Sexual Reproduction in Animals and Plants

Hitoshi Sawada • Naokazu Inoue
Megumi Iwano

Editors

Sexual Reproduction in Animals and Plants

 Springer Open

Editors
Hitoshi Sawada, Ph.D.
Director and Professor
Sugashima Marine Biological Laboratory
Graduate School of Science
Nagoya University
429-63 Sugashima, Toba
Mie 517-0004, Japan

Naokazu Inoue, Ph.D.
Associate Professor
Department of Cell Science
Institute of Biomedical Sciences
School of Medicine
Fukushima Medical University
1 Hikarigaoka, Fukushima
Fukushima 960-1295, Japan

Megumi Iwano, Ph.D.
Assistant Professor
Graduate School of Biological Sciences
Nara Institute of Science and Technology
8916-5 Takayama-cho, Ikoma
Nara 630-0192, Japan

ISBN 978-4-431-54588-0   ISBN 978-4-431-54589-7 (eBook)
DOI 10.1007/978-4-431-54589-7
Springer Tokyo Heidelberg New York Dordrecht London

Library of Congress Control Number: 2013955610

Printed on acid-free paper

Springer is part of Springer Science+Business Media (www.springer.com)

Preface

The International Symposium on the Mechanisms of Sexual Reproduction in Animals and Plants was held in Nagoya, Japan, as a joint meeting of the second Allo-authentication Meeting and the fifth International Symposium on the Molecular and Cell Biology of Egg- and Embryo-Coats (MCBEEC), November 12–16, 2012. This was the first international meeting where many plant and animal reproductive biologists gathered and discussed their recent progress.

Approximately 160 participants met in Nagoya from all over the world, and most of the oral presenters and several poster presenters contributed as authors of this book of proceedings. Although there are several books covering plant self-incompatibility and double-fertilization systems as well as animal fertilization, until now there has been no book covering recent progress in almost all fields of plant and animal fertilization.

This meeting was organized as part of the research project entitled "Elucidating common mechanisms of allogeneic authentication: mechanisms of sexual reproduction shared by animals and plants" supported by a Grant-in-aid for Scientific Research on Innovative Areas from MEXT, Japan. This project was established because self-sterile mechanisms in primitive chordates (ascidians) were found to be very similar to the self-incompatibility system in flowering plants and also because GCS1, a sperm-side factor responsible for gamete fusion, exists not only in plants but also in unicellular organisms and animals. These discoveries led us to speculate that there must be many other common mechanisms or molecules involved in sexual reproduction of animals and plants. We believe that to stimulate discussion in this innovative area, it is very important to summarize our current understanding of the mechanism of sexual reproduction. We hope that this book will be useful for many scientists, particularly those in the field of sexual reproduction—which we tentatively call "allo-authentication"—and in intercellular communications.

Toba, Japan   Hitoshi Sawada
Fukushima, Japan   Naokazu Inoue
Ikoma, Japan   Megumi Iwano

International Symposium on the Mechanisms of Sexual Reproduction in Animals and Plants
[Joint Meeting of the 2nd Allo-authentication Meeting and 5th Egg-Coat Meeting (MCBEEC)],
November 12–16, 2012, Nagoya Garden Palace, Nagoya, Japan

First Row (From Left to Right)
Katsuyuki Yamato, Gary Cherr, Kenji Murata, George Gerton, Luca Jovine, Harvey Florman, Timo Strünker, Miriam Sutovsky, Peter Sutovsky, Jurrien Dean, Masaru Okabe, Hitoshi Sawada, Alberto Darszon, Stefanie Sprunck, Noni Franklin-Tong, Emily Indriolo, Teh-hui Kao, Ravi Palanivelu, Kentaro K. Shimizu, Tetsuya Higashiyama, Hideo Mohri, Douglas Chandler

Second Row
Kenji Miyado, Hisayoshi Nozaki, Hiroyuki Sekimoto, Yasunori Sasakura, Makoto Hirai, Takako Saito, Ling Han, Tomoe Takahashi, Misato Yokoe, Hiroko Asano, Motoko Igarashi, Xintian Lao, Megumi Iwano, Kanae Ito, Mafumi Abiko, Chihiro Ono, Kaoru Yoshida, Ayumi Shiroshita, Yuki Takeshige, Miho Watanabe, Jiao Shi, Sheng Zhong, Jinging Liu, Naoko Hirano, Rinako Miyabayashi

Third Row
Tetsuyuki Entani, Chihiro Sato, Estelle Garenaux, Tomoko Koyano, Yoko Hattori, Midori Matsumoto, Eerdundagula Ebchuqin, Tomoko Igawa, Hitoshi Sugiyama, Masaaki Morisawa, Khaled Machaca, Yasuhiro Iwao, Akihiko Watanabe, Hideo Kubo, Tomohiro Sasanami, Ken Kitajima, Motonori Hoshi, Hanae Nodono, Kazuo Inaba, Luigia Santella, Kiyotaka Toshimori, Chizuru Ito, Maako Makino

Fourth Row
Naoto Yonezawa, Masahiro Suzuki, Toshiyuki Mori, Hiroki Okumura, Masatoshi Mita, Shigeyuki Kawano, Tatsuma Mohri, Keiichiro Kyozuka, Kenshou Furuta, Akihiro Nakamura, Katsuyuki Kakeda, Takashi Okamoto, Taizo Motomura, Gang Fu, Gary F. Clark, Yuki Hamamura, Kogiku Shiba, Noritaka Hirohashi, Lixy Yamada

Fifth Row
Takeru Kanazawa, Naoya Araki, Hideaki Fukushima, Masao Ueki, Jun Takayama, Youki Takezawa, Yuichirou Harada, Masato Uchiyama, Bayyinatul Muchtaromah, Retno Susilowati, Masako Mino, Jumpei Dake, Taku Kato, Akiko Konishi

Sixth Row
Yuhkoh Satouh, Haruhiko Miyata, Yoshitaka Fujihara, Masahito Ikawa, Tadashi Baba, Kenji Yamatoya, Takashi Ijiri, Ken-ichi Sato, Hidenori Takeuchi, Hiroki Tsutsui, Tsukasa Matsuda, Shunsuke Nishio, Naokazu Inoue, Masaya Morita, Shun Ohki, Gen Hiyama, Kazunori Yamamoto, Shinsaku Suetsugu, Shigemasa Hirada, Kei Otuska, Takuma Hirata

Contents

Part I
Sperm Attraction, Activation, and Acrosome Reaction

Chapter 1
Sperm Chemotaxis: The First Authentication Events Between Conspecific Gametes Before Fertilization

Manabu Yoshida

Abstract Sperm chemotaxis toward eggs before fertilization has been observed in many living organisms. Sperm chemotaxis is the first communication or signaling event between male and female gametes in the process of fertilization, and species-specific events occur in many cases. Thus, sperm chemotaxis may act as a safety process for authenticating that fertilization occurs between conspecific egg and sperm and helps to prevent crossbreeding. Here, we introduce mechanisms of sperm chemotaxis, focusing on cross-talk between gametes and species specificity. Furthermore, we discuss the interactions between sperm-activating and sperm-attracting factors (SAAFs) in the ascidian species and that SAAF receptors on sperm cells are not all-or-none responses. The SAAF receptors may accept SAAFs of related species (closely related molecules), with different affinities.

Keywords Fertilization • Species specificity • Sperm chemotaxis

1.1 Introduction

In all living organisms, male gametes are activated, with increase in their motility, and are subsequently attracted toward a female gamete in response to certain factors released from the female gametes or reproductive organs. Chemotactic behavior of male gametes toward the ovule was first described in Kingdom Plantae, bracken fern (Pfeffer 1884). Brown algae have also developed chemoattractants for male gametes, known as sexual pheromones (Maier and Müller 1986). In flowering plants, peptidic factors called LUREs attract the pollen tube toward the ovules, resulting in

M. Yoshida (✉)
Misaki Marine Biological Station, School of Science, Center for Marine Biology,
University of Tokyo, Miura, Kanagawa 238-0225, Japan
e-mail: yoshida@mmbs.s.u-tokyo.ac.jp

H. Sawada et al. (eds.), *Sexual Reproduction in Animals and Plants*,
DOI 10.1007/978-4-431-54589-7_1, © The Author(s) 2014

guiding the sperm cell to the ovule (Okuda et al. 2009). Gamete chemotaxis was also observed in Kingdom Fungi, and the aquatic fungus *Allomyces macrogynus* shows gamete chemotaxis (Machlis 1973).

In Kingdom Animalia, sperm chemotaxis toward the egg was first observed in the hydrozoan *Spirocodon saltatrix* (Dan 1950), and such an ability is now widely recognized in marine invertebrates, from cnidarians to ascidians (Miller 1966, 1985b; Cosson 1990), and in vertebrates, from fish to humans (Oda et al. 1995; Pillai et al. 1993; Suzuki 1958, 1959; Eisenbach 1999; Yanagimachi et al. 2013). In nematodes, spermatozoa are unflagellated but use an amoeboid movement to move from the bursa through the uterus to the spermatheca (Ward and Carrel 1979). A sperm-guiding factor present in the micropyle area of the egg of the teleost rosy barb has also been described (Amanze and Iyengar 1990).

In many cases, species specificity of sperm chemotaxis is present. Thus, these phenomena constitute the first communication event between the gametes during fertilization and prevent crossbreeding among different species. In this chapter, we review sperm chemotaxis and focus on the species specificity of this phenomenon.

1.2 Chemical Nature of Sperm Chemoattractants

Chemoattractant molecules for sperm in plants are low molecular weight organic compounds such as the bimalate ions in the bracken fern (Brokaw 1957, 1958) and unsaturated cyclic or linear hydrocarbons, such as ectocarpene, in the brown algae (Maier and Müller 1986). In the aquatic fungus *A. macrogynus*, the female gametes release a sesquiterpene "sirenin" as a attractant for male gametes (Machlis 1973), and interestingly, a different compound called "parisin" released by the male gametes is able to attract flagellated female gametes of the same species (Pommerville and Olson 1987).

In animals, sperm chemoattractants have been identified in several species, and most of these chemoattractant molecules are proteins or peptides. Chemoattractants such as "resact" in sea urchins (Ward et al. 1985; Guerrero et al. 2010), "sepsap" in cuttlefish (Zatylny et al. 2002), and "asterosap" in starfish (Böhmer et al. 2005) are peptides. A 21-kDa protein named "allurin" in the amphibian *Xenopus laevis* (Olson et al. 2001) and tryptophan in abalone (Riffell et al. 2002) act as sperm chemoattractants. In the hydrozoan *Hippopodius hippopus*, the attractant has not yet been identified but has been characterized as a small and thermoresistant protein with a molecular mass of 25 kDa and an isoelectric point of 3.5 (Cosson et al. 1986). Recently, a Coomassie Blue-affinity glycoprotein, "Micropyler Sperm Attractant" (MSA), around the opening and inside of the micropyle of herring and flounder eggs has been identified that guides ("attract") the spermatozoa into the micropyle (Yanagimachi et al. 2013).

On the other hand, nonproteinaceous chemoattractants have been identified in coral and ascidians: the chemoattractant of the coral *Montipora digitata* is an unsaturated fatty alcohol (Coll et al. 1994), and those of the ascidians *Ciona intestinalis* and *Ascidia sydneiensis* (Yoshida et al. 2002; Matsumori et al. 2013) are sulfated hydroxysterols. Mammalian spermatozoa also show chemotactic behavior, and many candidate chemoattractants for spermatozoa have been proposed (Eisenbach and Giojalas 2006). Recently, progesterone released from the cumulus oophorus was considered as a candidate of sperm attractant for human sperm (Guidobaldi et al. 2008). On the other hand, odorants such as bourgeonal (Spehr et al. 2003) and lyral (Fukuda et al. 2004), which are aromatic aldehydes used in perfumes, could also act as chemoattractants in human and mouse sperm, respectively.

Where are the sperm chemoattractants released? Fern sperms show a chemotactic response to secretions from the female reproductive structures (Pfeffer 1884). Sperm attractants of sea urchins and sea stars (starfish) are derived from the egg jelly (Ward et al. 1985; Nishigaki et al. 1996), and the source of sperm attractant of the hydrozoan, the siphonophore, is a cupule, the extracellular structure of the egg (Carré and Sardet 1981). Therefore, sperm attractants are released from the egg accessory organs or female gametes in these species. In contrast, in ascidians, sperm-attracting activity does not originate from the overall egg coat as a layer of jelly surrounding the eggs, but originates from the egg (Yoshida et al. 1993), indicating that the eggs themselves release the chemoattractant for the sperm.

1.3 Ca^{2+} Changes Mediate Sperm Chemotaxis

In all examples of well-characterized chemotaxis, the intracellular Ca^{2+} concentration ($[Ca^{2+}]_i$) appears to be a common element of absolute necessity in the attraction mechanism (Kaupp et al. 2008; Yoshida and Yoshida 2011). Ca^{2+} plays a key role in the regulation of flagellar beating, and in the case of sea urchin spermatozoa, the sperm attractant triggers $[Ca^{2+}]_i$ fluctuations (Böhmer et al. 2005; Wood et al. 2005) that appear to correlate with the asymmetrical beating of sperm flagella (Brokaw et al. 1974; Brokaw 1979). In the hydrozoan siphonophores, the diameters of the sperm trajectories decrease on approach of the sperm to the cupule (a sperm-attracting accessory organ of the egg), but the sperm trajectories are unchanged in the absence of Ca^{2+} (Cosson et al. 1984). A similar role for extracellular Ca^{2+} in mediating flagellar asymmetry of the spermatozoon during chemotactic behavior has been reported in hydrozoa (Miller and Brokaw 1970; Cosson et al. 1984). In ascidians, the spermatozoa normally exhibit circular movements, as just described, and maintain $[Ca^{2+}]_i$ at very low levels (Shiba et al. 2008). During chemotactic behavior the spermatozoa produce frequent and transient increases of $[Ca^{2+}]_i$ in the flagella (Ca^{2+} bursts) (Shiba et al. 2008). Interestingly, the Ca^{2+} bursts are consistently evoked at points at which the spermatozoon is around a temporally minimal value for a given sperm-activating and sperm-attracting factor (SAAF) concentration (Shiba et al. 2008) and to trigger a sequence of "turn-and-straight" movements. These data suggest that sperm attractants induce Ca^{2+} entry from extracellular spaces into the sperm cell, and the resultant increase in $[Ca^{2+}]_i$ mediates the beating of sperm flagella, resulting in chemotactic "turn-and-straight" movements.

1.4 Specificity of Sperm Chemotaxis in Species Other Than Ascidians

As described here, the molecular structures of sperm chemoattractants are different in different species, and factors from one species cannot attract the sperm of another species. This specificity ensures species-specific fertilization by preventing cross-breeding. Species or genus specificity in sperm chemotaxis has been observed in hydrozoa (Miller 1979) and in echinoderms, other than sea urchins (Miller 1985a, 1997). However, no chemotactic cross-reactivity exists in siphonophore species examined, and contact with seawater without attractants is enough to activate sperm motility, although the presence of Ca^{2+} ions in seawater is involved in the chemo-attraction process (Cosson 1990). Mammalian species also seem to share a common sperm attractant molecule (Sun et al. 2003; Guidobaldi et al. 2008; Teves et al. 2006), suggesting the lack of species specificity. In Mollusca, even though the abalone species seem to show species specificity in sperm chemotaxis (Riffell et al. 2004), there is a lack of species specificity of sperm chemotaxis among chitons (Miller 1977).

1.5 Species Specificity of Sperm Chemotaxis in Ascidians

In ascidians, species-specific sperm agglutination was reported in the early 1950s in five Mediterranean ascidians (Minganti 1951), and precise species-specificity tests of sperm attractants in egg ethanol extracts were also described in many ascidian species (Miller 1975, 1982) (Table 1.1). In these studies, ascidian sperm chemotaxis or agglutination tend to be species specific, but cross-reactivity among many species was also observed (Table 1.1). In particular, a lack of specificity was evident within the genus *Styela* (Miller 1975, 1982) (Table 1.1). However, the study on species specificity of ascidians contained both the order Phlebobranchiata, including the genus *Ciona*, and the order Stolidobranchiata, including the genus *Styela*, which are genetically distant, as per recent taxonomic data (Zeng et al. 2006; Tsagkogeorga et al. 2009).

We have previously identified the sperm chemoattractant released from the eggs of *Ciona intestinalis* as (25*S*)-3α,4β,7α,26-tetrahydroxy-5α-cholestane-3,26-disulfate, which was designated as the *Ciona* sperm-activating and -attracting factor (Ci-SAAF) (Yoshida et al. 2002; Oishi et al. 2004). The synthesized Ci-SAAF molecule possesses abilities to both activate motility and attract sperm (Yoshida et al. 2008; Oishi et al. 2004). The SAAF of another *Ciona* species, *C. savignyi* (Cs-SAAF), seems to be identical with Ci-SAAF and presents no specificity for the sperm activation of *C. savignyi* and vice versa (Yoshida et al. 1993, 2002). We have also recently identified As-SAAF from another phlebobranchian species, *A. sydneiensis*, as 3α,7α,8β,26-tetrahydroxy-5α-cholest-22-ene-3,26-disulfate (Matsumori et al. 2013); this was the first study leading to the identification of the chemoattractants of related species in Kingdom Animalia. Unexpectedly, Ci-SAAF and

Table 1.1 Species-specificity tests of sperm chemotaxis or agglutination in ascidians

			Sperm											
			1[a]	2	3	4	5	6	7	8	9	10	11	12
Egg extracts	Ciona intestinalis	1	++	++	–	/	–	–	–	–	–	–	–	–
	Ascidia callosa	2	+	++	/	/	++	–	/	/	/	/	/	/
	Corella inflata	3	–	–	–	–	–	–	/	/	–	–	–	–
	Corella willmeriana	4	±	+	±	++	++	/	/	/	/	/	/	/
	Chelyosoma productum	5	/	–	/	/	++	/	/	/	/	/	/	/
	Pyura haustor	6	–	–	/	/	–	±	/	/	/	/	/	/
	Styela plicata	7	–	/	/	/	/	–	++	–	–	/	/	/
	Styela clava	8	–	/	/	–	/	–	–	±	–	/	/	/
	Styela montereyensis	9	–	–	–	–	–	+	+	+	++	+	+	–
	Styela gibbsii	10	–	–	–	–	+	+	/	/	–	++	/	/
	Boltenia villosa	11	–	–	–	–	/	±	/	/	/	/	++	/
	Halocynthia igaboja	12	–	–	–	+	+	–	/	/	/	/	+	++

Species 1–5 are Phlebobranchia; species 6–12 are Stolidobranchia
++ strong activity, + weak activity, ± uncertain response, – negative response, / test not done
[a]Numbers show the same species shown in egg extracts
Source: Miller (1982)

a   Ci-SAAF   **b   As-SAAF**

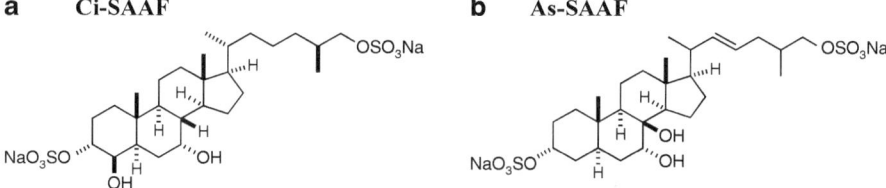

Fig. 1.1 Molecular structure of ascidian sperm attractants: *Ciona intestinalis* (Ci-SAAF) (**a**); *Ascidia sydneiensis* (As-SAAF) (**b**)

As-SAAF vary only by one double bond and the position of the OH group (Fig. 1.1). Even such a small difference in the sperm attractant molecules is enough to result in species-specific responses.

The cross-reactivity data of sperm chemotaxis for several ascidian species belonging to order Phlebobranchia show some specificity in the cross-reactivity between egg-conditioned seawater (ESW) and sperm response when comparing *Ciona* versus *Phallusia* and *Phallusia* versus *Ascidia*. However, this does not seem to be true in all cases in terms of "species" or "genus" specificity. For example, there is a "one-way" (no reciprocity) cross-reaction between *C. savignyi* and *A. sydneiensis* (Table 1.2) (Yoshida et al. 2013). Furthermore, even when a cross-reaction is observed, the level of activity is different. The interactions between the SAAFs in the ascidian species and the SAAF receptors on the sperm cells are not all-or-none responses. The SAAF receptor may accept SAAFs of related species, which are closely related molecules, with different affinity. Hence, sperm chemotaxis is neither a "species"- nor a "genus"-specific phenomenon among ascidians.

Table 1.2 Cross-reactivity in sperm chemotaxis elicited by egg-conditioned seawater (ESW) from different ascidian species

			Sperm				
			1[a]	2	3	4	5
Egg-conditioned seawater	*Ciona intestinalis*	1	++	++	/	/	/
	Ciona savignyi	2	++	++	–	/	–
	Phallusia nigra	3	/	–	++	/	–
	Phallusia mammillata	4	/	±	+	++	–
	Ascidia sydneiensis	5	/	+	–	/	++

++ active, + weakly active, ± uncertain response, – negative, / not examined
[a]Numbers show the same species shown in egg extracts
Source: Yoshida et al. (2013)

1.6 Conclusion

Sperm chemotaxis appears to be a much more specific phenomenon at the species or genus level in many animal species: cnidarians (Miller 1979), echinoderms other than sea urchins (Miller 1985a, 1997), and ascidians (Miller 1982; Yoshida et al. 2013). These results indicate that the specificity of sperm chemotaxis participates in the prevention of crossbreeding at fertilization. It is hypothesized that the interaction between sperm attractants from egg and attractant receptors on the sperm does not result in all-or-none responses, and that attractant receptors may accept some heterospecific sperm attractants having related chemical structures but with different binding or dissociation constants. Research into the precise chemical nature of sperm attractants and their corresponding receptors in different species may provide new horizons for studies of the fertilization system, especially on the mechanisms by which authentic interactions between conspecific eggs and spermatozoa occur.

Acknowledgments This work was supported in part by Grants-in-Aid for Scientific Research on Innovative Area "Elucidating Common Mechanisms of Allogeneic Authentication: Mechanisms of Sexual Reproduction Shared by Animals and Plants" from MEXT (#22112507 & 24112708).

References

Amanze D, Iyengar A (1990) The micropyle: a sperm guidance system in teleost fertilization. Development (Camb) 109:495–500
Böhmer M, Van Q, Weyand I, Hagen V, Beyermann M, Matsumoto M, Hoshi M, Hildebrand E, Kaupp UB (2005) Ca²⁺ spikes in the flagellum control chemotactic behavior of sperm. EMBO J 24(15):2741–2752
Brokaw CJ (1957) 'Electro-chemical' orientation of bracken spermatozoids. Nature (Lond) 179:525

Brokaw CJ (1958) Chemotaxis of bracken spermatozoids. The role of bimalate ions. J Exp Biol 35:192–196

Brokaw CJ (1979) Calcium-induced asymmetrical beating of triton-demembranated sea urchin sperm flagella. J Cell Biol 82:401–411

Brokaw CJ, Josslin R, Bobrow L (1974) Calcium ion regulation of flagellar beat symmetry in reactivated sea urchin spermatozoa. Biochem Biophys Res Commun 58:795–800

Carré D, Sardet C (1981) Sperm chemotaxis in siphonophores. Biol Cell 40:119–128

Coll JC, Bowden BF, Meehan GV, Konig GM, Carroll AR, Tapiolas DM, Alino PM, Heaton A, De Nys R, Leone PA, Maida M, Aceret TL, Willis RH, Babcock RC, Willis BL, Florian Z, Clayton MN, Miller RL (1994) Chemical aspects of mass spawning in corals. I. Sperm-attractant molecules in the eggs of the scleractinian coral *Montipora digitata*. Mar Biol 118:177–182

Cosson MP (1990) Sperm chemotaxis. In: Gagnon C (ed) Controls of sperm motility: biological and clinical aspects. CRC, Boca Raton, pp 104–135

Cosson MP, Carré D, Cosson J (1984) Sperm chemotaxis in siphonophores. II. Calcium-dependent asymmetrical movement of spermatozoa induced by attractant. J Cell Sci 68:163–181

Cosson J, Carré D, Cosson MP (1986) Sperm chemotaxis in siphonophores: identification and biochemical properties of the attractant. Cell Motil Cytoskeleton 6:225–228

Dan JC (1950) Fertilization in the medusan, *Spirocodon saltatrix*. Biol Bull 99:412–415

Eisenbach M (1999) Sperm chemotaxis. Rev Reprod 4:56–66

Eisenbach M, Giojalas LC (2006) Sperm guidance in mammals—an unpaved road to the egg. Nat Rev Mol Cell Biol 7(4):276–285. doi:10.1038/nrm1893

Fukuda N, Yomogida K, Okabe M, Touhara K (2004) Functional characterization of a mouse testicular olfactory receptor and its role in chemosensing and in regulation of sperm motility. J Cell Sci 117(pt 24):5835–5845

Guerrero A, Nishigaki T, Carneiro J, Tatsu Y, Wood CD, Darszon A (2010) Tuning sperm chemotaxis by calcium burst timing. Dev Biol 344(1):52–65. doi:10.1016/j.ydbio.2010.04.013

Guidobaldi HA, Teves ME, Unates DR, Anastasia A, Giojalas LC (2008) Progesterone from the cumulus cells is the sperm chemoattractant secreted by the rabbit oocyte cumulus complex. PLoS One 3(8):e3040. doi:10.1371/journal.pone.0003040

Kaupp UB, Kashikar ND, Weyand I (2008) Mechanisms of sperm chemotaxis. Annu Rev Physiol 70:93–117. doi:10.1146/annurev.physiol.70.113006.100654

Machlis L (1973) The chemotactic activity of various sirenins and analogues and the uptake of sirenin by the sperm of allomyces. Plant Physiol 52(6):527–530

Maier I, Müller DG (1986) Sexual pheromones in algae. Biol Bull 170:145–175

Matsumori N, Hiradate Y, Shibata H, Oishi T, Simma S, Toyoda M, Hayashi F, Yoshida M, Murata M, Morisawa M (2013) A novel sperm-activating and attracting factor from the ascidian *Ascidia sydneiensis*. Org Lett 15(2):294–297. doi:10.1021/ol303172n

Miller RL (1966) Chemotaxis during fertilization in the hydroid *Campanularia*. J Exp Zool 162:23–44

Miller RL (1975) Chemotaxis of the spermatozoa of *Ciona intestinalis*. Nature (Lond) 254:244–245

Miller RL (1977) Chemotactic behavior of chitons (Mollusca: Polyplacophora). J Exp Zool 202:203–212

Miller RL (1979) Sperm chemotaxis in the hydromedusae. I. Species-specificity and sperm behavior. Mar Biol 53:99–114

Miller RL (1982) Sperm chemotaxis in ascidians. Am Zool 22:827–840

Miller RL (1985a) Demonstration of sperm chemotaxis in echinodermata: Asteroidea, Holothuroidea, Ophiuroidea. J Exp Zool 234:383–414

Miller RL (1985b) Sperm chemo-orientation in Metazoa. In: Metz CB, Monroy A (eds) Biology of fertilization, vol 2. Academic Press, New York, pp 275–337

Miller RL (1997) Specificity of sperm chemotaxis among Great Barrier Reef shallow-water holothurians and ophiuroids. J Exp Biol 279:189–200

Miller RL, Brokaw CJ (1970) Chemotactic turning behaviour of *Tubularia* spermatozoa. J Exp Biol 52:699–706

Minganti A (1951) Esperienze sulle fertilizine nelle ascidie. Pubbl Staz Zool Napoli 23:58–65

Nishigaki T, Chiba K, Miki W, Hoshi M (1996) Structure and function of asterosaps, sperm-activating peptides from the jelly coat of starfish eggs. Zygote 4(3):237–245

Oda S, Igarashi Y, Ohtake H, Sakai K, Shimizu N, Morisawa M (1995) Sperm-activating proteins from unfertilized eggs of the Pacific herring *Clupea pallasii*. Dev Growth Differ 37:257–261

Oishi T, Tsuchikawa H, Murata M, Yoshida M, Morisawa M (2004) Synthesis and identification of an endogenous sperm activating and attracting factor isolated from eggs of the ascidian *Ciona intestinalis*; an example of nanomolar-level structure elucidation of a novel natural compound. Tetrahedron 60:6971–6980

Okuda S, Tsutsui H, Shiina K, Sprunck S, Takeuchi H, Yui R, Kasahara RD, Hamamura Y, Mizukami A, Susaki D, Kawano N, Sakakibara T, Namiki S, Itoh K, Otsuka K, Matsuzaki M, Nozaki H, Kuroiwa T, Nakano A, Kanaoka MM, Dresselhaus T, Sasaki N, Higashiyama T (2009) Defensin-like polypeptide LUREs are pollen tube attractants secreted from synergid cells. Nature (Lond) 458(7236):357–361. doi:10.1038/nature07882

Olson JH, Xiang X, Ziegert T, Kittelson A, Rawls A, Bieber AL, Chandler DE (2001) Allurin, a 21-kDa sperm chemoattractant from *Xenopus* egg jelly, is related to mammalian sperm-binding proteins. Proc Natl Acad Sci USA 98(20):11205–11210

Pfeffer W (1884) Locomotorische Richtungsbewegungen durch chemische Reize. Untersuch Botanisch Inst Tübingen 1:363–482

Pillai MC, Shields TS, Yanagimachi R, Cherr GN (1993) Isolation and partial characterization of the sperm motility initiation factor from eggs of the Pacific herring, *Clupea pallasi*. J Exp Zool 265:336–342

Pommerville J, Olson LW (1987) Evidence for a male-produced pheromone in *Allomyces macrogynus*. Exp Mycol 11(3):245–248. doi:dx.doi.org/10.1016/0147-5975(87)90012-0

Riffell JA, Krug PJ, Zimmer RK (2002) Fertilization in the sea: the chemical identity of an abalone sperm attractant. J Exp Biol 205(pt 10):1439–1450

Riffell JA, Krug PJ, Zimmer RK (2004) The ecological and evolutionary consequences of sperm chemoattraction. Proc Natl Acad Sci USA 101(13):4501–4506

Shiba K, Baba SA, Inoue T, Yoshida M (2008) Ca^{2+} bursts occur around a local minimal concentration of attractant and trigger sperm chemotactic response. Proc Natl Acad Sci USA 105(49):19312–19317. doi:10.1073/pnas.0808580105

Spehr M, Gisselmann G, Poplawski A, Riffell JA, Wetzel CH, Zimmer RK, Hatt H (2003) Identification of a testicular odorant receptor mediating human sperm chemotaxis. Science 299(5615):2054–2058

Sun F, Giojalas LC, Rovasio RA, Tur-Kaspa I, Sanchez R, Eisenbach M (2003) Lack of species-specificity in mammalian sperm chemotaxis. Dev Biol 255(2):423–427

Suzuki R (1958) Sperm activation and aggregation during fertilization in some fishes. I. Behavior of spermatozoa around the micropyle. Embryologia 4:93–102

Suzuki R (1959) Sperm activation and aggregation during fertilization in some fishes. II. Effect of distilled water on the sperm-stimulating capacity and fertilizability of eggs. Embryologia 4:359–367

Teves ME, Barbano F, Guidobaldi HA, Sanchez R, Miska W, Giojalas LC (2006) Progesterone at the picomolar range is a chemoattractant for mammalian spermatozoa. Fertil Steril 86(3):745–749. doi:10.1016/j.fertnstert.2006.02.080

Tsagkogeorga G, Turon X, Hopcroft RR, Tilak MK, Feldstein T, Shenkar N, Loya Y, Huchon D, Douzery EJ, Delsuc F (2009) An updated 18S rRNA phylogeny of tunicates based on mixture and secondary structure models. BMC Evol Biol 9:187. doi:10.1186/1471-2148-9-187

Ward S, Carrel JS (1979) Fertilization and sperm competition in the nematode *Caenorhabditis elegans*. Dev Biol 73(2):304–321

Ward GE, Brokaw CJ, Garbers DL, Vacquier VD (1985) Chemotaxis of *Arbacia punctulata* spermatozoa to resact, a peptide from the egg jelly layer. J Cell Biol 101:2324–2329

Wood CD, Nishigaki T, Furuta T, Baba SA, Darszon A (2005) Real-time analysis of the role of Ca^{2+} in flagellar movement and motility in single sea urchin sperm. J Cell Biol 169(5): 725–731

Yanagimachi R, Cherr G, Matsubara T, Andoh T, Harumi T, Vines C, Pillai M, Griffin F, Matsubara H, Weatherby T, Kaneshiro K (2013) Sperm attractant in the micropyle region of fish and insect eggs. Biol Reprod 88(2):47. doi:10.1095/biolreprod.112.105072

Yoshida M, Yoshida K (2011) Sperm chemotaxis and regulation of flagellar movement by Ca^{2+}. Mol Hum Reprod 17(8):457–465. doi:10.1093/molehr/gar041

Yoshida M, Inaba K, Morisawa M (1993) Sperm chemotaxis during the process of fertilization in the ascidians *Ciona savignyi* and *Ciona intestinalis*. Dev Biol 157:497–506

Yoshida M, Murata M, Inaba K, Morisawa M (2002) A chemoattractant for ascidian spermatozoa is a sulfated steroid. Proc Natl Acad Sci USA 99(23):14831–14836

Yoshida M, Shiba K, Yoshida K, Tsuchikawa H, Ootou O, Oishi T, Murata M (2008) Ascidian sperm activating and attracting factor: importance of sulfate groups for the activities and implication of its putative receptor. FEBS Lett 582(23-24):3429–3433. doi:10.1016/j.febslet.2008.09.006

Yoshida M, Hiradate Y, Sensui N, Cosson J, Morisawa M (2013) Species specificity of sperm motility activation and chemotaxis: a study on ascidian species. Biol Bull 224(3):156–165

Zatylny C, Marvin L, Gagnon J, Henry J (2002) Fertilization in *Sepia officinalis*: the first mollusk sperm-attracting peptide. Biochem Biophys Res Commun 296(5):1186–1193

Zeng L, Jacobs MW, Swalla BJ (2006) Coloniality has evolved once in stolidobranch ascidians. Integr Comp Biol 46(3):255–268. doi:10.1093/icb/icj035

Chapter 2
Respiratory CO_2 Mediates Sperm Chemotaxis in Squids

Noritaka Hirohashi, Yoko Iwata, Warwick H.H. Sauer, and Yasutaka Kakiuchi

Abstract The squid *Loligo* (*Heterololigo*) *bleekeri* uses two distinct insemination sites, inside or outside the female's body, which links to the mating behavior of two distinct types of males, consort or sneaker, respectively. We found that sperm release a self-attracting molecule, which causes only sneaker sperm to swarm. We identified respiratory CO_2 as the sperm chemoattractant and its sensor, membrane-bound flagellar carbonic anhydrase. Downstream signaling results from generation of an extracellular proton gradient, intracellular acidosis, and concomitant recovery from acidosis. This cycle in turn elicits Ca^{2+}-dependent flagellar turning/tumbling, resulting in chemotactic swarming.

Keywords Chemotaxis • CO_2 sensor • Sperm evolution • Spermatozoa • Squids

N. Hirohashi (✉)
Oki Marine Biological Station, Education and Research Center for Biological Resources,
Shimane University, 194 Kamo, Okinoshima-cho, Oki, Shimane 685-0024, Japan
e-mail: hiro@life.shimane-u.ac.jp

Y. Iwata
Atmosphere and Ocean Research Institute, University of Tokyo,
Kashiwa, Chiba 277-8564, Japan

W.H.H. Sauer
Department of Ichthyology and Fisheries Science, Rhodes University,
6140, Grahamstown, South Africa

Y. Kakiuchi
Graduate School of Humanities and Sciences, Ochanomizu University,
2-2-1 Otsuka, Tokyo 112-8610, Japan

H. Sawada et al. (eds.), *Sexual Reproduction in Animals and Plants*,
DOI 10.1007/978-4-431-54589-7_2, © The Author(s) 2014

2.1 Results

2.1.1 Sperm from Sneaker Males Swarm in Response to Respiratory CO_2 Emission

Sperm chemotaxis, widely recognized in metazoa (Miller 1975; Sun et al. 2009; Kaupp et al. 2008; Guerrero et al. 2010) and plants (Okuda et al. 2009), is the phenomenon in which sperm direct their movement in response to chemicals released from eggs or accessory cells, facilitating sperm–egg encounters. In addition to the well-known biological context in egg-derived chemical guidance for sperm attraction, spermatozoa often form motile conjugates that may be beneficial in competing with sperm from other males in polyandrous species (Moore et al. 2002; Fisher and Hoekstra 2010). Despite extended arguments on the evolutionary adaptation of sperm cooperation for reproductive success (Immler 2008; Foster and Pizzari 2010), little is known about how sperm form functional conjugates (Moore et al. 2002). Previously, we found that each male of the coastal squid *Loligo bleekeri* produces one of two types of morphologically distinct euspermatozoa, the two types being linked to distinctly different male mating behaviors (Iwata et al. 2011). Consort males produce spermatozoa with short flagella and transfer sperm capsules (spermatophores) to internal locations of the females, inside the oviduct, whereas sneaker males produce long-flagellum sperm and transfer spermatophores to the outer body wall of the same females (Fig. 2.1, Iwata et al. 2011). The evolutionary consequences by which such phenotypic dimorphism arose remain elusive; however, each

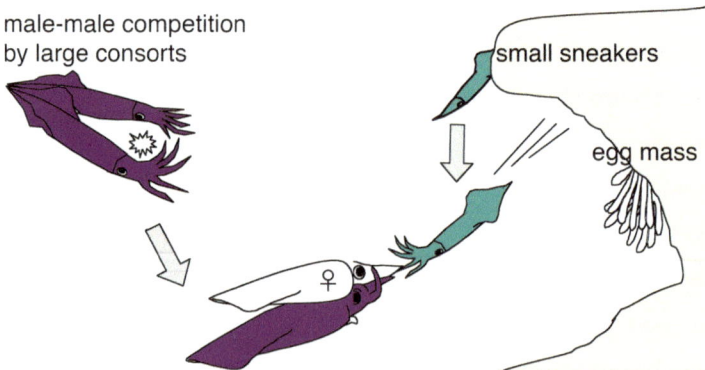

Fig. 2.1 Mating behaviors in *Loligo bleekeri*. In *L. bleekeri*, male individuals conduct one of two alternative reproductive tactics associated with body size. Consort males are relatively larger than females in body size, struggle physically with each other, and copulate predominately with the female to pass their sperm. Sneaker males, on the other hand, display maturity at a relatively smaller size than females and do not participate in male–male competition for mating. Instead, they access females by "sneaking" behavior to transfer their sperm in the course of the consort's mating

type of sperm is expected to face different sperm competitiveness and different fertilization modes.

We tested whether squid sperm show any swarming behavior by drawing sperm suspensions into glass capillary tubes. Within 3 min, sneaker, but not consort, sperm became concentrated and formed a regular striped pattern along the longitudinal axis of the capillary. The formation of this pattern was transient, although motility appeared unchanged for the duration of the experiment. To ascertain that swarming is an intrinsic trait specific to sneaker sperm, a mixture of sneaker and consort sperm, each labeled with different mitochondrial dyes, was introduced into the capillary tube. Only sneaker sperm formed swarms, and both mitochondrial dyes yielded the same result. Swarming did not involve reduced motility or physical binding among sperm, but rather each sperm in the swarm moved independently by actively swimming, a phenomenon similar to chemotactic swarming. A filter assay was used to determine whether sperm swarming resulted from a chemical cue. We transferred a small amount of labeled sperm and a large amount of nonlabeled sperm into the lower and upper chambers. We then observed changes in swim-up sperm numbers by confocal microscopy. Swim-up numbers doubled when sneaker, but not consort, sperm were placed in the lower and upper chambers, indicating that the sneaker sperm upward migration is caused by a chemical stimulus that elicits chemotaxis or chemokinesis or both. Unexpectedly, sneaker sperm swam up when consort or even starfish (*Asterina pectinifera*) sperm were placed in the upper chamber, suggesting that the sperm attractant may be a ubiquitous molecule generated by sperm respiration.

Because *Dictyostelium discoideum* (Kimmel and Parent 2003) and *Escherichia coli* (Budrene and Berg 1991) release the chemoattractants cAMP and L-aspartate, respectively, that promote self-organization, we tested each and found no effect on either type of squid sperm. Chemoattractants known in other cell types, such as dicarboxylic acids for fern spermatozoids (Brokaw 1957), L-tryptophan for abalone sperm (Riffell et al. 2002), and sugars or amino acids for bacteria, were all inactive for squid sperm. Finally, we tested various gases and found that only CO_2 attracts sneaker, but not consort, sperm. From these observations, we hypothesized that temporal swarming observed in the capillary tube was mediated by chemosensation in response to respiratory CO_2 (CO_2 taxis) emitted by sperm.

2.1.2 Flagellar Membrane-Localized Carbonic Anhydrase Serves as a Primary CO₂ Sensor

We reasoned that the transient character of the swarming could result from the instability of the formation of a chemical gradient and speculated that the nature of the chemical gradient could be CO_2 hydration products (protons or bicarbonate ions) rather than CO_2 itself. Because carbonic anhydrases (CAs) serve as primary CO_2 sensors in many biological systems (Wang et al. 2010; Chandrashekar et al. 2009; Sun et al. 2009; Ziemann et al. 2009), we tested broad and specific CA inhibitors and found an inhibitory effect on swarming by several of these compounds. We cloned a full-length cDNA encoding CA from sneaker testes and found that it is most similar to membrane-anchored CA isoforms. This transcript was also found in consort testis; therefore, we generated an antibody against a synthetic peptide to confirm protein expression in both types of sperm. Western blots of whole-cell extracts identified a ~31.7-kDa band (the calculated molecular mass of 28.8 kDa) in both sneaker and consort sperm. The flagella of both sneaker and consort sperm were equally stained by the antibody, and immunoreactivity was diminished by treating live sperm with proteinase K, indicating that CA localizes on the cell surface. We examined CO_2 metabolism and found that both sneaker and consort sperm converted their respiratory CO_2 to H^+ and HCO_3^- by CA and acidified the pH of the medium (pH_e). Notably, only sneaker sperm acidified intracellular pH (pH_i) concomitantly with pH_e acidification.

2.1.3   An Extracellular Proton Gradient Establishes and Maintains Swarming

We hypothesized that sneaker sperm sense a proton gradient by which swarming is enabled. We first measured pH_e using a pH-sensitive dye during swarm formation. Development of a proton gradient from the central part of the swarm was evident, although estimation of the precise pH_e values was precluded by the spatiotemporal alternation of sperm density that affects concentrations of the pH indicator. Next, when the pH_e gradient formation was interfered by buffering seawater (10 mM Tris or HEPES), no swarming was observed. Finally, we tested sperm behavior to acid-loaded pipettes. We found that both sneaker (below pH 5.0) and consort (below pH 4.0) sperm showed a chemotactic response to acid (acidotaxis) and kept swarming in the vicinity of the pipette for a longer period (~30 min). As expected, sperm did not respond to a pipette with 50 mM bicarbonate-containing agarose (pH 8.0), confirming a proton as the inducer of chemotaxis.

Why do consort sperm show acidotaxis, but not CO_2 taxis, despite the presence of CA? The acidotaxis assay revealed that the sensitivity of acid detection in sneaker sperm is ~1 pH unit higher than that in consort sperm. Given that no apparent pH_i decrease occurred in consort sperm, we hypothesized that only sneaker sperm have the acid-induced proton uptake system by which CO_2 taxis is driven. To explore this hypothesis, we first examined pH_i homeostasis at various pH_e using buffered seawater. We found that both sneaker and consort sperm were similar in maintaining their pH_i against alkalosis. However, only consort sperm showed pH_i homeostasis against acidosis. If swimming up or down the proton gradient is instantly reflected in the pH_i values, the pH_i changes could be a signaling component that mediates a chemotactic response. Sperm were placed in buffered seawater (pH 8.0 or 6.0) into which a pipette filled with 1 M sodium acetate (NaAc)-soaked agarose gel (pH 8.0 or 6.0) was inserted. In this setup, because NaAc crosses the plasma membrane and causes cytoplasmic acidosis, sperm are allowed to change pH$_i$ depending on their swimming direction: sperm swimming toward the pipette will become acidified and those swimming away from the pipette will recover from cytoplasmic acidosis in a constant pH$_e$ environment (pH 8.0 or 6.0). Sperm from sneaker males, but not consort males, showed directional movements toward the pipette when pH$_e$ was adjusted to 6.0. Conversely, when a pipette loaded with ammonium chloride (pH 5.0) (an alkalosis-inducing agent without pH$_e$ changes) was placed in seawater at pH 5.0, sneaker sperm showed chemorepellent behavior from the pipette. These results, together with other data, suggest that an environmental proton gradient enables synchronous changes in the pH$_i$ (acidic range) of sperm that facilitate directional movement to establish and maintain the swarm formation.

2.1.4 A Return from Intracellular Acidosis Evokes Calcium-Dependent Motor Responses for Turn/Tumbling

In sea urchins (Bohmer et al. 2005; Guerrero et al. 2010), ascidians (Shiba et al. 2008), and perhaps other animals (Cosson et al. 1984), sperm exhibit a coordinated transition of straight runs and quick turns, primarily regulated by calcium flux through the plasma membrane, enabling them to approach the chemoattractant source. Similarly, *L. bleekeri* sperm require extracellular Ca^{2+} for swarming in both experimental and natural conditions and for acidotaxis. We then analyzed the swimming trajectory of the sperm entering the border zone of the swarming region. We found that sperm ascending into a swarm tend to maintain straight trajectories, whereas sperm descending into a swarm make frequent turns. These results clearly demonstrated that sperm swarming is driven at least by chemotaxis but not by solely chemokinesis or trapping regardless of their possible existence. Two-dimensional swimming trajectory analysis showed that the reorientation consists of the initiation of the turn or tumbling motion followed by straight swimming directed toward the chemical source (straight–turn–straight). We asked whether this turn/tumbling initiation is caused by a pH$_i$-dependent calcium ion uptake. Unfortunately, we were unable to image flagellar [Ca^{2+}]$_i$; therefore, we took an alternative approach. Sperm preincubated in acidic (pH 5.0) or normal (pH 8.0) seawater were placed in Ca^{2+}-free seawater at pH 8.0 and tested to determine whether they would respond to a local Ca^{2+} release and, as a result, elicit turn/tumbling behavior. Both types of sperm, regardless of preincubation conditions, exhibited mostly straight swimming behavior in Ca^{2+}-free seawater. However, only sneaker sperm that were preincubated in the acidic environment evoked frequent high-turn swimming episodes in the vicinity of the Ca^{2+}-loaded pipette. These results indicate that intracellular acidosis primes the Ca^{2+} influx capacity in sneaker sperm, and the subsequent recovery from acidosis elicits Ca^{2+} uptake, which triggers transition of the swimming mode from straight to turn/tumbling (Fig. 2.2).

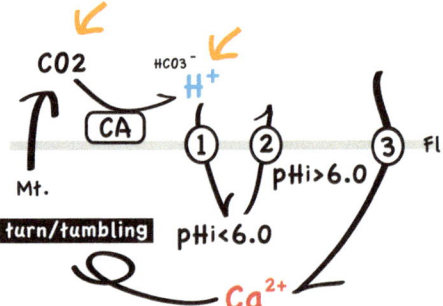

Fig. 2.2 Model of CO_2 chemotaxis in squid sperm. Respiratory CO_2 emitted from self and neighboring sperm is hydrated by the flagellar membrane-bound carbonic anhydrase (CA) into bicarbonate ions and protons. Only sneaker, but not consort, sperm influx contains extracellular protons generated from self and neighboring sperm by an unknown mechanism (*1*), resulting in intracellular acidosis (below pH_i 6.0). When sperm swim along a descending proton gradient, recovery from acidosis (above pH_i 6.0) occurs (*2*), which evokes calcium influx (*3*) that is necessary for turn/tumbling initiation in the flagellum. *Mt.* mitochondria, *Fl.* flagellum

2.2 Discussion

The spear squid *L. bleekeri* employs alternative mating tactics; large consort males take physical advantage in courtship with females and deposit their spermatophores inside the female's body. Therefore, fertilization is assumed to occur internally. In contrast, small sneaker males transfer their spermatophores by sneaking behavior at an external location just below the female's mouth, so that such sperm would encounter eggs when females hold the eggs in their arms during the egg-laying procedure. Although a factor that influences the male-type decision remains to be identified, this system offers extremely a unique situation where internal and external fertilization coexist within a single spawning episode. Previously, we found that sneaker sperm are ~50 % longer than consort sperm (Iwata et al. 2011). Although no such clear within-species dimorphic eusperm had been reported previously, there are many examples of sperm size differences among closely related species, which have largely been explained as the consequences of sperm competition (Gage 1994; Briskie and Montgomerie 1992; Gomendio and Roldan 1991). Unexpectedly, empirical data supported no evidence that larger sperm are favored in sperm competition in this species regarding the swimming velocity and sperm precedence at the storage site (in the seminal receptacle). Alternatively, different fertilization environments might be a prominent factor that could drive the evolution of sperm size (Iwata et al. 2011).

In this study, we found that sperm behavioral traits are also different between sneaker and consort spermatozoa. Sperm from sneaker, but not consort, males have a characteristic of forming motile conjugates when ejaculated into seawater and hold these in the close vicinity of the spermatophore. From an ecological aspect, retaining ability of ejaculates at the buccal region (site of egg deposition) would have a prominent effect on storing into the externally located seminal receptacle (female sperm storage organ) or fertilization success because mating and egg laying are temporally independent (Iwata et al. 2005). Especially, sperm should travel, either actively or passively, for the certain distance from the ejaculation site to the storage site by an unknown mechanism (Iwata et al. 2011; Sato et al. 2010; Lumkong 1992). Therefore, the sperm swarming trait together with the female arm crown architecture (Naud et al. 2005) would provide an effective diffusion-resistant situation against water movement.

The question of why only sneaker sperm have acquired the swarming trait would be intriguing to address in the light of postcopulatory sexual selection (Birkhead and Pizzari 2002) and natural selection (Foster and Pizzari 2010). Theoretically, a risk of sperm diffusion would be much greater on sneaker (externally deposited and stored) than on consort (internally deposited and stored) sperm, which could account for the evolution of complex adaptive traits on precopulatory (mating behavior) and postcopulatory (sperm function) sexual selection. We therefore carried out further investigations with other Loliginidae species that also employ alternative male mating behavior. In *Loligo reynaudii* and *Photololigo edulis*, sperm from sneaker individuals, as judged from the sperm mass morphology (Iwata and Sakurai 2007), exhibited self-swarming, whereas no swarming occurred for sperm from consort individuals. Moreover, in species employing only sneaker-type mating behavior, that is, males inseminate the external sites on females, such as *Idiosepius paradoxus* and *Todarodes pacificus*, sperm also showed swarming behavior, supporting our hypothesis that the swarming trait tightly associates with the fertilization mode rather than sperm competition between sneaker and consort (Parker 1990).

2.3 Perspectives

It remains unknown how changes in pH_i elicit $[Ca^{2+}]_i$ mobilization in this system. However, recent reports identified that CatSper, a mammalian sperm calcium channel essential for flagellum motility, can be activated by either progesterone (Strunker et al. 2011; Lishko et al. 2011) (a sperm chemoattractant) or intracellular alkalization (Kirichok et al. 2006). CO_2/acid detection in the mammalian gustatory system (Chandrashekar et al. 2009; Huang et al. 2006; Kawaguchi et al. 2010; Chang et al. 2010; Lahiri and Forster 2003) and central nervous system (Ziemann et al. 2009; Lahiri and Forster 2003) may represent molecular similarity to CO_2 taxis found in squid sperm in terms of intracellular acidosis via transcellular proton currents (Chang et al. 2010) and "off-response" (Kawaguchi et al. 2010). Because CO_2 emission is the cell's fundamental property, understanding the molecular pathway in the CO_2 taxis will provide a broad impetus to discover similar examples in biological systems.

Acknowledgments This study was supported by Narishige Zoological Science Award, Research Institute of Marine Invertebrates, Yamada Science Foundation, Grant-in-aid for Scientific Research on Innovative Areas from MEXT and Japanese Association for Marine Biology (JAMBIO) to N.H. and National Research Foundation to W.H.H.S.

References

Birkhead TR, Pizzari T (2002) Postcopulatory sexual selection. Nat Rev Genet 3(4):262–273
Bohmer M, Van Q, Weyand I, Hagen V, Beyermann M, Matsumoto M, Hoshi M, Hildebrand E, Kaupp UB (2005) Ca^{2+} spikes in the flagellum control chemotactic behavior of sperm. EMBO J 24(15):2741–2752
Briskie JV, Montgomerie R (1992) Sperm size and sperm competition in birds. Proc Biol Sci 247(1319):89–95
Brokaw CJ (1957) Electro-chemical orientation of bracken spermatozoids. Nature (Lond) 179(4558):525
Budrene EO, Berg HC (1991) Complex patterns formed by motile cells of *Escherichia coli*. Nature (Lond) 349(6310):630–633
Chandrashekar J, Yarmolinsky D, von Buchholtz L, Oka Y, Sly W, Ryba NJ, Zuker CS (2009) The taste of carbonation. Science 326(5951):443–445
Chang RB, Waters H, Liman ER (2010) A proton current drives action potentials in genetically identified sour taste cells. Proc Natl Acad Sci USA 107(51):22320–22325
Cosson MP, Carre D, Cosson J (1984) Sperm chemotaxis in siphonophores. II. Calcium-dependent asymmetrical movement of spermatozoa induced by the attractant. J Cell Sci 68:163–181
Fisher HS, Hoekstra HE (2010) Competition drives cooperation among closely related sperm of deer mice. Nature (Lond) 463(7282):801–803
Foster KR, Pizzari T (2010) Cooperation: the secret society of sperm. Curr Biol 20(7):R314–R316
Gage MJG (1994) Associations between body size, mating pattern, testis size and sperm lengths across butterflies. Proc R Soc Lond B Biol Sci 258:247–254
Gomendio M, Roldan ER (1991) Sperm competition influences sperm size in mammals. Proc Biol Sci 243(1308):181–185
Guerrero A, Nishigaki T, Carneiro J, Yoshiro T, Wood CD, Darszon A (2010) Tuning sperm chemotaxis by calcium burst timing. Dev Biol 344(1):52–65
Huang AL, Chen X, Hoon MA, Chandrashekar J, Guo W, Trankner D, Ryba NJ, Zuker CS (2006) The cells and logic for mammalian sour taste detection. Nature (Lond) 442(7105):934–938
Immler S (2008) Sperm competition and sperm cooperation: the potential role of diploid and haploid expression. Reproduction 135(3):275–283
Iwata Y, Sakurai Y (2007) Threshold dimorphism in ejaculate characteristics in the squid *Loligo bleekeri*. Mar Ecol Prog Ser 345:141–146
Iwata Y, Munehara H, Sakurai Y (2005) Dependence of paternity rates on alternative reproductive behaviors in the squid *Loligo bleekeri*. Mar Ecol Prog Ser 298:219–228
Iwata Y, Shaw P, Fujiwara E, Shiba K, Kakiuchi Y, Hirohashi N (2011) Why small males have big sperm: dimorphic squid sperm linked to alternative mating behaviours. BMC Evol Biol 11:236
Kaupp UB, Kashikar ND, Weyand I (2008) Mechanisms of sperm chemotaxis. Annu Rev Physiol 70:93–117
Kawaguchi H, Yamanaka A, Uchida K, Shibasaki K, Sokabe T, Maruyama Y, Yanagawa Y, Murakami S, Tominaga M (2010) Activation of polycystic kidney disease-2-like 1 (PKD2L1)-PKD1L3 complex by acid in mouse taste cells. J Biol Chem 285(23):17277–17281
Kimmel AR, Parent CA (2003) *Dictyostelium discoideum* cAMP chemotaxis pathway. Sci STKE 2003:cm1
Kirichok Y, Navarro B, Clapham DE (2006) Whole-cell patch-clamp measurements of spermatozoa reveal an alkaline-activated Ca^{2+} channel. Nature (Lond) 439(7077):737–740
Lahiri S, Forster RE 2nd (2003) CO_2/H^+ sensing: peripheral and central chemoreception. Int J Biochem Cell Biol 35(10):1413–1435
Lishko PV, Botchkina IL, Kirichok Y (2011) Progesterone activates the principal Ca^{2+} channel of human sperm. Nature (Lond) 471(7338):387–391
Lumkong A (1992) A histological study of the accessory reproductive-organs of female *Loligo forbesi* (Cephalopoda, Loliginidae). J Zool (Lond) 226:469–490

Miller RL (1975) Chemotaxis of the spermatozoa of Ciona intestinalis. Nature 254:244–245

Moore H, Dvorakova K, Jenkins N, Breed W (2002) Exceptional sperm cooperation in the wood mouse. Nature (Lond) 418(6894):174–177

Naud MJ, Shaw PW, Hanlon RT, Havenhand JN (2005) Evidence for biased use of sperm sources in wild female giant cuttlefish (*Sepia apama*). Proc Biol Sci 272(1567):1047–1051

Okuda S, Tsutsui H, Shiina K, Sprunck S, Takeuchi H, 1 Yui R, Kasahara RD, Hamamura Y, Mizukami A, Susaki D, Kawano N, Sakakibara T, Namiki S, Itoh K, Otsuka K, Matsuzaki M, Nozaki H, Kuroiwa T, Nakano A, Kanaoka MM, Dresselhaus T, Sasaki N, Higashiyama T (2009) Defensin-like polypeptide LUREs are pollen tube attractants secreted from synergid cells. Nature 458:357–361

Parker GA (1990) Sperm competition games: sneaks and extra-pair copulations. Proc Biol Sci 242:127–133

Riffell JA, Krug PJ, Zimmer RK (2002) Fertilization in the sea: the chemical identity of an abalone sperm attractant. J Exp Biol 205(pt 10):1439–1450

Sato N, Kasugai T, Ikeda Y, Munehara H (2010) Structure of the seminal receptacle and sperm storage in the Japanese pygmy squid. J Zool (Lond) 282:151–156

Shiba K, Baba SA, Inoue T, Yoshida M (2008) Ca²⁺ bursts occur around a local minimal concentration of attractant and trigger sperm chemotactic response. Proc Natl Acad Sci USA 105(49):19312–19317

Strunker T, Goodwin N, Brenker C, Kashikar ND, Weyand I, Seifert R, Kaupp UB (2011) The CatSper channel mediates progesterone-induced Ca²⁺ influx in human sperm. Nature (Lond) 471(7338):382–386

Sun L, Wang H, Hu J, Han J, Matsunami H, Luo M (2009) Guanylyl cyclase-D in the olfactory CO₂ neurons is activated by bicarbonate. Proc Natl Acad Sci USA 106(6):2041–2046

Wang YY, Chang RB, Liman ER (2010) TRPA1 is a component of the nociceptive response to CO₂. J Neurosci 30(39):12958–12963

Ziemann AE, Allen JE, Dahdaleh NS, Drebot II, Coryell MW, Wunsch AM, Lynch CM, Faraci FM, Howard MA 3rd, Welsh MJ, Wemmie JA (2009) The amygdala is a chemosensor that detects carbon dioxide and acidosis to elicit fear behavior. Cell 139(5):1012–1021

Chapter 3
Specific Mechanism of Sperm Storage in Avian Oviducts

Mei Matsuzaki, Gen Hiyama, Shusei Mizushima, Kogiku Shiba, Kazuo Inaba, and Tomohiro Sasanami

Abstract The capability for sperm storage in the female genital tract is frequently observed in vertebrates as well as in invertebrates. Because of the presence of a system that maintains the ejaculated sperm alive in the female reproductive tract in a variety of animals, including insects, fish, amphibians, reptiles, birds, and in mammals, this strategy appears to be advantageous for animal reproduction. Although the occurrence and physiological reasons for sperm storage have been reported extensively in many species, the mechanism for sperm storage in the female reproductive tract has been poorly understood until recently. In this chapter, we report our recent findings on the mechanism of sperm storage in avian oviducts, especially data obtained from the Japanese quail (*Coturnix japonica*), as an experimental model. Because sperm storage in birds occurs at body temperature (i.e. 41 °C), elucidation of the mechanism of sperm maintenance in the avian oviduct may open up new avenues for the development of novel strategies for sperm storage in vitro without cryopreservation.

Keywords Birds • Female reproductive tract • Sperm • Sperm storage tubules

M. Matsuzaki • G. Hiyama • S. Mizushima • T. Sasanami (✉)
Department of Applied Biological Chemistry, Faculty of Agriculture,
Shizuoka University, 836 Ohya, Shizuoka 422-8529, Japan
e-mail: atsasan@ipc.shizuoka.ac.jp

K. Shiba • K. Inaba
Shimoda Marine Research Center, University of Tsukuba,
5-10-1 Shimoda, Tsukuba 415-0025, Japan

H. Sawada et al. (eds.), *Sexual Reproduction in Animals and Plants*,
DOI 10.1007/978-4-431-54589-7_3, © The Author(s) 2014

23

3.1 Introduction

The timing of ovulation is known to be regulated by endocrine factors such as luteinizing hormone and progesterone in higher vertebrates, but the time of insemination into the female reproductive tract by natural mating or artificial insemination does not always synchronize with that of ovulation. To achieve efficient fertilization, sperm need to migrate to the site of fertilization when the ovulated oocytes are there. To increase the chance of fertilization, female animals frequently store sperm in their reproductive tract, and thus sperm storage works as a natural mechanism ensuring sperm encounter the ovulated oocytes at the right time and right place. For instance, some bat species mate in autumn, but fertilization does not take place immediately: the bats store the sperm in the oviduct for 5 months and fertilize them in the spring of the following year (Holt 2011). This phenomenon also optimizes the timing of the birth of their offspring until a suitable season for nursing arrives. Some reptiles, such as turtles, snakes, and lizards, have obvious potential for sperm storage in the oviduct for an extremely long period (maximum, 7 years). This long-term storage appears to work as an insurance against not finding mating partners in some breeding seasons (Holt and Lloyd 2010).

Because of the presence of specialized simple tubular invaginations in the oviduct, once ejaculated sperm have entered the female reproductive tract, they can survive up to 2–15 weeks in domestic birds, including chickens, turkeys, quail, and ducks, for various periods depending on the species (Bakst et al. 1994; Bakst 2011) in contrast to the relatively short lifespan in mammalian spermatozoa (i.e., several days). These specialized structures are generally referred to as sperm storage tubules (SST). SSTs are located in the uterovaginal junction (UVJ) and in the infundibulum, although the primary storage site for sperm is the SST in the UVJ (Burke and Ogasawara 1969; Brillard 1993). The spermatozoa are transported to the infundibulum, which is the site of fertilization and also serves as a secondary sperm storage site (Bakst 1981; Schindler et al. 1967). Although extensive investigations concerning the function of the SST in birds have been performed since its discovery in the 1960s by means of ultrastructural analysis (Bobr et al. 1964; Schuppin et al. 1984; Van Krey et al. 1967), the specific mechanisms involved in sperm uptake into the SST, sperm maintenance within it, and controlled sperm release from it remain to be elucidated. In this chapter, we report our recent findings on the mechanism of sperm storage in the avian oviduct, especially data obtained from the Japanese quail (*Coturnix japonica*), as an experimental model.

3.2 Sperm Release from the SST Is a Regulated Event in Birds

There are several reports indicating that sperm release from the SST is not regulated, but occurs in response to the mechanical pressures of a passing ovum, because no contractile elements associated with the SST were found (Van Krey et al. 1967; Tingari and

Fig. 3.1 Ultrastructural observation of the uterovaginal junction (UVJ) surface treated with progesterone. After mating, the animals were injected with vehicle alone (**a**) or 0.8 µg/ml progesterone (**b**). The UVJ was isolated 1 h after the injection, and the area of the entrance to the sperm storage tubules (SST) was observed by scanning electron microscopy. A representative photograph from those obtained from three different birds is shown. *Bar* = 10 µm

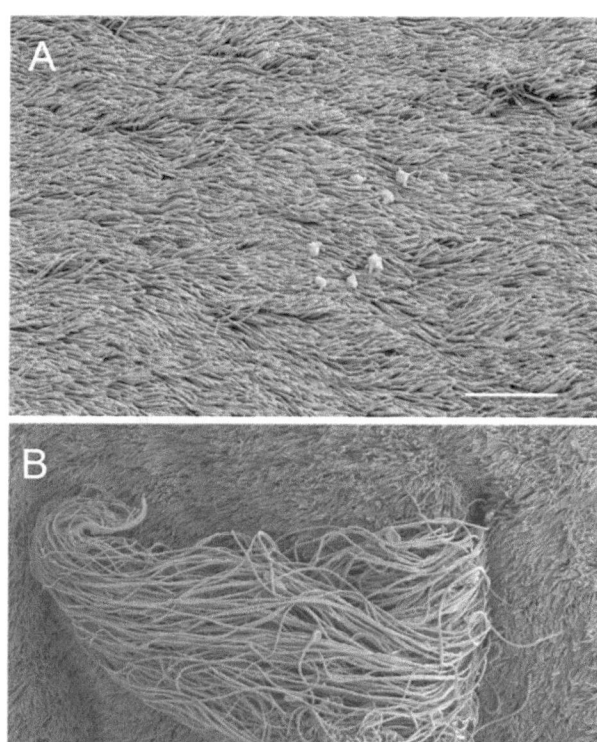

Lake 1973). In contrast, there is conflicting evidence showing that egress of the spermatozoa is regulated because resident spermatozoa were discharged from the SST close to the times of ovulation and oviposition (Bobr et al. 1964). To examine whether sperm release from the SST is regulated during the ovulatory cycle, female birds were mated 12 h after oviposition, and the SST in the UVJ at 2 or 13 h after mating (corresponding to a time 14 or 25 h after oviposition, respectively) was observed. The percentage of the SST containing sperm at 14 h after oviposition was high (i.e., approximately 50–60 %) and significantly decreased to approximately 40 % at 25 h. Also, a bundle of sperm extruding into the lumen of the UVJ from the SST was frequently seen at 20 h after oviposition, although no such sperm were observed at 8, 14, or 25 h. To test whether hormonal stimulation causes sperm release from the SST, the birds were injected with various steroid hormones and the SST filling rate was calculated. As a result, the percentage of the SST with sperm was only significantly decreased when the animals were treated with more than 0.8 µg/ml progesterone compared to that of the control birds, which were injected with a vehicle alone. Scanning electron microscopic observation revealed that the SST shrank in response to the injection of progesterone, and a bundle of the sperm tail extruded from the SST was observed (Fig. 3.1). This morphological change showed SST squeezed out the resident sperm into the lumen of the oviduct.

These results demonstrated that the release of the sperm from the SST is a regulated event during the ovulatory cycle, and progesterone acts as a sperm-releasing factor in birds (Ito et al. 2011). If the resident sperm are released from the SST without any regulation, most of the sperm ascending the oviduct may be trapped by the descending egg. It is reasonable to suppose that the sperm release from the SST is stimulated by progesterone because there is at least a 5-h grace period before the next ovulation and sperm released from the SST can reach the site of fertilization without hindrance from the descending egg. This process may be supported by a lubricant effect of cuticle materials, because the release of cuticle materials from the epithelial cells of the UVJ is also stimulated by progesterone injection (Ito et al. 2011).

3.3   Sperm Maintenance in the SST

Although the period of sperm storage is different in different species, once they have mated, female birds are able to produce fertilized eggs without repeated mating after a long period (e.g., maximum 3 months in the turkey hen). This is possible because these female birds store the resident sperm in the SST of the UVJ after mating, and the resident sperm are thought to be discharged from the SST by the stimulation of progesterone in each ovulatory cycle. This phenomenon suggests that the resident sperm can survive in the SST for extended periods at body temperature in birds (i.e., 41 °C). This surprising phenomenon was discovered more than half a century ago, but we currently know little about the mechanism that supports such long-term maintenance of the resident sperm in the SST of the female reproductive tract in birds. To clarify the SST functions, we first observed the resident sperm in the lumen of the SST by electron microscopy. It is reported that the sperm at the uterotubal junction (UTJ) in the bovine oviduct binds to the surface of epithelial cells and that this binding ensures the tethering of the sperm at the UTJ until the time of ovulation (Hunter 2008; Suarez 2010). In contrast to the situation in mammalian species, the resident sperm seems to be free from the epithelial cells of the SST in the quail oviduct (Fig. 3.2). This finding led us to hypothesize that unknown materials in the lumen of the SST may affect sperm mobility. To confirm this hypothesis, we incubated the ejaculated sperm in the presence or absence of prepared UVJ extracts. The flagellar movement of the sperm was recorded using a high-speed camera. When the sperm were incubated in the absence of the UVJ extracts, a vigorous flagellar movement was observed (Fig. 3.3a). However, in the presence of the UVJ extracts, we found that the flagellar movements were relatively quiescent, and that the amplitude of the flagellar movement, as well as the linear velocity of the sperm, decreased (Fig. 3.3b). More importantly, the addition of the UVJ extracts extended the sperm lifespan in vitro. In the presence of the UVJ extracts, sperm swam vigorously even after 48 h of incubation, whereas in the absence of the extracts, sperm usually died within 5 h (data not shown). These results indicate the possibility that unknown molecules responsible for sperm maintenance exist in the UVJ extracts. In the previous study, we also observed that the formation of secretory granules in the SST epithelial cells fluctuated during the

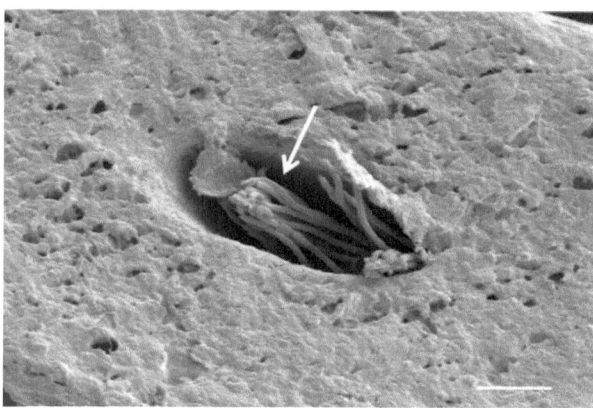

Fig. 3.2 Ultrastructural observation of the resident sperm in the SST. The UVJ was isolated at 1 h after mating, and the tissue was embedded in paraffin wax. Thick sections were prepared, and the surface of the cross section was observed by scanning electron microscopy. *Arrow* indicates the bundle of resident sperm in the SST. A representative photograph from those obtained from three different birds is shown. *Bar* = 5 μm

ovulatory cycle, and progesterone treatment mimicking the phenomena takes place during the ovulatory cycle. In SST cells, there are well-developed tight junctions among the cells in the apical region, and the SST epithelial cells appear to secrete their contents into the lumen of the SST, where the resident sperm are located. Although we did not elucidate the nature of the secretory granules, it is very likely that the contents of the granules in the UVJ extracts affect sperm physiology (i.e., sperm filling, storage, and release) in the SST.

3.4 Conclusion

In this chapter, we reported that sperm maintenance in the SST, as well as sperm release from the same location, are events regulated during the ovulatory cycle. For instance, we demonstrated that progesterone stimulates the release of resident sperm from the SST in the Japanese quail with a contraction-like morphological change of the SST. This process may be supported by the lubricant effect of cuticle materials secreted from the ciliated cells of the UVJ, as well as unknown materials supplied from the SST epithelial cells, in events coincidently triggered under progesterone control. In addition, we found secretory granules in SST epithelial cells, and the number of the secretory granules fluctuated during the ovulatory cycle, indicating that SST epithelial cells unknown materials into the lumen of the SST; these materials may affect sperm physiology (e.g., motility, respiration, and metabolism) (Ito et al. 2011). Although the nature of the molecules responsible for sperm maintenance for a long period of time remains to be clarified, we found extracts of UVJ possess the ability to reduce sperm motility and to extend sperm lifespan in vitro. Because sperm storage in

Fig. 3.3 Effects of UVJ
extracts on motility of
ejaculated sperm incubated in
the presence (**b**) or absence
(**a**) of the UVJ extracts
(300 µg/ml) at 39 °C for
10 min. The flagellar
movements of the sperm were
recorded by a high-speed
camera (200 frames per
second), and 10 images taken
at every 1/20 s were overlaid.
A representative photograph
from those obtained from
three different birds is shown.
Bar = 100 µm

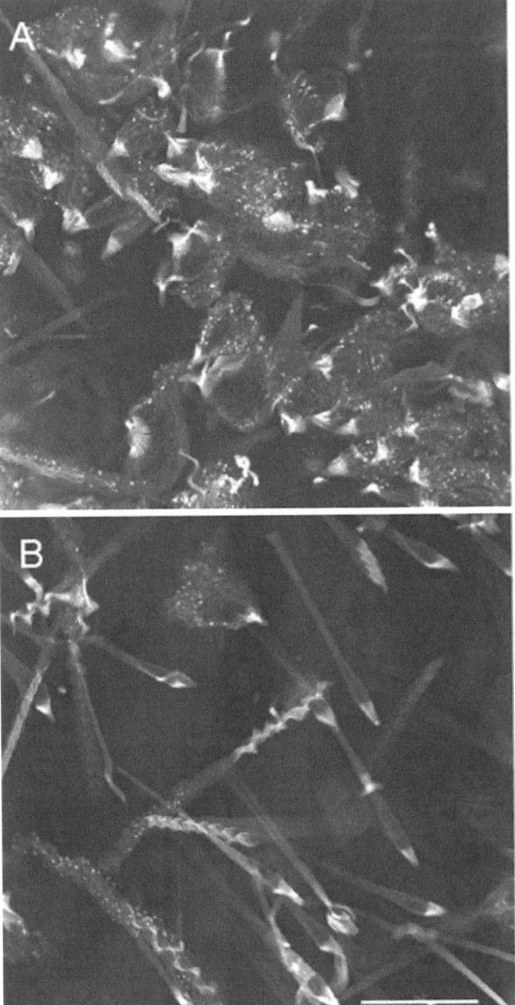

avian species occurs at high body temperature (i.e., 41 °C), elucidation of the
mechanism for sperm storage may lead to the development of new strategies for
sperm preservation at ambient temperatures.

Acknowledgments The authors are grateful to Ms. R. Hamano for technical assistance. This work
was supported in part by financial support from a Grant-in-Aid for Scientific Research on Innovative
Areas (24112710 to TS) and Grant-in-Aid for Scientific Research (B) (24380153 to T.S.).

References

Bakst MR (1981) Sperm recovery from oviducts of turkeys at known intervals after insemination and oviposition. J Reprod Fertil 62:159–164

Bakst MR (2011) Role of the oviduct in maintaining sustained fertility in hen. J Anim Sci. doi:10.2527/jas.2010-3663

Bakst MR, Wishart G, Brullard JP (1994) Oviductal sperm selection, transport, and storage in poultry. Poult Sci Rev 5:117–143

Bobr LW, Ogasawara FX, Lorenz FW (1964) Distribution of spermatozoa in the oviduct and fertility in domestic birds. II. Transport of spermatozoa in the fowl oviduct. J Reprod Fertil 8:49–58

Brillard JP (1993) Sperm storage and transport following natural mating and artificial insemination. Poult Sci 72:923–928

Burke WH, Ogasawara FX (1969) Presence of spermatozoa in uterovaginal fluids of the hen at various stages of the ovulatory cycle. Poult Sci 48:408–413

Holt WV (2011) Does apoptosis hold the key to long-term sperm storage mechanisms in vivo? Mol Reprod Dev 78:464–465

Holt WV, Lloyd RE (2010) Sperm storage in the vertebrate female reproductive tract: how does it work so well? Theriogenology 73:713–722

Hunter RHF (2008) Sperm release from oviduct epithelial binding is controlled hormonally by peri-ovulatory Graafian follicles. Mol Reprod Dev 75:167–174

Ito T, Yoshizaki N, Tokumoto T et al (2011) Progesterone is a sperm releasing factor from the sperm storage tubules in birds. Endocrinology 152:3952–3962

Schindler H, Ben-David E, Hurwits S, Kempenich O (1967) The relation of spermatozoa to the glandular tissue in the storage sites in the hen oviduct. Poult Sci 46:69–78

Schuppin GT, Van Krey HP, Denbow DM, Bakst MR, Meyer GB (1984) Ultrastructural analysis of uterovaginal sperm storage glands in fertile and infertile turkey breeder hens. Poult Sci 63:1872–1882

Suarez SS (2010) How do sperm get to the egg? Bioengineering expertise needed! Exp Mech 50:1267–1274

Tingari MD, Lake PE (1973) Ultrastructural studies on the uterovaginal sperm-host gland of the domestic hen, *Gallus domesticus*. J Reprod Fertil 34:423–431

Van Krey HP, Ogasawara FX, Pangborn J (1967) Light and electron microscope studies of possible sperm gland emptying mechanisms. Poult Sci 46:69–78

Chapter 4
Allurin: Exploring the Activity of a Frog Sperm Chemoattractant in Mammals

Lindsey Burnett, Hitoshi Sugiyama, Catherine Washburn, Allan Bieber, and Douglas E. Chandler

Abstract Allurin, a 21-kDa protein secreted by the oviduct of female *Xenopus* frogs, is incorporated into the jelly layers of eggs as they pass single file on their way to the uterus and subsequent spawning. Hydration of the egg jelly layers at spawning releases allurin as a chemoattractant that binds to the midpiece of *Xenopus* sperm in a dose-dependent manner. Gradients of allurin elicit directed swimming across a porous membrane in two-chamber assays and preferential, up-gradient swimming of sperm in video-microscopic assays. Allurin, purified from *X. laevis* or produced in recombinant form, also elicits chemotaxis by mouse sperm in two-chamber and video microscopic assays. Allurin binds to mouse sperm at the mid-piece and head, a pattern also seen in frog sperm. Western blots suggest the presence of an allurin-like protein in the follicular fluid of mice and humans and peptides that mimic subdomains within allurin elicit chemoattractive behavior in both mouse and human sperm. By sequence homology, allurin is a truncated member of the Cysteine-RIch Secretory Protein (CRISP) family whose members include Crisps 1, 2, and 4, which have been demonstrated to modulate mammalian sperm functions including

Dedication This chapter is dedicated to Allan L. Bieber, a long-time collaborator of ours who recently passed on. Allan was an expert biochemist who guided our purification of allurin and characterization of its disulfide bonding pattern using mass spectrometry. His long-term interest in venom proteins from snakes was culminated by his delight in finding that allurin is closely related to Crisp snake toxin proteins.

L. Burnett • C. Washburn • D.E. Chandler (✉)
School of Life Sciences, Arizona State University, Tempe, AZ 85287-4501, USA
e-mail: d.chandler@asu.edu

H. Sugiyama
Science and Technology Group, Okinawa Institute of Science and Technology, Okinawa 904-0495, Japan

A. Bieber
Department of Chemistry and Biochemistry, Arizona State University, Tempe, AZ 85287-4501, USA

H. Sawada et al. (eds.), *Sexual Reproduction in Animals and Plants*,
DOI 10.1007/978-4-431-54589-7_4, © The Author(s) 2014

capacitation, ion channel activity, and sperm–egg binding. Interestingly, allurin contains only two of the three domains found in these full-length CRISP proteins and in this respect is similar to the sperm self-recognition proteins HrUrabin and CiUrabin important in ascidian gamete interactions. These findings suggest that both full-length and truncated CRISP proteins play important reproductive roles in species widely separated in evolutionary time.

Keywords Crisp proteins • Egg jelly • Sperm chemotaxis • *Xenopus laevis*

4.1 Introduction

Sperm physiological responses are choreographed by a series of signaling ligands generated by female gametes and the organs in which they are housed. The signals conveyed inform the sperm of the location, direction, and status of nearby female gametes and elicit changes in sperm motility, sperm secretory status, and sperm metabolic activities required to successfully fertilize an egg. Some of these ligands are diffusible in nature and emanate from extracellular coatings surrounding the egg whereas other ligands form a more stable part of the extracellular coat structure and act only on sperm in contact with these matrices. Diffusible ligands include peptides, steroids, amino acids, and even proteins (Burnett et al. 2008a).

One class of protein ligands that have received increasing interest are the cysteine-rich secretory (Crisp) proteins whose members include snake venom toxins, sperm chemoattractants, sperm decapacitation factors, sperm self-identification markers, and sperm ion channel modulators (Burnett et al. 2008a; Gibbs et al. 2008; Kratzschmar et al. 1996; Koppers et al. 2011). Full-length Crisp proteins contain three domains, as indicated in Fig. 4.1. The largest "pathogenesis-related" (PR) domain at the N-terminal (cyan) is homologous to proteins expressed in a variety of plants in response to environmental stress or viral infection (Fernandez et al. 1997). Next is the Hinge domain (yellow), the smallest of the three, named for the fact that it links the first and third domain. Last is the ion channel regulatory (ICR) domain at the C-terminal (magenta), named for its ability to block or modulate a variety of voltage-dependent potassium and calcium channels as well as ryanodine receptors (Gibbs et al. 2006). The ICR domain is similar in tertiary structure to ion channel-blocking peptides from sea anenomes (Pennington et al. 1999; Alessandri-Haber et al. 1999; Cotton et al. 1997).

The mammalian Crisp proteins, 1 through 4, are reproductively important and full length. Earliest reports of sperm-associated Crisp proteins demonstrated that the rodent epididymis secretes Crisp 1, which binds to the post-acrosomal region of the sperm head (Cohen et al. 2011). This protein has been shown to migrate to the equatorial region of rat sperm during capacitation and to possibly participate in sperm–egg fusion (Cohen et al. 2000, 2007, 2008; Da Ros et al. 2004; Roberts et al. 2006, 2007, 2008; Ellerman et al. 2006). Antibodies raised against the protein inhibit sperm–egg fusion, and the protein has been demonstrated to bind to the egg

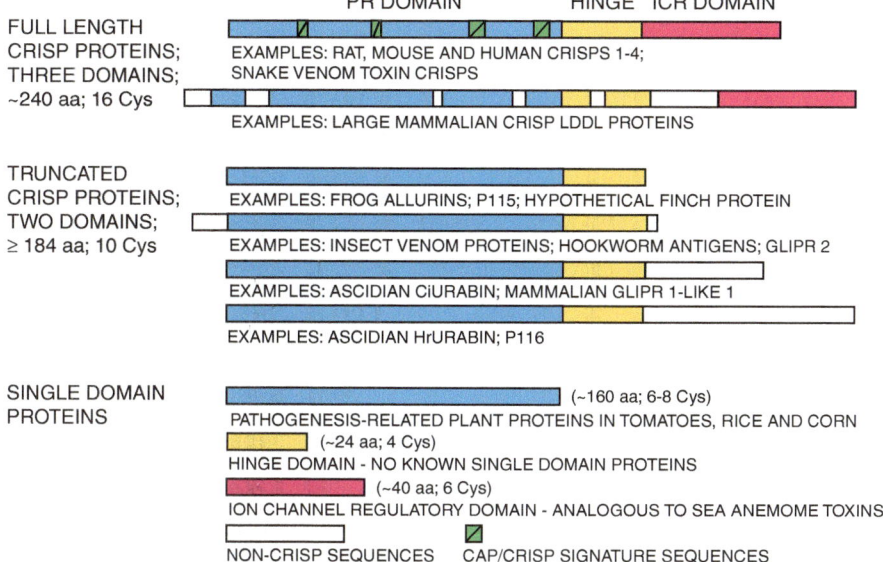

Fig. 4.1 Domain structure of full-length and truncated Crisp proteins. CAP/CRISP signature sequences are present in the pathogenesis-related (*PR*) domain of all Crisp proteins but are shown here only for full-length proteins. N-terminal signal sequences have been omitted for clarity

surface (Cohen et al. 2007; Ellerman et al. 2006). Although knockout of the protein does not produce sterility, it does reduce the efficiency of sperm–egg binding in vitro (Da Ros et al. 2008).

A second Crisp family member, TPX-1 or Crisp 2, discovered as an acrosomal granule constituent and as a major autoantigen in vasectomized rodents (Hardy et al. 1988; Foster and Gerton 1996), has also been shown to play a modulatory role in sperm–egg binding in rodents (Busso et al. 2007). Crisp 2 is thought to be released during the acrosome reaction, and the protein has been shown to help mediate binding of the sperm to the zona pellucida (Cohen et al. 2011).

Crisp 3, although widely expressed, is a prominent secreted protein in the prostate (Udby et al. 2005). This protein can be found in the circulation bound to a macroglobulin and has been proposed for use as a marker of prostatic hypertrophy (Bjartell et al. 2006; Udby et al. 2010). As a constituent of semen, the protein has been correlated with reproductive success in horses and cows (Schambony et al. 1998a, b; Topfer-Petersen et al. 2005).

Crisp 4, present in rodents but not in other mammals, has been the first Crisp family protein demonstrated to have a molecular role in sperm physiology. Darszon and coworkers (Gibbs et al. 2010a) have shown that Crisp 4 inhibits the opening of TRP-M8 channels in mouse sperm that are required to initiate capacitation. Channel inhibition by Crisp 4 or possibly by its homologue Crisp 1 may play a role in preventing premature capacitation, that is, acting as a "decapacitation factor." In addition, Crisp 4 has been shown to modulate sperm binding to the zona pellucida

(Turunen et al. 2012). Found in the epididymis, this protein is considered to be a functional orthologue to human Crisp 1 (Nolan et al. 2006).

The complexity of the CRISP family has been further increased by the discovery of "truncated" members missing the "ion channel regulatory" domain characteristic of full-length Crisp proteins such as Crisp 1 and Crisp 4. As shown in Fig. 4.1, truncated members typically contain the PR and Hinge domains along with amino-acid sequences at the C-terminal that vary in length from a few residues to several hundred. Members of this class include allurin, the frog sperm chemoattractant whose characteristics are described in the current chapter, insect venom and hookworm antigen proteins (Burnett et al. 2008a), the protease inhibitor-like proteins P115 and P116 (Gibbs et al. 2008), *Ciona intestinalis* (Ci) Urabin and *Halocynthia roretzi* (Hr) Urabin, recently discovered self-identity ligands on ascidian sperm (Urayama et al. 2008; Yamaguchi et al. 2011), and the glioma pathogenesis related-like proteins, whose functions are currently unknown, but which include members that are expressed in a testis-specific manner (Gibbs et al. 2010b).

Commonly, all these proteins are grouped into the Cysteine Rich–Antigen–Pathogenesis related (CAP) superfamily of proteins that are characterized by a homologous PR domain, with four regions of particularly high sequence homology referred to as "CAP signatures" or "CRISP signatures" (green in Fig. 4.1) (Gibbs et al. 2008).

4.2   Characterization of Allurin as a Frog Sperm Chemoattractant

Amphibian eggs undergo a marked decrease in fertilizability if their outer jelly layers are removed. This observation was initially made in *Bufo japonicas* and *Bufo arenaras* and more recently in *Xenopus laevis* (Katagiri 1987; Krapf et al. 2009; Olson and Chandler 1999). However, reintroduction of diffusible jelly components (referred to as "egg water") restores the fertilizability of these jellyless eggs almost to their original level of 80 % or greater (Olson and Chandler 1999). For this reason, our laboratory tested the possibility that a diffusible jelly component from *X. laevis* eggs might be acting as a sperm chemoattractant by using two types of assays (Burnett et al. 2011a). The first uses a modified Boyden chamber that has a porous polycarbonate filter separating an upper sperm chamber and a lower chemoattractant chamber. Sperm passing through the filter in response to a chemotactic gradient of egg water were increased fivefold over those in control experiments in which egg water was not present (Al-Anzi and Chandler 1998). This response was dose dependent, egg water specific, and was not seen in the presence of egg water mixed uniformly throughout the lower chamber. This assay guided the subsequent purification of allurin from egg water using anion-exchange chromatography in conjunction with a NaCl step gradient for elution (Olson et al. 2001; Sugiyama et al. 2009).

A second chemoattraction assay, developed by Sally Zigmond to study neutrophils (Zigmond 1977) and modified by Giojalas and coworkers for use with sperm

(Fabro et al. 2002), employs video microscopy to record the directional movements of individual sperm. Trajectories are recorded of sperm swimming on an observational platform that lies between two troughs, one acting as a reservoir of sperm and the other as a reservoir of chemoattractant (see Fig. 4.2a). The sperm on the platform swim within a chemotactic gradient formed by diffusion. Their movement at 7 frame/s was analyzed for velocity and directionality using X–Y Cartesian coordinates, the gradient axis being along the X coordinate. Values in successive frames as well as average values over entire trajectories were compared.

Data analysis revealed that in the presence of an egg-water gradient (red/dark bars, Fig. 4.2b) average sperm velocity was about 43 μm/s and not significantly different from sperm velocity in the absence of a gradient (white bars, Fig. 4.2b). In addition, similar sperm velocities were observed in sperm traveling up the gradient and down the gradient, suggesting that egg water does not contain a chemokinetic agent that alters sperm swimming speed. In contrast, net sperm travel along the X coordinate (gradient axis) was increased threefold by the presence of an egg-water gradient (first bar set, Fig. 4.2c). In comparison, sperm movement along the Y coordinate perpendicular to the gradient axis was random and not increased by egg water (second bar set, Fig. 4.2c). Sperm movement along the X coordinate, when analyzed in greater detail, showed a marked change in trajectory distribution from relatively random movement (white bars, Fig. 4.2d) to those favoring positive, upgradient movement (red/dark bars, Fig. 4.2d). These findings demonstrated that X. *laevis* (Xl) sperm prefer to swim toward the reservoir of egg water, indicating the presence of a chemoattractant.

4.3 Allurin Is a Chemoattractant for Mammalian Sperm

Because rodents express a number of Crisp proteins that have striking roles in sperm physiology and sperm–egg interactions, it was of interest to determine whether a truncated Crisp protein such as allurin had effects on mammalian sperm. Initially we used the two-chamber assay to demonstrate that a gradient of X. *laevis* egg water did indeed have the ability increase mouse sperm passage across a porous membrane by 2.5 fold (Burnett et al. 2011b, 2012). Subsequently, we tested the ability of mouse sperm to bind Alexa 488-conjugated allurin in a region-specific manner. Dye-conjugated allurin bound strongly to the subequatorial region of the mouse sperm head and in a punctuate pattern to the midpiece of the sperm flagellum (see Fig. 4.3c, d). Remarkably, this pattern of binding was similar to that observed when frog sperm were exposed to the same allurin conjugate (Fig. 4.3a, b).

These findings suggested the need for a more detailed study of allurin-induced chemotaxis in mouse sperm using the Zigmond chamber. In these studies we compared the effects of allurin purified from egg water with those of mouse follicular fluid, a known source of chemoattractants for mouse sperm. Mouse sperm, in contrast to Xl sperm, are strong swimmers and can swim against gravity. Thus, the observational platform of the Zigmond chamber can face upward when evaluating

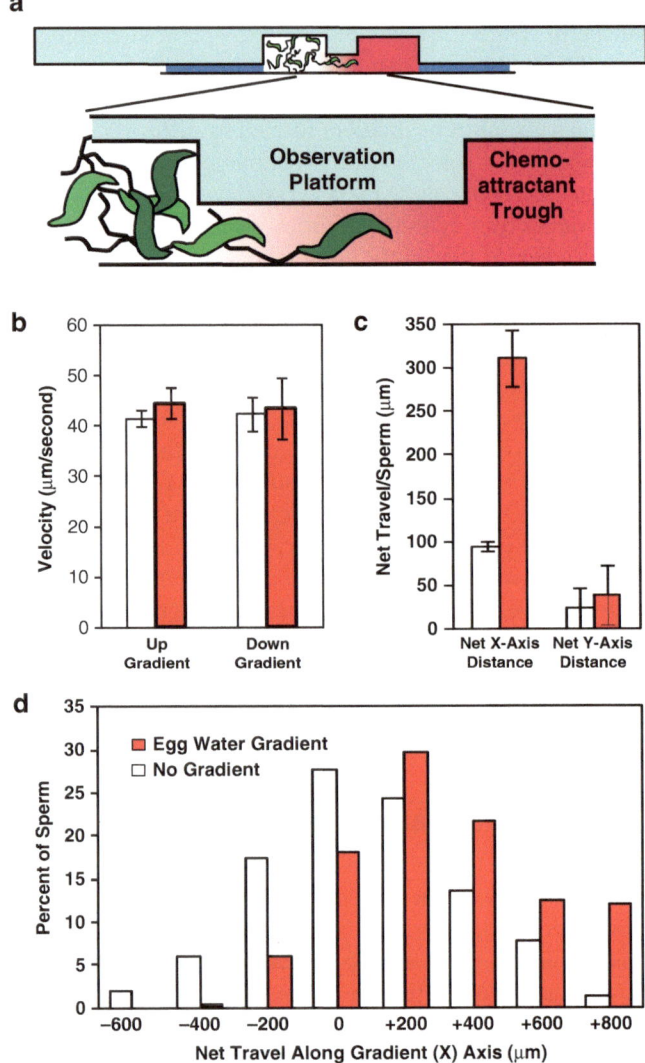

Fig. 4.2 Swimming behavior of *Xenopus laevis* sperm in a gradient of egg water that contains allurin. (**a**) Cut-away diagram of an inverted Zigmond chamber. Sperm trajectories are tracked under the observation platform of the chamber by video microscopy. The chamber is inverted because these sperm cannot swim against gravity. *Inset*: Sperm in the observation area are exposed to a gradient of chemoattractant. (**b**) Sperm velocity is unchanged in the presence of an egg-water gradient (*red/dark bars*) regardless of whether the sperm a swimming up or down the gradient. (**c**) The average net travel of sperm along the gradient (X) axis is increased threefold in the presence of an egg-water gradient but is not increased along the Y axis. (**d**) Distribution of net travel along the gradient (X) axis for individual sperm is shifted to large positive values in the presence of an egg-water gradient (*red/dark bars*) providing evidence of directed sperm movement toward the chemoattractant trough. *Error bars* represent the mean ± SE of 150 sperm in three experiments. (**a** reproduced with permission from Burnett et al. 2012; **b–d** reproduced from Burnett et al. 2011c)

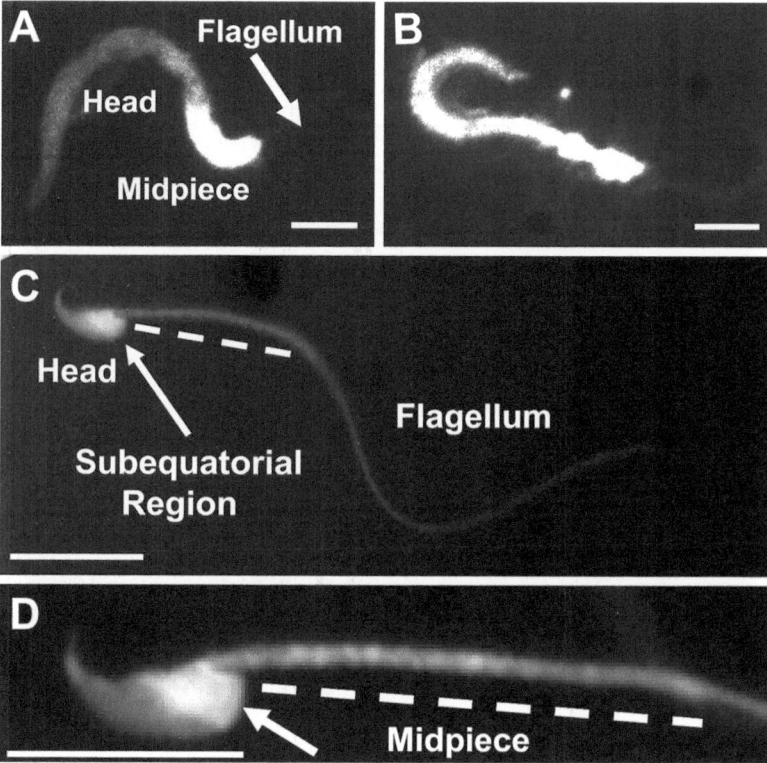

Fig. 4.3 Fluorescence micrographs of Oregon green 488-conjugated allurin bound to frog and mouse sperm. (**a, b**) Allurin binds to *X. laevis* sperm at the midpiece and to a variable extent at the head. No binding is observed at the flagellum. *Bars* = 2 μm. (**c, d**) Allurin binds to mouse sperm at the subequatorial region of the head (*arrows*) and at the midpiece of the flagellum (*dashed lines*). *Bars* = 10 μm. (**a** and **b** reproduced with permission from Burnett et al. 2011c; **c** and **d** reproduced with permission from Burnett et al. 2011b)

mouse sperm (Fig. 4.4a, inset), whereas the observational platform faces downward when observing Xl sperm (Fig. 4.2a, inset). As shown in Fig. 4.4b, both purified allurin (red/dark bar) and mouse follicular fluid (yellow/light bar) gradients produced no marked change in sperm swimming velocity when compared to sperm swimming in the absence of a gradient (white bar, Fig. 4.4b). In contrast, the net distance swum by sperm along the *X* (gradient) axis was increased dramatically in the presence of either an allurin or a follicular fluid gradient (first bar set, Fig. 4.4c). Such an increase was not seen along the *Y*-axis but instead random movement having a net displacement near zero, much like controls in which no chemotactic agent is present (second bar set, Fig. 4.4c). Differences in movement along the gradient (*X*)-axis between chemotaxing sperm and control sperm are clearly seen in distribution plots such as that in Fig. 4.4d. The presence of either an allurin (red/dark bars) or follicular fluid (yellow/light bars) gradient markedly decreases the percentage of sperm swimming down the gradient while markedly increasing the percentage of sperm swimming up gradient toward the chemoattractant trough.

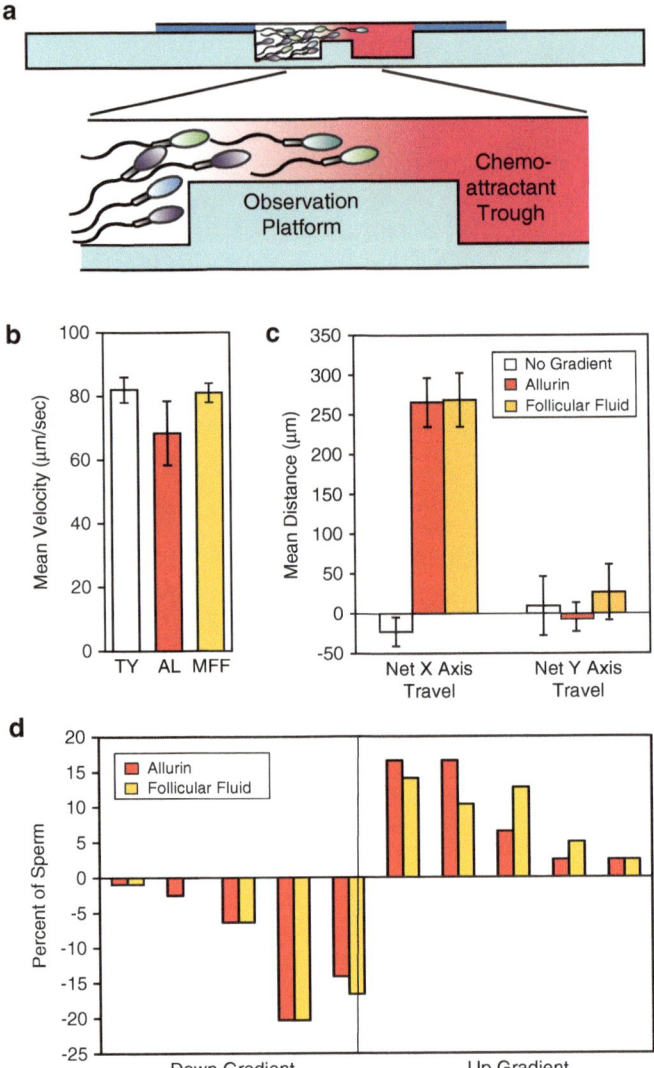

Fig. 4.4 Swimming behavior of mouse sperm in allurin and follicular fluid gradients. (**a**) Sperm movement is tracked on the observation platform of a Zigmond chamber. *Inset*: Sperm in the observation area are exposed to a gradient of chemoattractant emanating from the trough. (**b**) Average forward velocity of sperm in an allurin gradient (*red/dark bar*) and in a mouse follicular fluid gradient (*yellow/light bar*) is similar to that in controls (*open bar*). (**c**) The average net travel of sperm along the gradient (*X*) axis is substantial in the presence of an allurin or a follicular fluid gradient. (**d**) Distribution of differences in sperm travel along the gradient (*X*) axis (presence of chemoattractant - absence). Data is binned according to magnitude of *X* axis travel (vertical line is zero). The percentage of sperm traveling down gradient away from the chemoattractant was markedly reduced while the percentage of sperm traveling up gradient was markedly increased. *Error bars* represent the mean ± SE for 80 sperm in four experiments. *AL* allurin, *MFF* mouse follicular fluid. (**a** reproduced with permission from Burnett et al. 2012; **b, c, d** reproduced from Burnett et al. 2011b)

Fig. 4.5 Fluorescence immunocytochemical staining of a secondary follicle in the mouse ovary using anti-allurin antibodies. A strong signal is seen in the cytoplasm of mural granulosa cells and in the cumulus cells surrounding the oocyte. The antibodies used cross-react with full-length Crisp proteins, thus demonstrating their expression in developing follicles. Draq 5 was used as a nuclear counterstain. *Bar* = 25 μm. (Reproduced with permission from Burnett et al. 2012)

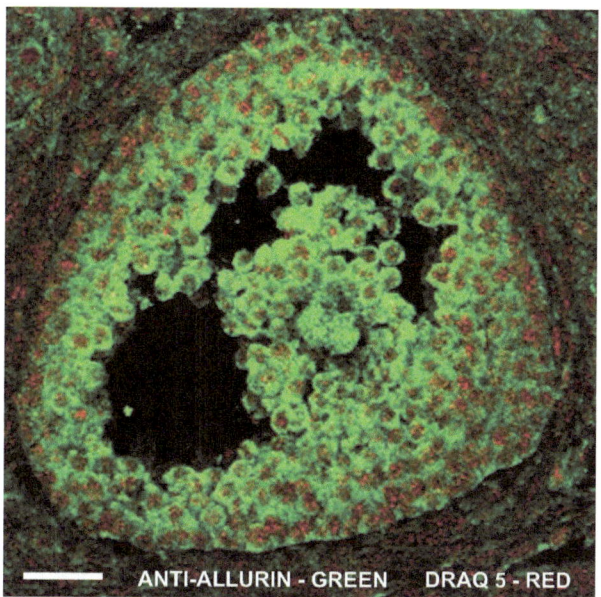

ANTI-ALLURIN - GREEN   DRAQ 5 - RED

These data provide strong evidence that allurin binds to and produces sperm chemoattractant behavior in both *Xenopus* and mouse sperm. Given that the identity of sperm chemoattractants in follicular fluid is still a matter of debate, these data also provide impetus for asking whether there are allurin-like chemoattractant proteins in follicular fluid. Two observations suggest that further study is warranted. First, immunocytochemistry using anti-allurin antibodies strongly labels the mural granulosa and cumulus cells surrounding the oocyte in secondary ovarian follicles of the mouse (Fig. 4.5). The localization appears to be cytoplasmic and is consistent with granulosa cell secretion of an allurin-like protein. Second, the secretion of such a protein into the follicular fluid can be verified by Western blot. As shown in Fig. 4.6, mouse follicular fluid, although containing hundreds of proteins of varying molecular weights (left, FF, Commassie stained), reveals just one band at 20 kDa that labels with anti-allurin antibodies (right, FF, double asterisk). This finding suggests the presence of a truncated Crisp protein in follicular fluid, albeit one whose identity and chemoattractant activity is presently unknown.

4.4 The Future of Crisp Protein Relationships in Reproduction

In retrospect, the discovery that allurin is also a chemoattractant for mammalian sperm should not have been surprising. CAP superfamily and Crisp subfamily proteins have entered into a broad array of reproductive mechanisms across both

Fig. 4.6 Western blot detection of an allurin-like protein in mouse follicular fluid. Follicular fluid (*FF*) contains numerous Coomassie blue-stained proteins spanning a wide range of sizes as shown by sodium dodecyl sulfate (SDS)-polyacrylamide gel electrophoresis (*left*, *FF*). Western blotting of the gel with anti-allurin antibodies reveals one labeled band at a relative mobility of 20 kDa (*right, FF, double asterisk*). This band is similar in mobility to the monomer band in purified *X. laevis* allurin (*right, AL, single asterisk*). (Reproduced with permission from Burnett et al. 2011b)

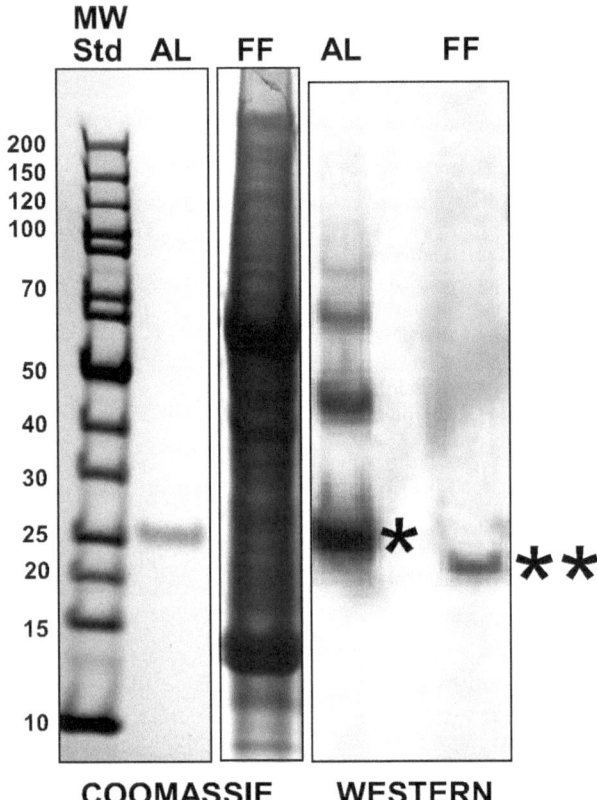

vertebrate and invertebrate animal phyla. Evolutionary relationships can be brought out by comparing domain organization and the positioning of the highly conserved cysteines which, through formation of disulfide bonds, play a vital role in tertiary structure and stability.

The PR domain, found as a stand-alone domain in both plants and animals, may have been the earliest domain of the CAP superfamily to evolve. It presumed origin before to the plant–animal bifurcation is hinted at by the existence of current-era yeast proteins that exhibit substantial homology to this domain Choudhary and Schneiter 2012). This domain, consisting of a β-pleated sheet sandwiched between α-helices on each face, is stabilized by three disulfide bonds (Fernandez et al. 1997). The PR domain has taken on a number of suspected roles during evolution including cation binding and proteolytic and disintegrin activities, as well as abilities to bind to sperm and oocyte surfaces.

Metal binding most notably includes high-affinity coordination with cadmium and zinc, abilities that could be related to sequestering of potentially lethal metals in the case of cadmium or a basis for metalloprotease activity in the case of zinc. The coordinating residues for cadmium or zinc binding have been clearly predicted from the X-ray crystallographic structures of Crisp snake toxins (Wang et al. 2005;

Guo et al. 2005; Shikamoto et al. 2005). Similarly, the structure of the Crisp-related protein glioma pathogenesis-related protein 1 in its zinc bound form has been determined (Asojo et al. 2011). Indeed, the PR domain structure bears some similarity to Zn^{2+}-requiring metalloproteases although only a few Crisp-related proteins have proven activity (e.g., Milne et al. 2003). However, these structural similarities may underlie the fact that the PR domain is capable of docking with a number of protein partners in other contexts.

A more detailed inspection of the sequence of this domain in a variety of CAP/CRISP proteins shows the presence of a PR domain (blue shading, Fig. 4.7) having at least four CAP/CRISP signature regions of high amino-acid identity (green shading, Fig. 4.7) and six to eight cysteines that are generally disulfide bonded (red on yellow, Fig. 4.7). Four of these cysteines [residues 73, 148, 153, and 164 in *Xenopus tropicalis* (Xt) allurin] are essentially invariant. In contrast, the positioning of two other cysteines shows variation between two subgroups of Crisp proteins. One subgroup represented by mammalian Crisps, snake toxin crisps, and a lamprey Crisp protein have these cysteines within the CAP/CRISP 2 and -4 signature sequences (residues 93 and 167 in Xt allurin). A second subgroup represented by plant PR-1 and ascidian sperm self-recognition proteins have these two cysteines near the CAP/CRISP signature 1 sequence at residues 109 and 115 in the PR-1 protein of tomato (arrows, Fig. 4.7). Xt allurin and a hypothetical Crisp protein in finch both conform perfectly to the first subgroup and thus might be expected to have a tertiary structure much like that delineated in the X-ray diffraction structures of snake toxin Crisp proteins. If so, one can expect all six of the cysteines in the PR domains of these proteins to be disulfide bonded and the pattern of bonding to be overlapping as in the toxin proteins (Wang et al. 2005; Guo et al. 2005).

Unexpectedly, some truncated Crisp proteins, including *X. laevis* (Xl) allurin and wasp venom proteins, do not conform to either pattern of cysteines and in fact have been shown to have two free cysteines that are not disulfide bonded (Sugiyama et al. 2009; Henriksen et al. 2001). What this means for the tertiary structure of Xl allurin is not clear, although its structure in the PR domain is likely to be quite different from that of Xt allurin. It comes as a surprise then that Xl allurin and Xt allurin show similar chemoattractant activity, even across species (Burnett et al. 2008b); this would seem to imply that the chemoattractant activity of the allurins is mediated by a different region of the PR domain or by the hinge domain as hypothesized next.

Indeed, the PR domain appears to have come into greater use when coupled with the smaller hinge domain. Although the origin of the hinge domain is uncertain, its amino-acid sequence is highly homologous in all CAP/CRISP proteins and notably features four invariant cysteines (yellow/light shading, Fig. 4.7). In X-ray diffraction structures of CAP proteins these cysteines form a pair of overlapping disulfide bonds that maintain the consistent "chair-like" tertiary structure of this domain (Wang et al. 2005; Guo et al. 2005). The widespread use of this PR domain–Hinge domain combination in Crisp proteins ranges from ascidians, worms, and snails in the invertebrates to amphibians, snakes, and mammals in the vertebrates. Given this, it may pertinent to ask what kinds of increased functionality have been gained by the use of this domain together with the PR domain.

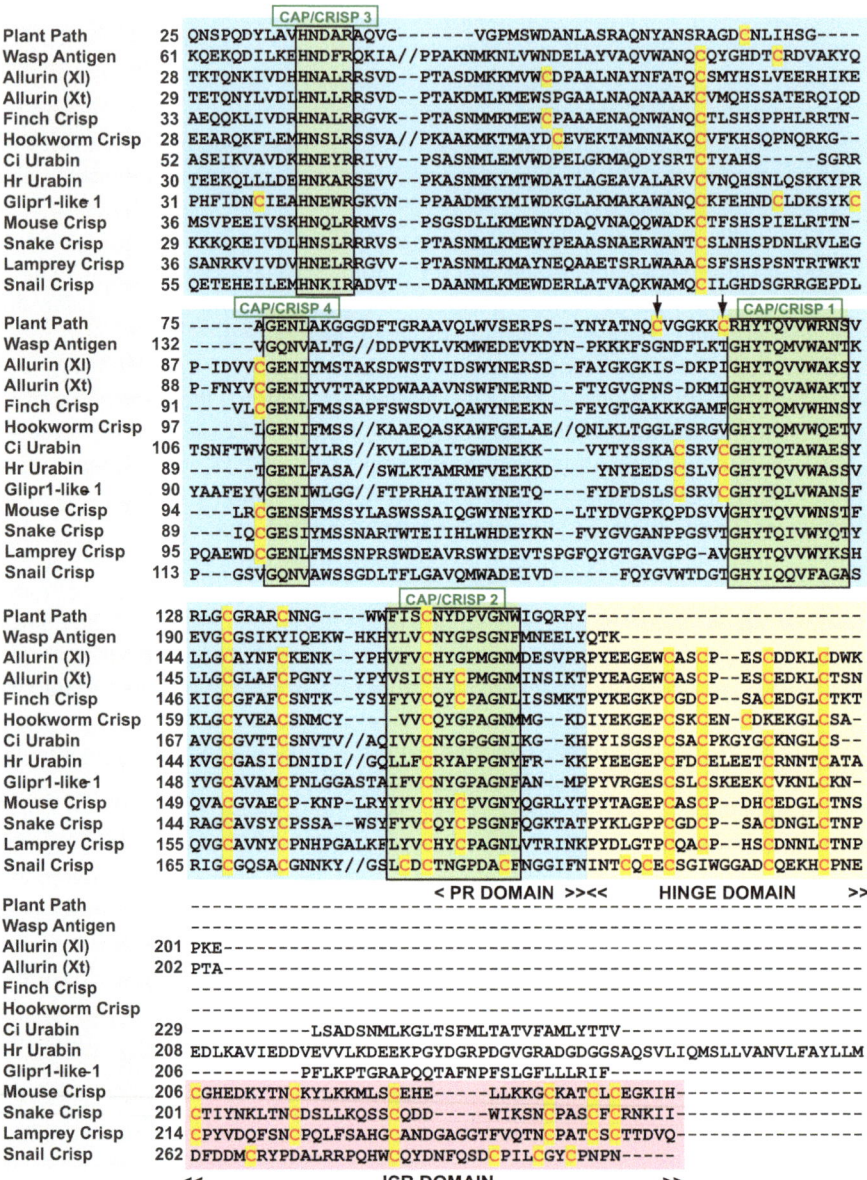

Fig. 4.7 Clustal W-aligned amino-acid sequences of homologous CAP/CRISP proteins. Signal sequences and initial nonhomologous N-terminal regions have been omitted for clarity. The PR, Hinge, and ICR (ion channel regulatory) domains are shaded in *blue, yellow,* and *pink* respectively. The PR domain includes four CAP/CRISP signature sequences of high homology (*boxed* and *shaded* in *green*). Cysteines are highlighted in *red* on *yellow*. Initial residue positions are indicated at the *left* of each sequence line. *Double slashes* indicate short internal sequences omitted for clarity. Genbank acquisition numbers for the sequences are: AJ011520.1/GI:3660528, M98858.1/GI:162550, AF393653.2/GI:317185854, NM001201342.1/GI:318065112, XM002191406.1/GI:224048894, AY288089.1/GI:31075034, NM001246286.1/GI:350534677, BAG68488.1/GI:195972735, NP689992.1/GI:22749527, BC011150.1/GI:15029853, AY324325.1/GI:32492058, BAF56484.1/GI:145046200, AJ491318.1/GI:32187774

One possibility is that rather than acting independently, these two domains might both contribute surfaces that together allow new protein–protein interactions. An interesting piece of data that hints at this comes from the X-ray crystal structure of pseudecin, one of the Crisp protein snake toxins from the viper *Pseudecis porphyriacus* that inhibits cyclic nucleotide gated channels (Suzuki et al. 2008). In this protein, the surface formed by the C-terminal-most region of the PR domain and a surface formed by the hinge domain face one another to provide a deep cleft that could be a site of protein docking. Indeed, binding of Zn^{2+} by this protein results in a marked shift in the positions of these two surfaces, and the authors have postulated that this may regulate the interaction of this protein with the channel (Suzuki et al. 2008). Similar tantalizing results come from our studies of peptides representing sequences within allurin that mimic this region of the PR and Hinge domains. Synthetic peptides representing the Hinge region and the C-terminal of the PR domain have been found to have dose-dependent chemoattractant activity for both frog and mouse sperm in two-chamber assays (Washburn et al. 2011; Washburn, unpublished observations). In contrast, peptides of random sequence or having homologous sequences from snake toxin Crisp proteins do not show activity. Of further interest is that these peptides bind to sperm at the midpiece, the same spatial pattern seen in binding of allurin to sperm.

As indicated in Fig. 4.1, full-length Crisp proteins contain the PR domain–Hinge domain combination just discussed as well as the ion channel regulatory (ICR) domain. The ICR domain, as previously mentioned, has been demonstrated to modulate the actions of voltage-dependent potassium channels, voltage-dependent calcium channels, and ryanodine receptors (Gibbs et al. 2006; Brown et al. 1999; Morrissette et al. 1995; Nobile et al. 1994, 1996). These properties target channels in muscle and nerve during the actions of Crisp protein toxins found in snakes and lizards. Although the presence of Crisp proteins in insect venoms would seem to be a parallel use it should be pointed out that the ICR domain is not present in the truncated Crisp proteins found in these venoms.

In mammals, full-length Crisp proteins are associated with sperm production in the testis, sperm maturation in the epididymis, and sperm capacitation and oocyte binding in the female reproductive tract (Gibbs et al. 2008; Burnett et al. 2008a). One might speculate that the ICR domain of these proteins may be modifying the behavior of ion channels in sperm. Ion channel activity is important in sperm capacitation, and there is evidence that Crisp 1 may be acting as a decapacitation factor that keeps certain channels quiescent in the male tract, but which upon removal during capacitation in the female tract, allows new channel activity (Nixon et al. 2006). Indeed, Crisp 4, a rodent orthologue of Crisp 1, has been shown to modify TRP-M8 channels in mouse sperm that may be involved in sperm capacitation and chemotaxis (Gibbs et al. 2010a).

Interestingly, the ICR domain has structural similarities to sea anemone toxin peptides that have been shown to block voltage-dependent potassium and calcium channels. In vertebrates, however, the ICR does not appear as an independent domain but is virtually always within the context of a full-length Crisp protein.

This observation would seem to imply that the PR domain–Hinge domain combination associated with it must play some role in enhancing the actions of the ICR domain. One possibility is that these domains are important for cell-surface binding and localization so as to target ICR domain actions. Evidence for this capacity comes from the fact that Crisp 1 binds to the mouse sperm surface in the epididymis and remains bound to the sperm in the female tract (Cohen et al. 2000; Da Ros et al. 2004; Roberts et al. 2002, 2008). Its localization on the sperm and its binding lifetime are dependent on its glycosylation state. The "D" form of Crisp 1 binds to the sperm head in the postacrosomal region and remains bound until ejaculated into the female tract. The "D" form inhibits premature acrosome reaction in the male tract, but once in the female tract appears to unbind from the sperm during capacitation (Roberts et al. 2008; Nixon et al. 2006). In contrast, the "E" form of Crisp 1, having a different glycosylation pattern, binds to sperm in the epididymis, but is not easily removed even in the female tract (Da Ros et al. 2004; Roberts et al. 2008); rather, it migrates to the equatorial region of the sperm where it is thought to take part in sperm binding at the egg plasma membrane (Cohen et al. 2008, 2011).

Indeed, binding of Crisp 1 and Crisp 2 to the egg plasma membrane of rats has been characterized (Ellerman et al. 2006; Cohen et al. 2011). The structural feature of Crisp 1 responsible for egg binding appears to lie in the CAP/CRISP 2 signature region of the PR domain. This signature is common to both Crisp 1 and Crisp 2, and the fact that these two Crisp proteins compete for the same site on the egg surface suggests that this signature is likely the binding partner in both proteins (Cohen et al. 2011). If peptides representing this signature are incubated with eggs, peptide binding is not only detected but is accompanied by block of sperm–egg plasma membrane adhesion and subsequent fertilization (Ellerman et al. 2006). Similarly, we have shown that allurin, absent an ICR domain, binds to the sperm head at the equatorial region in a manner similar to Crisp 1, but different in binding to the midpiece of the flagellum as well (see Fig. 4.3).

Of additional interest is that the "E" form of Crisp 1, during production in the epididymis, is proteolytically processed in a manner that eliminates the ICR domain, leaving a truncated Crisp 1 that bears a lot of similarity to allurin, albeit highly glycosylated (Roberts et al. 2002). This point raises the possibility that proteolytic processing of a full-length Crisp protein in the ovarian follicle could lead to the allurin-like protein detected in mouse follicular fluid that we have hypothesized might bind to sperm and act as a chemoattractant (see Fig. 4.7).

Although the binding partners of mammalian Crisp proteins on the egg surface are not yet known, binding partners of Crisp proteins have been studied in sperm. In mouse sperm, a cell-surface protein designated SHTAP has been shown to bind Crisp 2, be expressed exclusively in the testis, and to relocalize during sperm capacitation (Jamsai et al. 2009). In addition, binding of Crisp 2 to SHTAP in a yeast two-hybrid system seems to require both the PR and Hinge domains as well as the ICR domain to maximize binding. Again, the region in the PR domain required for binding was that of the CAP/CRISP 2 signature sequence at the C-terminal of the domain (Jamsai et al. 2009).

Parallel data come from truncated Crisp proteins found on the surface of ascidian sperm. As demonstrated by Sawada and coworkers, ascidian sperm display GPI-anchored Crisp proteins that play a role in mediating self-recognition between sperm and egg so as to prevent self-fertilization (Urayama et al. 2008; Yamaguchi et al. 2011). In two different species, *H. roretzi* and *C. intestinalis*, a GPI-anchored Urabin on the surface of the sperm head binds to a specific partner protein in the vitelline coat of the conspecific egg (VC70 and VC57 in the two respective species). Initial binding of the sperm Urabin to the vitelline coat supports subsequent binding of a pair of polycystin-1-like proteins on the sperm surface (s-Themis A and B) with binding partners on the vitelline coat (v-Themis A and B) (Yamaguchi et al. 2011). Both these specific sperm–vitelline coat interactions are then thought to lead to weakening of the interaction and removal of the sperm, thus blocking fertilization.

In summary, data from multiple sperm species support the hypothesis that the CAP/CRISP signatures 1 and 2 mediate binding and localization of Crisp proteins to target cells and this binding is enhanced by the Hinge domain. In some cases, as in the mammalian and snake venom Crisp proteins, such binding may be a requirement for actions of the ICR domain on ion channels in the membrane of the target cell. In the case of allurin this binding appears to be to a yet-uncharacterized receptor leading to signal transduction events that modulate sperm flagellar activity and chemotaxis.

4.5   Conclusion

Allurin, a 21-kDa protein synthesized in the *Xenopus* oviduct, is incorporated into the jelly surrounding spawned *Xenopus* eggs. Upon release into the surrounding medium the protein binds to an uncharacterized receptor on the *Xenopus* sperm midpiece and elicits chemotactic behavior; that is, sperm preferentially swim up an allurin gradient toward higher concentrations of the protein. Surprisingly, allurin elicits a similar response in mouse sperm. The amino acid sequence of allurin indicates that it is a truncated member of the Crisp family having two domains. Our results suggest that either the pathogenesis-related domain or Hinge domain of allurin activates an evolutionarily conserved signaling system that controls flagellar waveform and directional swimming. Future research is needed to determine whether Crisp protein signaling systems play a role in mammalian sperm chemotaxis in vivo.

References

Al-Anzi B, Chandler DE (1998) A sperm chemoattractant is released from *Xenopus* egg jelly during spawning. Dev Biol 198:366–375

Alessandri-Haber N, Lecoq A, Gasparini S et al (1999) Mapping the functional anatomy of BgK on Kv1.1, Kv1.2, and Kv1.3. Clues to design analogs with enhanced selectivity. J Biol Chem 274:35653–35661

Asojo OA, Koski RA, Bonafé N (2011) Structural studies of human glioma pathogenesis-related protein 1. Acta Crystallogr D Biol Crystallogr 67:847–855

Bjartell A, Johansson R, Bjork T et al (2006) Immunohistochemical detection of cysteine-rich secretory protein 3 in tissue and in serum from men with cancer or benign enlargement of the prostate gland. Prostate 66:591–603

Brown RL, Haley TL, West KA et al (1999) Pseudechetoxin: a peptide blocker of cyclic nucleotide-gated ion channels. Proc Natl Acad Sci USA 96:754–759

Burnett LA, Xiang X, Bieber AL et al (2008a) Crisp proteins and sperm chemotaxis: discovery in amphibians and explorations in mammals. Int J Dev Biol 52:489–501

Burnett LA, Boyles S, Spencer C et al (2008b) *Xenopus tropicalis* allurin: expression, purification and characterization of a sperm chemoattractant that exhibits cross-species activity. Dev Biol 316:408–416

Burnett LA, Tholl N, Chandler DE (2011a) Two types of assays for detecting frog sperm chemoattraction. J Vis Exp 58(1-8):e3407. doi:10.3791/3407

Burnett L, Anderson D, Rawls A et al (2011b) Mouse sperm exhibit chemotaxis to allurin, a truncated member of the cysteine-rich secretory protein family. Dev Biol 360:318–328

Burnett L, Sugiyama H, Bieber A et al (2011c) Egg jelly proteins stimulate directed motility in *Xenopus laevis* sperm. Mol Reprod Dev 78:450–462

Burnett LA, Washburn CA, Sugiyama H et al (2012) Allurin, an amphibian sperm chemoattractant having implications for mammalian sperm physiology. Int Rev Cell Mol Biol 295:1–61

Busso D, Goldweic N, Hayashi M et al (2007) Evidence for the involvement of testicular protein CRISP2 in mouse sperm–egg fusion. Biol Reprod 76:701–708

Choudhary V, Schneiter R (2012) Pathogen-related yeast (PRY) proteins and members of the CAP superfamily are secreted sterol-binding proteins. Proc Natl Acad Sci USA 109:16882–16887

Cohen DJ, Rochwerger L, Ellerman DA et al (2000) Relationship between the association of rat epididymal protein "DE" with spermatozoa and the behavior and function of the protein. Mol Reprod Dev 56:180–188

Cohen DJ, Da Ros VG, Busso D et al (2007) Participation of epididymal cysteine-rich secretory proteins in sperm–egg fusion and their potential use for male fertility regulation. Asian J Androl 9:528–532

Cohen DJ, Busso D, Da Ros V et al (2008) Participation of cysteine-rich secretory proteins (CRISP) in mammalian sperm–egg interaction. Int J Dev Biol 52:737–742

Cohen DJ, Maldera JA, Vasen G et al (2011) Epididymal protein CRISP1 plays different roles during the fertilization process. J Androl 32:672–678

Cotton J, Crest M, Couet F et al (1997) A potassium-channel toxin from the sea anemone *Bunodosoma granulifera*, an inhibitor for Kv1 channels. Revision of the amino acid sequence, disulfide-bridge assignment, chemical synthesis, and biological activity. Eur J Biochem 244:192–202

Da Ros VG, Munice MJ, Cohen DJ et al (2004) Bicarbonate is required for migration of sperm epididymal protein DE (CRISP 1) to the equatorial segment and expression of rat sperm fusion ability. Biol Reprod 70:1325–1332

Da Ros VG, Maldera JA, Willis WD et al (2008) Impaired sperm fertilizing ability in mice lacking Cysteine-RIch Secretory Protein 1 (CRISP1). Dev Biol 320:12–18

Ellerman DA, Cohen DJ, Da Ros VG et al (2006) Sperm protein "DE" mediates gamete fusion through an evolutionarily conserved site of the CRISP family. Dev Biol 297:228–237

Fabro G, Rovasio RA, Civalero S et al (2002) Chemotaxis of capacitated rabbit spermatozoa to follicular fluid revealed by a novel directionality-based assay. Biol Reprod 67:1565–1571

Fernandez C, Szyperski T, Bruyere T et al (1997) NMR solution structure of the pathogenesis-related protein P14a. J Mol Biol 266:576–593

Foster GA, Gerton JA (1996) Autoantigen 1 of the guinea pig sperm acrosome is the homologue of mouse Tpx-1 and human TPX1 and is a member of the cysteine-rich secretory protein (CRISP) family. Mol Reprod Dev 44:221–229

Gibbs GM, Scanlon MJ, Swarbrick J et al (2006) The cysteine-rich secretory protein domain of Tpx-1 is related to ion channel toxins and regulates ryanodine receptor calcium signaling. J Biol Chem 281:4156–4163

Gibbs GM, Roelants K, O'Bryan MK (2008) The CAP superfamily: cysteine-rich secretory proteins, antigen 5, and pathogenesis-related 1 proteins–roles in reproduction, cancer, and immune defense. Endocr Rev 29:865–897

Gibbs GM, Orta G, Reddy T et al (2010a) Cysteine-rich secretory protein 4 is an inhibitor of transient receptor potential M8 with a role in establishing sperm function. Proc Natl Acad Sci USA 108:7034–7039

Gibbs GM, Lo JCY, Nixon B et al (2010b) Glioma pathogenesis-related 1-like 1 is enriched, dynamically modified, and redistributed during male germ cell maturation and has a potential role in sperm–oocyte binding. Endocrinology 151:2331–2342

Guo M, Teng M, Niu L et al (2005) Crystal structure of the cysteine-rich secretory protein stecrisp reveals that the cysteine-rich domain has a K$^+$ channel inhibitor-like fold. J Biol Chem 280:12405–12412

Hardy DM, Huang TT, Driscoll WJ et al (1988) Purification and characterization of the primary acrosomal autoantigen of guinea pig epididymal spermatozoa. Biol Reprod 38:423–437

Henriksen A, King TP, Mirza O et al (2001) Major venom allergen of yellow jackets, Ves v 5: structural characterization of a pathogenesis-related protein superfamily. Proteins 45:438–448

Jamsai D, Rijal S, Bianco DM et al (2009) A novel protein, sperm head and tail associated protein (SHTAP), interacts with cysteine-rich secretory protein 2 (CRISP2) during spermatogenesis in the mouse. Biol Cell 102:93–106

Katagiri C (1987) Role of oviductal secretions in mediating gamete fusion in anuran amphibians. Zool Sci 4:1–14

Koppers AJ, Reddy T, O'Bryan MK (2011) The role of cysteine-rich secretory proteins in male fertility. Asian J Androl 13:111–117

Krapf D, O'Brien ED, Cabada MO et al (2009) Egg water from the amphibian *Bufo arenarum* modulates the ability of homologous sperm to undergo the acrosome reaction in the presence of the vitelline envelope. Biol Reprod 80:311–319

Kratzschmar J, Haendler B, Eberspaecher U et al (1996) The human cysteine-rich secretory protein (CRISP) family. Primary structure and tissue distribution of CRISP 1, CRISP 2 and CRISP 3. Eur J Biochem 236:827–836

Milne TJ, Abbenante G, Tyndall JD et al (2003) Isolation and characterization of a cone snail protease with homology to CRISP proteins of the pathogenesis-related protein superfamily. J Biol Chem 278:31105–31110

Morrissette J, Kratzschmar J, Haendler B et al (1995) Primary structure and properties of helothermine, a peptide toxin that blocks ryanodine receptors. Biophys J 68:2280–2288

Nixon B, MacIntyre DA, Mitchell LA et al (2006) The identification of mouse sperm-surface-associated proteins and characterization of their ability to act as decapacitation factors. Biol Reprod 74:275–287

Nobile M, Magnelli V, Lagostena L et al (1994) The toxin helothermine affects potassium currents in newborn rat cerebellar granule cells. J Membr Biol 139:49–55

Nobile M, Noceti F, Prestipino G et al (1996) Helothermine, a lizard venom toxin, inhibits calcium current in cerebellar granules. Exp Brain Res 110:15–20

Nolan MA, Wu L, Bang HJ et al (2006) Identification of rat cysteine-rich secretory protein 4 (Crisp 4) as the ortholog to human CRISP 1 and mouse Crisp 4. Biol Reprod 74:984–991

Olson JH, Chandler DE (1999) *Xenopus laevis* egg jelly contains small proteins that are essential for fertilization. Dev Biol 210:401–410

Olson J, Xiang X, Ziegert T et al (2001) Allurin, a 21 kD sperm chemoattractant from *Xenopus* egg jelly, is homologous to mammalian sperm-binding proteins. Proc Natl Acad Sci USA 98:11205–11210

Pennington MW, Lanigan MD, Kalman K et al (1999) Role of disulfide bonds in the structure and potassium channel blocking activity of ShK toxin. Biochemistry 38:14549–14558

Roberts KP, Ensrud KM, Hamilton DW (2002) A comparative analysis of expression and processing of the rat epididymal fluid and sperm-bound forms of proteins D and E. Biol Reprod 67:525–533

Roberts KP, Ensrud KM, Wooters JL et al (2006) Epididymal secreted protein Crisp 1 and sperm function. Mol Cell Endocrinol 250:122–127

Roberts KP, Johnston D, Nolan MA et al (2007) Structure and function of epididymal protein cysteine-rich secretory protein-1. Asian J Androl 9:508–514

Roberts KP, Ensrud-Bowlin KM, Piehl LB et al (2008) Association of the protein D and protein E forms of rat CRISP1 with epididymal sperm. Biol Reprod 79:1046–1053

Schambony A, Hess O, Gentzel M et al (1998a) Expression of CRISP proteins in the male equine genital tract. J Reprod Fertil Suppl 53:67–72

Schambony A, Gentzel M, Wolfes H et al (1998b) Equine CRISP 3: primary structure and expression in the male genital tract. Biochim Biophys Acta 1387:206–216

Shikamoto Y, Suto K, Yamazaki Y et al (2005) Crystal structure of a CRISP family Ca^{2+}-channel blocker derived from snake venom. J Mol Biol 350:735–743

Sugiyama H, Burnett L, Xiang X et al (2009) Purification and multimer formation of allurin, a sperm chemoattractant from *Xenopus laevis* egg jelly. Mol Reprod Dev 76:527–536

Suzuki N, Yamazaki Y, Brown RL et al (2008) Structures of pseudechetoxin and pseudecin, two snake-venom cysteine-rich secretory proteins that target cyclic nucleotide-gated ion channels: implications for movement of the C-terminal cysteine-rich domain. Acta Crystallogr D 64:1034–1042

Topfer-Petersen E, Ekhlasi-Hundrieser M, Kirchhoff C et al (2005) The role of stallion seminal proteins in fertilisation. Anim Reprod Sci 89:159–170

Turunen H, Sipila P, Krutskikh A et al (2012) Loss of cysteine-rich secretory protein 4 (Crisp4) leads to deficiency in sperm–zona pellucida interaction in mice. Biol Reprod 86:121–128

Udby L, Bjartell A, Malam J et al (2005) Characterization and localization of cysteine-rich secretory protein 3 (CRISP 3) in the human male reproductive tract. J Androl 26:333–342

Udby L, Johnsen AH, Borregaard N (2010) Human CRISP-3 binds serum alpha(1)B-glycoprotein across species. Biochim Biophys Acta 1800:481–485

Urayama S, Harada Y, Nakagawa Y et al (2008) Ascidian sperm glycosylphosphatidylinositol-anchored CRISP-like protein as a binding partner for an allorecognizable sperm receptor on the vitelline coat. J Biol Chem 283:21725–21733

Wang J, Shen B, Guo M et al (2005) Blocking effect and crystal structure of natrin toxin, a cysteine-rich secretory protein from *Naja atra* venom that targets the BKCa channel. Biochemistry 44:10145–10152

Washburn CA, Bieber AL, Tubbs K et al (2011) Mammalian sperm chemotaxis is elicited by peptide mimics of cysteine rich secretory proteins. In: Abstracts, National meeting of the Society for the Study of Reproduction, Portland

Yamaguchi A, Saito T, Yamada L et al (2011) Identification and localization of the sperm CRISP family protein CiUrabin involved in gamete interaction in the ascidian *Ciona intestinalis*. Mol Reprod Dev 78:488–497

Zigmond SH (1977) Ability of polymorphonulear leukocytes to orient in gradients of chemical factors. J Cell Biol 75:606–616

Chapter 5
Structure, Function, and Phylogenetic Consideration of Calaxin

Kazuo Inaba, Katsutoshi Mizuno, and Kogiku Shiba

Abstract Sperm chemotaxis is widely seen both in animals and plants and is considered to be necessary for efficient success of fertilization. Although intracellular Ca^{2+} is known to play important roles in sperm chemotaxis, the molecular mechanism causing the change in flagellar waveform that drives sperm directed toward the egg is still unclear. Several Ca^{2+}-binding proteins, especially calmodulin, have been discussed as an important regulator of the molecular motor dynein in flagellar motility during chemotactic movement of sperm. However, there has been no experimental evidence to show the binding of calmodulin to dyneins. Recently, we found a novel Ca^{2+}-binding protein, termed calaxin, in the axonemes of sperm flagella in the ascidian *Ciona intestinalis*. Calaxin binds to the outer arm dynein in a Ca^{2+}-dependent manner and suppresses its activity to slide microtubules at high Ca^{2+} concentration. Inhibition of calaxin results in significant loss of chemotactic behavior of sperm, indicating that calaxin is essential for sperm chemotaxis. In this chapter, we describe the finding history, molecular nature, and the roles in sperm chemotaxis of calaxin, as well as its phylogenetic consideration.

Keywords Axonemal dynein • Calaxin • Opisthokont • Sperm chemotaxis

5.1 Ca^{2+} and Flagellar Motility

The flagellar wave is composed of a bend with larger angle called the principal bend (P-bend) and a bend with a smaller angle called the reverse bend (R-bend). Sperm showing the same extent of both bends show symmetrical waveform and swim

K. Inaba (✉) • K. Mizuno • K. Shiba
Shimoda Marine Research Center, University of Tsukuba,
5-10-1 Shimoda, Shizuoka 415-0025, Japan
e-mail: kinaba@kurofune.shimoda.tsukuba.ac.jp

H. Sawada et al. (eds.), *Sexual Reproduction in Animals and Plants*,
DOI 10.1007/978-4-431-54589-7_5, © The Author(s) 2014

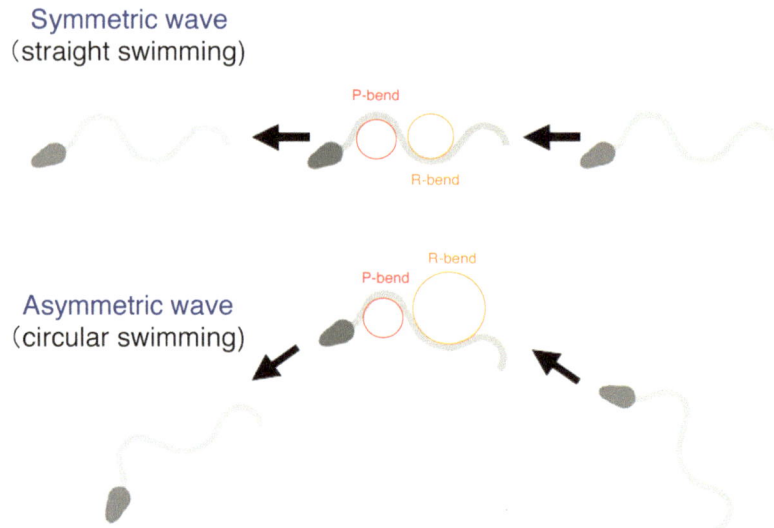

Fig. 5.1 Flagellar waveform and the direction of sperm movement. The flagellar wave is composed of a large principal bend (P-bend) and a smaller reverse bend (R-bend). Sperm with almost the same extent of P-bend and R-bend move straight, whereas those with larger R-bend move in circular fashion. Inverse of the radius of inscribed *circle* represents curvature, which is a parameter to express the extent of flagellar bending

straight, whereas a decrease in R-bend results in asymmetry of waveform and the circular swimming of sperm (Fig. 5.1). This waveform conversion is conducted by the regulation of dynein-driven axonemal motility. It has been well known that Ca^{2+} plays an important role in the regulation of the axonemal motility in eukaryotic flagella and cilia (Kamiya and Witman 1984; Gibbons and Gibbons 1980; Sale 1986). In fact, increasing the concentration of Ca^{2+} in Triton-demembranated sperm induces conversion of the flagellar waveform from symmetry to asymmetry (Brokaw 1979). An extremely asymmetrical waveform is induced at very high calcium concentrations (Gibbons and Gibbons 1980; Sale 1986). Sperm with this waveform show cane-shaped "quiescence," which is observed by electric or mechanical stimulation of sperm flagella in the sea urchin (Shingyoji and Takahashi 1995; Kambara et al. 2011). On the self-nonself recognition of sperm and egg in *Ciona*, sperm show quiescence with a straight flagellum in response to the increase in intracellular Ca^{2+} (Saito et al. 2012). The basic regulation of flagellar waveform by Ca^{2+} is thought to be performed by the specific activation of dynein arms, which results in the changes in flagellar waveforms (Brokaw 1979; Lindemann and Goltz 1988). Accumulating evidence indicates that the activity of inner arm dyneins is regulated by signals from the radial spoke/central pair in a Ca^{2+}-dependent manner (Smith 2002; Nakano et al. 2003). On the other hand, independent regulation of the outer arm dynein by Ca^{2+} is also pointed out (Mitchell and Rosenbaum 1985; Wakabayashi et al. 1997; Sakato and King 2003).

Calmodulin has been a strong candidate to regulate the conversion of flagellar and ciliary waveform. In fact, several studies have discussed the presence and potential roles of calmodulin in *Tetrahymena* cilia (Jamieson et al. 1979; Blum et al. 1980), *Chlamydomonas* flagella (Gitelman and Witman 1980), and sperm flagella (Tash and Means 1983; Brokaw and Nagayama 1985; Lindemann et al. 1991). It is well demonstrated that calmodulin is present in radial spokes and central pair and regulates the function of these structures in the modulation of axonemal dyneins in *Chlamydomonas* (Smith and Yang 2004). Another Ca^{2+}-binding protein, centrin, is known to be a component of inner arm dynein (Piperno et al. 1992).

In contrast, analysis of *Chlamydomonas* mutants indicates that outer arm dyneins are essential for conversion of waveform asymmetry in response to changes in Ca^{2+} concentration (Kamiya and Okamoto 1985; Wakabayashi et al. 1997; Sakato and King 2003). In fact, a Ca^{2+}-binding protein is contained in the outer arm dynein as a light chain (LC4) (King and Patel-King 1995). Another Ca^{2+}-binding protein associated with the outer arm dynein in *Chlamydomonas* flagella is DC3, a component of outer arm dynein docking complex (ODA-DC) (Casey et al. 2003). DC3 protein is structurally distinct from other Ca^{2+}-binding proteins in *Chlamydomonas* flagella, such as calmodulin, centrin, and LC4. Intriguingly this protein shows sequence similarity to a protein predicted in the Apicomplexa *Plasmodium yoelii* and *Plasmodium falciparum* (Casey et al. 2003), but its orthologue has not been found in the genome of the ascidian *Ciona intestinalis* (Hozumi et al. 2006). Thus, the Ca^{2+}-binding protein that regulates axonemal dyneins had not been fully characterized in sperm flagella. Calmodulin was reported to regulate flagellar motility and could be extracted from axonemes with outer arm dynein, but it has not been clarified whether calmodulin can bind directly to outer arm dynein (Tash et al. 1988). In fact, isolated outer arm dynein does not contain calmodulin as a subunit in *Ciona* and sea urchin (Inaba 2007).

5.2   Finding Calaxin

During the course of immunoscreening-based cDNA screening for axonemal proteins in *C. intestinalis*, we isolated multiple clones from testis cDNA library encoding a protein with sequence similarity to calcineurin B (Padma et al. 2003). Phylogenetic analysis revealed that this protein is grouped not into calcineurin B but into a family of neuronal calcium sensor (NCS). We named this novel Ca^{2+}-binding NCS family protein in *Ciona* as "calaxin," for calcium-binding axonemal protein (Mizuno et al. 2009).

NCS proteins have been identified in many organisms ranging from yeast to human (Burgoyne and Weiss 2001). Major five classes of NCS have been well studied in human: NCS-1 (frequenin), neurocalcin and its related proteins (visinin-like protein VILIP and hippocalcin), recoverin, GCAP (guanylyl cyclase-activating protein), and KChIP (Kv channel-interacting protein) (Burgoyne 2004). In mammals, recoverin and GCAPs are expressed only in the retina and regulate phototransduction and others are expressed in neuronal tissues. NCS-1 is also expressed in many nonneuronal cell types, and its orthologue is present in yeast. The NCS proteins contain four EF hand motifs but only three (or two in the case of recoverin and KChIP1) are able to bind Ca^{2+}. Eleven of 15 mammalian NCS proteins are N-terminally myristoylated, which are important in Ca^{2+}-dependent interaction with the plasma membrane. In contrast to these NCS proteins, calaxin does not possess the N-terminal consensus motif for myristoylaion and belongs to a class distinct from these five NCS classes (Mizuno et al. 2009).

Immunolocalization reveals that calaxin is localized at the vicinity of the outer arm dyneins (Mizuno et al. 2009). Sucrose density gradient centrifugation clearly indicates that calaxin directly interacts with the outer arm dynein in a Ca^{2+}-dependent manner. Far Western blotting and a cross-linking experiment show that calaxin binds to the β-heavy chain in the presence of Ca^{2+}, whereas it binds to β-tubulin in both the presence and absence of Ca^{2+}. Preliminary experiments showed that calaxin binds to the N-terminal stem region of β-heavy chain (Mizuno, unpublished observation).

Although a phylogenetic analysis shows that *Chlamydomonas* LC4 and *Ciona* calaxin are grouped into different classes of Ca^{2+}-binding protein, there are many similarities between them (Sakato and King 2003; Sakato et al. 2007; Mizuno et al. 2009; 2012). First, they appear to undergo dynamic conformational change in response to Ca^{2+} binding. Second, their binding sites are the stems of specific dynein heavy chains: γ-heavy chain of *Chlamydomonas* outer arm dynein for LC4 and its orthologue in *Ciona*, β-heavy chain for calaxin. Third, they mediate binding between dynein and microtubules. Regardless of these common properties, calaxin exhibits characteristic features: Ca^{2+}-dependent binding to dynein heavy chain and Ca^{2+}-independent binding to β-tubulin (and possibly to intermediate chain 2 [IC2], ortho-logue of *Chlamydomonas* IC1). Because *Ciona* lacks both LC4 and DC3, calaxin might be evolved to play double roles of LC4/DC3 in Ca^{2+}-dependent regulation of outer arm dynein (also see next section).

5.3 Mechanism of Calaxin-Mediated Modulation of Flagellar Movements During Sperm Chemotaxis

During chemotaxis in *Ciona*, sperm repeat straight and turn movements to come toward the egg (Fig. 5.2). The turn movement accompanies transient increase in intracellular Ca^{2+} concentration and asymmetry of flagellar waveform (Shiba et al. 2008). It was not been elucidated how the increase in intracellular Ca^{2+} concentration induced the modulation of flagellar bending during sperm chemotaxis. As we previously showed that calaxin was directly bound to the outer arm dynein in a Ca^{2+}-dependent manner, it was a strong candidate for direct Ca^{2+}-dependent modulator for flagellar waveform of sperm (Mizuno et al. 2009). Localization of calaxin to epithelial cilia also suggests a possibility that calaxin is a general Ca^{2+} sensor to modulate ciliary and flagellar motility (Mizuno et al. 2009).

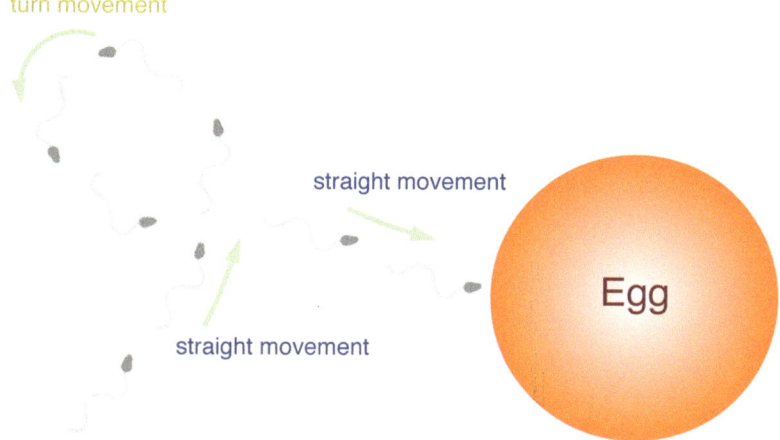

Fig. 5.2 Sperm trajectory during chemotaxis to the egg. During chemotactic movements, *Ciona* sperm show a unique turning movement associated with a flagellar change to an asymmetrical waveform, followed by a straight movement with symmetrical waveform

An antidiabetic compound, repaglinide, is a specific inhibitor for NCS and is also effective on calaxin (Okada et al. 2003; Mizuno et al. 2012). In the presence of repaglinide, sperm do not show the unique turn movement, resulting in less effective chemotaxis. Flagellar bending with strong asymmetry continues for ~0.1 ms during one turn in normal sperm. Repaglinide-treated sperm exhibit transient flagellar asymmetry, but the asymmetry is not sustained for long, and sperm exhibit only incomplete turning (Mizuno et al. 2012). Thus, it is suggested that calaxin plays a key role in sustaining the asymmetrical waveform, not in its formation (Mizuno et al. 2012).

In a demembranated sperm model treated with repaglinide, flagellar bending becomes attenuated at a high concentration of Ca^{2+}. This attenuation is not observed at low concentration of Ca^{2+}, suggesting that calaxin regulates dynein-driven microtubule sliding for asymmetrical bending at higher Ca^{2+} concentration ($<10^{-6}$ M). By using an in vitro assay system with purified dynein, microtubules, and calaxin, the roles of calaxin in the regulation of dynein-driven microtubule sliding can be directly examined (Mizuno et al. 2012). Increasing the concentration of Ca^{2+} has a small effect on the velocity of microtubule sliding by *Ciona* outer arm dynein. Addition of calaxin gives no significant change in the sliding. On the other hand, at higher Ca^{2+} concentrations ($<10^{-6}$ M), addition of calaxin significantly reduced the velocity of microtubule translocation. Thus, calaxin is thought to bind and suppress outer arm dynein at high concentrations of Ca^{2+}. This suppression is thought necessary for the propagation of asymmetrical bending.

The mechanism of calaxin-mediated chemotactic turn is summarized in Fig. 5.3. Before the chemotactic turn, Ca^{2+} concentration in sperm is low and the calaxin is dissociated from dynein. Symmetric P- and R-bends are properly propagated, resulting in straight swimming of sperm. When intracellular Ca^{2+} concentration is

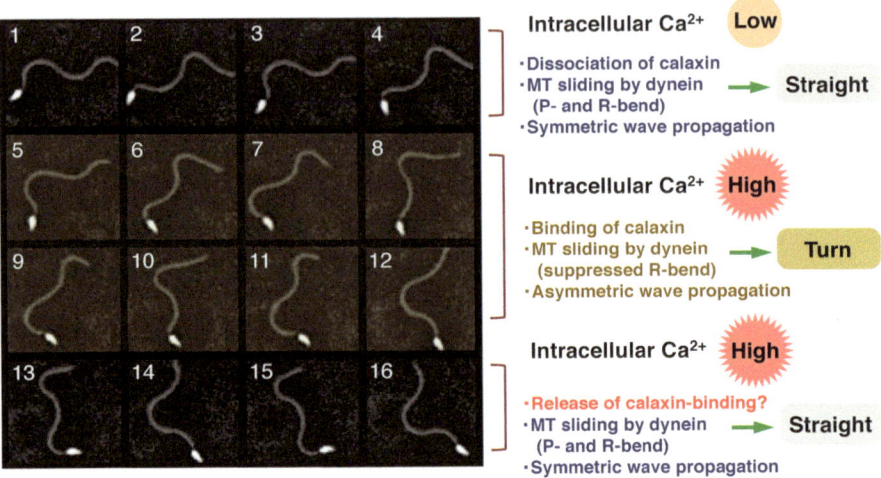

Fig. 5.3 Molecular events during chemotactic turn movement of *Ciona* sperm shown by 16 sequential images of sperm waveform during chemotactic turn. The 8 images highlighted in the *center* represent turn with asymmetrical waveform. The molecular events during this process are driven by the changes of intracellular Ca^{2+} and interaction between calaxin and dynein (see text for more detail)

raised by Ca^{2+} influx, calaxin suppresses dynein-driven microtubule sliding, resulting in propagation of asymmetrical bending and turn movement of sperm. After the chemotactic turn, sperm show straight movement. Therefore, calaxin is thought to be again dissociated from dynein and sperm swim straight with a symmetrical flagellar waveform. Ca^{2+} imaging of live *Ciona* sperm, however, demonstrates that intracellular Ca^{2+} concentration is still high just after the chemotactic turn (Shiba et al. 2008). It is possible that binding of calaxin to dynein is controlled not by absolute Ca^{2+} concentration but by the difference in Ca^{2+} concentration. Alternatively, calaxin may be downregulated by some factors after the chemotactic turn. The mechanism of calaxin after the chemotactic turn is still to be elucidated.

5.4 A Phylogenetic Consideration of Calaxin

Homology search against databases of other organisms demonstrates that calaxin orthologues are present in vertebrates, such as human, mouse and *Xenopus*. Calaxin is also found in invertebrates, both deuterostome (*Ciona*, lancet, and sea urchin) and protostome (*Drosophila*). Search against genome databases of the sea anemone *Nematostella vectensis* and the choanoflagellate *Monosiga brevicollis* also identifies calaxin orthologues in these organisms (Mizuno et al. 2009). However, calaxin has not been found in yeast, *Volvox*, *Trypanosoma*, or *Arabidopsis*, implying that calaxin is metazoan specific (Mizuno et al. 2009). Recently, high-throughput next-generation sequencing enables us to determine draft sequences from a number of other organisms.

```
Hs    MNRKKLQKLTDTLTKN--CKHFNKFEVNCLIKLFYDLVGGVERQGLVVGLDRNAFRNILH
Ci    -MSKKNQKLAEELYKTSCQKHFTKTEVESLIICYKNLLEGLK-------MDRNLFRDILH
Bd    ---------------MLAKTCVSRAEIDILGRSFTLTAQTSDK------IDRSRFRDMLA
      .  ..:  *::  *     :             .         :**.  **::*

Hs    VTFGMTDDMIMDRVFRGFDKDNDGCVNVLEWIHGLSLFLRGSLEEKMKYCFEVFDLNGDG
Ci    QKFNMTEDLLMDRVFRAFDKDSDSYISLTEWVEGLSVFLRGTLDEKMEYTFTVFDLNGDG
Bd    DTFGVDDSLIMD-------RDADNYISFDEYIKGMSVFLNGRYEERLKFCFRVYDLNGDR
      .*.:  :.::**        :*  *.  :..  *::.*:*:**.*   :*:::  *  *:*****

Hs    FISKEEMFHMLKNSLLKQPSEEDPDEGIKDLVEITLKKMDHDHDGKLSFADYELAVREET
Ci    YISREEMFQMLKTCLVKQPTEEDPDEGIKDLVEIALKKMDHDHDSRLSKKDFKDAVLIEP
Bd    YISKEEMFQMLKNCLVKGAVEEDED-GVKDLVDLVLKKLDEDRDGRVSEADWAGAIAKET
      :**:****:***..*:*  .  *** *  *:****:::.***:*.*:*.::*   *:  *:  *.

Hs    LLLEAFGPCLPDPKSQMEFEAQVFKDPNEFNDM
Ci    LLLEAFGKCLPDEKSSEIFEYHVLGVKQCRG--
Bd    LLMEAFGHCLPDAKVDEYD-------------
      **:**** **** *  .
```

Fig. 5.4 Multiple alignment of calaxin. Sequences of calaxin from *Homo sapiens* (Hs), *Ciona intestinalis* (Ci), and *Batrachochytrium dendrobatidi*s (Bd) are aligned by ClustalW. *Asterisks, colons,* or *dots* indicate identical residues in all sequences in the alignment, conserved substitutions, or semi-conserved substitutions, respectively

We have recently found a calaxin orthologue in the chytrid fungus *Batrachochytrium dendrobatidis* (Fig. 5.4). Further search against genome databases of other organisms supports the idea that calaxin is an opisthokont-specific protein to regulate axonemal dyneins (Inaba et al., manuscript in preparation).

Ca^{2+}-dependent regulation of flagellar waveform is important for responses of organisms to several stimuli. For example, *Chlamydomonas* exhibits several light-induced behavioral responses, including phototaxis, photophobic response, and photokinesis (Witman 1993; Wakabayashi and King 2006). *Paramecium* swims both backward and forward according to the changes in intracellular Ca^{2+} (Naitoh and Kaneko 1972). The outer arm dynein of *Paramecium* cilia has not been well characterized, but a gene for the orthologue of *Chlamydomonas* LC4 or DC3 is found in the *Paramecium* genome. On the other hand, neither LC4 nor DC3 is found in *Ciona*, as already described. Considering the regulation of the outer arm dynein by Ca^{2+} commonly seen in *Chlamydomonas*, *Paramecium*, and *Ciona*, it is possible to consider that calaxin is an opisthokont-specific innovation for Ca^{2+}-dependent regulation of axonemal dyneins.

5.5   Perspectives

Calaxin was first identified in *Ciona* sperm but was found to be distributed all through opisthokonts. Considering the presence of Ca^{2+}-dependent regulator for dynein, LC4, in *Chlamydomonas* and other bikont species, what does this "innovation of calaxin" in opisthokonts mean? It is possible that an unknown mechanism for motility regulation might have been innovated in the supergroup of opisthokonts as well as structural diversification in the axonemes at the base of bikonts and opisthokonts. Further phylogenetic or structural evidence is necessary to conclude the evolutional and reproductive significance of calaxin.

Acknowledgments We thank Y. Shikata, K. Oiwa, H. Sakakibara, H. Kojima, S.A. Baba, O. Kutomi, K. Hirose, M. Okai, Y. Takahashi, M. Tanokura, K. Seto, and Y. Degawa for useful advice in the present study. We are grateful to all staff members of the Education and Research Center of Marine Bio-Resources, Tohoku University, and to the National Bio Resource Project (NBRP) for supplying *C. intestinalis*. This work was supported in part by a grant from MEXT (Ministry of Education, Culture, Sports, Science and Technology), Japan, and by JST-BIRD (Japan Science and Technology Agency-Institute for Bioinformatics Research and Development), Japan, to K.I.

References

Blum JJ, Hayes A, Jamieson GA Jr, Vanaman TC (1980) Calmodulin confers calcium sensitivity on ciliary dynein ATPase. J Cell Biol 87:386–397

Brokaw CJ (1979) Calcium-induced asymmetrical beating of triton-demembranated sea urchin sperm flagella. J Cell Biol 82:401–411

Brokaw CJ, Nagayama SM (1985) Modulation of the asymmetry of sea urchin sperm flagellar bending by calmodulin. J Cell Biol 100:1875–1883

Burgoyne RD (2004) The neuronal calcium-sensor proteins. Biochim Biophys Acta 1742:59–68

Burgoyne RD, Weiss JL (2001) The neural calcium-sensor family of Ca^{2+}-binding proteins. Biochem J 353:1–12

Casey DM, Inaba K, Pazour GJ, Takada S, Wakabayashi K, Wilkerson CG, Kamiya R, Witman GB (2003) DC3, the 21-kDa subunit of the outer dynein arm-docking complex (ODA-DC), is a novel EF-hand protein important for assembly of both the outer arm and the ODA-DC. Mol Biol Cell 14:3650–3663

Gibbons BH, Gibbons IR (1980) Ca^{2+}-induced quiescence in reactivated sea urchin sperm. J Cell Biol 84:13–27

Gitelman SE, Witman GB (1980) Purification of calmodulin from *Chlamydomonas*: calmodulin occurs in cell bodies and flagella. J Cell Biol 8:764–770

Hozumi A, Satouh Y, Makino Y, Toda T, Ide H, Ogawa K, King SM, Inaba K (2006) Molecular characterization of *Ciona* sperm outer arm dynein reveals multiple components related to outer arm docking complex protein 2. Cell Motil Cytoskeleton 63:591–603

Inaba K (2007) Molecular basis of sperm flagellar axonemes: structural and evolutionary aspects. Ann N Y Acad Sci 1101:506–526

Jamieson GA Jr, Vanaman TC, Blum JJ (1979) Presence of calmodulin in *Tetrahymena*. Proc Natl Acad Sci USA 76:6471–6475

Kambara Y, Shiba K, Yoshida M, Sato C, Kitajima K, Shingyoji C (2011) Mechanism regulating Ca^{2+}-dependent mechanosensory behaviour in sea urchin spermatozoa. Cell Struct Funct 36:69–82

Kamiya R, Okamoto M (1985) A mutant of *Chlamydomonas* reinhardtii that lacks the flagellar outer dynein arm but can swim. J Cell Sci 74:181–191

Kamiya R, Witman GB (1984) Submicromolar levels of calcium control the balance of beating between the two flagella in demembranated models of *Chlamydomonas*. J Cell Biol 98:97–107

King SM, Patel-King RS (1995) Identification of a Ca^{2+}-binding light chain within *Chlamydomonas* outer arm dynein. J Cell Sci 108:3757–3764

Lindemann CB, Goltz JS (1988) Calcium regulation of flagellar curvature and swimming pattern in triton X-100-extracted rat sperm. Cell Motil Cytoskeleton 10:420–431

Lindemann CB, Gardner TK, Westbrook E, Kanous KS (1991) The calcium-induced curvature reversal of rat sperm is potentiated by cAMP and inhibited by anti-calmodulin. Cell Motil Cytoskeleton 20:316–324

Mitchell DR, Rosenbaum JL (1985) A motile *Chlamydomonas* flagellar mutant that lacks outer dynein arms. J Cell Biol 100:1228–1234

Mizuno K, Padma P, Konno A, Satouh Y, Ogawa K, Inaba K (2009) A novel neuronal calcium sensor family protein, calaxin, is a potential Ca^{2+}-dependent regulator for the outer arm dynein of metazoan cilia and flagella. Biol Cell 101:91–103

Mizuno K, Shiba K, Okai M, Takahashi Y, Shitaka Y, Oiwa K, Tanokura M, Inaba K (2012) Calaxin drives sperm chemotaxis by Ca^{2+}-mediated direct modulation of a dynein motor. Proc Natl Acad Sci USA 109:20497–20502

Naitoh Y, Kaneko H (1972) Reactivated triton-extracted models of *Paramecium*: modification of ciliary movement by calcium ions. Science 176:523–524

Nakano I, Kobayashi T, Yoshimura M, Shingyoji C (2003) Central-pair-linked regulation of microtubule sliding in flagellar axonemes. J Cell Sci 116:1627–1636

Okada M, Takezawa D, Tachibanaki S, Kawamura S, Tokumitsu H, Kobayashi R (2003) Neuronal calcium sensor proteins are direct targets of the insulinotropic agent repaglinide. Biochem J 375:87–97

Padma P, Satouh Y, Wakabayashi K, Hozumi A, Ushimaru Y, Kamiya R, Inaba K (2003) Identification of a novel leucine-rich repeat protein as a component of flagellar radial spoke in the ascidian *Ciona intestinalis*. Mol Biol Cell 14:774–785

Piperno G, Mead K, Shestak W (1992) The inner dynein arms I2 interact with a "dynein regulatory complex" in *Chlamydomonas* flagella. J Cell Biol 118:1455–1463

Saito T, Shiba K, Inaba K, Yamada L, Sawada H (2012) Self-incompatibility response induced by calcium increase in sperm of the ascidian *Ciona intestinalis*. Proc Natl Acad Sci USA 109:4158–4162

Sakato M, King SM (2003) Calcium regulates ATP-sensitive microtubule binding by *Chlamydomonas* outer arm dynein. J Biol Chem 278:43571–43579

Sakato M, Sakakibara H, King SM (2007) *Chlamydomonas* outer arm dynein alters conformation in response to Ca^{2+}. Mol Biol Cell 18:3620–3634

Sale WS (1986) The axonemal axis and Ca^{2+}-induced asymmetry of active microtubule sliding in sea urchin sperm tails. J Cell Biol 102:2042–2052

Shiba K, Baba SA, Inoue T, Yoshida M (2008) Ca^{2+} bursts occur around a local minimal concentration of attractant and trigger sperm chemotactic response. Proc Natl Acad Sci USA 105:19312–19317

Shingyoji C, Takahashi K (1995) Flagellar quiescence response in sea urchin sperm induced by electric stimulation. Cell Motil Cytoskeleton 31:59–65

Smith EF (2002) Regulation of flagellar dynein by calcium and a role for an axonemal calmodulin and calmodulin dependent kinase. Mol Biol Cell 13:3303–3313

Smith EF, Yang P (2004) The radial spokes and central apparatus: mechano-chemical transducers that regulate flagellar motility. Cell Motil Cytoskeleton 57:8–17

Tash JS, Means AR (1983) Cyclic adenosine 3',5' monophosphate, calcium and protein phosphorylation in flagellar motility. Biol Reprod 28:75–104

Tash JS, Krinks M, Patel J, Means RL, Klee CB, Means AR (1988) Identification, characterization, and functional correlation of calmodulin-dependent protein phosphatase in sperm. J Cell Biol 106:1625–1633

Wakabayashi K, Yagi T, Kamiya R (1997) Ca^{2+}-dependent waveform conversion in the flagellar axoneme of *Chlamydomonas* mutants lacking the central-pair/radial spoke system. Cell Motil Cytoskeleton 38:22–28

Wakabayashi K, King SM (2006) Modulation of *Chlamydomonas reinhardtii* flagellar motility by redox poise. J Cell Biol 173:743–754

Witman GB (1993) *Chlamydomonas* phototaxis. Trends Cell Biol 3:403–408

Chapter 6
Cl⁻ Channels and Transporters in Sperm Physiology

C.L. Treviño, G. Orta, D. Figueiras-Fierro, J.L. De la Vega-Beltran,
G. Ferreira, E. Balderas, O. José, and A. Darszon

Abstract Spermatozoa must decode environmental and cellular cues to succeed in fertilization, and this process relies heavily on ion channels. New observations bring to light the relevant participation of Cl⁻ channels and anion transporters in some of the main sperm functions. Here we review the evidence that indicates the participation of Cl⁻ channels in motility, maturation, and the acrosome reaction (AR), and what is known about their molecular identity and regulation. Our better understanding of sperm anion transport will yield tools to handle some infertility problems, improve animal breeding and preserve biodiversity, and develop selective and secure male contraceptives.

Keywords Capacitation • Chloride • Hyperpolarization • Ionic currents • Sperm

6.1 Introduction

Spermatozoa must find and fertilize the egg to deliver their genetic information and generate a unique individual in organisms of sexual reproduction. Ion pumps and transporters are used by cells to build up and maintain ion concentration gradients across their membranes. These ionic gradients allow cells to respond to their changing environment, to signals from other cells and to perform secondary transport. Ion

C.L. Treviño • G. Orta • D. Figueiras-Fierro • J.L. De la Vega-Beltran
E. Balderas • O. José • A. Darszon (✉)
Departamento de Genética del Desarrollo y Fisiología Molecular, Instituto de Biotecnología,
Universidad Nacional Autónoma de México, Cuernavaca, Morelos 62210, Mexico
e-mail: darszon@ibt.unam.mx

G. Ferreira
Laboratorio de Canales Iónicos, Departamento de Biofísica, Facultad de Medicina,
Universidad de la República, Montevideo 11800, Uruguay

H. Sawada et al. (eds.), *Sexual Reproduction in Animals and Plants*,
DOI 10.1007/978-4-431-54589-7_6, © The Author(s) 2014

channels can transport millions of ions per second, which enables them to rapidly modify the cell electric potential and the concentrations of internal second messengers within a wide time range, depending on their mode of regulation (Hille 2001).

Spermatozoa are small, differentiated, and morphologically complex cells (Yanagimachi 1994). They must be endowed with decoding systems for multiple signals along their journey to reach the egg to succeed in fertilizing it. During their maturational convoluted journey through the epididymis and the female reproductive tract, mammalian sperm encounter many environmental changes (Dacheux et al. 2012; Hung and Suarez 2010; Visconti et al. 2011). Although significant advances have been made in recent years, the set of ion channels and transporters needed for sperm to achieve fertilization is still not fully known (Darszon et al. 2011; Lishko et al. 2012; Publicover and Barratt 2012). Ion channel inhibition and knockout experiments have clearly revealed the major role these transporters play in sperm maturation, the regulation of motility, and the acrosome reaction (AR) (Darszon et al. 2011; Lishko et al. 2012). As gene transcription and protein synthesis seem not to occur in mature sperm, their proteins are generated during spermatogenesis (Baker 2011).

Considering that ion channels are minor membrane protein components, demonstrating the functional presence of a particular ion channel in spermatozoa needs controlled immunological or proteomic detection combined with electrophysiological, ion-sensitive fluorescent functional assays and pharmacology. When possible, eliminating the specific ion channel from spermatozoa might unravel its function (Kirichok et al. 2006; Santi et al. 2010). Initial glimpses of the properties of some sperm ion channels were derived from planar bilayers with incorporated sperm plasma membranes (reviewed by Darszon et al. 1999). Notably, some of the first sperm single-channel recordings were of K^+ and Cl^- channels obtained in planar bilayers with incorporated sea urchin sperm plasma membranes (Labarca et al. 1996; Lievano et al. 1985; Morales et al. 1993) and of Ca^{2+} channels from boar sperm plasma membranes (Cox and Peterson 1989).

Obtaining electrophysiological recordings directly in sperm to study their ion channels was exceedingly difficult for a long time (Darszon et al. 1999; Guerrero et al. 1987; Jimenez-Gonzalez et al. 2006; Kirichok and Lishko 2011; Ren and Xia 2010; Weyand et al. 1994). Achieving whole-cell patch-clamp recordings became more feasible when Kirichok et al. (2006) were able to seal the cytoplasmic droplet of mouse epididymal spermatozoa and then mature human spermatozoa (Kirichok and Lishko 2011). This novel strategy is allowing the characterization of sperm-specific channels such as CatSper (Kirichok et al. 2006) and SLO3 (Navarro et al. 2007; Santi et al. 2010; Schreiber et al. 1998; Zeng et al. 2011), and of sperm anion channels that are present in other cell types (Orta et al. 2012; Ferrera et al. 2010). Additionally, a voltage-sensitive H^+ channel involved in the intracellular pH (pH_i) regulation in human sperm and less importantly in mouse sperm (Kirichok and Lishko 2011), and ATP-gated channels of the purinergic family, P2X2, in mouse epididymal sperm have been recorded (Navarro et al. 2011). Ionic currents with properties consistent with TRPM8 channels were recorded in testicular sperm (Gibbs et al. 2011; Martinez-Lopez et al. 2011). Intracellular Ca^{2+} concentration $[Ca^{2+}]_i$

imaging experiments at different temperatures and membrane potential (E_m) measurements suggested TRPM8-like channels are present in more mature mouse and human sperm (De Blas et al. 2009; Martinez-Lopez et al. 2011).

Alternatively, whole-cell recordings have now been obtained directly patching the head of human spermatozoa using a modification of the perforated patch-clamp strategy. Employing this approach, Cl⁻ currents displaying characteristics associated with Ca²⁺-dependent Cl⁻ channels were documented in human spermatozoa (Orta et al. 2012).

Research in past years has established that in most cells Cl⁻ is actively transported and not at electrochemical equilibrium. As a result, Cl⁻ can participate in signaling and perform work, an important matter when considering the functional roles of anion channels and transporters, in addition to their more established participation in fluid secretion and volume regulation (Duran et al. 2012). In spermatozoa, as in other cells, Cl⁻ is the main anion and is involved in volume regulation and osmotic stress protection (Cooper and Yeung 2007; Furst et al. 2002; Yeung et al. 2005). Capacitation, a maturational process and the AR, a unique exocytotic event essential for fertilization, both discussed next, are significantly affected in mouse and human spermatozoa when external Cl⁻ concentrations are lowered (Chen et al. 2009; Figueiras-Fierro et al. 2013; Orta et al. 2012; Wertheimer et al. 2008; Yeung and Cooper 2008). Although these findings suggest that Cl⁻ plays a relevant role in sperm physiology, not much is known about its transport across the membrane of this fundamental cell. This review summarizes information about Cl⁻ channels and transporters for which evidence suggests their presence in sperm and their involvement in important sperm functions such as epididymal maturation, capacitation, motility, and the AR.

6.2 Maturation During Epididymal Transit

The epididymis is a specialized duct of the male reproductive system that fulfills four important functions for spermatozoa: transport, concentration, maturation, and storage (Turner 2008). The movement of sperm through the epididymis involves hydrostatic pressure and smooth muscle contractions. Depending on the species, sperm transit through the epididymis ranges from 2 to 13 days (Turner 2008). The epididymis is divided into three main sections: caput, corpus, and cauda, proximal to distal to the exit from the testis. Sperm from the caput section are immotile and lack important characteristics required for fertilization. In contrast, sperm obtained from the cauda have the highest fertilizing capacity. The osmolarity encountered by sperm during transit into the epididymis increases from 280 (in the rete testis fluid) to up to 400 mmol/kg (in the cauda epididymis fluid) (Yeung et al. 2006). On ejaculation into the female reproductive tract, spermatozoa experience hypo-osmotic stress, which is counterbalanced through the process known as regulatory volume decrease (RVD) involving influx and efflux of water and osmolytes (Yeung et al. 2006). RVD capability is acquired during epididymal transit, and sperm from the cauda exhibit the greatest RVD capacity. Results indicating the role of K$^+$ channels during sperm RVD at the time of epididymal maturation suggest a parallel involvement of Cl$^-$ channels to compensate the positive charges and maintain electroneutrality (Cooper and Yeung 2007). The molecular identity of the Cl$^-$ channels involved in volume regulation is not established. Candidates such as ClC-2 (CLCN2) and ClC-3 (CLCN3) have been proposed to play a role in somatic cells (Furst et al. 2002; Nilius and Droogmans 2003); however, their function is still controversial (Sardini et al. 2003). Interestingly, CLCN3 was detected by Western blot and localized to the sperm tail by immunofluorescence (Yeung et al. 2005). Although the function of K$^+$ and Cl$^-$ channels in RVD is still under study, the expression of such channels in sperm from several species suggests they may play an important role during epididymal maturation, a matter that awaits further research.

6.3 Motility

Sperm motility is activated when spermatozoa enter the female tract. Motility is one of the most important functions carried out by the sperm because it is essential to achieve fertilization. Indeed, several sperm motility defects can cause male sterility (e.g., sAC, PKA, sNHE, GAPDHs, CatSper, PMCA4, SLO3) (Esposito et al. 2004; Miki et al. 2004; Nolan et al. 2004; Okunade et al. 2004; Quill et al. 2001; Ren et al. 2001; Santi et al. 2010; Wang et al. 2007; Zeng et al. 2011). Sperm motility is driven by the flagellum, an appendage with an ultrastructure very similar to that of cilia. The axoneme is the principal structure that propels a flagellum (Lindemann and Goltz 1988); it is composed of a particular arrangement of microtubules, usually in the configuration of nine doublets surrounding a central pair. Movement results by repetitive cycles of flagellar bending, arising from microtubule sliding using the force generated by dynein ATPases whose activity is modulated by pH, ATP, ADP, Ca^{2+}, and phosphorylation (Christen et al. 1983; Lindemann and Goltz 1988). Ion transport that supports and controls flagellar beating plays key roles in sperm motility regulation (Guerrero et al. 2011; Kaupp et al. 2008).

Upon ejaculation, sperm initiate motility with a relatively low-amplitude flagellar beat known as activated motility (Wennemuth et al. 2003). The stimulation of the sperm soluble adenylate cyclase by HCO$_3^-$ and the consequent cAMP/PKA stimulation is required for the activated motility (Carlson et al. 2007; Esposito et al. 2004; Hess et al. 2005; Nolan et al. 2004; Xie et al. 2006). After some time (varying according to species) in the female tract, spermatozoa become hyperactivated, displaying vigorous asymmetrical flagellar beating with large amplitude and high curvature. Hyperactivation helps sperm to detach from temporary binding sites along the female genital tract and to penetrate the extracellular matrix of cumulus cells and the zona pellucida (ZP) surrounding the oocyte (Suarez 2008). The mechanisms involved in hyperactivation are not well understood; however, it is known that a [Ca^{2+}]$_i$ rise mediated by CatSper channels is required for hyperactivation. This Ca^{2+} channel, only present in the sperm flagella, is weakly voltage dependent and activated by an increase in pH_i (Kirichok et al. 2006; Ren et al. 2001). CatSper null male mice are infertile mainly because of a failure to hyperactivate (Carlson et al. 2005; Carlson et al. 2003; Quill et al. 2001; Ren et al. 2001). It has been proposed that the hyperpolarization of the sperm plasma associated with capacitation increases the driving force for Ca^{2+}, facilitating Ca^{2+} influx through CatSper channels during cytosolic alkalinization (Navarro et al. 2007).

6.4 Capacitation

A defined period of time in the female genital tract is necessary for mammalian spermatozoa to acquire their ability to fertilize (Austin 1952; Chang 1951). Altogether the set of changes required for this maturation process is called capacitation and involves the development of a distinctive sperm motility pattern known as hyperactivation, and the sperm capacity to undergo the AR, an exocytotic event that allows the sperm to fertilize the egg. During sperm capacitation PKA is activated (Harrison 2004), leading to tyrosine phosphorylation increases (Visconti et al. 1995a, b); pH_i (Zeng et al. 1995) and $[Ca^{2+}]_i$ elevate (Baldi et al. 1991; Breitbart 2003; DasGupta et al. 1993; Suarez et al. 1993; Xia and Ren 2009); plasma membrane composition and organization are modified (Cross 1998; Davis 1981; Gadella and Harrison 2000; Go and Wolf 1983; Travis and Kopf 2002; Visconti et al. 1999); and the cell E_m is hyperpolarized in the mouse and other species (Arnoult et al. 1999; Demarco et al. 2003; Munoz-Garay et al. 2001; Zeng et al. 1995).

6.4.1 Membrane Potential Changes During Sperm Capacitation

It is yet not fully understood how and why E_m is hyperpolarized in some mammalian sperm species. Indeed, hyperpolarization is important in mouse, bovine, and equine sperm capacitation (Arnoult et al. 1996; Arnoult et al. 1999; De La Vega-Beltran et al. 2012; Demarco et al. 2003; McPartlin et al. 2011; Munoz-Garay et al. 2001; Zeng et al. 1995), although it has not been demonstrated in human sperm. The sperm resting E_m is relatively depolarized in most mammalian sperm (between −30 and −40 mV) (De La Vega-Beltran et al. 2012; Demarco et al. 2003; Espinosa and Darszon 1995; Hernandez-Gonzalez et al. 2006; McPartlin et al. 2011; Munoz-Garay et al. 2001; Santi et al. 2010; Zeng et al. 1995). Sperm populations are very heterogeneous and only a fraction of the cells capacitate in vitro (~30 %); therefore, the average E_m values must be cautiously considered. Capacitated mouse spermatozoa display an average E_m approximately −60 mV. Indeed, when Arnoult et al. (Arnoult et al. 1999) measured E_m in individual spermatozoa using di-8-AN-EPPS, a voltage-sensitive dye, they documented that capacitated sperm populations consisted of at least two groups: one hyperpolarized (~−80 mV), possibly representing capacitated sperm, and another of noncapacitated sperm with a resting E_m approximately -43 mV. Various findings suggest that hyperpolarization is essential for sperm to acquire the ability to undergo a physiological AR. For instance, carrying out capacitation in the presence of high external KCl significantly reduces the ZP-induced mouse sperm AR (Arnoult et al. 1999; De La Vega-Beltran et al. 2012; Zeng et al. 1995). These observations lead to the proposal that an E_m hyperpolarization is important for capacitation and thus is required for the AR.

Initially the role of the capacitation-associated hyperpolarization was thought to be needed to remove inactivation from T-type voltage-dependent Ca^{2+} channels (Ca_V3), which could then be activated by physiological agonists (e.g., ZP) to culminate with the induction of the AR (Arnoult et al. 1996; Arnoult et al. 1999; Santi et al. 1996; Zeng et al. 1995). Recent evidence suggests that hyperpolarization of the sperm plasma membrane is necessary and sufficient to prepare sperm for the AR (De la Vega et al. 2012). Even though the molecular entities and the mechanisms responsible for the hyperpolarization are not yet established, it could be the result of (1) an increase in K^+ permeability caused by the activation of K^+-selective channels, and (2) a reduction of Na^+ permeability by decreasing the activity of Na^+ channels. In this context, the regulation and activity of Cl^- permeability through Cl^- channels and transporters could also play a direct or indirect role in the regulation of the sperm plasma E_m.

6.5 The Acrosome Reaction

The sperm head contains a large secretory vesicle located at its posterior end called the acrosome (Yanagimachi 1998). Spermatozoa of many species need to undergo the fusion of their single acrosome to the plasma membrane to be able to fertilize the female gamete. This process, called the AR, is now believed to occur in multiple steps. As multiple fusion points between the acrosomal and plasma membrane are involved in the AR, plasma membrane–outer acrosome hybrid vesicles are liberated, resulting in the release of the acrosomal content (Buffone et al. 2012). The fusion machinery involved in this reaction is regulated by Ca^{2+} and similar to that found in many neuroendocrinal secretory cells (Bello et al. 2012; Castillo Bennett et al. 2010). Notably, where and what triggers the physiologically relevant AR is being reevaluated (Inoue et al. 2011; Jin et al. 2011; Visconti and Florman 2010; Yanagimachi 1998).

In this context ZP and progesterone, as well as other AR inducers, require reexamination to establish their physiological relevance.

Various transduction pathways are required to converge for the ZP-induced AR to occur, and complex $[Ca^{2+}]_i$ changes are involved (for review, see Mayorga et al. 2007). The physiologically relevant AR changes in $[Ca^{2+}]_i$ include external and internal Ca^{2+} sources (Breitbart et al. 2010; Costello et al. 2009; Darszon et al. 2011; Florman et al. 2008). At the present time three different Ca^{2+} channels are thought to mediate the $[Ca^{2+}]_i$ responses associated to the AR. They appear to be functionally linked in a manner that is not fully understood (Darszon et al. 2011; Florman et al. 2008; Publicover et al. 2007). Early on it was thought that voltage-dependent Ca^{2+} (Ca_V) channels were involved in the initial $[Ca^{2+}]_i$ increase detected during the AR induced by ZP in mouse sperm, taking into account their pharmacology and that of this reaction (Darszon et al. 2011). $Ca_V3.2$ was the most likely Ca_V candidate to participate in the mouse AR (Arnoult et al. 1996; Escoffier et al. 2007; Lievano et al. 1996; Trevino et al. 2004). However, $Ca_V3.1$ and 3.2 knockout mice are fertile (Stamboulian et al. 2004), and Ca_V currents, although recorded in testicular sperm, were not detected in epididymal sperm (Martinez-Lopez et al. 2009; Ren and Xia 2010). These results raised doubts about the participation of Ca_V3 channels in the mouse sperm AR, although solid immunological data demonstrate their presence (Escoffier et al. 2007; Trevino et al. 2004).

A sustained $[Ca^{2+}]_i$ increase lasting up to minutes is associated with the AR. Evidence indicates internal Ca^{2+} stores (i.e., the acrosome) participate releasing Ca^{2+} through the IP3 receptor, the second type of Ca^{2+} channel involved in the AR, which is activated as a consequence of IP3 production during the AR (Darszon et al. 2011; Florman et al. 2008; Mayorga et al. 2007; Publicover et al. 2007). Plasma membrane Ca^{2+} channels (SOCS) activate as Ca^{2+} store emptying occurs, causing the sustained $[Ca^{2+}]_i$ increase. Several components such as STIM, ORAI, and TRPCs may constitute SOCS (Moreno and Vaca 2011). STIM and ORAI have been suggested to be present in human and mouse sperm and contribute to the sustained $[Ca^{2+}]_i$ elevation leading to the AR (Costello et al. 2009; Darszon et al. 2012).

As discussed, Ca^{2+} transport plays a fundamental role in the AR. Much less is known about how Cl⁻ movements influence this event. Interestingly, niflumic acid (NFA), best known as a Ca^{2+}-activated Cl⁻ channel inhibitor, was reported a long time ago to block the first Cl⁻ single-channel activity recorded in mammalian sperm as well as the AR induced by solubilized ZP, progesterone, and GABA in mouse sperm (Espinosa et al. 1998). NFA also partially inhibited a Ca^{2+}-induced hyperpolarization partially driven by Cl⁻ in mouse spermatozoa (Espinosa et al. 1998). Anion channel blockers such as NFA, for example, DIDS, also inhibit the mouse and human sperm AR, as well as Cl⁻ channels detected in these cells (Espinosa and Darszon 1995; Espinosa et al. 1998; Figueiras-Fierro et al. 2013; Orta et al. 2012).

Considering the reevaluation of the site(s) and mechanisms that trigger the AR requires also a reexamination of the evidence involving neurotransmitter receptors in this process. These matters are discussed in the section on the $GABA_A$ and glycine receptors, which are particularly relevant to this review as they mediate Cl⁻ fluxes.

6.6 Cl⁻ Channels and Transporters Linked to Sperm Physiology

In a recent paper our group reported that when sperm are incubated in Cl⁻-free media, most of the capacitation-associated processes are impaired (Hernandez-Gonzalez et al. 2007; Wertheimer et al. 2008). In this condition, increase in tyrosine phosphorylation and hyperpolarization of the sperm E_m are not observed. Consistently, sperm did not hyperactivate (discussed below), were unable to undergo the AR, and failed to fertilize in an in vitro assay. Although in the absence of Cl⁻ cAMP agonists rescued phosphorylation events, this condition was not sufficient to allow the sperm to fertilize in vitro. These findings highlight the importance of Cl⁻ homeostasis in sperm during capacitation, and suggest that one or more Cl⁻ transport systems are present in sperm. The identity of the specific sperm Cl⁻ transporters involved in capacitation is still unclear.

The activity of all Cl⁻ transporters present in a particular cell type defines its $[Cl^-]_i$ levels. These transporters can be divided in two categories: Cl⁻ channels and specialized Cl⁻ carriers (Jentsch et al. 2005; Nilius and Droogmans 2003). Cl⁻ channels are distributed in four groups: (1) CFTR channels; (2) the γ-aminobutyric (GABA)-gated and related glycine-gated neurotransmitter receptors; (3) Ca^{2+}-activated Cl⁻ channels; and (4) CLC channels. Cl⁻ carriers couple the transport of Cl⁻ to the movement of another ion in either opposite direction (antiporter) or in the same direction (cotransporter or symporter). Cl⁻ carriers are classified in two main families: (1) the electro-neutral cation Cl⁻ cotransporter family and (2) the electro-neutral Cl^-/HCO_3^- exchanger family.

6.6.1   CFTR Channels

The ABC transporter family has a unique member, the cystic fibrosis transmembrane conductance regulator (CFTR), an anion channel modulated by cAMP/PKA and ATP. The CFTR consists of two membrane-spanning domains (MSDs), two nucleotide-binding domains (NBDs), and a regulatory (R) domain. The MSDs have six transmembrane helices each, which are linked by the regulatory domain. In cells possessing either endogenous or recombinant CFTR, the channel displays the following anionic selectivity sequence: $Br^- > Cl^- > I^- > F^-$ (Anderson et al. 1991). The channel pore is formed by the MSDs, whereas the gating activity is related to ATP hydrolysis by the NBDs and phosphorylation of the R domain (Sheppard and Welsh 1999). Mutations in CFTR cause cystic fibrosis (CF), an autosomic recessive genetic disease characterized by abnormal transport of Cl⁻ and HCO_3^- leading to viscous secretions in many epithelial cells, but especially the lungs, pancreas, liver, and intestine.

CFTR mutations affect male fertility; as on average 95 % of male CF patients have congenital bilateral absence of the vas deferens, making them infertile. In addition, other mechanisms related to sperm physiology may be affected by CFTR mutations, leading to infertility in CF (Popli and Stewart 2007). Supporting this notion, it has been found that fertility in uremic patients may be reduced through alterations in CFTR expression (Xu et al. 2012). CFTR has been shown to be present in both human and mouse sperm by our group and others, using specific antibodies (Chan et al. 2006; Hernandez-Gonzalez et al. 2007; Li et al. 2010; Xu et al. 2007).

Xu et al. (2007) showed, using CFTR inhibitors and specific antibodies, as well as heterozygous CFTR mutants, that sperm capacitation and the associated HCO_3^- transport are significantly reduced when compared to mice in normal fertilizing conditions. CFTR function has also been identified electrophysiologically as ATP-dependent Cl^- currents that are stimulated by cAMP, cGMP, and genistein, and inhibited by DPC and $CFTR_{inh}$-172 in whole-cell clamp recordings from testicular and epididymal mouse sperm (Figueiras-Fierro et al. 2013). The biophysical and pharmacological properties of CFTR recorded from epididymal mice spermatozoa, as well as its role in the AR, are shown in Fig. 6.1. Moreover, this particular Cl^- current is absent in testicular sperm from mice displaying the most common variant of the CF mutation, the loss-of-function mutation of the CFTR gene known as $\Delta F508$. All these findings support the idea that CFTR is present in mature spermatozoa and that it is involved in the capacitation events.

How E_m is regulated by Cl^- and other anions is still poorly understood. As CFTR is mainly a Cl^- channel, it possibly participates in regulation of the resting E_m; however, Cl^- substitution by nonpermeable anions (e.g., gluconate or methanesulfonate) does not modify it. On the other hand, this procedure inhibits hyperpolarization development during capacitation (Hernandez-Gonzalez et al. 2007). Findings supporting this idea in mouse sperm are (1) adding 250 μM of the CFTR antagonist DPC (diphenylamine-2-carboxylic acid) inhibits hyperpolarization associated with capacitation and also the AR induced by sZP, without modifying tyrosine phosphorylation levels; (2) the CFTR agonist genistein (5–10 μM) hyperpolarizes noncapacitated spermatozoa; and (3) addition of permeable analogues of cAMP to noncapacitated sperm elevates $[Cl^-]_i$ (Hernandez-Gonzalez et al. 2007). CFTR channels are also known to interact with and regulate other ion channels (i.e., epithelial Na^+ channels, ENaCs) and transporters (i.e., the Cl^-/HCO_3^- exchanger, SCL family) (Berdiev et al. 2009; Konig et al. 2001; Kunzelmann and Schreiber 1999; Perez-Cornejo and Arreola 2004). The interaction of CFTR with the Cl^-/HCO_3^- exchangers is important to explain its role in pH_i regulation (discussed in Sect. 6.6.6). Of particular relevance to understand how CFTR influences the resting E_m is its interaction with ENaC. As the sperm resting E_m is mildly depolarized (~ -35 mV) and thus shifted from the K^+ equilibrium potential, it cannot be explained mainly by a high K^+ permeability through K^+ channels, as it usually happens in many cells. Just a 10 % contribution of Na^+ permeability would explain the observed resting E_m in noncapacitated sperm, and their inhibition during capacitation would contribute to the observed hyperpolarization (see Hernandez-Gonzalez et al. 2006; Hernandez-Gonzalez et al. 2007). CFTR can downregulate ENaCs through mechanisms still not fully understood (Konig et al. 2001), and the CFTR agonist genistein hyperpolarizes sperm E_m (Hernandez-Gonzalez et al. 2007) and diminishes $[Na^+]_i$ (Escoffier et al. 2012). In summary, CFTR is important in sperm physiology regulating pH_i and E_m in the resting and capacitating conditions, directly through its Cl^- and anion permeability or indirectly through its interaction with ENaC and Cl^-/HCO_3^- exchangers.

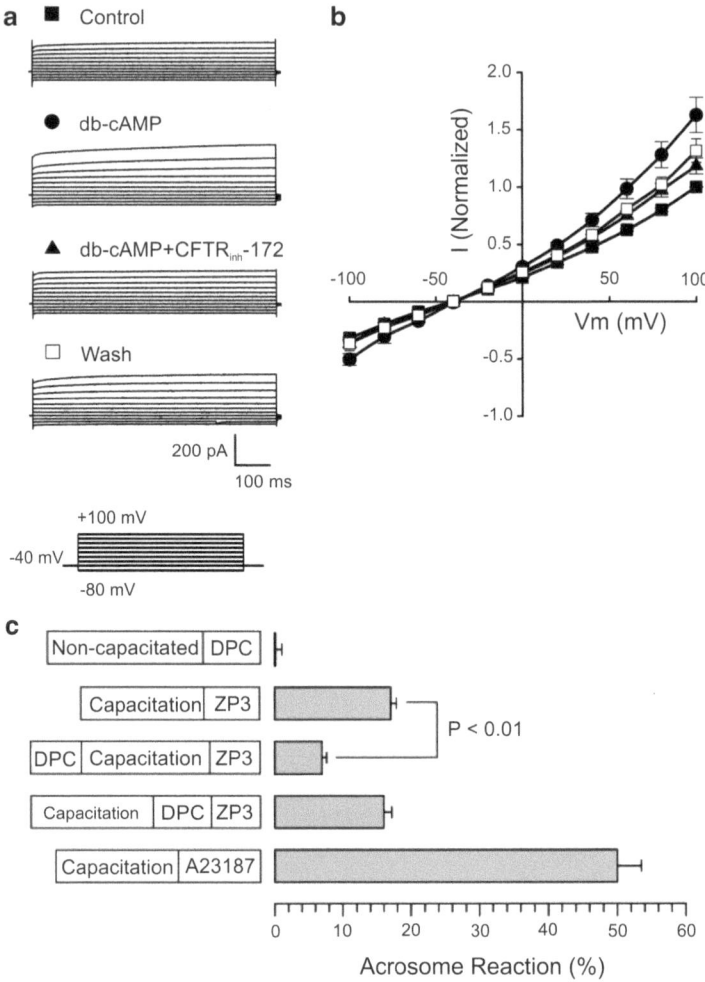

Fig. 6.1 Cystic fibrosis transmembrane conductance regulator (CFTR) channels are functional in epididymal mouse spermatozoa. (**a**) Representative whole-cell patch-clamp currents recorded by applying voltage pulses from a holding potential of −40 mV to test potentials ranging from 100 to −80 mV in 20-mV steps. The protocol used for eliciting CFTR currents is shown between **a** and **c** traces. Sperm were exposed to recording solutions where the primary ion is Cl⁻. Control Cl⁻ currents (*filled squares*) were significantly stimulated by extracellular db-cAMP (100 μM; *filled circles*) and inhibited following the addition to the external media of the specific inhibitor, CFTR$_{inh}$-172 (5 μM; *filled triangles* db-cAMP+CFTR$_{inh}$-172). CFTR$_{inh}$-172 inhibition was partially reversible (*hollow squares*). The effects of agonist and CFTR$_{inh}$-172 were recorded in the presence of niflumic acid (NFA) (50 μM), to eliminate the contribution of other Cl⁻ channels. (**b**) *I–V* relationship of the currents in **a**. CFTR currents similar to those detected in epididymal sperm were also present in testicular sperm. (**c**) CFTR inhibition during capacitation significantly inhibited acrosome reaction (AR). Mouse sperm were capacitated in medium supplemented with bovine serum albumin and NaHCO$_3$, and AR was induced with ZP or A23187 (Ca²⁺ ionophore). The presence of DPC, a classic CFTR inhibitor, during sperm capacitation decreased the ZP-induced AR if added during capacitation but not if added during AR induction. For **a** and **b**, symbols represent the mean ± SEM of four experiments; some SE bars were smaller than the symbols. The currents were normalized with respect to the stimulated Cl⁻ current at 100 mV. For **c**, data are normalized with respect to spontaneous AR and presented as the mean ± SEM with $n \geq 3$

6.6.2 GABA and Glycine Receptors

$GABA_A$ receptors are ligand-gated anion channels selective for Cl⁻ ions, which mediate inhibitory neurotransmission in the central nervous system (CNS). These receptors are pentameric and composed of different subunits. They were first identified pharmacologically as being activated by GABA and the selective agonist muscimol, blocked by bicuculline and picrotoxin, and modulated by benzodiazepines, barbiturates, and certain other central nervous system (CNS) depressants (Macdonald and Olsen 1994; Sieghart 1995). $GABA_A$ receptors are also found outside the CNS, in tissues such as liver, lung, and immune cells (Sigel and Steinmann 2012). Polymerase chain reaction (PCR) and immunocytochemistry studies revealed the presence of $GABA_A$ and $GABA_B$ receptors in spermatogenic cells, and patch-clamp studies showed that GABA application to round spermatogenic cells induced an inward Cl⁻ current (Kanbara et al. 2011). The presence and function of $GABA_A$ receptors in sperm from different species have been explored by several laboratories (Jin et al. 2009; Meizel 1997; Puente et al. 2011; Wistrom and Meizel 1993). For example, Ritta et al. established that this neurotransmitter plays a role in the regulation of motility in bovine and human sperm (Ritta et al. 2004; Ritta et al. 1998). In rat sperm, GABA and progesterone accelerated the process of capacitation and hyperactivated motility, effects that were inhibited by bicuculline and picrotoxin (Jin et al. 2009). Additionally, it has been shown that GABA can induce the AR (Puente et al. 2011; Shi et al. 1997) and also that $GABA_A$ receptors can modulate the response to progesterone in these cells (Hu et al. 2002; Ritta et al. 1998; Shi and Roldan 1995; Turner et al. 1994).

Glycine receptors are also heteropentameric Cl⁻ channels that mediate inhibitory transmission in the CNS, although they are also found in retina and macrophages (Webb and Lynch 2007). The alkaloid strychnine has been used as a very specific antagonist for these receptors. The presence and functionality of glycine receptors in sperm have been studied especially in Meizel's laboratory. They reported the expression of different glycine receptor isoforms in sperm from human, porcine, mouse, and hamster using both immunocytochemistry and Western blot analysis (Bray et al. 2002; Kumar and Meizel 2008; Llanos et al. 2001; Meizel and Son 2005; Melendrez and Meizel 1995; Sato et al. 2000). These receptors were detected in the head and flagellum, suggesting distinct roles at different cell locations. For example, antibodies against glycine receptors A1 and A2 inhibited the ZP3-induced AR in human sperm (Bray et al. 2002).

Compounds such as glycine, GABA, and acetylcholine were reported many years ago to induce the AR; however, the physiological significance of these observations was unclear, considering that the accepted physiological inductor for this reaction was ZP3. Recent findings have challenged this paradigm, and where, when, and what induces the AR are open questions again (Inoue et al. 2011; Jin et al. 2011; Kunzelmann et al. 2011).

6.6.3   Ca²⁺-Activated Cl⁻ Channels (CaCCs)

In many cell types, volume control and secretion are critical (i.e., reproductive tract smooth muscle cells, oviduct and ductus epididymis cells, spermatids, epithelial cells in exocrine glands and trachea, airway, and vascular smooth muscle cells), and spermatozoa are not an exception. Usually these cells possess Ca^{2+}-activated Cl^- channels (CaCCs) that display similar biophysical and molecular features (Hartzell et al. 2005; Huang et al. 2009). Such currents were initially documented in *Xenopus* oocytes (Miledi 1982) but have now been recorded in many cell types.

An elevation of $[Ca^{2+}]_i$ resulting from release from intracellular stores or influx through plasma membrane channels activates CaCCs. In spite of the frequent presence of CaCCs in cells, their molecular identity is not fully known. Candidates considered for CaCCs are bestrophins, tweety, and CLCA. However, all of them failed to reproduce the native behavior of CaCCs when expressed in heterologous expression systems (Hartzell et al. 2009). Recently, using bioinformatics approaches, different research teams succeeded isolating and expressing genes in heterologous expression systems that elicit almost identical CaCCs currents to those reported in native cells. The identity of these molecules that seem to be the molecular basis of CaCCs are anoctamins (TMEM16) (Caputo et al. 2008; Ferrera et al. 2010; Schroeder et al. 2008; Yang et al. 2008). The term anoctamin refers to the property of being anion selective and having subunits with a putative topology consisting of eight transmembrane segments and cytosolic N- and C-termini (Galietta 2009). One of the main TMEM16 proteins whose activity strongly resembles that of CaCCs is TMEM16A, corresponding to anoctamin-1. The TMEM16/anoctamin family has nine more members named TMEM16B to -K, or anoctamin 2–10 (Galindo and Vacquier 2005). The biophysical characteristics among the members of this family of ion channels such as voltage dependence, selectivity, and conductance differ (Pifferi et al. 2009; Scudieri et al. 2012). Almost all show enzymatic properties in addition to the transport function, which has led to the speculation that they are multifunctional proteins, one of them being permeable to Cl^- (Tian et al. 2012).

As mentioned, earlier electrophysiological evidence for the presence of Cl^- channels in sperm had been gathered. The initial patch-clamp recordings directly on epididymal mouse sperm revealed an anion channel displaying biophysical properties and sensitivity to NFA, resembling to the Ca^{2+}-dependent Cl^- channels (Espinosa et al. 1998; Hogg et al. 1994). Thereafter, a Cl^--permeable channel showing long stable openings was documented in patch-clamp studies in the cell-attached mode in the human sperm. Distinct channel clustering and activity was detected in different sperm head regions whose functional significance awaits determination (Jimenez-Gonzalez et al. 2007).

More recently, using a modified perforated patch-clamp technique, whole-cell recordings of the human mature spermatozoa head revealed that an important component of their Cl^- currents is in fact caused by CaCCs, possibly of the TMEM16A type (Orta et al. 2012). Supporting evidence was based on the biophysical properties and in the pharmacology profile of the recorded currents. The typical results of such experiments characterizing biophysically and pharmacologically CaCCs in epididymal human spermatozoa are shown in Fig. 6.2. One of the most specific

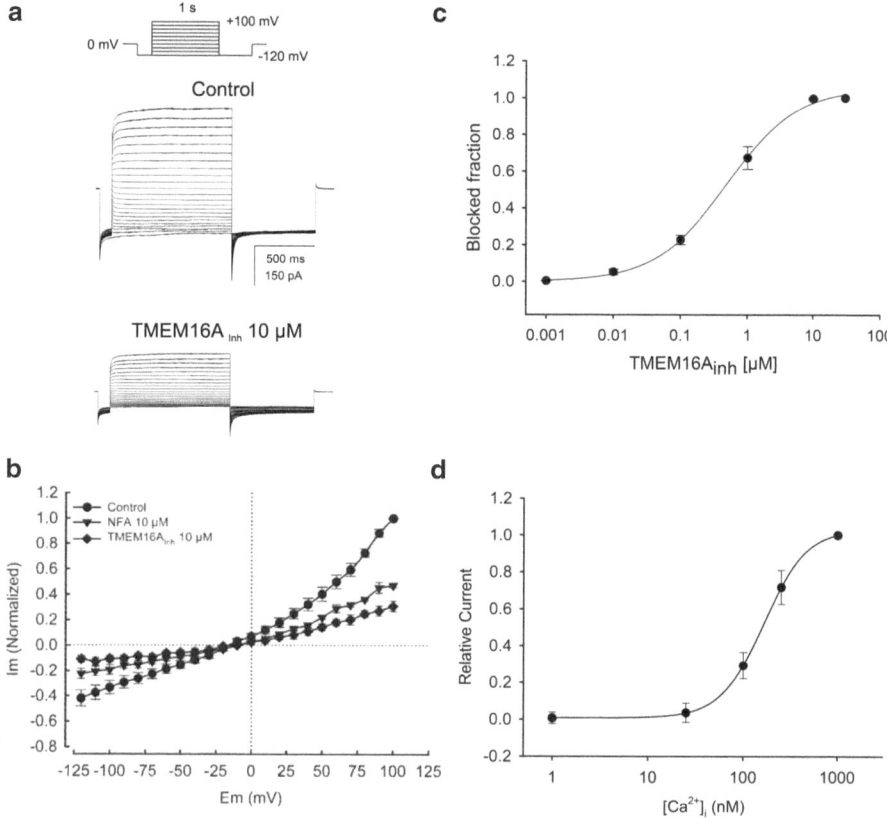

Fig. 6.2 Ca^{2+}-activated Cl^- channel (CaCC) currents in epididymal human spermatozoa. (**a**) Whole-cell CaCC currents recorded from a human spermatozoa. Currents were obtained at a holding potential of 0 mV with the indicated voltage step protocol (*top panel*). The control currents (Control) were recorded before and after exposure to 10 μM TMEM16A$_{inh}$ (*bottom panel*). (**b**) I–V plot of the blockade of the CaCC currents ([Ca^{2+}]$_i$=250 nM) by 10 μM NFA and 10 μM TMEM16A$_{inh}$. Currents were normalized with respect to maximal current in control measurements at the end of each voltage pulse. (**c**) Dose-dependent blockade of CaCC currents by TMEM16A$_{inh}$. Current amplitudes were measured at +100 mV by averaging seven to nine original current traces and normalized with respect to the maximal blocked fraction. (**d**) Concentration–response curve at different [Ca^{2+}]$_i$ on the macroscopic inward and outward Cl^- currents obtained from seals on the cytoplasmic droplet against the current obtained at 1,000 nM [Ca^{2+}]$_i$ recorded at +100 mV. The continuous lines represent the data fitted to the Hill equation with the following parameters: K_d=163 nM and n_H=1.9 at +100 mV. For **b** and **c**, data represent the mean±SEM with n=6; for **d**, data represent mean±SEM with n=7

antagonists of TMEM16A channels, the drug TMEM16A$_{inh}$ (20 μM), inhibited these currents and reduced ~80 % of the AR. These results suggest the critical participation of these channels during the AR. Results supporting this hypothesis are shown in Fig. 6.3. It is known that during the AR there are large [Ca^{2+}]$_i$ changes leading to profound modifications in the sperm head morphology involving acrosome swelling and a decrease in regulatory volume; therefore, CaCCs channels may

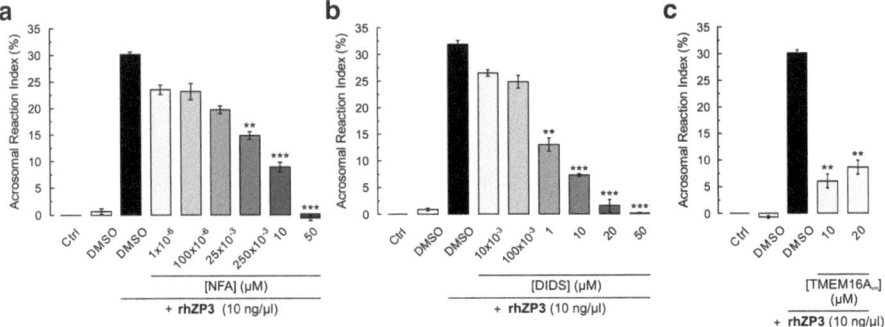

Fig. 6.3 The rhZP3-induced AR in human spermatozoa is inhibited by CaCC/TMEM16A$_{inh}$. Motile human spermatozoa were obtained by the swim-up technique and capacitated during 5 h. Sperm populations were preincubated for 15 min with different NFA, DIDS, and TMEM16A$_{inh}$ concentrations. The AR was induced with rhZP3 (10 ng μl^{-1}). Cells were fixed with cold methanol and the acrosomal status evaluated after staining sperm with FITC-PSA. Acrosomal reaction was expressed as an index (ARI = percentage of AR normalized against the maximum AR with ionomycin) and was used to estimate the percentage of AR inhibition. NFA (**a**) and DIDS (**b**), two CaCCs blockers, inhibited 90 % of the AR. TMEM16A$_{inh}$ (**c**) blocked approximately 80 %, indicating that TMEM16A channels may have an essential contribution in the AR. For all data, *error bars* represent mean ± SEM with *n* = 4–6. Statistical comparisons according to Student's unpaired *t* test indicated *$P < 0.05$; **$P < 0.01$; ***$P < 0.001$ versus spermatozoa incubated with 0.1 % DMSO + 10 ng μl^{-1} of rhZP3

participate in this event (Zanetti and Mayorga 2009). Blockage of CaCCs by NFA, DIDS, and TMEM16A$_{inh}$ may affect the reduction in regulatory volume, which in turn seems to be important to regulate acrosomal–plasma membrane distance, essential for acrosomal exocytosis (Zanetti and Mayorga 2009).

Evidence gathered recently has shown that CaCCS and voltage and Ca^{2+}-gated K$^+$ channels (BK$_{Ca}$, Slo1, or K$_{Ca}$1.1) have some unexpected similarities regarding their pharmacological properties (Greenwood and Leblanc 2007; Sones et al. 2009). Classical Cl$^-$ channel inhibitors with different structures such as anthracene-9-carboxylate, NFA, and ethacrynic acid also behave as BK$_{Ca}$ agonists (Greenwood and Large 1995; Ottolia and Toro 1994; Toma et al. 1996). The potent ability to inhibit the AR by classical anion channel blockers such as NFA may be explained by a combination of the effects of this drug on BK$_{Ca}$ and CaCCs, as BK$_{Ca}$ channels are also present in mammalian sperm (Rossato et al. 2001; Wu et al. 1998).

Notably, it has also been shown that TMEM16A channels may increase their permeability to HCO$_3^-$ on raising [Ca^{2+}]$_i$ through Ca^{2+}/calmodulin modulation (Jung et al. 2013). This finding is relevant to sperm physiology because HCO$_3^-$ plays a key role in cAMP production and pH$_i$ regulation (Visconti et al. 2011). It has also been shown that CaCCa can be modulated through the activity of other ion channels also present in spermatozoa (i.e., purinergic receptors) (Wang et al. 2013). Taking these facts into account, it is likely that new evidence of their critical relevance will be gathered in the future.

In summary, CaCCs significantly influence sperm physiology and are likely players in the solubilized ZP-induced AR. It will be interesting to further investigate their role in sperm motility, as inklings of their involvement in sea urchin sperm chemotaxis have been reported (Alvarez et al. 2012; Guerrero et al. 2013; Ingermann et al. 2008; Wood et al. 2003; Wood et al. 2007).

6.6.4 Voltage-Dependent Anion Channels (VDACs)

VDACs are porins, usually located on the outer mitochondrial membrane (Liu et al. 2010). They seem to be critical for mitochondria metabolism and regulation of apoptosis as they are permeant to small hydrophilic molecules. In this sense, they have been described as molecules able to "regulate cell life and death" (Shoshan-Barmatz et al. 2010). They seem to be involved in the pathogenesis of a variety of processes such as cancer, neurodegenerative diseases, and ischemic-reperfusion injuries in the heart (Peixoto et al. 2012). VDAC2 is the main variant present in the male germline, although VDAC3 has also been found (Liu et al. 2010). VDACs have been localized not only in the mitochondrial sheath but also in the plasma membrane and outer dense fiber (Hinsch et al. 2004; Liu et al. 2009; Triphan et al. 2008). It has been reported recently that they might play a role in human fertility (Kwon et al. 2013). Consistent with this proposal, VDAC2 was found to be one of the putative sperm head membrane proteins that bind ZP (Petit et al. 2013).

6.6.5 Secondary Active Cl⁻ Transporters

A $[Cl^-]_i$ increase has been reported to occur during capacitation (Hernandez-Gonzalez et al. 2007; Meizel and Turner 1996), and it is possible that several Cl⁻ transport systems participate in this process. Under physiological conditions electro-neutral carriers such as NCC, NKCC support transport of Cl⁻ into the cell and KCC outside the cell (Russell 2000). Pharmacological approaches have been used to test the participation of these carriers during sperm capacitation. Stilbenes such as DIDS and SITS, which are general Cl⁻ transport blockers, reduced sperm capacitation parameters to similar levels as those observed in the absence of Cl⁻. Two specific inhibitors (bumetanide and furosemide) for NKCC blocked the increase in tyrosine phosphorylation, hyperactivation, and the ability of the sperm to fertilize in vitro, but the concentration required to observe these effects was higher than that specifically reported to inhibit NKCC (Garg et al. 2007; Russell 2000). NKCC function requires the presence of Na⁺, K⁺, and Cl⁻, but tyrosine phosphorylation is affected in the absence of Na⁺ and Cl⁻ but not K⁺, suggesting that bumetanide and furosemide may be acting through a target other than NKCC. It is worth noting that the ZP-induced AR depended on the presence of the three ions and was inhibited at much lower concentrations of bumetanide, suggesting that NKCC might have a role in the preparation of the sperm for the physiologically induced AR. NKCC1 transcripts were detected in spermatids, and male mouse null mutants of this protein have defects in spermatogenesis and are infertile (Pace et al. 2000). Thiazide, a specific NCC inhibitor, did not interfere with capacitation-associated processes.

6.6.6 Cl⁻/HCO₃⁻ Exchangers

Cl^-/HCO_3^- transporters are proteins that exchange Cl^- for HCO_3^- in either direction. HCO_3^- regulation is particularly important in spermatozoa because it activates cAMP synthesis by the atypical soluble adenylyl cyclase present in these cells (Hess et al. 2005; Okamura et al. 1985). The specific carriers responsible for HCO_3^- transport have not yet been fully defined. Experiments from our group suggest that Na^+/HCO_3^- cotransporters allow HCO_3^- influx into mouse sperm (Demarco et al. 2003). However, other HCO_3^- transport systems can also play a role in the control of HCO_3^- levels in sperm. In particular, Cl^-/HCO_3^- exchangers have been proposed to be involved in HCO_3^- homeostasis. These exchangers also influence pH_i, cell volume, and E_m through their contribution to determine the Cl^- gradient. Two superfamilies group the Cl^-/HCO_3^- exchangers, SLC4 and SLC26; they have different anion selectivity and unique tissue distribution. The SLC4 superfamily is composed of 3 genes (AE1, AE2, and AE3), each of them represented by more than one alternative spliced sequence. The SLC26 gene superfamily consists of 11 genes but only SLC26A3, SLC26A4, and SLC26A6 have Cl^-/HCO_3^- exchange activity.

The AE2 gene is the only member from the SLC4 superfamily reported in spermatogenic cells. This gene has five splice isoforms (AE2a, AE2b1, AE2b2, AE2c1, AE2c2), and mice lacking their expression die before weaning of severely retarded development (Gawenis et al. 2004). Mice expressing only the AE2c isoform survive but they are infertile and exhibit testicular dysplasia, consistent with the observation that AE2 is highly expressed in the testis (Medina et al. 2003).

Recent Western blot and immunofluorescence results from several laboratories, including ours, indicate the presence of SLC26A3 and SLC26A6 in the sperm midpiece (Chan et al. 2009; Chavez et al. 2011; Chen et al. 2009). Chavez et al. (2011), also provided evidence that these transporters co-precipitate with CFTR and that tenidap, a specific SLC26A3 inhibitor, blocked the capacitation-associated hyperpolarization and the ZP-induced AR. However, tenidap did not affect the activation of a cAMP pathway or the increase in tyrosine phosphorylation, suggesting that these transporters are not directly involved in the regulation of the soluble adenylate cyclase.

6.7 Final Remarks

Work reported in recent years corroborates the significant repercussions of ion channels in sperm maturation, capacitation, and the AR. However, our knowledge regarding the molecular mechanisms regulating these processes is still limited. The study of ion transport in spermatozoa, specially the use of electrophysiological techniques, has been a difficult enterprise because of their small size and the stiffness of their membrane. Fortunately, new patch-clamp strategies have improved our ability to study sperm ion channels. Although the inability of these cells to perform transcription and translation significantly complicates knocking down or expressing exogenous proteins, advances in the production of genetically modified mice have enhanced the identification of key proteins, processes, and mechanisms essential to sperm function. The increasing sensitivity and speed of single-cell ion-imaging strategies is helping to unravel sperm signaling networks and how ion channels participate in them. We hope the readers have become aware that the study of sperm anion transport requires our full attention, as it is deeply implicated in the physiology of this important cell. Enhancing our understanding of sperm ion transport will impact our capacity to preserve animal species and improve our control of fertility.

Acknowledgments This work was supported by DGAPA: IN202312 (to A.D.) and IN202212-3 (to C.T.), CONACyT: 128566 (to A.D.) and 99333 (to C.T.), CSIC-UdelaR international collaboration grants to G.F. with A.D. and C.T. NIH: R01 HD038082 (to Pablo Visconti) and CSIC p944 (to Gonzalo Ferreira).

References

Alvarez L, Dai L, Friedrich BM, Kashikar ND, Gregor I, Pascal R, Kaupp UB (2012) The rate of change in Ca(2+) concentration controls sperm chemotaxis. J Cell Biol 196:653–663

Anderson MP, Gregory RJ, Thompson S, Souza DW, Paul S, Mulligan RC, Smith AE, Welsh MJ (1991) Demonstration that CFTR is a chloride channel by alteration of its anion selectivity. Science 253:202–205

Arnoult C, Cardullo RA, Lemos JR, Florman HM (1996) Activation of mouse sperm T-type Ca^{2+} channels by adhesion to the egg zona pellucida. Proc Natl Acad Sci USA 93:13004–13009

Arnoult C, Kazam IG, Visconti PE, Kopf GS, Villaz M, Florman HM (1999) Control of the low voltage-activated calcium channel of mouse sperm by egg ZP3 and by membrane hyperpolarization during capacitation. Proc Natl Acad Sci USA 96:6757–6762

Austin CR (1952) The capacitation of the mammalian sperm. Nature (Lond) 170:326

Baker MA (2011) The omics revolution and our understanding of sperm cell biology. Asian J Androl 13:6–10

Baldi E, Casano R, Falsetti C, Krausz C, Maggi M, Forti G (1991) Intracellular calcium accumulation and responsiveness to progesterone in capacitating human spermatozoa. J Androl 12:323–330

Bello OD, Zanetti MN, Mayorga LS, Michaut MA (2012) RIM, Munc13, and Rab3A interplay in acrosomal exocytosis. Exp Cell Res 318:478–488

Berdiev BK, Qadri YJ, Benos DJ (2009) Assessment of the CFTR and ENaC association. Mol Biosyst 5:123–127

Bray C, Son JH, Kumar P, Harris JD, Meizel S (2002) A role for the human sperm glycine receptor/Cl(−) channel in the acrosome reaction initiated by recombinant ZP3. Biol Reprod 66:91–97

Breitbart H (2003) Signaling pathways in sperm capacitation and acrosome reaction. Cell Mol Biol (Noisy-le-grand) 49:321–327

Breitbart H, Rotman T, Rubinstein S, Etkovitz N (2010) Role and regulation of PI3K in sperm capacitation and the acrosome reaction. Mol Cell Endocrinol 314:234–238

Buffone MG, Ijiri TW, Cao W, Merdiushev T, Aghajanian HK, Gerton GL (2012) Heads or tails? Structural events and molecular mechanisms that promote mammalian sperm acrosomal exocytosis and motility. Mol Reprod Dev 79:4–18

Caputo A, Caci E, Ferrera L, Pedemonte N, Barsanti C, Sondo E, Pfeffer U, Ravazzolo R, Zegarra-Moran O, Galietta LJ (2008) TMEM16A, a membrane protein associated with calcium-dependent chloride channel activity. Science 322:590–594

Carlson AE, Hille B, Babcock DF (2007) External Ca²⁺ acts upstream of adenylyl cyclase SACY in the bicarbonate signaled activation of sperm motility. Dev Biol 312:183–192

Carlson AE, Quill TA, Westenbroek RE, Schuh SM, Hille B, Babcock DF (2005) Identical phenotypes of CatSper1 and CatSper2 null sperm. J Biol Chem 280:32238–32244

Carlson AE, Westenbroek RE, Quill T, Ren D, Clapham DE, Hille B, Garbers DL, Babcock DF (2003) CatSper1 required for evoked Ca²⁺ entry and control of flagellar function in sperm. Proc Natl Acad Sci USA 100:14864–14868

Castillo Bennett J, Roggero CM, Mancifesta FE, Mayorga LS (2010) Calcineurin-mediated dephosphorylation of synaptotagmin VI is necessary for acrosomal exocytosis. J Biol Chem 285:26269–26278

Chan HC, Ruan YC, He Q, Chen MH, Chen H, Xu WM, Chen WY, Xie C, Zhang XH, Zhou Z (2009) The cystic fibrosis transmembrane conductance regulator in reproductive health and disease. J Physiol 587:2187–2195

Chan HC, Shi QX, Zhou CX, Wang XF, Xu WM, Chen WY, Chen AJ, Ni Y, Yuan YY (2006) Critical role of CFTR in uterine bicarbonate secretion and the fertilizing capacity of sperm. Mol Cell Endocrinol 250:106–113

Chang MC (1951) Fertilizing capacity of spermatozoa deposited into the fallopian tubes. Nature (Lond) 168:697–698

Chavez JC, Hernandez-Gonzalez EO, Wertheimer E, Visconti PE, Darszon A, Trevino CL (2011) Participation of the Cl⁻/HCO₃⁻ exchangers SLC26A3 and SLC26A6, the Cl⁻ channel CFTR and the regulatory factor SLC9A3R1 in mouse sperm capacitation. Biol Reprod 86:1–14

Chen WY, Xu WM, Chen ZH, Ni Y, Yuan YY, Zhou SC, Zhou WW, Tsang LL, Chung YW, Hoglund P, Chan HC, Shi QX (2009) Cl⁻ is required for HCO₃⁻ entry necessary for sperm capacitation in guinea pig: involvement of a Cl⁻/HCO₃⁻ exchanger (SLC26A3) and CFTR. Biol Reprod 80:115–123

Christen R, Schackmann RW, Shapiro BM (1983) Metabolism of sea urchin sperm. Interrelationships between intracellular pH, ATPase activity, and mitochondrial respiration. J Biol Chem 258:5392–5399

Cooper TG, Yeung CH (2007) Involvement of potassium and chloride channels and other transporters in volume regulation by spermatozoa. Curr Pharm Des 13:3222–3230

Costello S, Michelangeli F, Nash K, Lefievre L, Morris J, Machado-Oliveira G, Barratt C, Kirkman-Brown J, Publicover S (2009) Ca²⁺-stores in sperm: their identities and functions. Reproduction 138:425–437

Cox T, Peterson RN (1989) Identification of calcium-conducting channels in isolated boar sperm plasma membranes. Biochem Biophys Res Commun 161:162–168

Cross NL (1998) Role of cholesterol in sperm capacitation. Biol Reprod 59:7–11

Dacheux JL, Belleannee C, Guyonnet B, Labas V, Teixeira-Gomes AP, Ecroyd H, Druart X, Gatti JL, Dacheux F (2012) The contribution of proteomics to understanding epididymal maturation of mammalian spermatozoa. Syst Biol Reprod Med 58:197–210

Darszon A, Labarca P, Nishigaki T, Espinosa F (1999) Ion channels in sperm physiology. Physiol Rev 79:481–510

Darszon A, Nishigaki T, Beltran C, Trevino CL (2011) Calcium channels in the development, maturation, and function of spermatozoa. Physiol Rev 91:1305–1355

Darszon A, Sanchez-Cardenas C, Orta G, Sanchez-Tusie AA, Beltran C, Lopez-Gonzalez I, Granados-Gonzalez G, Trevino CL (2012) Are TRP channels involved in sperm development and function? Cell Tissue Res 349(3):749–764

DasGupta S, Mills CL, Fraser LR (1993) Ca(2+)-related changes in the capacitation state of human spermatozoa assessed by a chlortetracycline fluorescence assay. J Reprod Fertil 99:135–143

Davis BK (1981) Timing of fertilization in mammals: sperm cholesterol/phospholipid ratio as a determinant of the capacitation interval. Proc Natl Acad Sci USA 78:7560–7564

De Blas GA, Darszon A, Ocampo AY, Serrano CJ, Castellano LE, Hernandez-Gonzalez EO, Chirinos M, Larrea F, Beltran C, Trevino CL (2009) TRPM8, a versatile channel in human sperm. PLoS One 4:e6095

De La Vega-Beltran JL, Sanchez-Cardenas C, Krapf D, Hernandez-Gonzalez EO, Wertheimer E, Trevino CL, Visconti PE, Darszon A (2012) Mouse sperm membrane potential hyperpolarization is necessary and sufficient to prepare sperm for the acrosome reaction. J Biol Chem 287:44384–44393

Demarco IA, Espinosa F, Edwards J, Sosnik J, De La Vega-Beltran JL, Hockensmith JW, Kopf GS, Darszon A, Visconti PE (2003) Involvement of a Na^+/HCO^{-3} cotransporter in mouse sperm capacitation. J Biol Chem 278:7001–7009

Duran C, Qu Z, Osunkoya AO, Cui Y, Hartzell HC (2012) ANOs 3-7 in the anoctamin/Tmem16 Cl⁻ channel family are intracellular proteins. Am J Physiol Cell Physiol 302:C482–C493

Escoffier J, Boisseau S, Serres C, Chen CC, Kim D, Stamboulian S, Shin HS, Campbell KP, De Waard M, Arnoult C (2007) Expression, localization and functions in acrosome reaction and sperm motility of Ca(V)3.1 and Ca(V)3.2 channels in sperm cells: an evaluation from Ca(V)3.1 and Ca(V)3.2 deficient mice. J Cell Physiol 212:753–763

Escoffier J, Krapf D, Navarrete F, Darszon A, Visconti PE (2012) Flow cytometry analysis reveals a decrease in intracellular sodium during sperm capacitation. J Cell Sci 125:473–485

Espinosa F, Darszon A (1995) Mouse sperm membrane potential: changes induced by Ca^{2+}. FEBS Lett 372:119–125

Espinosa F, de la Vega-Beltran JL, Lopez-Gonzalez I, Delgado R, Labarca P, Darszon A (1998) Mouse sperm patch-clamp recordings reveal single Cl⁻ channels sensitive to niflumic acid, a blocker of the sperm acrosome reaction. FEBS Lett 426:47–51

Esposito G, Jaiswal BS, Xie F, Krajnc-Franken MA, Robben TJ, Strik AM, Kuil C, Philipsen RL, van Duin M, Conti M, Gossen JA (2004) Mice deficient for soluble adenylyl cyclase are infertile because of a severe sperm-motility defect. Proc Natl Acad Sci USA 101:2993–2998

Ferrera L, Caputo A, Galietta LJ (2010) TMEM16A protein: a new identity for Ca(2+)-dependent Cl(-) channels. Physiology (Bethesda) 25:357–363

Figueiras-Fierro D, Acevedo JJ, Martinez-Lopez P, Escoffier J, Sepulveda FV, Balderas E, Orta G, Visconti PE, Darszon A (2013) Electrophysiological evidence for the presence of cystic fibrosis transmembrane conductance regulator (CFTR) in mouse sperm. J Cell Physiol 228:590–601

Florman HM, Jungnickel MK, Sutton KA (2008) Regulating the acrosome reaction. Int J Dev Biol 52:503–510

Furst J, Gschwentner M, Ritter M, Botta G, Jakab M, Mayer M, Garavaglia L, Bazzini C, Rodighiero S, Meyer G, Eichmuller S, Woll E, Paulmichl M (2002) Molecular and functional aspects of anionic channels activated during regulatory volume decrease in mammalian cells. Pflugers Arch 444:1–25

Gadella BM, Harrison RA (2000) The capacitating agent bicarbonate induces protein kinase A-dependent changes in phospholipid transbilayer behavior in the sperm plasma membrane. Development (Camb) 127:2407–2420

Galietta LJ (2009) The TMEM16 protein family: a new class of chloride channels? Biophys J 97:3047–3053

Galindo BE, Vacquier VD (2005) Phylogeny of the TMEM16 protein family: some members are overexpressed in cancer. Int J Mol Med 16:919–924

Garg P, Martin CF, Elms SC, Gordon FJ, Wall SM, Garland CJ, Sutliff RL, O'Neill WC (2007) Effect of the Na-K-2Cl cotransporter NKCC1 on systemic blood pressure and smooth muscle tone. Am J Physiol Heart Circ Physiol 292:H2100–H2105

Gawenis LR, Ledoussal C, Judd LM, Prasad V, Alper SL, Stuart-Tilley A, Woo AL, Grisham C, Sanford LP, Doetschman T, Miller ML, Shull GE (2004) Mice with a targeted disruption of the AE2 Cl⁻/HCO₃⁻ exchanger are achlorhydric. J Biol Chem 279:30531–30539

Gibbs GM, Orta G, Reddy T, Koppers AJ, Martinez-Lopez P, Luis de la Vega-Beltran J, Lo JC, Veldhuis N, Jamsai D, McIntyre P, Darszon A, O'Bryan MK (2011) Cysteine-rich secretory protein 4 is an inhibitor of transient receptor potential M8 with a role in establishing sperm function. Proc Natl Acad Sci USA 108:7034–7039

Go KJ, Wolf DP (1983) The role of sterols in sperm capacitation. Adv Lipid Res 20:317–330

Greenwood IA, Large WA (1995) Comparison of the effects of fenamates on Ca-activated chloride and potassium currents in rabbit portal vein smooth muscle cells. Br J Pharmacol 116: 2939–2948

Greenwood IA, Leblanc N (2007) Overlapping pharmacology of Ca^{2+}-activated Cl^- and K^+ channels. Trends Pharmacol Sci 28:1–5

Guerrero A, Carneiro J, Pimentel A, Wood CD, Corkidi G, Darszon A (2011) Strategies for locating the female gamete: the importance of measuring sperm trajectories in three spatial dimensions. Mol Hum Reprod 17:511–523

Guerrero A, Espinal J, Wood CD, Rendon JM, Carneiro J, Martinez-Mekler G, Darszon A (2013) Niflumic acid disrupts marine spermatozoan chemotaxis without impairing the spatiotemporal detection of chemoattractant gradients. J Cell Sci 126:1477–1487

Guerrero A, Sanchez JA, Darszon A (1987) Single-channel activity in sea urchin sperm revealed by the patch-clamp technique. FEBS Lett 220:295–298

Harrison RA (2004) Rapid PKA-catalysed phosphorylation of boar sperm proteins induced by the capacitating agent bicarbonate. Mol Reprod Dev 67:337–352

Hartzell C, Putzier I, Arreola J (2005) Calcium-activated chloride channels. Annu Rev Physiol 67:719–758

Hartzell HC, Yu K, Xiao Q, Chien LT, Qu Z (2009) Anoctamin/TMEM16 family members are Ca^{2+}-activated Cl- channels. J Physiol 587(Pt 10):2127–2139

Hernandez-Gonzalez EO, Sosnik J, Edwards J, Acevedo JJ, Mendoza-Lujambio I, Lopez-Gonzalez I, Demarco I, Wertheimer E, Darszon A, Visconti PE (2006) Sodium and epithelial sodium channels participate in the regulation of the capacitation-associated hyperpolarization in mouse sperm. J Biol Chem 281:5623–5633

Hernandez-Gonzalez EO, Trevino CL, Castellano LE, de la Vega-Beltran JL, Ocampo AY, Wertheimer E, Visconti PE, Darszon A (2007) Involvement of cystic fibrosis transmembrane conductance regulator in mouse sperm capacitation. J Biol Chem 282:24397–24406

Hess KC, Jones BH, Marquez B, Chen Y, Ord TS, Kamenetsky M, Miyamoto C, Zippin JH, Kopf GS, Suarez SS, Levin LR, Williams CJ, Buck J, Moss SB (2005) The "soluble" adenylyl cyclase in sperm mediates multiple signaling events required for fertilization. Dev Cell 9:249–259

Hille B (2001) Ion channels of excitable membranes. Sinauer Associates, Sunderland

Hinsch KD, De Pinto V, Aires VA, Schneider X, Messina A, Hinsch E (2004) Voltage-dependent anion-selective channels VDAC2 and VDAC3 are abundant proteins in bovine outer dense fibers, a cytoskeletal component of the sperm flagellum. J Biol Chem 279:15281–15288

Hogg RC, Wang Q, Large WA (1994) Action of niflumic acid on evoked and spontaneous calcium-activated chloride and potassium currents in smooth muscle cells from rabbit portal vein. Br J Pharmacol 112:977–984

Hu JH, He XB, Wu Q, Yan YC, Koide SS (2002) Biphasic effect of GABA on rat sperm acrosome reaction: involvement of GABA(A) and GABA(B) receptors. Arch Androl 48:369–378

Huang F, Rock JR, Harfe BD, Cheng T, Huang X, Jan YN, Jan LY (2009) Studies on expression and function of the TMEM16A calcium-activated chloride channels. Proc Natl Acad Sci USA 106(50):21413–21418

Hung PH, Suarez SS (2010) Regulation of sperm storage and movement in the ruminant oviduct. Soc Reprod Fertil Suppl 67:257–266

Ingermann RL, Holcomb M, Zuccarelli MD, Kanuga MK, Cloud JG (2008) Initiation of motility by steelhead (*Oncorhynchus mykiss*) sperm: membrane ion exchangers and pH sensitivity. Comp Biochem Physiol A Mol Integr Physiol 151:651–656

Inoue N, Satouh Y, Ikawa M, Okabe M, Yanagimachi R (2011) Acrosome-reacted mouse spermatozoa recovered from the perivitelline space can fertilize other eggs. Proc Natl Acad Sci USA 108:20008–20011

Jentsch TJ, Neagoe I, Scheel O (2005) CLC chloride channels and transporters. Curr Opin Neurobiol 15:319–325

Jimenez-Gonzalez C, Michelangeli F, Harper CV, Barratt CL, Publicover SJ (2006) Calcium signalling in human spermatozoa: a specialized 'toolkit' of channels, transporters and stores. Hum Reprod Update 12:253–267

Jimenez-Gonzalez MC, Gu Y, Kirkman-Brown J, Barratt CL, Publicover S (2007) Patch-clamp 'mapping' of ion channel activity in human sperm reveals regionalisation and co-localisation into mixed clusters. J Cell Physiol 213:801–808

Jin JY, Chen WY, Zhou CX, Chen ZH, Yu-Ying Y, Ni Y, Chan HC, Shi QX (2009) Activation of GABA$_A$ receptor/Cl⁻ channel and capacitation in rat spermatozoa: HCO$_3$⁻ and Cl⁻ are essential. Syst Biol Reprod Med 55:97–108

Jin M, Fujiwara E, Kakiuchi Y, Okabe M, Satouh Y, Baba SA, Chiba K, Hirohashi N (2011) Most fertilizing mouse spermatozoa begin their acrosome reaction before contact with the zona pellucida during in vitro fertilization. Proc Natl Acad Sci USA 108:4892–4896

Jung J, Nam JH, Park HW, Oh U, Yoon JH, Lee MG (2013) Dynamic modulation of ANO1/TMEM16A HCO$_3$(−) permeability by Ca^{2+}/calmodulin. Proc Natl Acad Sci USA 110:360–365

Kanbara K, Mori Y, Kubota T, Watanabe M, Yanagawa Y, Otsuki Y (2011) Expression of the GABA$_A$ receptor/chloride channel in murine spermatogenic cells. Histol Histopathol 26:95–106

Kaupp UB, Kashikar ND, Weyand I (2008) Mechanisms of sperm chemotaxis. Annu Rev Physiol 70:93–117

Kirichok Y, Lishko PV (2011) Rediscovering sperm ion channels with the patch-clamp technique. Mol Hum Reprod 17:478–499

Kirichok Y, Navarro B, Clapham DE (2006) Whole-cell patch-clamp measurements of spermatozoa reveal an alkaline-activated Ca^{2+} channel. Nature (Lond) 439:737–740

Konig J, Schreiber R, Voelcker T, Mall M, Kunzelmann K (2001) The cystic fibrosis transmembrane conductance regulator (CFTR) inhibits ENaC through an increase in the intracellular Cl⁻ concentration. EMBO Rep 2:1047–1051

Kumar P, Meizel S (2008) Identification and spatial distribution of glycine receptor subunits in human sperm. Reproduction 136:387–390

Kunzelmann K, Kongsuphol P, Chootip K, Toledo C, Martins JR, Almaca J, Tian Y, Witzgall R, Ousingsawat J, Schreiber R (2011) Role of the Ca^{2+}-activated Cl⁻ channels bestrophin and anoctamin in epithelial cells. Biol Chem 392:125–134

Kunzelmann K, Schreiber R (1999) CFTR, a regulator of channels. J Membr Biol 168:1–8

Kwon WS, Park YJ, Mohamed-el SA, Pang MG (2013) Voltage-dependent anion channels are a key factor of male fertility. Fertil Steril 99:354–361

Labarca P, Santi C, Zapata O, Morales E, Beltr'an C, Li'evano A, Darszon A (1996) A cAMP regulated K⁺-selective channel from the sea urchin sperm plasma membrane. Dev Biol 174:271–280

Li CY, Jiang LY, Chen WY, Li K, Sheng HQ, Ni Y, Lu JX, Xu WX, Zhang SY, Shi QX (2010) CFTR is essential for sperm fertilizing capacity and is correlated with sperm quality in humans. Hum Reprod 25:317–327

Lievano A, Sanchez JA, Darszon A (1985) Single-channel activity of bilayers derived from sea urchin sperm plasma membranes at the tip of a patch-clamp electrode. Dev Biol 112:253–257

Lievano A, Santi CM, Serrano CJ, Trevino CL, Bellve AR, Hernandez-Cruz A, Darszon A (1996) T-type Ca^{2+} channels and alpha1E expression in spermatogenic cells, and their possible relevance to the sperm acrosome reaction. FEBS Lett 388:150–154

Lindemann CB, Goltz JS (1988) Calcium regulation of flagellar curvature and swimming pattern in triton X-100–extracted rat sperm. Cell Motil Cytoskeleton 10:420–431

Lishko PV, Kirichok Y, Ren D, Navarro B, Chung JJ, Clapham DE (2012) The control of male fertility by spermatozoan ion channels. Annu Rev Physiol 74:453–475

Liu B, Wang Z, Zhang W, Wang X (2009) Expression and localization of voltage-dependent anion channels (VDAC) in human spermatozoa. Biochem Biophys Res Commun 378:366–370

Liu B, Zhang W, Wang Z (2010) Voltage-dependent anion channel in mammalian spermatozoa. Biochem Biophys Res Commun 397:633–636

Llanos MN, Ronco AM, Aguirre MC, Meizel S (2001) Hamster sperm glycine receptor: evidence for its presence and involvement in the acrosome reaction. Mol Reprod Dev 58:205–215

Macdonald RL, Olsen RW (1994) GABA$_A$ receptor channels. Annu Rev Neurosci 17:569–602

Martinez-Lopez P, Santi CM, Trevino CL, Ocampo-Gutierrez AY, Acevedo JJ, Alisio A, Salkoff LB, Darszon A (2009) Mouse sperm K$^+$ currents stimulated by pH and cAMP possibly coded by Slo3 channels. Biochem Biophys Res Commun 381:204–209

Martinez-Lopez P, Trevino CL, de la Vega-Beltran JL, Blas GD, Monroy E, Beltran C, Orta G, Gibbs GM, O'Bryan MK, Darszon A (2011) TRPM8 in mouse sperm detects temperature changes and may influence the acrosome reaction. J Cell Physiol 226(6):1620–1631

Mayorga LS, Tomes CN, Belmonte SA (2007) Acrosomal exocytosis, a special type of regulated secretion. IUBMB Life 59:286–292

McPartlin LA, Visconti PE, Bedford-Guaus SJ (2011) Guanine-nucleotide exchange factors (RAPGEF3/RAPGEF4) induce sperm membrane depolarization and acrosomal exocytosis in capacitated stallion sperm. Biol Reprod 85:179–188

Medina JF, Recalde S, Prieto J, Lecanda J, Saez E, Funk CD, Vecino P, van Roon MA, Ottenhoff R, Bosma PJ, Bakker CT, Elferink RP (2003) Anion exchanger 2 is essential for spermiogenesis in mice. Proc Natl Acad Sci USA 100:15847–15852

Meizel S (1997) Amino acid neurotransmitter receptor/chloride channels of mammalian sperm and the acrosome reaction. Biol Reprod 56:569–574

Meizel S, Son JH (2005) Studies of sperm from mutant mice suggesting that two neurotransmitter receptors are important to the zona pellucida-initiated acrosome reaction. Mol Reprod Dev 72:250–258

Meizel S, Turner KO (1996) Chloride efflux during the progesterone-initiated human sperm acrosome reaction is inhibited by lavendustin A, a tyrosine kinase inhibitor. J Androl 17:327–330

Melendrez CS, Meizel S (1995) Studies of porcine and human sperm suggesting a role for a sperm glycine receptor/Cl$^-$ channel in the zona pellucida-initiated acrosome reaction. Biol Reprod 53:676–683

Miki K, Qu W, Goulding EH, Willis WD, Bunch DO, Strader LF, Perreault SD, Eddy EM, O'Brien DA (2004) Glyceraldehyde 3-phosphate dehydrogenase-S, a sperm-specific glycolytic enzyme, is required for sperm motility and male fertility. Proc Natl Acad Sci USA 101:16501–16506

Miledi R (1982) A calcium-dependent transient outward current in *Xenopus laevis* oocytes. Proc R Soc Lond B Biol Sci 215:491–497

Morales E, de la Torre L, Moy GW, Vacquier VD, Darszon A (1993) Anion channels in the sea urchin sperm plasma membrane. Mol Reprod Dev 36:174–182

Moreno C, Vaca L (2011) SOC and now also SIC: store-operated and store-inhibited channels. IUBMB Life 63:856–863

Munoz-Garay C, De la Vega-Beltran JL, Delgado R, Labarca P, Felix R, Darszon A (2001) Inwardly rectifying K(+) channels in spermatogenic cells: functional expression and implication in sperm capacitation. Dev Biol 234:261–274

Navarro B, Kirichok Y, Clapham DE (2007) KSper, a pH-sensitive K$^+$ current that controls sperm membrane potential. Proc Natl Acad Sci USA 104:7688–7692

Navarro B, Miki K, Clapham DE (2011) ATP-activated P2X2 current in mouse spermatozoa. Proc Natl Acad Sci USA 108:14342–14347

Nilius B, Droogmans G (2003) Amazing chloride channels: an overview. Acta Physiol Scand 177:119–147

Nolan MA, Babcock DF, Wennemuth G, Brown W, Burton KA, McKnight GS (2004) Sperm-specific protein kinase A catalytic subunit Calpha2 orchestrates cAMP signaling for male fertility. Proc Natl Acad Sci USA 101:13483–13488

Okamura N, Tajima Y, Soejima A, Masuda H, Sugita Y (1985) Sodium bicarbonate in seminal plasma stimulates the motility of mammalian spermatozoa through direct activation of adenylate cyclase. J Biol Chem 260:9699–9705

Okunade GW, Miller ML, Pyne GJ, Sutliff RL, O'Connor KT, Neumann JC, Andringa A, Miller DA, Prasad V, Doetschman T, Paul RJ, Shull GE (2004) Targeted ablation of plasma membrane Ca^{2+}-ATPase (PMCA) 1 and 4 indicates a major housekeeping function for PMCA1 and a critical role in hyperactivated sperm motility and male fertility for PMCA4. J Biol Chem 279:33742–33750

Orta G, Ferreira G, Jose O, Trevino CL, Beltran C, Darszon A (2012) Human spermatozoa possess a calcium-dependent chloride channel that may participate in the acrosomal reaction. J Physiol (Lond) 590:2659–2675

Ottolia M, Toro L (1994) Potentiation of large conductance K/Ca channels by niflumic, flufenamic, and mefenamic acids. Biophys J 67:2272–2279

Pace AJ, Lee E, Athirakul K, Coffman TM, O'Brien DA, Koller BH (2000) Failure of spermatogenesis in mouse lines deficient in the Na(+)-K(+)-2Cl(-) cotransporter. J Clin Invest 105:441–450

Peixoto PM, Dejean LM, Kinnally KW (2012) The therapeutic potential of mitochondrial channels in cancer, ischemia-reperfusion injury, and neurodegeneration. Mitochondrion 12:14–23

Perez-Cornejo P, Arreola J (2004) Regulation of Ca(2+)-activated chloride channels by cAMP and CFTR in parotid acinar cells. Biochem Biophys Res Commun 316:612–617

Petit FM, Serres C, Bourgeon F, Pineau C, Auer J (2013) Identification of sperm head proteins involved in zona pellucida binding. Hum Reprod 28:852–865

Pifferi S, Dibattista M, Menini A (2009) TMEM16B induces chloride currents activated by calcium in mammalian cells. Pflugers Arch 458:1023–1038

Popli K, Stewart J (2007) Infertility and its management in men with cystic fibrosis: review of literature and clinical practices in the UK. Hum Fertil (Camb) 10:217–221

Publicover S, Harper CV, Barratt C (2007) [Ca²⁺]ᵢ signalling in sperm: making the most of what you've got. Nat Cell Biol 9:235–242

Publicover SJ, Barratt CL (2012) Chloride channels join the sperm 'channelome'. J Physiol (Lond) 590:2553–2554

Puente MA, Tartaglione CM, Ritta MN (2011) Bull sperm acrosome reaction induced by gamma-aminobutyric acid (GABA) is mediated by GABAergic receptors type A. Anim Reprod Sci 127:31–37

Quill TA, Ren D, Clapham DE, Garbers DL (2001) A voltage-gated ion channel expressed specifically in spermatozoa. Proc Natl Acad Sci USA 98:12527–12531

Ren D, Navarro B, Perez G, Jackson AC, Hsu S, Shi Q, Tilly JL, Clapham DE (2001) A sperm ion channel required for sperm motility and male fertility. Nature (Lond) 413:603–609

Ren D, Xia J (2010) Calcium signaling through CatSper channels in mammalian fertilization. Physiology (Bethesda) 25:165–175

Ritta MN, Bas DE, Tartaglione CM (2004) In vitro effect of gamma-aminobutyric acid on bovine spermatozoa capacitation. Mol Reprod Dev 67:478–486

Ritta MN, Calamera JC, Bas DE (1998) Occurrence of GABA and GABA receptors in human spermatozoa. Mol Hum Reprod 4:769–773

Rossato M, Di Virgilio F, Rizzuto R, Galeazzi C, Foresta C (2001) Intracellular calcium store depletion and acrosome reaction in human spermatozoa: role of calcium and plasma membrane potential. Mol Hum Reprod 7:119–128

Russell JM (2000) Sodium-potassium-chloride cotransport. Physiol Rev 80:211–276

Santi CM, Darszon A, Hernandez-Cruz A (1996) A dihydropyridine-sensitive T-type Ca²⁺ current is the main Ca²⁺ current carrier in mouse primary spermatocytes. Am J Physiol 271:C1583–C1593

Santi CM, Martinez-Lopez P, de la Vega-Beltran JL, Butler A, Alisio A, Darszon A, Salkoff L (2010) The SLO3 sperm-specific potassium channel plays a vital role in male fertility. FEBS Lett 584:1041–1046

Sardini A, Amey JS, Weylandt KH, Nobles M, Valverde MA, Higgins CF (2003) Cell volume regulation and swelling-activated chloride channels. Biochim Biophys Acta 1618:153–162

Sato Y, Son JH, Meizel S (2000) The mouse sperm glycine receptor/chloride channel: cellular localization and involvement in the acrosome reaction initiated by glycine. J Androl 21:99–106

Schreiber M, Wei A, Yuan A, Gaut J, Saito M, Salkoff L (1998) Slo3, a novel pH-sensitive K⁺ channel from mammalian spermatocytes. J Biol Chem 273:3509–3516

Schroeder BC, Cheng T, Jan YN, Jan LY (2008) Expression cloning of TMEM16A as a calcium-activated chloride channel subunit. Cell 134:1019–1029

Scudieri P, Sondo E, Ferrera L, Galietta LJ (2012) The anoctamin family: TMEM16A and TMEM16B as calcium-activated chloride channels. Exp Physiol 97:177–183

Sheppard DN, Welsh MJ (1999) Structure and function of the CFTR chloride channel. Physiol Rev 79:S23–S45

Shi QX, Roldan ER (1995) Evidence that a GABAA-like receptor is involved in progesterone-induced acrosomal exocytosis in mouse spermatozoa. Biol Reprod 52:373–381

Shi QX, Yuan YY, Roldan ER (1997) Gamma-aminobutyric acid (GABA) induces the acrosome reaction in human spermatozoa. Mol Hum Reprod 3:677–683

Shoshan-Barmatz V, De Pinto V, Zweckstetter M, Raviv Z, Keinan N, Arbel N (2010) VDAC, a multi-functional mitochondrial protein regulating cell life and death. Mol Aspects Med 31:227–285

Sieghart W (1995) Structure and pharmacology of gamma-aminobutyric acidA receptor subtypes. Pharmacol Rev 47:181–234

Sigel E, Steinmann ME (2012) Structure, function, and modulation of GABA(A) receptors. J Biol Chem 287:40224–40231

Sones WR, Leblanc N, Greenwood IA (2009) Inhibition of vascular calcium-gated chloride currents by blockers of KCa1.1, but not by modulators of KCa2.1 or KCa2.3 channels. Br J Pharmacol 158:521–531

Stamboulian S, Kim D, Shin HS, Ronjat M, De Waard M, Arnoult C (2004) Biophysical and pharmacological characterization of spermatogenic T-type calcium current in mice lacking the CaV3.1 (alpha1G) calcium channel: CaV3.2 (alpha1H) is the main functional calcium channel in wild-type spermatogenic cells. J Cell Physiol 200:116–124

Suarez SS (2008) Regulation of sperm storage and movement in the mammalian oviduct. Int J Dev Biol 52:455–462

Suarez SS, Varosi SM, Dai X (1993) Intracellular calcium increases with hyperactivation in intact, moving hamster sperm and oscillates with the flagellar beat cycle. Proc Natl Acad Sci USA 90:4660–4664

Tian Y, Schreiber R, Kunzelmann K (2012) Anoctamins are a family of Ca^{2+}-activated Cl^- channels. J Cell Sci 125:4991–4998

Toma C, Greenwood IA, Helliwell RM, Large WA (1996) Activation of potassium currents by inhibitors of calcium-activated chloride conductance in rabbit portal vein smooth muscle cells. Br J Pharmacol 118:513–520

Travis AJ, Kopf GS (2002) The role of cholesterol efflux in regulating the fertilization potential of mammalian spermatozoa. J Clin Invest 110:731–736

Trevino CL, Felix R, Castellano LE, Gutierrez C, Rodriguez D, Pacheco J, Lopez-Gonzalez I, Gomora JC, Tsutsumi V, Hernandez-Cruz A, Fiordelisio T, Scaling AL, Darszon A (2004) Expression and differential cell distribution of low-threshold Ca(2+) channels in mammalian male germ cells and sperm. FEBS Lett 563:87–92

Triphan X, Menzel VA, Petrunkina AM, Cassara MC, Wemheuer W, Hinsch KD, Hinsch E (2008) Localisation and function of voltage-dependent anion channels (VDAC) in bovine spermatozoa. Pflugers Arch 455:677–686

Turner KO, Garcia MA, Meizel S (1994) Progesterone initiation of the human sperm acrosome reaction: the obligatory increase in intracellular calcium is independent of the chloride requirement. Mol Cell Endocrinol 101:221–225

Turner TT (2008) De Graaf's thread: the human epididymis. J Androl 29:237–250

Visconti PE, Bailey JL, Moore GD, Pan D, Olds-Clarke P, Kopf GS (1995a) Capacitation of mouse spermatozoa. I. Correlation between the capacitation state and protein tyrosine phosphorylation. Development (Camb) 121:1129–1137

Visconti PE, Florman HM (2010) Mechanisms of sperm–egg interactions: between sugars and broken bonds. Sci Signal 3:pe35

Visconti PE, Galantino-Homer H, Ning X, Moore GD, Valenzuela JP, Jorgez CJ, Alvarez JG, Kopf GS (1999) Cholesterol efflux-mediated signal transduction in mammalian sperm. beta-cyclodextrins initiate transmembrane signaling leading to an increase in protein tyrosine phosphorylation and capacitation. J Biol Chem 274:3235–3242

Visconti PE, Krapf D, de la Vega-Beltran JL, Acevedo JJ, Darszon A (2011) Ion channels, phosphorylation and mammalian sperm capacitation. Asian J Androl 13:395–405

Visconti PE, Moore GD, Bailey JL, Leclerc P, Connors SA, Pan D, Olds-Clarke P, Kopf GS (1995b) Capacitation of mouse spermatozoa. II. Protein tyrosine phosphorylation and capacitation are regulated by a cAMP-dependent pathway. Development (Camb) 121:1139–1150

Wang D, Hu J, Bobulescu IA, Quill TA, McLeroy P, Moe OW, Garbers DL (2007) A sperm-specific Na⁺/H⁺ exchanger (sNHE) is critical for expression and in vivo bicarbonate regulation of the soluble adenylyl cyclase (sAC). Proc Natl Acad Sci USA 104:9325–9330

Wang J, Haanes KA, Novak I (2013) Purinergic regulation of CFTR and Ca²⁺-activated Cl⁻ channels and K⁺ channels in human pancreatic duct epithelium. Am J Physiol Cell Physiol 304(7):C673–C684

Webb TI, Lynch JW (2007) Molecular pharmacology of the glycine receptor chloride channel. Curr Pharm Des 13:2350–2367

Wennemuth G, Carlson AE, Harper AJ, Babcock DF (2003) Bicarbonate actions on flagellar and Ca²⁺ -channel responses: initial events in sperm activation. Development (Camb) 130:1317–1326

Wertheimer EV, Salicioni AM, Liu W, Trevino CL, Chavez J, Hernandez-Gonzalez EO, Darszon A, Visconti PE (2008) Chloride is essential for capacitation and for the capacitation-associated increase in tyrosine phosphorylation. J Biol Chem 283:35539–35550

Weyand I, Godde M, Frings S, Weiner J, Muller F, Altenhofen W, Hatt H, Kaupp UB (1994) Cloning and functional expression of a cyclic-nucleotide-gated channel from mammalian sperm. Nature (Lond) 368:859–863

Wistrom CA, Meizel S (1993) Evidence suggesting involvement of a unique human sperm steroid receptor/Cl⁻ channel complex in the progesterone-initiated acrosome reaction. Dev Biol 159:679–690

Wood CD, Darszon A, Whitaker M (2003) Speract induces calcium oscillations in the sperm tail. J Cell Biol 161:89–101

Wood CD, Nishigaki T, Tatsu Y, Yumoto N, Baba SA, Whitaker M, Darszon A (2007) Altering the speract-induced ion permeability changes that generate flagellar Ca²⁺ spikes regulates their kinetics and sea urchin sperm motility. Dev Biol 306:525–537

Wu WL, So SC, Sun YP, Chung YW, Grima J, Wong PY, Yan YC, Chan HC (1998) Functional expression of P2U receptors in rat spermatogenic cells: dual modulation of a Ca(2+)-activated K⁺ channel. Biochem Biophys Res Commun 248:728–732

Xia J, Ren D (2009) The BSA-induced Ca²⁺ influx during sperm capacitation is CATSPER channel-dependent. Reprod Biol Endocrinol 7:119

Xie F, Garcia MA, Carlson AE, Schuh SM, Babcock DF, Jaiswal BS, Gossen JA, Esposito G, van Duin M, Conti M (2006) Soluble adenylyl cyclase (sAC) is indispensable for sperm function and fertilization. Dev Biol 296:353–362

Xu HM, Li HG, Xu LG, Zhang JR, Chen WY, Shi QX (2012) The decline of fertility in male uremic patients is correlated with low expression of the cystic fibrosis transmembrane conductance regulator protein (CFTR) in human sperm. Hum Reprod 27:340–348

Xu WM, Shi QX, Chen WY, Zhou CX, Ni Y, Rowlands DK, Yi Liu G, Zhu H, Ma ZG, Wang XF, Chen ZH, Zhou SC, Dong HS, Zhang XH, Chung YW, Yuan YY, Yang WX, Chan HC (2007) Cystic fibrosis transmembrane conductance regulator is vital to sperm fertilizing capacity and male fertility. Proc Natl Acad Sci USA 104:9816–9821

Yanagimachi R (1994) Mammalian fertilization. In: Knobile E, Neill JD (eds) The physiology of reproduction. Raven, New York, pp 189–317

Yanagimachi R (1998) Intracytoplasmic sperm injection experiments using the mouse as a model. Hum Reprod 13(Suppl 1):87–98

Yang YD, Cho H, Koo JY, Tak MH, Cho Y, Shim WS, Park SP, Lee J, Lee B, Kim BM, Raouf R, Shin YK, Oh U (2008) TMEM16A confers receptor-activated calcium-dependent chloride conductance. Nature (Lond) 455:1210–1215

Yeung CH, Barfield JP, Cooper TG (2005) Chloride channels in physiological volume regulation of human spermatozoa. Biol Reprod 73:1057–1063

Yeung CH, Barfield JP, Cooper TG (2006) Physiological volume regulation by spermatozoa. Mol Cell Endocrinol 250:98–105

Yeung CH, Cooper TG (2008) Potassium channels involved in human sperm volume regulation–quantitative studies at the protein and mRNA levels. Mol Reprod Dev 75:659–668

Zanetti N, Mayorga LS (2009) Acrosomal swelling and membrane docking are required for hybrid vesicle formation during the human sperm acrosome reaction. Biol Reprod 81:396–405

Zeng XH, Yang C, Kim ST, Lingle CJ, Xia XM (2011) Deletion of the Slo3 gene abolishes alkalization-activated K^+ current in mouse spermatozoa. Proc Natl Acad Sci USA 108: 5879–5884

Zeng Y, Clark EN, Florman HM (1995) Sperm membrane potential: hyperpolarization during capacitation regulates zona pellucida-dependent acrosomal secretion. Dev Biol 171:554–563

Chapter 7
Equatorin-Related Subcellular and Molecular Events During Sperm Priming for Fertilization in Mice

Chizuru Ito, Kenji Yamatoya, and Kiyotaka Toshimori

Abstract Spermatozoa must undergo a priming process that renders them competent for fertilization. This priming process involves the initiation of acrosomal exocytosis, remodeling of the acrosomal substructure, and biochemical modification of related molecules. However, the mechanism underlying sperm priming process has remains unclear because of the number of molecules involved in sperm capacitation and the ensuing acrosome reaction (acrosomal exocytosis). Here, we focus on the acrosomal type 1 membrane protein equatorin and the related subcellular and molecular events that occur during the sperm priming process in mice.

Keywords Acrosome • Acrosome reaction • Equatorin • Fertilization

7.1 Introduction

Sperm priming involves the specific subcellular and molecular changes occurring in capacitated spermatozoa that are inevitable for sperm–egg fusion (Yanagimachi 1994; Toshimori 2009). The priming events occur at the periacrosomal plasma membrane of spermatozoa that approach or arrive at the zona pellucida, and the changes appear to continue in a fertilizing spermatozoon that penetrates into the perivitelline space. However, the underlying mechanism remains unresolved

C. Ito • K. Toshimori (✉)
Department of Reproductive Biology and Medicine, Graduate School of Medicine, Chiba University, Chiba 260-8670, Japan
e-mail: ktoshi@faculty.chiba-u.jp

K. Yamatoya
Department of Reproductive Biology and Medicine, Graduate School of Medicine, Chiba University, Chiba 260-8670, Japan

Biomedical Research Center, Chiba University, Chiba 260-8670, Japan

H. Sawada et al. (eds.), *Sexual Reproduction in Animals and Plants*,
DOI 10.1007/978-4-431-54589-7_7, © The Author(s) 2014

because the priming process is rapid and complicated and because the precise localization and chemical nature of the molecules involved are still unclear. As the initial changes are thought to initiate in the apical region of the anterior acrosome (Yanagimachi 1994; Eddy 2006; Toshimori 2009), the essential sperm molecules involved in the priming process are presumed to be localized at the plasma membrane, in the periacrosomal space, on the outer acrosomal membrane, and in the anterior acrosome. However, there are unknown pathway linkages between the apical region and other sperm domains during capacitation (Buffone et al. 2012). To date, proteins anchored to the plasma membrane that are important for mediating the capacitation include calcium-related proteins, such as CatSper1, CatSper 2, and PMCA (Eddy 2006). SNARE regulatory proteins, such as α-SNAP, NSF, rab3a (small GTPase), and synaptotagmins, are detected around the acrosomal region, suggesting their involvement in the regulation of acrosomal exocytosis (Tomes et al. 2002; Michaut et al. 2001; Yunes et al. 2002). The outer acrosomal membrane-anchored proteins could be involved in the priming process, including the signaling pathway to activate internal molecules, and could also function as components of the platform to maintain the acrosomal structure and other crucial fertilization events. These proteins include ADAM3/cyritestin (Forsbach and Heinlein 1998), ZPBP1/SP38/IAM38 (Yu et al. 2006; Ferrer et al. 2012), SPACA1/SAMP32 (Hao et al. 2002; Ferrer et al. 2012), SPACA4/SAMP14 (Shetty et al. 2003), OBF13/Izumo1 (Okabe et al. 1987; Inoue et al. 2005), SPESP1/ESP (Wolkowicz et al. 2003), and MN9/equatorin (Toshimori et al. 1992, 1998). Because the proteins proposed to be essential for sperm–egg fusion, sperm Izumo1 (Inoue et al. 2005) and egg CD9 (Kaji et al. 2000; Le Naour et al. 2000; Miyado et al. 2000), are discussed in other chapters; here, we focus on the nature of the acrosomal protein equatorin (Eqtn) and its related subcellular and molecular events during the sperm priming process in mice.

7.2   Equatorin and Its Chemical Nature

Equatorin was first detected as an MN9 antigen in the mouse, rat, hamster, and human mature spermatozoa using the mouse monoclonal antibody MN9 (Toshimori et al. 1992). Equatorin is composed of a complex of 48-kDa and 38-kDa proteins located in the acrosome. The antigen was renamed "equatorin" because the MN9 antibody showed a strong affinity for the equatorial segment and the MN9 antigen was found to be enriched in the equatorial segment (Toshimori et al. 1998). Purified MN9 antibody suppresses the events that occur from acrosome reaction to egg activation under in vitro (Toshimori et al. 1998) and in vivo conditions (Yoshinaga et al. 2001). The MN9 antigen is relocated from the acrosome to the plasma membrane over the equatorial segment during the acrosome reaction (Manandhar and Toshimori 2003). Equatorin was purified by immunoprecipitation using the MN9 antibody, and purified equatorin was identified by LC-MS/MS analysis (Fig. 7.1) (Yamatoya et al. 2009). A Mascot search revealed a single significant candidate,

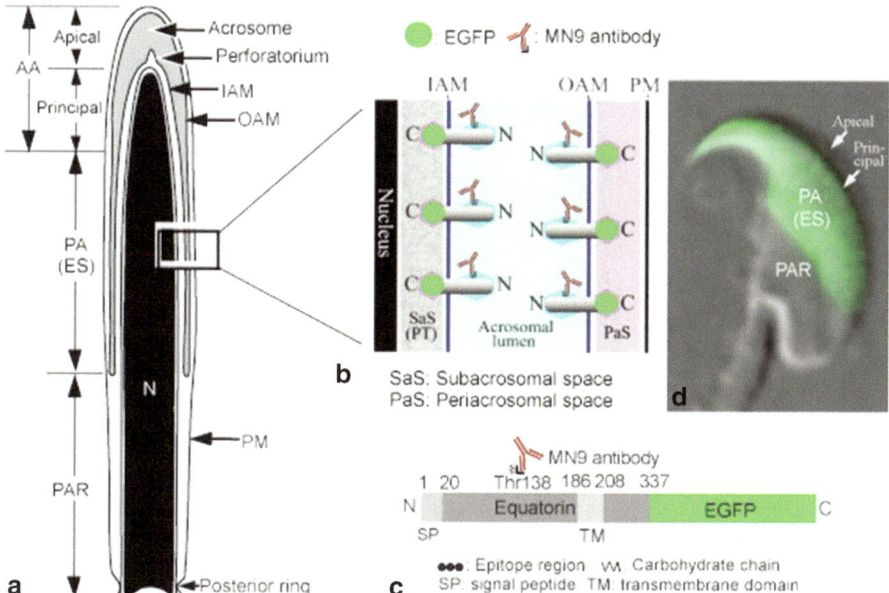

Fig. 7.1 Sperm domains, localization of equatorin, topology of the epitope recognized by the MN9 antibody and Eqtn-EGFP, and a generated Eqtn-EGFP transgenic mouse spermatozoon. (**a, b**) The MN9 antibody recognizes the N-terminus in the acrosomal lumen, whereas EGFP is present at the C-terminus side in both the periacrosomal and subacrosomal spaces. (**c, d**) The molecular structure of the mouse Eqtn-EGFP chimeric protein (**c**) used to generate the Eqtn-EGFP transgenic spermatozoa (**d**). The conserved domains are 1–20 amino acids (aa) for the potential signal peptide (SP) and 186–208 aa for the transmembrane (TM) domain; the epitope for the MN9 antibody is present in the region containing a carbohydrate chain branched from the threonine at aa 138 (Thr138). *AA* anterior acrosome, *ES* equatorial segment, *IAM* inner acrosomal membrane, *OAM* outer acrosomal membrane, *PA* posterior acrosome, *PAR* postacrosomal region, *PM* plasma membrane

4930579C15Rik, the human orthologue of which is reported to be frequently deleted in cancer (Ruiz et al. 2000). The equatorin gene and its chemical nature were clarified with the aid of recently developed techniques for carbohydrates, such as the Pro-Q Emerald 300 glycoprotein gel staining kit and SYPRO Ruby protein gel stain. Equatorin is a highly glycosylated and sialylated protein, an *N,O*-sialoglycoprotein, and it is insoluble in mild detergents. There are long and short forms of equatorin encoded by the *Eqtn* gene, which is located on mouse chromosome 4 (the EQTN gene on human chromosome 9, 9p21). The core protein size is approximately 27 kDa, and the *O*-sialylated carbohydrate region branching from the threonine 138 is involved in the epitope for the MN9 antibody (Fig. 7.2). Equatorin is also called Afaf (Li et al. 2006) or C9orf11 (Ruiz et al. 2000).

Information on the equatorin epitope region recognized by the MN9 antibody is based on *Galnt3* gene deletion experiments. Galnt3-null mouse homozygotic testicular germ cells exhibit drastically reduced reactivity with the MN9 antibody, and the mice are infertile because of oligoashthenoteratozoospermia (Miyazaki et al. 2013).

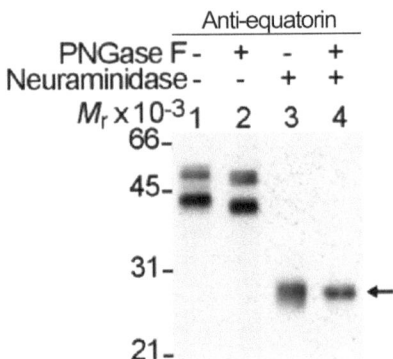

Fig. 7.2 Glycosylation status of equatorin by mobility shift assay using glycosidase treatment. *Lanes 1–4*: cauda sperm extract analyzed by Western blotting (15% gel) with the MN9 antibody. *Lane 1*, without glycosidase treatment. *Lane 2*, treatment with PNGase F only. *Lane 3*, treatment with neuraminidase only. *Lane 4*, treatment with PNGase F and neuraminidase. Mobility shifts are observed with both the PNGase and neuraminidase treatments, but the MN9 antigenicity remains on the molecule at approximately 27 kDa (*arrow*). (Modified from Yamatoya et al. 2009)

In these mutant mice, spermatozoa are rare in the cauda epididymides, most of which exhibit deformed rounded heads. Galnt3 is a GalNAc transferase family protein involved in the initiation of mucin-type O-glycosylation (Bennett et al. 1996). The VVA (*Vicia villosa* agglutinin) lectin, which can recognize the Tn antigen (GalNAc-O-Ser/Thr) generated by GalNac transferase, also recognizes the acrosomal regions of spermatids and spermatozoa, whereas VVA binding is drastically reduced in Galnt3-null mice (Miyazaki et al. 2013). Thus, GlcNAcβ1-3-GalNAcα1-Ser/Thr mediated by Galnt3 (GalNAc transferase) is involved in the formation of the equatorin epitope region recognized by the MN9 antibody.

7.3 Expression and Molecular Size of Equatorin in the Testis

When examined by in situ hybridization, mouse *Eqtn* mRNA is specifically expressed in early- to mid-round spermatids (step 1–6) and decreases in late-round (steps 7–8) spermatids (Ito et al. 2013). The molecular size of mouse equatorin is approximately 65 kDa in testicular germ cells (Ito et al. 2013), but the size decreases to 40–50 kDa in cauda epididymal spermatozoa when examined by Western blotting (Yoshida et al. 2010).

7.4 Localization of Equatorin in Mature Spermatozoa

Equatorin is anchored to the acrosomal membrane (type 1 membrane protein) in mature spermatozoa, but the amount in the acrosomal membrane varies among the domains. When examined by immunogold electron microscopy, the immunogold particles are abundant on the inner acrosomal membrane in the principal region and equatorial segment (Ito et al. 2013). However, on the outer acrosomal membrane, the gold particles are depleted in the principal region but abundant at the equatorial segment, whereas gold particles are not detected on the outer acrosomal membrane in the apical region, suggesting that equatorin is absent or very poor on the outer acrosomal membrane in the apical region. In addition, electron microscopic data suggest that equatorin is embedded in electron-dense acrosomal matrix substances, locations where immunogold particles are generally 5–70 nm away from the acrosomal membranes, with the N-terminus in the acrosomal lumen (Ito et al. 2013). Such three-dimensional structures suggest that the N-terminus of equatorin is embedded in the complex of the inner acrosomal membrane and acrosomal matrix (CIMAM) and the complex of the outer acrosomal membrane and acrosomal matrix (COMAM) (Toshimori 2011; Ito et al. 2013). Similarly, the C-terminus of equatorin appears to associate with the perinuclear theca substances in the subacrosomal space and with the periacrosomal substances in the periacrosomal space.

These lines of subcellular evidence provide some clues to propose the function of equatorin in spermatozoa, as deficiency in the acrosomal membrane-binding matrix protein SPESP1 (ESP) causes embrittlement of the equatorial segment but increases the expression of equatorin (Fujihara et al. 2012). Because SPESP-1 is known to be localized to the equatorial segment of ejaculated human sperm (Wolkowicz et al. 2003), SPESP1 may work redundantly with equatorin. Antisera raised against recombinant SPESP-1 inhibit the binding and fusion of human sperm to the hamster eggs (Wolkowicz et al. 2008), but SPESP-1 cannot be detected by a rabbit anti-mouse SPESP1 polyclonal antiserum that was produced by immunization with mouse SPESP1 polypeptide (MYGSNVFPEGRTSD) after the acrosome reaction (Fujihara et al. 2012). SPESP-1-deficient mice are fertile, although Spesp1+/– and Spesp1–/– spermatozoa have a lower fusing ability compared to wild-type spermatozoa (Fujihara et al. 2010). The protein(s) associated with equatorin have not been identified to date.

7.5 Behavior of Equatorin During the Acrosome Reaction

7.5.1 Before and the Very Initial Stage of the Acrosome Reaction

Equatorin is not detected before or at the very initial stage of the acrosome reaction by indirect immunofluorescence (IIF) microscopy using the MN9 antibody because the antibody cannot reach the epitope region on the N-terminus of equatorin, which is in the acrosomal lumen. However, as the acrosome reaction proceeds, the epitope region is gradually exposed at the area near the principal regions and becomes detected by the MN9 antibody (Fig. 7.3). When examined by IIF microscopy using an antibody against Izumo1 (#125), the MN9 epitope region appears later than Izumo1 (unpublished data), which is localized on the outer and inner acrosomal membranes of the apical segment (Inoue et al. 2005; Satouh et al. 2012).

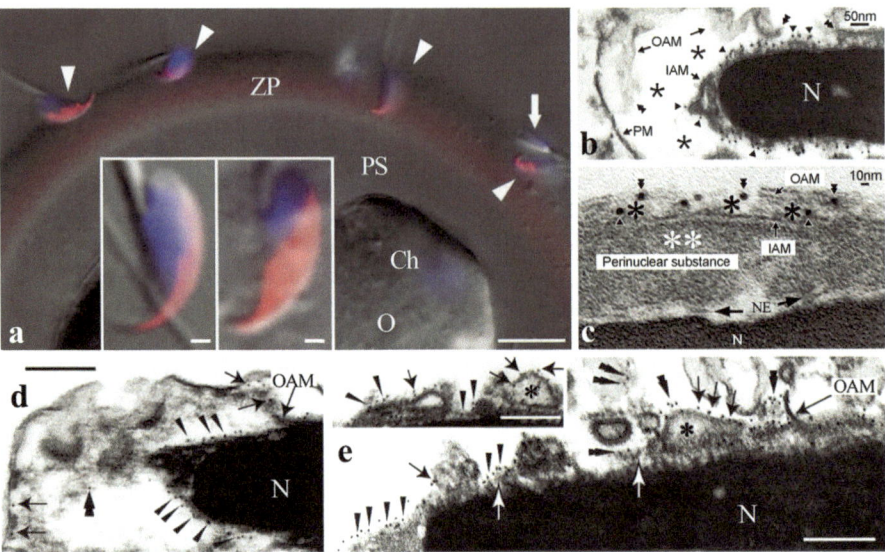

Fig. 7.3 Subcellular behavior of equatorin during the acrosome reaction. Indirect immunofluorescence with the MN9 antibody (**a**) and immunogold electron microscopy with the MN9 antibody (**b–e**). (**a**) Various stages of the acrosome reaction are found in spermatozoa that are attached to the zona pellucida (*arrowheads*). An acrosome-intact spermatozoon is indicated with an *arrow*. *Left and right insets:* High-magnification images of different types of MN9-immunostaining patterns. *Ch* chromosome, *O* ooplasm, *PS* perivitelline space, *ZP* zona pellucida. MN9 *red*, nucleus or chromosome *blue* (Hoechst). *Bars* **a** 10 μm; *insets* 1 μm. (**b, c**) Immunogold particles (10-nm gold particles, *arrowheads*) are abundant on the inner acrosomal membrane facing the acrosome lumen, as indicated with *asterisks*, but depleted (no gold particles in this photograph) on the outer acrosomal membrane in the anterior acrosomal region before or at the very initial stage of the acrosome reaction; gold particles are not present on the plasma membrane. In the posterior acrosome (equatorial segment), the gold particles are present on both the inner (*arrowheads*) and outer (*double arrowheads*) acrosomal membranes; the particles appear to associate with the electron-dense matrix substances facing the narrowed internal lumen (*). Electron-dense perinuclear substances (** in this figure; Ito et al. 2013) are found in the space between the inner acrosomal membrane and the nuclear envelope (*NE*). (**d, e**) Early stage (**d**) and advanced stage (**e**) of the acrosome reaction. Immunogold particles (5 nm) are dense on the inner acrosomal membrane (*arrowheads*) and weak on the outer acrosomal membrane (*arrows*) and in the amorphous substance (*double arrowhead*). In the advanced stage (**e**), some gold particles are found in association with the hybrid vesicles (*asterisks*) formed by the plasma membrane and outer acrosomal membrane (*arrows* in **e** and *inset*). Some other gold particles are present on the inner acrosomal membrane (*arrowheads*). *Bar* **e** 200 nm. *IAM* inner acrosomal membrane, *N* nucleus, *OAM* outer acrosomal membrane, *PM* plasma membrane. (**a–c** and **d, e** reproduced with slight modifications from Yoshida et al. 2010) and Yamatoya et al. 2009, respectively)

7.5.2 Early to Middle Stages of the Acrosome Reaction

Some amount of equatorin is relocated on the surface of the plasma membrane over the equatorial segment but not to the plasma membrane over the postacrosomal region (Yoshida et al. 2010). This fact suggests that the epitope region recognized

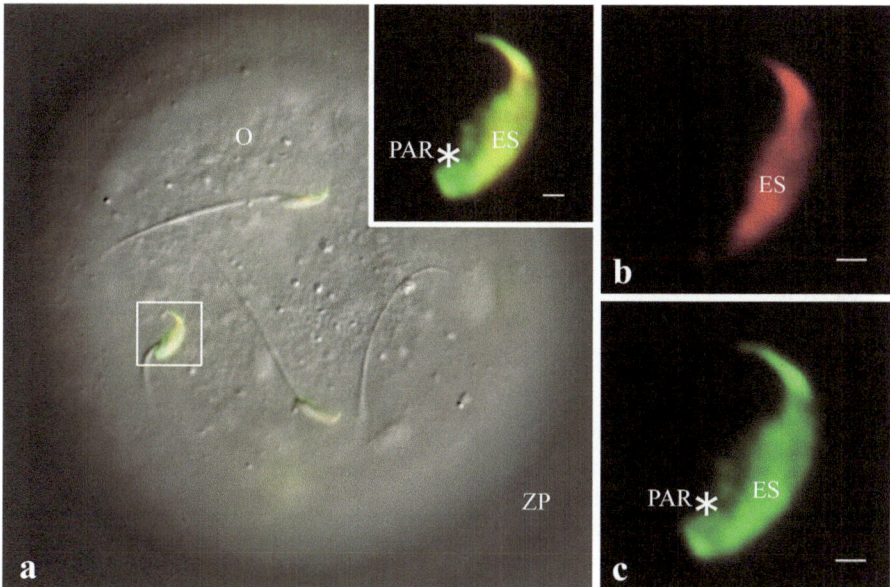

Fig. 7.4 Localization of equatorin in the perivitelline spermatozoa. Indirect immunofluorescence double staining with the anti-equatorin antibody MN9 antibody (*red*) and anti-Izumo1 antibody (*green*). Perivitelline spermatozoa accumulated in the perivitelline space in CD9-null eggs are the subjects of the immunofluorescence study. The square area at the lower magnification (**a**) of the merged DIC image is enlarged in the *insets* of **a**, **b**, and **c**. Equatorin is localized at the equatorial segment region (*ES* in *inset* and **b**), whereas Izumo1 is found at both the equatorial segment and postacrosomal region (* in **c**). *O* oocyte surface, *ZP* zona pellucida. *Bar* 1 m

by the MN9 antibody has a specific affinity for the equatorial plasma membrane; however, a sufficient amount of equatorin remains in the acrosome-reacted spermatozoa. Immunofluorescence double-staining data using the anti-Izumo1 antibody and anti-MN9 antibody suggest that Izumo1 can spread rapidly and widely to the postacrosomal region in the acrosome-reacting spermatozoa, whereas the equatorin epitope region recognized by MN9 antibody remains confined to the anterior acrosomal region (unpublished data). The molecular size of equatorin appears to be reduced to 35 kDa at the end of the acrosome reaction (Yoshida et al. 2010).

7.5.3 Advanced Stage and After the Acrosome Reaction

A sufficient amount of equatorin remains in the acrosomal membranes in the principal region and at the equatorial segment in the advanced stage of the acrosome reaction; this pattern is similar in a fertilizing spermatozoon that reaches the perivitelline space (i.e., perivitelline spermatozoon). In fact, the perivitelline spermatozoa accumulated in the perivitelline space of CD9-deficient eggs display both equatorin and Izumo1 (Fig. 7.4).

The equatorial segment is stabilized by the extensive scaffolding network that connects the outer acrosomal membrane to the inner acrosomal membrane (Fawcett 1975), a site thought to contain many enzymes and matrix molecules (Yanagimachi 1994; Eddy 2006; Toshimori 2009). The equatorin protein that is stably embedded in this network is carried into the ooplasm where associated enzymes and matrix molecules are fully released. A study investigating this process by employing Eqtn-EGFP transgenic mouse spermatozoa is currently being performed in our laboratory.

7.6 Possible Roles of Equatorin

At present, there are several lines of evidence invoked to propose the roles for equatorin. First, the evidence that the protein is produced and distributed in the acrosomal membrane in early spermatids and is retained in mature spermatozoa (Figs. 7.1, 7.3, 7.5) suggests that equatorin serves as a platform component to stabilize the acrosomal membrane and to maintain the shape of the acrosome. This role is presumed to continue until the sperm–egg interaction stage. Within this context, the nature of the branching sialylated carbohydrate chains (MN9 epitope region) on the threonine 138 (Figs. 7.1, 7.2) at the N-terminus will be important because the protein is maintained in the acidic environment of the acrosome lumen before the acrosome reaction, whereas the C-terminus is under neutral conditions, that is, in the periacrosomal space or in the perinuclear space. Equatorin may also function to anchor other molecules, although supporting evidence is not available thus far. Second, some amount of the equatorin that is relocated to the equatorial segment plasma membrane (Fig. 7.4) may play additional roles, as previously suggested by the findings that the MN9 antibody could suppress the events from acrosome reaction to egg activation under in vitro and in vivo conditions (Toshimori et al. 1998; Yoshinaga et al. 2001).

7.7 Perspective

Our next purpose is to determine the precise role of equatorin in the course of fertilization process and during spermatogenesis. To pursue this purpose, the *equatorin* gene-deleted mouse line has been established, and the analyses are in progress in our laboratory. Also, to visualize the behavior of equatorin, we are performing varieties of imaging studies employing Eqtn-EGFP transgenic mouse spermatozoa.

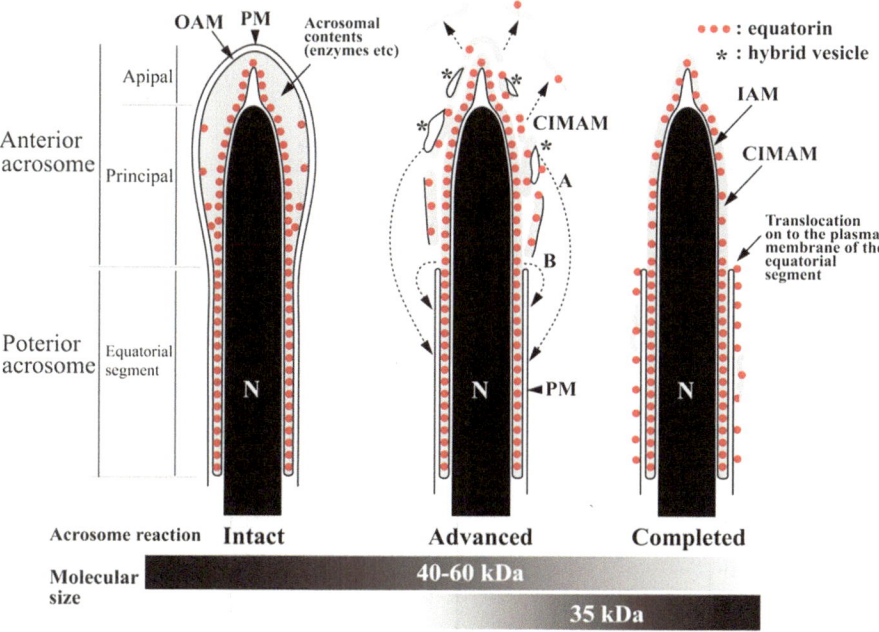

Fig. 7.5 Change of the localization and molecular size of equatorin during the acrosome reaction. Before the acrosome reaction (intact), equatorin (*red*) is enriched on the whole inner acrosomal membrane (*IAM*) and the outer acrosomal membrane (*OAM*) at the equatorial segment but is absent in the apical region and sparse on the outer acrosomal membrane in the principal region. There are several possible routes for the relocation of equatorin onto the sperm surface. First, according to the reassociation model indicated in **A**, as the acrosome reaction proceeds, equatorin (*) detaches from OAM in the principal region or from other acrosomal membranes and reassociates with the sperm surface; a sufficient amount of equatorin remains associated with the amorphous substance of the acrosomal matrix (*gray color*). Second, according to the lateral diffusion model indicated in **B**, equatorin in the acrosomal lumen diffuses to the sperm surface along the acrosomal membrane and plasma membranes (*PM*), and some of the equatorin becomes localized at the plasma membrane over the equatorial segment (*ES*); a sufficient amount of equatorin remains on IAM in acrosome reaction-completed spermatozoa. As the acrosome reaction progresses, low molecular mass equatorin (35 kDa) appears and increases in signal intensity. *N* nucleus

Acknowledgments The authors thank Ms. K. Kamimura and Mr. T. Mutoh for their excellent technical assistance. This work was supported by a grant from the Grant-in-Aid for Scientific Research on Innovation Areas to K.T. (22112504, 24112706), the Grant-in-Aid for Scientific Research to K.T. (22390033), in part to C.I. (24592441) and to K.Y. (23791809).

References

Bennett EP, Hassan H, Clausen H (1996) cDNA cloning and expression of a novel human UDP-N-acetyl-alpha-D-galactosamine. Polypeptide *N*-acetylgalactosaminyltransferase, GalNAc-t3. J Biol Chem 271:17006–17012

Buffone MG, Ijiri TW, Cao W et al (2012) Heads or tails? Structural events and molecular mechanisms that promote mammalian sperm acrosomal exocytosis and motility. Mol Reprod Dev 79:4–18

Eddy EM (2006) The spermatozoon. In: Knobil E, Neill JD (eds) The physiology of reproduction, vol 1. Raven, New York, pp 3–54

Fawcett DW (1975) The mammalian spermatozoon. Dev Biol 44:394–436

Ferrer M, Rodriguez H, Zara L et al (2012) MMP2 and acrosin are major proteinases associated with the inner acrosomal membrane and may cooperate in sperm penetration of the zona pellucida during fertilization. Cell Tissue Res 349:881–895

Forsbach A, Heinlein UA (1998) Intratesticular distribution of cyritestin, a protein involved in gamete interaction. J Exp Biol 201:861–867

Fujihara Y, Murakami M, Inoue N et al (2010) Sperm equatorial segment protein 1, SPESP1, is required for fully fertile sperm in mouse. J Cell Sci 123:1531–1536

Fujihara Y, Satouh Y, Inoue N et al (2012) SPACA1-deficient male mice are infertile with abnormally shaped sperm heads reminiscent of globozoospermia. Development (Camb) 139: 3583–3589

Hao Z, Wolkowicz MJ, Shetty J et al (2002) SAMP32, a testis-specific, isoantigenic sperm acrosomal membrane-associated protein. Biol Reprod 66:735–744

Inoue N, Ikawa M, Isotani A et al (2005) The immunoglobulin superfamily protein Izumo is required for sperm to fuse with eggs. Nature (Lond) 434:234–238

Ito C, Yamatoya K, Yoshida K et al (2013) Integration of the mouse sperm fertilization-related protein equatorin into the acrosome during spermatogenesis as revealed by super-resolution and immuno-electron microscopy. Cell Tissue Res 352:739–750. doi:10.1007/s00441-00013-01605-y

Kaji K, Oda S, Shikano T et al (2000) The gamete fusion process is defective in eggs of Cd9-deficient mice. Nat Genet 24:279–282

Le Naour F, Rubinstein E, Jasmin C et al (2000) Severely reduced female fertility in CD9-deficient mice. Science 287:319–321

Li YC, Hu XQ, Zhang KY et al (2006) Afaf, a novel vesicle membrane protein, is related to acrosome formation in murine testis. FEBS Lett 580:4266–4273

Manandhar G, Toshimori K (2003) Fate of postacrosomal perinuclear theca recognized by monoclonal antibody MN13 after sperm head microinjection and its role in oocyte activation in mice. Biol Reprod 68:655–663

Michaut M, De Blas G, Tomes CN et al (2001) Synaptotagmin VI participates in the acrosome reaction of human spermatozoa. Dev Biol 235:521–529

Miyado K, Yamada G, Yamada S et al (2000) Requirement of CD9 on the egg plasma membrane for fertilization. Science 287:321–324

Miyazaki T, Mori M, Yoshida CA et al (2013) Galnt3 deficiency disrupts acrosome formation and leads to oligoasthenoteratozoospermia. Histochem Cell Biol 139:339–354

Okabe M, Adachi T, Takada K et al (1987) Capacitation-related changes in antigen distribution on mouse sperm heads and its relation to fertilization rate in vitro. J Reprod Immunol 11:91–100

Ruiz A, Pujana MA, Estivill X (2000) Isolation and characterisation of a novel human gene (C9orf11) on chromosome 9p21, a region frequently deleted in human cancer. Biochim Biophys Acta 1517:128–134

Satouh Y, Inoue N, Ikawa M et al (2012) Visualization of the moment of mouse sperm–egg fusion and dynamic localization of IZUMO1. J Cell Sci 125:4985–4990

Shetty J, Wolkowicz MJ, Digilio LC et al (2003) SAMP14, a novel, acrosomal membrane-associated, glycosylphosphatidylinositol-anchored member of the Ly-6/urokinase-type plasminogen activator receptor superfamily with a role in sperm–egg interaction. J Biol Chem 278:30506–30515

Tomes CN, Michaut M, De Blas G et al (2002) SNARE complex assembly is required for human sperm acrosome reaction. Dev Biol 243:326–338

Toshimori K (2009) Dynamics of the mammalian sperm head: modifications and maturation events from spermatogenesis to egg activation. Adv Anat Embryol Cell Biol 204:5–94

Toshimori K (2011) Dynamics of the mammalian sperm membrane modification leading to fertilization: a cytological study. J Electron Microsc (Tokyo) 60(Suppl 1):S31–S42

Toshimori K, Tanii I, Araki S et al (1992) Characterization of the antigen recognized by a monoclonal antibody MN9: unique transport pathway to the equatorial segment of sperm head during spermiogenesis. Cell Tissue Res 270:459–468

Toshimori K, Saxena DK, Tanii I et al (1998) An MN9 antigenic molecule, equatorin, is required for successful sperm–oocyte fusion in mice. Biol Reprod 59:22–29

Wolkowicz MJ, Shetty J, Westbrook A et al (2003) Equatorial segment protein defines a discrete acrosomal subcompartment persisting throughout acrosomal biogenesis. Biol Reprod 69:735–745

Wolkowicz MJ, Digilio L, Klotz K et al (2008) Equatorial segment protein (ESP) is a human alloantigen involved in sperm–egg binding and fusion. J Androl 29:272–282

Yamatoya K, Yoshida K, Ito C et al (2009) Equatorin: identification and characterization of the epitope of the MN9 antibody in the mouse. Biol Reprod 81:889–897

Yanagimachi R (1994) Fertilization. In: Knobil E, Neill JD (eds) The physiology of reproduction, vol 1. Raven, New York, pp 189–317

Yoshida K, Ito C, Yamatoya K et al (2010) A model of the acrosome reaction progression via the acrosomal membrane-anchored protein equatorin. Reproduction 139:533–544

Yoshinaga K, Tanii I, Oh-oka T et al (2001) Changes in distribution and molecular weight of the acrosomal protein acrin2 (MC41) during guinea pig spermiogenesis and epididymal maturation. Cell Tissue Res 303:253–261

Yu Y, Xu W, Yi YJ et al (2006) The extracellular protein coat of the inner acrosomal membrane is involved in zona pellucida binding and penetration during fertilization: characterization of its most prominent polypeptide (IAM38). Dev Biol 290:32–43

Yunes R, Tomes C, Michaut M et al (2002) Rab3A and calmodulin regulate acrosomal exocytosis by mechanisms that do not require a direct interaction. FEBS Lett 525:126–130

Chapter 8
Acrosome Reaction-Mediated Motility Initiation That Is Critical for the Internal Fertilization of Urodele Amphibians

Eriko Takayama-Watanabe, Tomoe Takahashi,
Misato Yokoe, and Akihiko Watanabe

Abstract The reproductive modes of extant amphibians are highly diversified in their adaptation to species-specific reproductive environments. In the external fertilization of most amphibians, fertilization begins from motility initiation of the sperm by the hyposmolality of freshwater at spawning, whereas in internal fertilization, the sperm initiate motility without any change in osmolality. Acrosome reaction (AR)-mediated motility initiation is an initial event of the internal fertilization of the urodele amphibian *Cynops pyrrhogaster*, also known as the Japanese fire belly newt. It initiates motility of female-stored sperm on the surface of the jelly layer in the cloaca. This unique mechanism of motility initiation is based on the fine structure of the jelly matrix, in which AR-inducing substance (ARIS) and sperm motility-initiating substance (SMIS) are localized in the sheet-like structure covering the outer surface and in the granules beneath it. The ARIS and the SMIS repeatedly increase the sperm intracellular Ca^{2+} level and result in the sporadic initiation of motility within 4 min. The SMIS activity is also present in the jelly layer of the externally fertilizing anuran amphibian *Discoglossus pictus* (the Mediterranean painted frog), but a thick layer of matrix simply covers the outer surface of the jelly layer. The AR-mediated mechanism may be established in the reproductive strategies for the internal fertilization of urodeles.

Keywords Amphibians • Acrosome reaction • Extracellular matrix • Fertilization • Sperm motility

E. Takayama-Watanabe
Institute of Arts and Sciences, Yamagata University, 1-4-12 Kojirakawa,
Yamagata 990-8560, Japan

T. Takahashi • M. Yokoe • A. Watanabe (✉)
Department of Biology, Yamagata University, 1-4-12 Kojirakawa, Yamagata 990-8560, Japan
e-mail: watan@sci.kj.yamagata-u.ac.jp

H. Sawada et al. (eds.), *Sexual Reproduction in Animals and Plants*,
DOI 10.1007/978-4-431-54589-7_8, © The Author(s) 2014

8.1 Diversity of Reproductive Modes in Amphibians

Amphibian reproduction adapts to various environments in freshwater and on land. The variety of reproductive modes among amphibians has been expected to provide models for the study of unknown mechanisms underlying reproduction (Wake and Dickie 1998). Duellman and Trueb (1994) proposed the diversification path of reproductive modes in amphibian evolution. Fertilization occurs in lentic freshwater in the primitive mode of amphibian reproduction, and eggs and sperm are simultaneously spawned in freshwater. The sperm initiate motility in response to the hyposmolality in the freshwater and fertilize the eggs.

Amphibian eggs are surrounded by an oviduct-secreted matrix called the jelly layer. A variety of modifications are seen in the morphology and physiology of the oviductal secretion, which assure reproduction under a species-specific environment. For example, the fertilization of some anurans occurs in lotic freshwater. For the adaptation to water flow, the jelly layer is modified to form network-forming morphology, as seen in the frog *Rana tagoi* (Fig. 8.1a). In a more evolved mode of anuran reproduction, fertilization occurs arboreally. In this case, a foam nest is often formed with an oviductal secretion, providing the fertilization environment on the tree, as seen in the treefrog *Rhacophorus arboreus* (Fig. 8.1b).

Internal fertilization is the most evolved mode of amphibian reproduction. It is seen in a few anuran species, more than 90 % of urodele species, and all caecilian species. In urodeles, the internal fertilization evolved from external fertilization (Duellman and Trueb 1994). In the internal fertilization of urodeles, sperm are quiescently stored in the spermatheca, the female sperm reservoir in the cloaca, for months (Duellman and Trueb 1994; Greven 1998). At the beginning of fertilization, the sperm are inseminated on the surface of the jelly layer. The jelly layer is responsible for sperm motility initiation without a hyposmotic condition, as seen in the newt *Cynops pyrrhogaster* (Ukita et al. 1999).

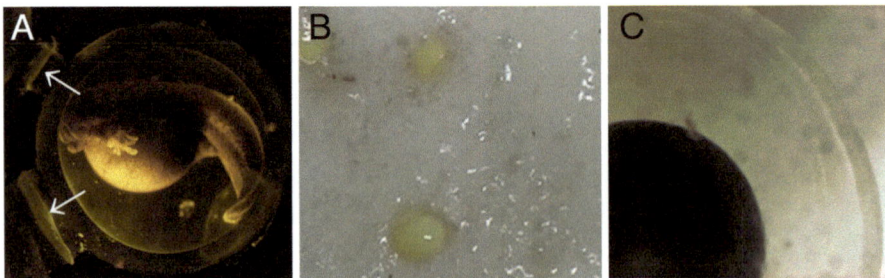

Fig. 8.1 Modification of oviductal secretion in amphibians. (**a**) Jelly layer of *Rana tagoi*. *Arrows* indicate joint regions between the eggs. (**b**) Foam nest of *Rhacophorus arboreus*. (**c**) Jelly layer of *Cynops pyrhogaster*

8.2 The Jelly Layer of Amphibian Eggs

The jelly layer of amphibian eggs is a gelatinous, sugar-rich matrix surrounding the eggs as the outermost egg coat. It is a multifunctional structure for attachment of fertilized eggs to the substrates, the physical barrier for developing embryos, defense against bacterial invasion, and fertilization (Wake and Dickie 1998). The jelly layer is composed of several sublayers, each of which contains qualitatively and quantitatively different components (Fig. 8.1c; Greven 1998; Okimura et al. 2001). The components are secreted in a specific compartment of the pars convoluta, a distal portion of the oviduct, and sometimes in the most proximal portion of the oviduct (called the uterus or ovisac). They are accumulated on the surface of the vitelline envelope to form each sublayer (from the inner to outer) in ovulated eggs passing through the oviduct.

The role of the jelly layer in amphibian fertilization has been investigated for many years. It is a source of divalent cations for the sperm–egg interaction (Katagiri 1987; Takayama-Watanabe et al., in press), and proteinaceous components also have significant roles in the interaction (Arranz and Cabada 2000; Olson et al. 2001; Watanabe et al. 2009, 2010). Interestingly, some components such as AR-inducing substance (ARIS) seem to localize in a species-specific manner (Watanabe and Onitake 2002), suggesting that the function of the jelly layer in fertilization is modified among amphibians.

In urodeles, fertilization is strongly dependent on the outer sublayer(s) of the jelly layer (MacLaughlin and Humphries 1978; Takahashi et al. 2006). Activities for sperm AR induction and motility initiation localize in the outermost sublayer of the newt *C. pyrrhogaster* (Sasaki et al. 2002; Watanabe et al. 2003). Takahashi et al. (2006) reported that sperm AR on the surface of the jelly layer is critical for the fertilization of *C. pyrrhogaster* because the insemination of AR-induced sperm, in contrast to that of AR-intact sperm, remarkably reduced fertilization rates. In the anurans *Bufo japonicus* and *Xenopus laevis*, because the AR is induced after sperm passing through the jelly layer (Yoshizaki and Katagiri 1982; Ueda et al. 2002), a specific mechanism for the sperm–egg interaction may be present on the surface of the jelly layer of *C. pyrrhogaster,* and it may be responsible for the success of internal fertilization.

8.3 Acrosome Reaction-Mediated Motility Initiation

Acrosome reaction-mediated motility initiation is the mechanism underlying the induction of motility of female-stored sperm in the internal fertilization of *C. pyrrhogaster* (Watanabe et al. 2010). This initiation is based on the fine structures of the jelly surface, where a sheet-like structure covers the outer surface of the egg jelly and sequesters many granules from outside (Watanabe et al. 2010; Fig. 8.2a). By immunological localization, sperm motility-initiating substance (SMIS) and ARIS

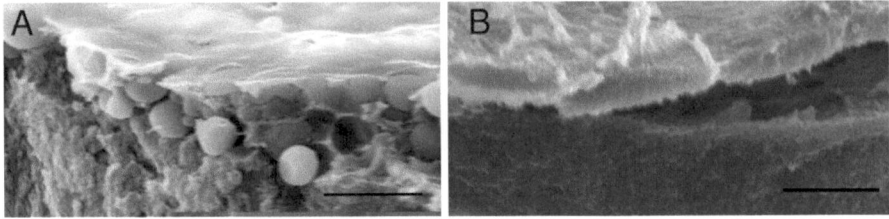

Fig. 8.2 Fine structures on outer surface of amphibian jelly layer. (**a**) *Cynops pyrrhogaster*. (**b**) *Discoglossus pictus*

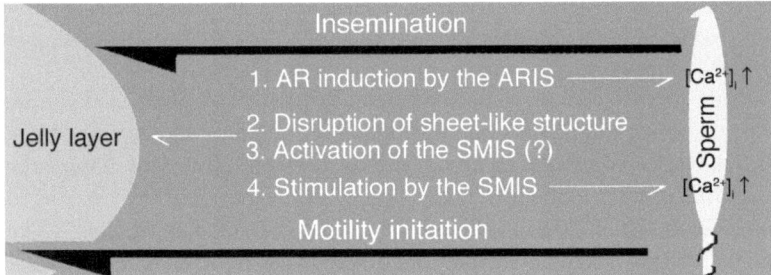

Fig. 8.3 Sperm–egg interaction on the surface of the jelly layer in *Cynops pyrrhogaster*. *ARIS* acrosome reaction-inducing substance, *SMIS* sperm motility-initiating substance

exist in the sheet-like structure and the granules, respectively. At the beginning of internal fertilization, the large egg size of urodeles results in outspread the external opening of the spermatheca in the cloaca, and the quiescently stored sperm are mechanically inseminated onto the surface of the egg jelly (Fig. 8.3). The sperm immediately (<30 s) undergo the AR in response to the ARIS in the sheet-like structure (Watanabe et al. 2011). The enzymes released from the acrosomal vesicle are suggested to disrupt the sheet-like structure and expose the SMIS in the granules to the jelly surface (Watanabe et al. 2010). The SMIS in the jelly layer is largely inactivated by the association with jelly substances (Mizuno et al. 1999), and recent data from our laboratory indicate that a sperm protease may mediate the activation of the SMIS (Yokoe et al., unpublished data). The active SMIS, in turn, stimulates sperm to initiate motility. In contrast to the AR, motility initiation occurs sporadically among sperm within 1–4 min after insemination (Watanabe et al. 2003, 2011).

The isolation of the SMIS in the jelly layer is unique to AR-mediated motility initiation. Although sperm motility is controlled by a variety of extracellular cues such as osmotic conditions, steroids, peptides, and proteases (Morisawa et al. 1999; Miyata et al. 2012) in animal species, all are ready to affect sperm by themselves at fertilization. The exposure of the SMIS and its activation fully depends on the inseminated sperm, which should be one of the reasons why sporadic motility initiation occurs through the AR-mediated mechanism. Sporadic motility initiation is also caused by valid levels of intracellular Ca^{2+} of the quiescently stored sperm (Watanabe et al. 2011).

In the sperm of *C. pyrrhogaster*, intracellular Ca^{2+} is increased in both AR induction and motility initiation, as in the sperm of other animal species (Darszon et al. 2006), and immediately drops down to the initial level (Watanabe et al. 2011). In addition, we recently found that subsequent influxes of Ca^{2+} in the motility-initiated sperm cause a high motility state (Takahashi et al. 2013). Activation of the motility state is needed for sperm to propel through the viscous jelly matrix. Conversely, AR-mediated motility initiation and the subsequent activation of the motility state are potent to exclude sperm, diminishing their ability to appropriately control the intracellular Ca^{2+} level. Actually, many sperm are left on the surface of the jelly layer in naturally spawned eggs of *C. pyrrhogaster* (Takahashi et al. 2006). It is interesting that the insemination of too many sperm can achieve fertilization without AR-mediated motility initiation (Takahashi et al. 2006). Because AR induction is essential for sperm to interact with the vitelline envelope (Nakai et al. 1999), fertilization is thought to occur by abnormal sperm that spontaneously miss the acrosome and initiate motility (Watanabe and Onitake 2003). In that case, most of the fertilized eggs fail in normal development because of polyspermy, although the eggs of *C. pyrrhogaster* are physiologically polyspermic (Street 1940). Such abnormalities of sperm physiology are commonly seen in sperm from every male (Watanabe and Onitake 2003) and are thought to be increased during the months of female sperm storage. AR-mediated motility initiation may have a role in controlling the amount of sperm participating in fertilization to ensure the success of embryonic development.

8.4 SMIS Activity in the Amphibian Jelly Layer

Sperm motility-initiating substance (SMIS) activity in egg jelly is crucial to initiate motility in the internal fertilization of urodele amphibians. How SMIS-triggering motility initiation is originated in the amphibian fertilization system may be significant to understand the diversification mechanism for the amphibian mode of reproduction. To address this question, we performed comparative studies using the anti-SMIS monoclonal antibody (Ohta et al. 2011; Takayama-Watanabe et al. 2012). SMIS activity is present in the jelly layers of the externally fertilizing urodele *Hynobius lichenatus* and the anuran *Discoglossus pictus*, indicating that SMIS activity is widely conserved among urodeles and anurans. In the primitive anuran *D. pictus*, the acrosome reaction (AR) is induced in the outermost jelly sublayer (Campanella et al. 1997), although a thick layer of jelly matrix covers the outer surface without locating granules (Takayama-Watanabe et al. 2012; Fig. 8.2b). In addition, no sporadic feature is seen in sperm motility initiation by the jelly substances, suggesting that the SMIS activity has a distinct role in the fertilization of *D. pictus*. Because *D. pictus* sperm initiate motility in response to hyposmolality, the SMIS activity seems to be redundant for the motility initiation at their external fertilization. It is well known that mammalian sperm show a specific motility state to propel through the oviductal matrix (Yanagimachi 1994). It is suspected that the SMIS activity in the jelly layer of *D. pictus* is for the activation of the sperm motility state in relationship to the penetration of the jelly matrix.

8.5 Perspective

The SMIS is a key for the AR-mediated motility initiation critical for the success of internal fertilization of urodeles. It acts based on the fine structure of the egg jelly surface at the beginning of fertilization. Modifications of the jelly matrix-based mechanism may be an engine for the adaptation of motility control of amphibian sperm to various environments. Diversification of reproductive mode widely occurs among animal species and contributes to the success of fertilization under a specific condition. Every mode of reproduction sometimes looks so specific that we know little about the molecular mechanism for its diversification. The SMIS-triggering mechanism is potent to reveal the plasticity of sperm motility control that correlates with the diversification of reproductive modes in amphibian species.

Acknowledgments This work was supported by a Grant-in-Aid for Scientific Research (C) 24570246 and (A) 24240062 from the Japan Society for the Promotion of Science.

References

Arranz LE, Cabada MO (2000) Diffusible high glycosylated protein from *Bufo arenarum* egg-jelly coat: biological activity. Mol Reprod Dev 56:392–400

Campanella C, Carotenuto R, Infante V et al (1997) Sperm–egg interaction in the painted frog (*Discoglossus pictus*): an ultrastructural study. Mol Reprod Dev 47:323–333

Darszon A, Acevedo JJ, Galindo BE et al (2006) Sperm channel diversity and functional multiplicity. Reproduction 131:977–988

Duellman WE, Trueb L (1994) Biology of amphibians. Johns Hopkins University Press, Baltimore

Greven H (1998) Survey of the oviduct of salamandrids with special reference to the viviparous species. J Exp Zool 282:507–525

Katagiri C (1987) Role of oviductal secretions in mediating gamete fusion in anuran amphibians. Zool Sci 4:1–14

MacLaughlin EW, Humphries AAJ (1978) The jelly envelopes and fertilization of eggs of the newt, *Notophthalmus viridescens*. J Morphol 158:73–90

Miyata H, Thaler CD, Haimo LT et al (2012) Protease activation and the signal transduction pathway regulating motility in sperm from the water strider *Aquarius remigis*. Cytoskeleton 69:207–220

Mizuno J, Watanabe A, Onitake K (1999) Initiation of sperm motility of the newt, *Cynops pyrrhogaster*, is induced by the heat-stable component of egg-jelly. Zygote 7:329–334

Morisawa M, Oda S, Yoshida M et al (1999) Transmembrane signal transduction for the regulation of sperm motility in fishes and ascidians. In: Gagnon C (ed) The male gamete. Cache River Press, Vienna, pp 149–160

Nakai S, Watanabe A, Onitake K (1999) Sperm surface heparin/heparan sulfate is responsible for sperm binding to the uterine envelope in the newt, *Cynops pyrrhogaster*. Dev Growth Differ 41:101–107

Ohta M, Kubo H, Nakauchi Y et al (2011) Sperm motility-initiating activity in the egg jelly of the externally-fertilizing urodele amphibian, *Hynobius lichenatus*. Zool Sci 27:875–879

Okimura M, Watanabe A, Onitake K (2001) Organization of carbohydrate components in the egg-jelly layers of the newt, *Cynops pyrrhogaster*. Zool Sci 18:909–918

Olson JH, Xiang X, Ziegert T et al (2001) Alluirn, a 21-kDa sperm chemoattractant from *Xenopus* egg jelly, is related to mammalian sperm-binding proteins. Proc Natl Acad Sci USA 98:11205–11210

Sasaki T, Kamimura S, Takai H et al (2002) The activity for the induction of the sperm acrosome reaction localizes in the outer layer and exists in the high-molecular-weight components of the egg-jelly of the newt, *Cynops pyrrhogaster*. Zygote 10:1–9

Street JC (1940) Experiments on the organization of the unsegmented egg of *Triturus pyrrhogaster*. J Exp Zool 85:383–408

Takahashi S, Nakazawa H, Watanabe A et al (2006) The outermost layer of egg-jelly is crucial to successful fertilization in the newt, *Cynops pyrrhogaster*. J Exp Zool 305A:1010–1017

Takahashi T, Kutsuzawa M, Shiba K et al (2013) Distinct Ca^{2+} channels maintain a high motility state of the sperm that may be needed for penetration of egg jelly of the newt, *Cynops pyrrhogaster*. Dev Growth Differ 55:657–667

Takayama-Watanabe E, Campanella C, Kubo H et al (2012) Sperm motility initiation by egg jelly of the anuran *Discoglossus pictus* may be mediated by sperm motility-initiating substance of the internally-fertilizing newt, *Cynops pyrrhogaster*. Zygote 20:417–422

Takayama-Watanabe E, Ochiai H, Tanino S et al (2014) Contribution of different Ca^{2+} channels to the acrosome reaction-mediated initiation of sperm motility in the newt, *Cynops pyrrhogaster*. Zygote (in press)

Ueda Y, Yoshizaki N, Iwao Y (2002) Acrosome reaction in sperm of the frog, *Xenopus laevis*: its detection and induction by oviductal pars recta secretion. Dev Biol 243:55–64

Ukita M, Itoh T, Watanabe T et al (1999) Substances for the initiation of sperm motility in egg-jelly of the Japanese newt, *Cynops pyrrhogaster*. Zool Sci 16:793–802

Wake MH, Dickie R (1998) Oviductal structure and function and reproductive modes in amphibians. J Exp Zool 282:477–506

Watanabe A, Onitake K (2002) The urodele egg-coat as the apparatus adapted for the internal fertilization. Zool Sci 19:1341–1347

Watanabe A, Onitake K (2003) Sperm activation. In: Sever DM (ed) Reproductive biology and phylogeny of urodele (amphibian). Science Publishers, Enfield, pp 423–455

Watanabe T, Ito T, Watanabe A et al (2003) Characteristics of sperm motility induced on the egg-jelly in the internal fertilization of the newt, *Cynops pyrrhogaster*. Zool Sci 20:345–352

Watanabe A, Fukutomi K, Kubo H et al (2009) Identification of egg-jelly substances triggering sperm acrosome reaction in the newt, *Cynops pyrrhogaster*. Mol Reprod Dev 76:399–406

Watanabe T, Kubo H, Takeshima S et al (2010) Identification of the sperm motility-initiating substance in the newt, *Cynops pyrrhogaster*, and its possible relationship with the acrosome reaction during internal fertilization. Int J Dev Biol 54:591–597

Watanabe A, Takayama-Watanabe E, Vines CA et al (2011) Sperm motility-initiating substance in newt egg-jelly induces differential initiation of sperm motility based on sperm intracellular calcium levels. Dev Growth Differ 53:9–17

Yanagimachi R (1994) Mammalian fertilization. In: Knobil E, Neill JD (eds) The physiology of reproduction, 2nd edn. Raven, New York, pp 189–317

Yoshizaki N, Katagiri C (1982) Acrosome reaction in sperm of the toad, *Bufo bufo japonicus*. Gamete Res 6:343–352

Chapter 9
Analysis of the Mechanism That Brings Protein Disulfide Isomerase-P5 to Inhibit Oxidative Refolding of Lysozyme

Miho Miyakawa, Shuntaro Shigihara, Gosuke Zukeran,
Tetsutaro Tomioka, Tasuku Yoshino, and Kuniko Akama

Abstract Mammalian sperm acquire fertilizing capacity during epididymal transit. During this process, the expression of protein disulfide isomerase-P5 (PDI-P5, P5) decreases. Almost all members of the PDI family accelerate the formation, reduction, and isomerization of disulfide bonds, whereas boar P5, which is composed of two active thioredoxin domains, a and a', and an inactive b domain, inhibits the oxidative refolding of reduced and denatured lysozyme due to nonproductively folded lysozyme resulting from intermolecular isomerization. We investigated the reductive activities of cysteine mutants of boar P5 and of the a and $a'b$ domains, using an insulin turbidity assay, and their ability to inhibit the oxidative refolding of reduced and denatured lysozyme. We also analyzed refolded products by Western blotting. The reductive activities of the C-terminal variants, C171A, C174A, and C171/174A, were lower than those of the N-terminal variants C36A, C39A, C36/39A, and $a'b$. The inhibitory activities of N- and C-terminal cysteine mutants were ~60 % and ~75 %, respectively, whereas the inhibitory activities of C36/39A and of $a'b$ were ~60 % and 30 %, respectively. The properties of a mixture of a and $a'b$ were similar to those of $a'b$ alone. These results suggested that two active domains of P5 cooperatively inhibited refolding by intermolecular isomerization and that this effect was increased when the a and $a'b$ domains were linked. The findings were supported by Western blotting analysis of the refolded products. These results provide new insight into the molecular mechanism by which P5 inhibits protein refolding.

M. Miyakawa • G. Zukeran • T. Tomioka • T. Yoshino
Department of Chemistry, Graduate School of Science,
Chiba University, 1-33 Yayoi-cho, Inage-ku, Chiba, Chiba 263-8522, Japan

S. Shigihara
Department of Chemistry, Faculty of Science, Chiba University,
1-33 Yayoi-cho, Inage-ku, Chiba, Chiba 263-8522, Japan

K. Akama (✉)
Department of Chemistry, Graduate School of Science, Center for General Education,
Chiba University, 1-33 Yayoi-cho, Inage-ku, Chiba, Chiba 263-8522, Japan
e-mail: akama@faculty.chiba-u.jp

H. Sawada et al. (eds.), *Sexual Reproduction in Animals and Plants*,
DOI 10.1007/978-4-431-54589-7_9, © The Author(s) 2014

Keywords Isomerase activity • Lysozyme-refolding inhibition • PDI • PDI-P5

9.1 Introduction

Caput epididymal sperm released from the testes are biologically incompetent, acquiring functionality during transit from the epididymal caput to the cauda via the corpus. Ejaculated sperm are able to move vigorously and participate in the complex cascades of interactions that culminate in fertilization of the oocyte (Yanagimachi et al. 1994). Various kinds of studies on epididymal sperm maturation have been reported (McLaughlin et al. 1997; Frayne et al. 1998; Zhu et al. 2001; Saxena et al. 2002; Baker et al. 2005, 2012; Ellerman et al. 2006; Toshimori et al. 2006; Inoue et al. 2008; Ijiri et al. 2011). Protein disulfide isomerase-P5 (PDI-P5, P5) has been reported present in mouse sperm membrane fraction (Stein et al. 2006) and to be involved in asymmetrical organogenesis in zebrafish (Hoshijima et al. 2002), in regulating human platelet function (Jordan et al. 2005), and in promoting the ability of human tumors to evade the immune system (Kaiser et al. 2007; Gumireddy et al. 2007). We have reported that boar P5 is downregulated during transit from the epididymal corpus to the cauda and inhibits oxidative refolding of reduced and denatured lysozyme (Akama et al. 2010). To date, however, the mechanism by which P5 inhibits protein refolding has not been elucidated.

P5 is composed of two active thioredoxin domains, a and a', and an inactive b domain, with each active thioredoxin domain containing an active CGHC site. We investigated the mechanism by which P5 inhibits protein refolding by assaying the inhibitory activities of highly purified domain-deleted or cysteine variants of P5. Our results suggest that the a domain inhibits refolding by promoting the intermolecular isomerization of lysozyme with the a' domain, with the a-$a'b$ attachment essential for complete inhibition.

9.2 Materials and Methods

9.2.1 Expression and Purification of PDI-P5 Variants

The cDNAs corresponding to P5 deletion variants, the a and $a'b$ fragments, were amplified by polymerase chain reaction (PCR) and inserted into the expression vector pET30a(+) (Novagen). The cysteine mutants of P5 cDNA were generated by amplification of wild-type P5 cDNA cloned into pET30a(+) vector using pfu polymerase (Promega). The domain structure of P5 and its variants are shown in Fig. 9.1. Expression and purification of pET-PDI-P5 variants were performed as described (Akama et al. 2010), with ammonium sulfate precipitation and hydrophobic chromatography on phenyl-Sepharose (CL-4B) (GE Healthcare Bio-Science) performed before affinity chromatography.

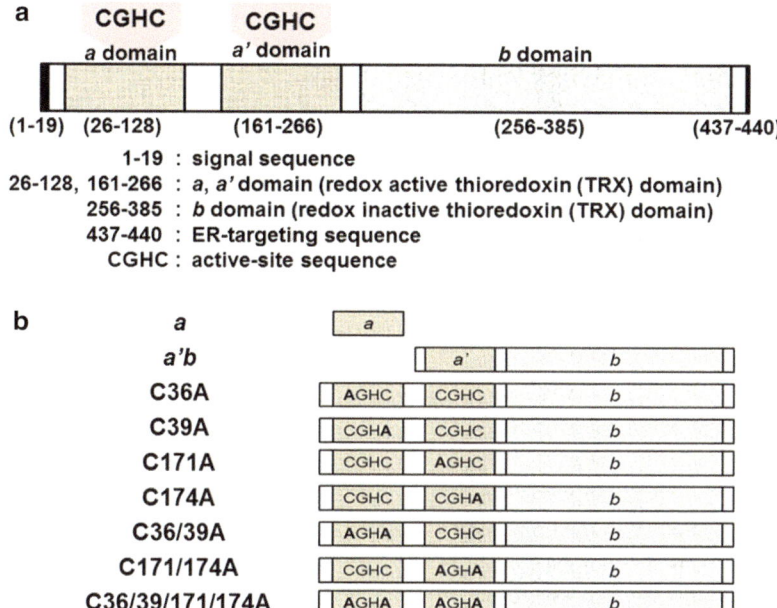

Fig. 9.1 Domain structure of boar P5 and its variants. (**a**) Domain organization of boar P5. (**b**) P5 variants: active domains are shown in light brown and the inactive domain in *light gray*. Replacement of cysteines by alanines is shown in *boldface*

9.2.2 Insulin Turbidity and Lysozyme Refolding Assays

Insulin turbidity (Holmgren 1979) and lysozme refolding (Puig and Gilbert 1994) assays were performed as described.

9.2.3 Western Blotting

Reaction products of lysozyme refolding assay were Western blotted as described (Akama et al. 2010).

9.3 Results

9.3.1 Reductive Activity of a′ Domain

Purified recombinant mature P5 showed thiol-dependent reductase activity, catalyzing the reduction of insulin disulfides with DTT as previously reported (Akama et al. 2010). All cysteine-to-alanine mutations decreased the reductive activity of P5.

C-terminal mutants (C171A, C174A, and C171/174A) had about ~50 % of the activity of wild type, whereas N-terminal mutants (C36A, C39A, C36/39A and $a'b$) had about ~75 % of the activity, suggesting that the disulfide reducing activity of the a' domain was greater than that of the other domains. The quadruple mutant C36/39/171/174A, in which all four active cysteines were mutated to alanine, had no reducing activity, indicating that cysteines located in the inactive thioredoxin domain, b, were not involved in disulfide reduction.

9.3.2 Chaperone Activities of the P5 Mutants

Reduced and denatured lysozyme is commonly used as a substrate to determine whether PDIs have chaperone activity and promote oxidative refolding in vitro. Most PDI family members possess chaperone activity, catalyzing the formation and rearrangement of disulfide bonds during the correct folding of nascent proteins. We have found, however, that P5 possesses anti-chaperone activity, inhibiting the oxidative refolding of the denatured lysozyme (Akama et al. 2010). Western blot analysis has suggested that the anti-chaperone activity of P5 is the result of its intermolecular isomerization with lysozyme, yielding lysozyme aggregates (Akama et al. 2010).

To clarify this mechanism, we compared the ability of P5 and P5 mutants to inhibit oxidative refolding. The a domain enhanced lysozyme refolding, behaving like thioredoxin alone, whereas the $a'b$ fragment without the N-terminal a domain inhibited folding ~30 % compared with the wild type (100 %). The inhibitory activity of equal concentrations of the a and $a'b$ fragments was similar to that of $a'b$ alone, suggesting that the enhancement of refolding by the a domain did not occur when $a'b$ was present. The double N-terminal mutant C36/39A had greater inhibitory activity (60 %) than $a'b$. These results suggested that the presence of the a domain enhanced the intermolecular isomerization of P5 with denatured lysozyme, and that the covalent bonding of a to $a'b$ was essential for complete inhibition. The other N-terminal active domain mutants (C36A, C39A) had ~60 % and the C-terminal active domain mutants (C171A, C174A and C171/174A) had ~75 % of the activity of the wild type, indicating that the a domain inhibited refolding when combined with the a' domain.

9.3.3 Detection of Lysozyme Aggregates by Western Blotting

Our previous study has indicated that the P5 inhibition of lysozyme refolding is caused by lysozyme aggregates resulting from intermolecular isomerization. Using Western blotting, we found that the quantity of lysozyme aggregates was reduced when refolding was catalyzed by C36A and C171A, compared with native P5 (Fig. 9.2). Spontaneous crosslinking of reduced and denatured lysozyme alone was observed in refolding buffer (Fig. 9.2), explaining why its refolding rate was no higher than 40 %.

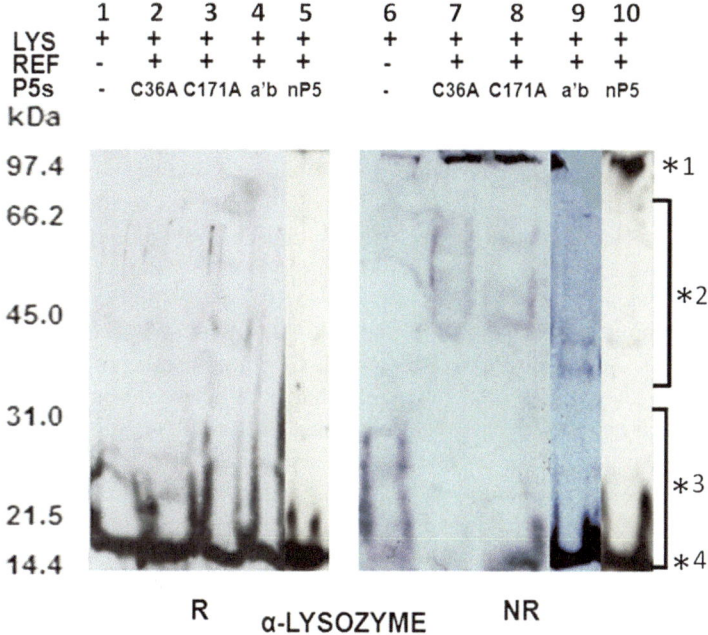

Fig. 9.2 Western blotting of refolded products of lysozyme. *1, intermolecular crosslinked aggregates. *2, crosslinked intermediates (highly crosslinked, lanes 7 and 8). *3, intermediates of lysozyme produced during spontaneous refolding. *4l, lysozyme monomers. *P5s*, P5 and its variants; *nP5*, native P5; *R*, reduced; *NR*, nonreduced. P5 concentration, 0.6 μM

In the presence of native P5, lysozyme aggregates were detected in the high molecular region, while lysozyme intermediates were not, similar to our earlier findings (Akama et al. 2010). In the presence of C36A or C171A, lysozyme intermediates had higher molecular weights than those of lysozyme alone, with the amounts of aggregates being smaller than for P5. No aggregates were observed in the presence of *a'b*, but the sizes of the intermediates were smaller than those observed with the cysteine mutants. These results further indicated that the *a* and *a'* domains cooperatively contributed to the production of lysozyme aggregates and that the *a* domain accelerated the aggregate production.

9.4   Discussion

9.4.1   Collaborative Isomerization by Two Active Domains

The results of the lysozyme refolding assay showed that N-terminal active cysteines contribute more to the inhibition of refolding than C-terminal active cysteines, similar to findings with rat and human P5 (Kramer et al. 2001; Kikuchi et al. 2002).

Using human P5 variants, a single active domain, particularly the *a* domain, has been shown sufficient for isomerase activity (Kikuchi et al. 2002). When both cysteines in either the *a* or *a'* domain are replaced with serine, the isomerase activities of C36/39S and C171/174S, using insulin as a substrate, are 50 % and 75 %, respectively, relative to the wild type, suggesting that the *a* domain can compensate for the replacement of cysteines in the *a'* domain. However, a study of yeast PDI has shown that the isomerase activity of mutants, in which cysteines residues were replaced with alanine, is less than that of the serine mutants, a finding thought to be the result of the reducing activity remaining in the hydroxyl group of serine (Tian et al. 2006).

Using a lysozyme refolding assay, we found that the inhibitory activity of the N-terminal mutants (C36A, C39A, and C36/39A) was ~60 % of that of the wild type, whereas the inhibitory activity of the C-terminal mutants (C171A, C174A, and C171/174A) was ~75 %, indicating that both active domains collaborate in intermolecular isomerization. Western blotting showed that comparable amounts of lysozyme aggregates were present in the presence of C36A or C171A, indicating that the *a* and *a'* domains cooperatively inhibited refolding. The relatively stronger reducing activity of the *a'* relative to the *a* domain observed in the insulin turbidity assay also suggested collaborative isomerization. The refolding inhibitory activity of both *a'b* and the mixture of *a* and *a'b* was 30 % of that of the wild type, suggesting that a covalent bond between the two was essential for complete inhibition. This finding was supported by our Western blotting results, with *a'b* alone showing lysozyme intermediates but no aggregates. The collaborative inhibition by the *a* and *a'* domains suggests that they have comparable access to substrates, although these domains are adjacent to each other, possibly because the amino acid chain is relatively long (36 residues) between these domains, enabling the *a* domain to move widely. The more efficient inhibition by C36/39A, in which both cysteines in the *a* domain were mutated to inactive alanine, than by *a'b*, which lacked an *a* domain, suggests that the mobile *a* domain holds and stabilizes large lysozyme aggregates.

9.4.2 Importance of Thioredoxin Domain Order

Evolution has given rise to the PDI family of proteins, with 1–6 thioredoxin domain(s). The ancestral PDIs may have been characterized by active domains packed together, with these domains structurally separated in descendant PDIs (Pedone et al. 2010). Protein disulfide oxidoreductases (PDOs), composed of only two thioredoxin domains, can be classified as ancestral types. P5 may also be classified as an ancestral PDI because it contains three thioredoxin domains, in the order *a*, *a'*, and *b*, similar to PDOs. The domain structure of P5, consisting of two contiguous active domains and a third inactive domain, may be responsible for its potent isomerization activity and unique function; this in turn may be the result of the nonspecialized roles of both the *a* and *a'* domains. Rather, they cooperatively isomerize disulfides. The downregulation of P5 from the epididymal corpus to the cauda suggests that its potent isomerization activity may be involved in sperm maturation.

9.5   Conclusion

We investigated the properties of domain-deleted and cysteine variants of P5 having an active a and a' domain and an inactive b domain. The reductive activities of C171A, C174A, and C171/174A (a' domain variants) were ~50 % of the wild type, whereas those of C36A, C39A, C36/39A (a domain variants), and $a'b$ were ~75 %. The inhibitory activities of lysozyme refolding of a and a' domain variants were ~60 % and ~75 % of wild type, respectively, whereas the inhibitory activity of a mixture of a and $a'b$ was 30 % as well as that of $a'b$. Together with the results of Western blotting, these findings suggested that two active domains of P5 cooperatively inhibited lysozyme refolding by intermolecular isomerization, and that the isomerase activity was enhanced by a domain and the covalent bond between a and $a'b$ was essential for the "potent" isomerase activity. Our finding on its isomerization mechanism may contribute for further understanding of molecular basis of sperm maturation, asymmetrical organogenesis, and tumor immune evasion.

References

Akama K et al (2010) Protein disulfide isomerase-P5, downregulated in the final stage of boar epididymal sperm maturation, catalyzes disulfide formation to inhibit protein function in oxidative refolding of reduced denatured lysozyme. Biochem Biophys Acta 1804:1272–1284

Baker MA et al (2005) Identification of post-translational modifications that occur during sperm maturation using difference in two-dimensional gel electrophoresis. Proteomics 5:1003–1012

Baker MA et al (2012) Analysis of phosphopeptide changes as spermatozoa acquire functional competence in the epididymis demonstrates changes in the post-translational modification of Izumo1. J Proteome Res 11:5252–5264

Ellerman DA et al (2006) A role for sperm surface protein disulfide isomerase activity in gamete fusion: evidence for the participation of ERp57. Dev Cell 10:831–837

Frayne J et al (1998) The MDC family of proteins and their processing during epididymal transit. J Reprod Fertil Suppl 53:149–155

Gumireddy K et al (2007) In vivo selection for metastasis promoting genes in the mouse. Proc Natl Acad Sci USA 104:6696–6701

Holmgren A (1979) Thioredoxin catalyzes the reduction of insulin disulfides by dithiothreitol and dihydrolipoamide. J Biol Chem 254:9627–9632

Hoshijima K et al (2002) A protein disulfide isomerase expressed in the embryonic midline is required for left/right asymmetries. Genes Dev 16:2518–2529

Ijiri TW et al (2011) Identification and validation of mouse sperm proteins correlated with epididymal maturation. Proteomics 11:4047–4062

Inoue N et al (2008) Putative sperm fusion protein IZUMO and the role of N-glycosylation. Biochem Biophys Res Commun 377:910–914

Jordan PA et al (2005) A role for the thiol isomerase protein ERP5 in platelet function. Blood 105:1500–1507

Kaiser BK et al (2007) Disulphide-isomerase-enabled shedding of tumour-associated NKG2D ligands. Nature (Lond) 447:482–486

Kikuchi M et al (2002) Functional analysis of human P5, protein disulfide isomerase homologue. J Biochem 132:451–455

Kramer B et al (2001) Functional roles and efficiencies of the thioredoxin boxes of calcium-binding proteins 1 and 2 in protein folding. Biochem J 357:83–95

McLaughlin EA et al (1997) Cloning and sequence analysis of rat fertilin alpha and beta developmental expression, processing and immunolocalization. Mol Hum Reprod 3:801–809

Pedone E et al (2010) Multiple catalytically active thioredoxin folds: a winning strategy for many functions. Cell Mol Life Sci 67:3797–3814

Puig A, Gilbert HF (1994) Protein disulfide isomerase exhibits chaperone and anti-chaperone activity in the oxidative refolding of lysozyme. J Biol Chem 269:7764–7771

Saxena DK et al (2002) Behaviour of a sperm surface transmembrane glycoprotein basigin during epididymal maturation and its role in fertilization in mice. Reproduction 123:435–444

Stein KK et al (2006) Proteomic analysis of sperm regions that mediate sperm–egg interactions. Proteomics 6:3533–3543

Tian G et al (2006) The crystal structure of yeast protein disulfide isomerase suggests cooperativity between its active sites. Cell 124:61–73

Toshimori K et al (2006) The involvement of immunoglobulin superfamily proteins in spermatogenesis and sperm–egg interaction. Reprod Med Biol 5:87–93

Yanagimachi R et al (1994) Mammalian fertilization. In: Knobil K, Neil JD (eds) The physiology of reproduction. Raven, New York, pp 189–317

Zhu GZ et al (2001) Testase 1 (ADAM 24), A plasma membrane-anchored sperm protease implicated in sperm function during epididymal maturation or fertilization. J Cell Sci 114: 1787–1794

Part II
Gametogenesis, Gamete Recognition, Activation, and Evolution

Chapter 10
Effect of Relaxin-Like Gonad-Stimulating Substance on Gamete Shedding and 1-Methyladenine Production in Starfish Ovaries

Masatoshi Mita, Yuki Takeshige, and Masaru Nakamura

Abstract As in lower vertebrates, starfish oocyte maturation and ovulation are induced by a hormonal substance. Gonad-stimulating substance (GSS) is the first mediator in inducing these maturation and ovulation. GSS secreted from the nervous system acts on the ovary to produce 1-methyladenine (1-MeAde) as a maturation-inducing hormone. 1-MeAde induces germinal vesicle breakdown (GVBD) and follicular envelope breakdown (FEBD), releasing the oocyte from the ovaries, that is, spawning. Recently, GSS was purified from the radial nerves of the starfish *Asterina pectinifera* and its chemical structure was determined to be a relaxin-like peptide. To further elucidate the physiological roles of GSS on oocyte maturation and ovulation, this study examined the effect of synthetic GSS on 1-MeAde production in intact ovaries, folliculated oocytes, isolated follicle cells, and spawned ovaries containing follicle cells of the starfish (sea star) *A. pectinifera*. Spawning was induced by synthetic GSS as a dose-dependent manner. However, a high concentration of GSS failed to induce spawning. 1-MeAde production in an ovarian fragment also declined at a high concentration of GSS. Similar dual effects of GSS were observed in folliculated oocytes. In contrast, 1-MeAde production in isolated follicle cells and spawned ovaries did not decrease at a high concentration of GSS. Interestingly, egg jelly inhibited GSS-induced 1-MeAde production in follicle cells. These results may suggest that egg jelly disturbs GSS action on 1-MeAde production and spawning in ovaries.

Keywords 1-Methyladenine • Egg jelly • Gonad-stimulating substance • Spawning • Starfish

M. Mita (✉) • Y. Takeshige
Department of Biology, Faculty of Education, Tokyo Gakugei University, Tokyo 184-8501, Japan
e-mail: bio-mita@u-gakugei.ac.jp

M. Nakamura
General Research Center, Churashima Foundation, Ishikawa 888, Motobu-cho, Kunigami-gun, Okinawa 905-0206, Japan

H. Sawada et al. (eds.), *Sexual Reproduction in Animals and Plants*,
DOI 10.1007/978-4-431-54589-7_10, © The Author(s) 2014

10.1   Introduction

In most starfish, fertilization occurs in seawater, outside the female's body. A ripe ovary contains a huge number of fully grown oocytes of almost equal size. Each oocyte still possesses a large nucleus (germinal vesicle), which is arrested in late prophase of the first maturation division. The oocyte is surrounded by a single follicular layer. The follicles adhere firmly to each other and to the ovarian wall (Kanatani and Shirai 1969). Resumption of meiosis in immature oocytes and spawning from the ovary are induced by 1-methyladenine (1-MeAde) (Kanatani et al. 1969). 1-MeAde is produced by ovarian follicle cells on stimulation by gonad-stimulating substance (GSS) released from radial nerves (Kanatani 1985). However, the action of 1-MeAde is indirect; it acts on the oocyte membrane receptor (Yoshikuni et al. 1988; Tadenuma et al. 1992) to activate a cytoplasmic maturation or M-phase promoting factor (MPF) (Okumura et al. 2002; Hiraoka et al. 2004), which is the direct trigger of germinal vesicle breakdown (GVBD) (Kishimoto and Kanatani 1976) and follicular envelope breakdown (FEBD) (Kishimoto et al. 1984). Once FEBD occurs, the denuded oocytes become freely movable within the ovary and are forced out by contraction of the ovarian wall (Shirai et al. 1981).

Previous studies have shown that an inhibitor of spawning, "shedhibin," is contained in the nerve extract (Cheat 1966a, b). However, crude GSS inhibits spawning at a high concentration (Kanatani and Shirai 1970). Recently, GSS was purified from the radial nerves of the starfish *Asterina pectinifera* and its chemical structure was determined to be a relaxin-like peptide (Mita et al. 2009). In this chapter, hormonal action of GSS on spawning is described from the aspect of 1-MeAde production within ovaries.

10.2   Effect of GSS on Spawning in Ovarian Fragments

Starfish (the sea star *Asterina pectinifera*) were collected from several locations in Japan in the breeding season. An experiment was carried out to examine the effect of synthetic GSS on spawning. Ovarian fragments were placed in modified van't Hoff's artificial seawater (ASW) at pH 8.2 containing synthetic GSS at various concentrations. After 1 h incubation at 20 °C, spawning was induced by GSS at concentrations exceeding 0.2 nM GSS (Fig. 10.1). The median effective concentration (EC_{50}) was obtained with about 0.5 nM GSS. However, GSS-induced spawning decreased at concentrations above 10 nM, and above 20 nM GSS failed to induce spawning. These oocytes within the ovary were matured in a high concentration of GSS. It appears that GSS at a high concentration acts in some toxic way on the ovarian wall to inhibit its contraction, resulting in failure of spawning.

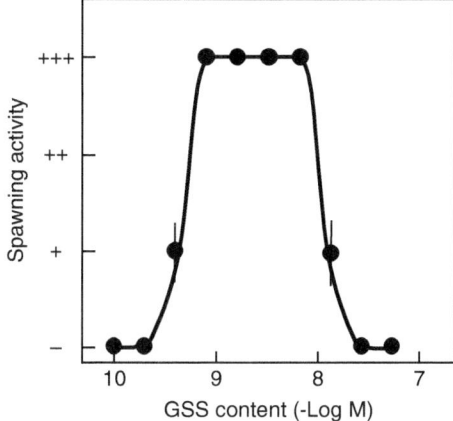

Fig. 10.1   Effect of gonad-stimulating substance (*GSS*) on gamete shedding from ovaries in starfish *Asterina pectinifera*. Ovarian fragments were placed in artificial seawater (ASW) containing GSS at concentrations indicated. After 1 h incubation at 20 °C, the sample was examined to determine whether spawning had occurred. (+++), spawning occurred and most oocytes matured; (++), about 50 % oocytes matured; (+), a few oocytes were matured; and (−), no spawning occurred. Values are mean ± SEM of four separate assays using different animals

10.3   Effect of GSS on 1-MeAde Production

Because GSS mediates oocyte maturation by acting on the ovary to produce 1-MeAde (Kanatani et al. 1969), we tested the action of synthetic GSS on 1-MeAde production in the whole ovary. When ovarian fragments of *A. pectinifera* were incubated in ASW containing synthetic GSS at various concentrations, the 1-MeAde produced was found in the incubation medium. The amount of 1-MeAde produced increased as the GSS concentration was raised, and reached maximum at 2 nM GSS (Fig. 10.2a). However, it declined at concentrations above 2 nM GSS. In contrast, GSS-induced 1-MeAde production in isolated follicle cells increased in a dose-dependent manner and reached a plateau at concentrations above 2 nM GSS (Fig. 10.2b), which agrees with previous reports (Mita et al. 1987, 2009). Because GSS stimulates follicle cells as a target in starfish ovaries (Hirai and Kanatani 1971; Hirai et al. 1973), it is possible that unknown matter within the ovaries affects GSS action on follicle cells to produce 1-MeAde.

The ovaries in starfish *A. pectinifera* are filled with immature oocytes, each surrounded by an envelope consisting of follicle cells (Fig. 10.3a). Following treatment of ovary with 1-MeAde, not only GVBD but also FEBD occurs, and denuded maturing oocytes are spawned from the ovary. The follicle cells become clustered within the ovaries (Mita and Nakamura 1994). After spawning, follicle cells exit the ovaries. Thus, we examined the effect of GSS on 1-MeAde production in the spawned ovaries. Following incubation of the spawned ovaries with GSS at various

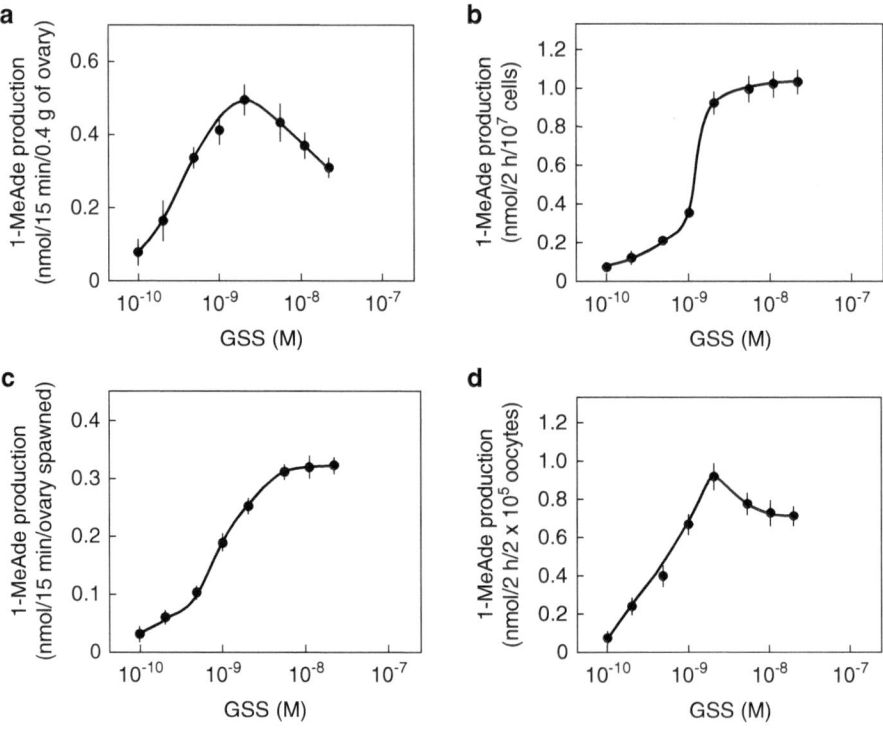

Fig. 10.2 Effect of GSS on 1-methyladenine (1-MeAde) production by an ovarian fragment (**a**), isolated follicle cells (**b**), folliculated oocytes (**c**), and spawned ovary (**d**) of starfish *A. pectinifera*. After incubation in ASW at 20 °C containing GSS at concentrations as indicated, 1-MeAde concentration was determined by biological assay. Values are mean ± SEM of four separate assays using different animals

concentrations, 1-MeAde was produced in the incubation medium (Fig. 10.2c). The production of 1-MeAde was markedly dependent on GSS concentration. A decline in 1-MeAde production was not observed at a high concentration of GSS, in agreement with the result obtained in isolated follicle cells. Thus, effects of a high concentration of GSS on 1-MeAde production in the spawned ovaries are different from those in the intact ovaries. To elucidate this, we used folliculated oocytes to examine GSS action on 1-MeAde production. The folliculated oocytes were prepared as described previously (Mita 1985). GSS stimulated 1-MeAde production in folliculated oocytes (Fig. 10.2d). The amount of 1-MeAde produced by folliculated oocytes increased in a dose-dependent manner of GSS, but it decreased slightly at a high concentration of GSS. This result suggests that oocytes or other related matter disrupt 1-MeAde production in follicle cells.

On the other hand, it has been shown that L-glutamate inhibits GSS-induced spawning in *A. pectinifera* (Ikegami et al. 1967). Thus, we examined the effect of L-glutamate on GSS-induced 1-MeAde production in follicle cells. When isolated follicle cells were incubated with GSS at concentrations of 2 and 20 nM in the

Fig. 10.3 Cross sections of ovary in starfish *A. pectinifera.* (**a**) Light micrograph shows folliculated oocytes. *Bar* 50 μm. Association of follicle cells with oocyte (**b**) was observed by ultramicroscopy. *Bars* 1 mm. *Ct* cortex, *EJ* egg jelly, *Fc* follicle cells, *GV* germinal vesicle, *Oc* oocyte

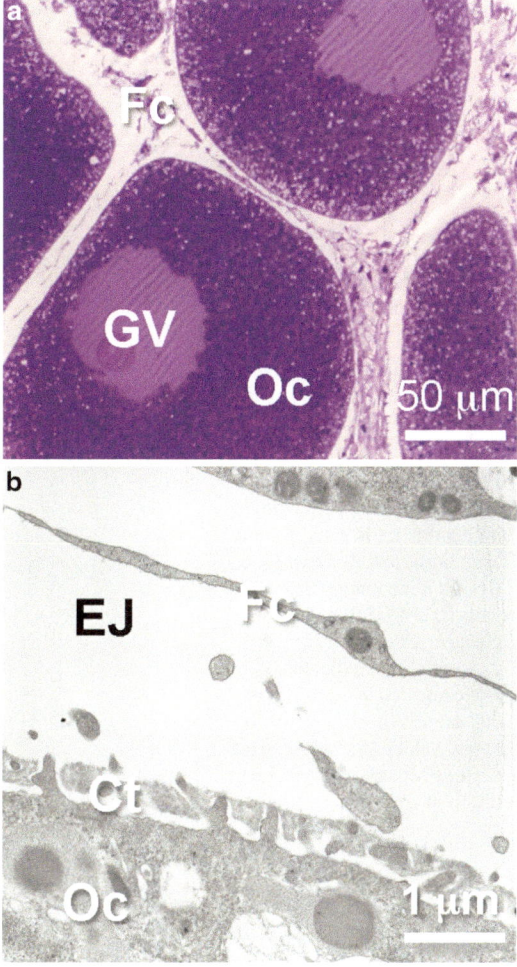

absence or presence of 10 mM L-glutamate, 1-MeAde production occurred regardless of presence of L-glutamate (Table 10.1). This result suggests that L-glutamate has no effect on GSS-induced 1-MeAde production in follicle cells.

10.4   Effect of Egg Jelly on GSS-Induced 1-MeAde Production

Although the follicle cells are covered with the oocyte surface (Fig. 10.3a), there is an interspace between the follicular envelope and oocyte surface (Fig. 10.3b). It has been shown that egg jelly is distributed in this interspace (Shirai et al. 1981). Thus, we examined whether the egg jelly affects GSS-induced 1-MeAde production by follicle cells.

Table 10.1 Effect of L-glutamate on gonad-stimulating substance (GSS)-induced 1-methyladenine (1-MeAde) production in isolated follicle cells of the starfish *Asterina pectinifera*

Condition	1-MeAde production (nmoles/2 h/10^7 cells)
Control	Undetectable
L-Glutamate (10 mM)	Undetectable
GSS (2 nM)	0.92±0.11
GSS (20 nM)	1.08±0.17
GSS (2 nM)+L-glutamate (10 mM)	0.88±0.15
GSS (20 nM)+L-glutamate (10 mM)	0.98±0.15

After 2 h incubation of follicle cells with 2 or 10 nM GSS in the absence or presence of 10 mM L-glutamate, 1-MeAde concentration was determined by biological assay
Values are mean±SEM of three separate assays using different animals

Fig. 10.4 Effect of egg jelly on GSS-induced 1-MeAde production by isolated follicle cells in starfish *A. pectinifera*. After 2 h incubation at 20 °C with 10 nM GSS in the presence of egg jelly at concentrations indicated, 1-MeAde concentration was determined by biological assay. Values are mean±SEM of three separate assays using different animals

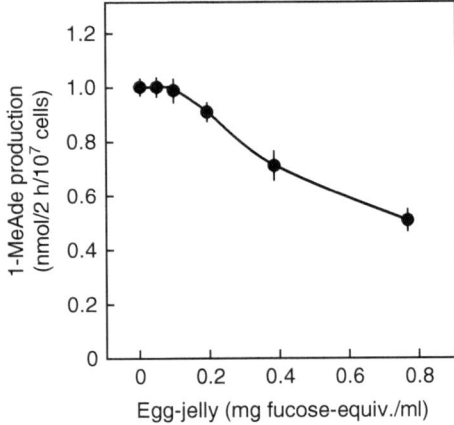

Egg jelly was prepared by acidifying a 20 % defolliculated oocyte suspension in ASW with HCl to pH 5.5 for 4–5 min. The oocytes were removed by centrifugation at $500\,g$ for 5 min, and the supernatant was adjusted to pH 8.2 with NaOH. Undissolved matter was removed by centrifugation at $10,000\,g$ for 30 min at 4 °C. The concentration of egg-jelly solution was estimated by carbohydrate determination using the method of Dische and Shettles (1951) with D-fucose as the standard. When isolated follicle cells were incubation with 10 nM GSS in the presence of egg jelly at various concentrations, 1-MeAde production decreased gradually as the egg-jelly concentration was raised (Fig. 10.4), suggesting that the egg jelly inhibits GSS-induced 1-MeAde production in follicle cells. The decrease in 1-MeAde production at a high concentration of GSS may be caused by the egg jelly. However, it is unexpected that egg jelly has an inhibitory effect on GSS action to stimulate 1-MeAde production within the ovaries. It has also been shown that egg jelly acts directly to induce contraction of the ovarian walls (Shirai et al. 1981). Further studies on ovarian wall contraction should provide useful insights into the hormonal control of ovulation and spawning of starfish.

10.5   Conclusion

1. The oocytes released from ovaries of starfish *Asterina (A.) pectinifera* were induced by GSS in a dose-dependent manner.
2. However, a high concentration of GSS failed to induce the spawning.
3. 1-MeAde production by ovarian fragments declined at a high concentration of GSS.
4. Similar dual effects of GSS were observed in folliculated oocytes, whereas 1-MeAde production in isolated follicle cells and spawned ovaries did not decrease at a high concentration of GSS.
5. Because egg jelly inhibited GSS-induced 1-MeAde production in follicle cells, it may be possible that the egg jelly disturbs GSS action on 1-MeAde production and spawning in ovaries.

References

Cheat AB (1966a) Neurochemical control of gamete release in starfish. Biol Bull 130:43–58
Cheat AB (1966b) The gamete shedding substance of starfishes: a physiological-biochemical study. Am Zool 6:263–271
Dische Z, Shettles LB (1951) A new spectrophotometric test for the detection of methyl pentose. J Biol Chem 192:579–582
Hirai S, Kanatani H (1971) Site of production of meiosis-inducing substance in ovary of starfish. Exp Cell Res 57:224–227
Hirai S, Chida K, Kanatani H (1973) Role of follicle cells in maturation of starfish oocytes. Dev. Growth & Differ 15:21–31
Hiraoka D, Hori-Oshima S, Fukuhara T, Tachibana K, Okumura E, Kishimoto T (2004) PDK1 is required for the hormonal signaling pathway leading to meiotic resumption in starfish oocytes. Dev Biol 276:330–336
Ikegami S, Tamura S, Kanatani H (1967) Starfish gonad: Action and chemical identification of spawning inhibitor. Science 158:1052–1053
Kanatani H, Shirai H (1969) Mechanism of starfish spawning. II. Some aspects of action of a neural substance obtained from radial nerve. Biol Bull 137:297–311
Kanatani H, Shirai H, Nakanishi K, Kurokawa T (1969) Isolation and identification of meiosis-inducing substance in starfish, *Asterias amurensis*. Nature (Lond) 221:273–274
Kanatani H, Shirai H (1970) Mechanism of starfish spawning. III. Properties and action of meiosis-inducing substance produced in gonad under influence of gonad-stimulating substance. Dev Growth Differ 12:119–140
Kanatani H (1985) Oocyte growth and maturation in starfish. In: Metz CB, Monroy A (eds) Biology of fertilization, vol 1. Academic, New York, pp 119–140
Kishimoto T, Kanatani H (1976) Cytoplasmic factor responsible for germinal vesicle breakdown meiotic maturation in starfish oocyte. Nature (Lond) 260:321–322
Kishimoto T, Usui N, Kanatani H (1984) Breakdown of starfish ovarian follicle induced by maturation-promoting factor. Dev Biol 101:28–34

Mita M (1985) Effect of cysteine and its derivatives on 1-methyladenine production by starfish follicle cells. Dev Growth Differ 27:563–572

Mita M, Ueta N, Nagahama Y (1987) Regulatory functions of cyclic adenosine 3',5'-monophosphate in 1-methyladenine production by starfish follicle cells. Biochem Biophys Res Commun 147(1):8–12

Mita M, Nakamura M (1994) Influence of gonad-stimulating substance on 1-methyladenine level responsible for germinal vesicle breakdown and spawning in the starfish *Asterina pectinifera*. J Exp Zool 269:140–145

Mita M, Yoshikuni M, Ohno K, Shibata Y, Paul-Prasanth B, Pichayawasin S, Isobe M, Nagahama Y (2009) A relaxin-like peptide purified from radial nerves induces oocyte maturation and ovulation in the starfish, *Asterina pectinifera*. Proc Natl Acad Sci USA 106:9507–9512

Mita M, Yamamoto K, Nagahama Y (2011) Interaction of relaxin-like gonad-stimulating substance with ovarian follicle cells of the starfish *Asterina pectinifera*. Zoolog Sci 28:764–769

Okumura E, Fukuhara T, Yoshida H, Hanada Si S, Kozutsumi R, Mori M, Tachibana K, Kishimoto T (2002) Akt inhibits Myt1 in the signaling pathway that leads to meiotic G2/M-phase transition. Nat Cell Biol 4:111–116

Shirai H, Yoshimoto Y, Kanatani H (1981) Mechanism of starfish spawning. IV. Tension generation in the ovarian wall by 1-methyladenine at the time of spawning. Biol Bull 161:172–179

Tadenuma H, Takahashi K, Chiba K, Hoshi M, Katada T (1992) Properties of 1-methyladenine receptors in starfish oocyte membranes: involvement of pertussis toxin-sensitive GTP-binding protein in the receptor-mediated signal transduction. Biochem Biophys Res Commun 186: 114–121

Yoshikuni M, Ishikawa K, Isobe M, Goto T, Nagahama Y (1988) Characterization of 1-methyladenine binding in starfish oocyte corties. Proc Natl Acad Sci USA 85:1874–1877

Chapter 11
Incapacity of 1-Methyladenine Production to Relaxin-Like Gonad-Stimulating Substance in Ca²⁺-Free Seawater-Treated Starfish Ovarian Follicle Cells

Miho Watanabe, Kazutoshi Yamamoto, and Masatoshi Mita

Abstract Previous studies have shown that brief treatment of follicle cells from ovaries of the starfish *Asterina pectinifera* with Ca^{2+}-free seawater (CaFSW) deprived the follicle cells of their capacity to respond to gonad-stimulating substance (GSS). To elucidate the failure of GSS, this study examined the hormonal action of GSS on CaFSW-treated follicle cells of *A. pectinifera*, particularly the mode of signal transduction. GSS failed to stimulate the production of 1-methyladenine (1-MeAde) and cyclic AMP (cAMP) in CaFSW-treated follicle cells and the incapacity was irreversible. According to competitive experiments using radioiodinated and radioinert GSS, highly specific binding was observed in follicle cells, although their affinities and binding sites in CaFSW-treated follicle cells were absolutely inferior to those in intact cells. Furthermore, GSS did not stimulate adenylyl cyclase in membrane preparations of CaFSW-treated follicle cells. Both Gsα and Giα were detected immunologically in membranes of CaFSW-treated follicle cells as well as those of nontreated cells. These results suggest that signal transduction for GSS in CaFSW-treated follicle cells does not flow readily from GSS receptors to G proteins.

Keywords 1-Methyladenine • Cyclic AMP • G protein • Relaxin-like gonad-stimulating substance • Starfish

M. Watanabe • M. Mita (✉)
Department of Biology, Faculty of Education, Tokyo Gakugei University,
Tokyo 184-8501, Japan
e-mail: bio-mita@u-gakugei.ac.jp

K. Yamamoto
Department of Biology, School of Education, Waseda University, Tokyo 162-8480, Japan

H. Sawada et al. (eds.), *Sexual Reproduction in Animals and Plants*,
DOI 10.1007/978-4-431-54589-7_11, © The Author(s) 2014

11.1 Introduction

In starfish, fully grown oocytes in the ripe ovary remain arrested at the prophase of the first maturation division. Resumption of meiosis in these immature oocytes is induced by 1-MeAde, which is produced by ovarian follicle cells on stimulation by GSS released from radial nerves (Kanatani et al. 1969; Kanatani 1985). Recently, GSS was purified from the radial nerves of the starfish *Asterina pectinifera* and its chemical structure was determined to be a relaxin-like peptide (Mita et al. 2009). In this chapter, 1-MeAde production in follicle cells as a role of GSS is described from the aspect of signal transduction mode, based on incapacity of follicle cells to produce 1-MeAde after washing with CaFSW.

11.2 Irreversible Incapacity of 1-MeAde Production in CaFSW-Treated Follicle Cells

Starfish (*A. pectinifera*) were collected from several locations in Japan in the breeding season. Ovarian follicle cells were separated from folliculated oocytes as described previously (Mita 1985). The isolated follicle cells were suspended in modified van't Hoff's artificial seawater (ASW) at pH 8.2 (Kanatani and Shirai 1970). In the case of treatment with CaFSW, where NaCl was substituted for CaCl$_2$, follicle cells were washed three times with CaFSW, then suspended in normal ASW (Fig. 11.1). Previous studies have shown that the action of GSS on 1-MeAde

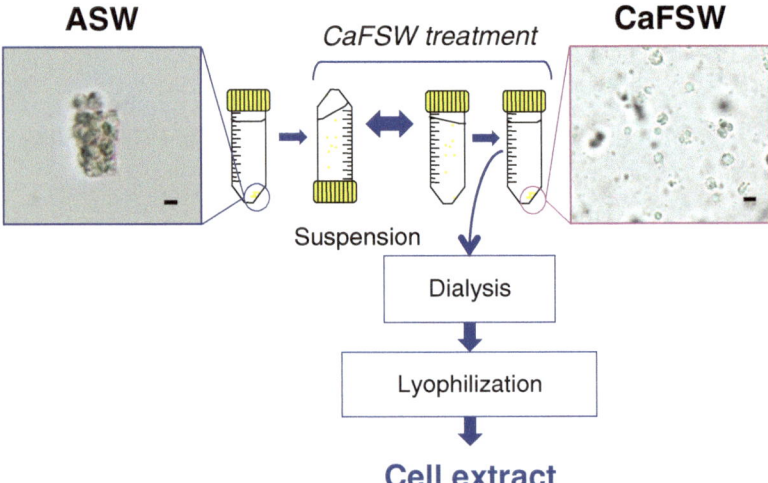

Fig. 11.1 Treatment of starfish ovarian follicle cells with Ca^{2+}-free seawater (CaFSW) and preparation of cell extracts: follicle cell suspension in artificial seawater (*ASW*) (*left*) and *CaFSW* (*right*) by light microscopy. *Bars* 10 μm

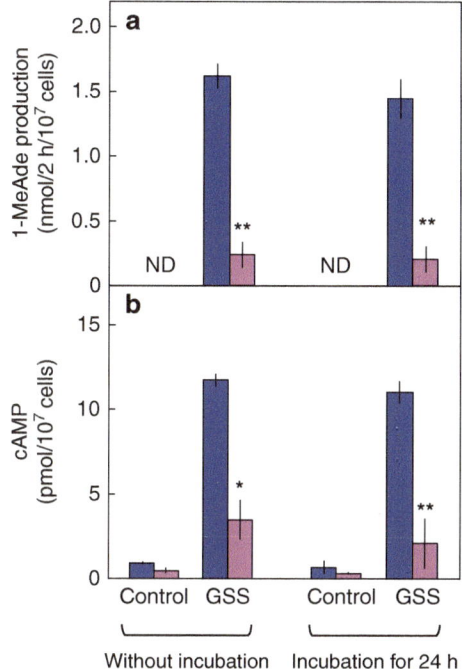

Fig. 11.2 Effect of CaFSW treatment on gonad-stimulating substance (GSS)-induced 1-methyl-adenine (1-MeAde) (**a**) and cyclic adenosine monophosphate (cAMP) (**b**) production in starfish ovarian follicle cells. After treatment with either ASW (*blue*) or CaFSW (*purple*), follicle cells were suspended in ASW. With or without for 24 h at 4 °C, these cells were incubated in the absence or presence of GSS (20 nM) for 2 h at 20 °C. Each *column* and *vertical line* show the mean for three independent samples and SEM, respectively. *ND* not detectable. *Asterisks* indicate significantly lower than value in ASW (*$P < 0.05$, **$P < 0.01$)

production by follicle cells is mediated through the production of cAMP (Mita et al. 1987; Mita and Nagahama 1991). However, after washing the follicle cells with CaFSW, the GSS-dependent production of 1-MeAde and cAMP decreased to a large degree (Fig. 11.2), in agreement with the previous report (Mita 1994). It is considered that the decrease of 1-MeAde production by follicle cells treated with CaFSW is caused by the low level of cAMP. However, it is unclear whether the failure of GSS on 1-MeAde and cAMP production in CaFSW-treated follicle cells is irreversible. Thus, follicle cells were incubated in ASW for 24 h at 4 °C after CaFSW treatment. In spite of incubation, these cells did not produce 1-MeAde and cAMP with 20 nM GSS (Fig. 11.2): the incapacity was irreversible. These results suggest that, once washed with CaFSW, follicle cells lose the ability of 1-MeAde and cAMP production by GSS. On the other hand, the CaFSW-treated follicle cells were capable of producing 1-MeAde in the presence of 1-methyladenosine (1-MeAdo) (Fig. 11.3). Because 1-MeAdo ribohydrolase is present in follicle cells (Shirai and Kanatani 1972), 1-MeAdo is converted to 1-MeAde by 1-MeAdo ribo-hydrolase independently of the signal transduction pathway for GSS. Presumably, CaFSW treatment damages follicle cells for signal transduction for GSS.

Fig. 11.3 Effect of CaFSW treatment on GSS- and 1-MeAdo-induced 1-MeAde (**a**) and cAMP (**b**) production in starfish ovarian follicle cells. After treatment with either ASW (*blue*) or CaFSW (*purple*), follicle cells were suspended in ASW and incubated in the absence or presence of GSS (20 nM) or 1-MeAdo (1 mM) for 2 h at 20 °C. Each *column* and *vertical line* show mean for four independent samples and SEM, respectively. *ND* not detectable. *Significantly lower than value in ASW ($P<0.05$); **significantly higher than value in ASW ($P<0.05$)

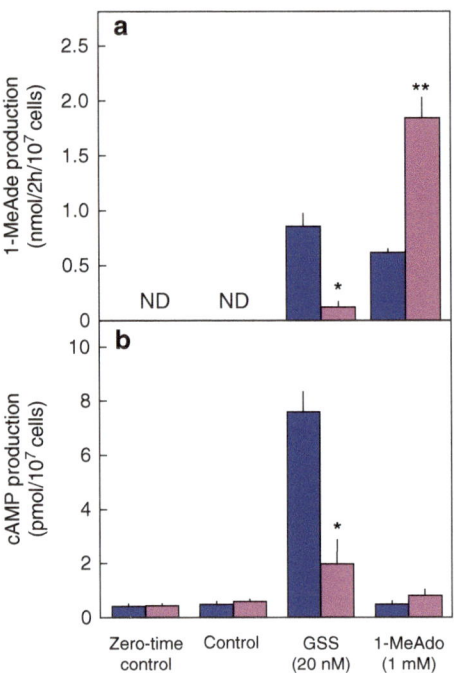

11.3 Signal Transduction for GSS in CaFSW-Treated Follicle Cells

Previous studies have shown that the action of GSS on 1-MeAde production in follicle cells is effected through its receptors, G proteins and adenylyl cyclase (Mita and Nagahama 1991). To obtain more information on the failure of GSS to produce 1-MeAde in CaFSW-treated follicle cells, properties of GSS receptors were examined. Membrane preparations from follicle cells after treatment with or without CaFSW were incubated with radioiodinated and radioinert GSS for 2 h at 20 °C. On the basis of the competitive binding experiment, Scatchard plots (Scatchard 1949) were used to estimate the dissociation constant (K_d) and the number of binding sites (NBS). Specific binding by GSS was found in CaFSW-treated follicle cells. K_d values in CaFSW-treated follicle cells were higher than those in nontreated cells (Table 11.1). NBS also decreased in CaFSW-treated follicle cells, suggesting that the affinity and number of receptors in CaFSW-treated follicle cells were absolutely inferior to those in intact cells.

Previous experiments on immunoblot analysis have shown the presence of two types of G proteins in follicle cell membranes (Mita et al. 2011). To identify the G proteins, immunoblotting using antibodies against the α-subunit of Gs (Gsα) and Gi-3 (Giα) and the β-subunit (Gβ) was performed to determine the presence of G proteins in a membrane preparation of CaFSW-treated follicle cells. In follicle cells after CaFSW treatment, the Gsα and Giα antibodies recognized 45- and 41-kDa

Table 11.1 Dissociation constant (K_d) and number of binding sites (NBS) for gonad-stimulating substance (GSS) in membrane preparations of starfish ovarian follicle cells treated with either ASW or CaFSW

Condition	K_d (nM)	NBS (pmol/mg protein)
ASW	0.93 ± 0.08	4.51 ± 0.88
CaFSW	1.24 ± 0.12	2.13 ± 0.39

Values show mean for six independent samples and SEM, respectively

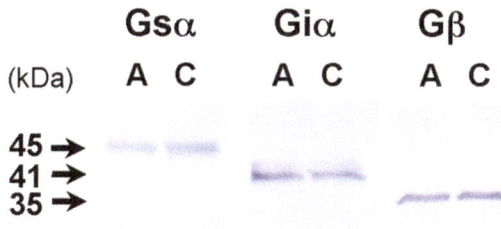

Fig. 11.4 Immunoblotting after SDS-PAGE of crude membrane preparations obtained from starfish ovarian follicle cells treated with either ASW (A) or CaFSW (C) with anti-Gsα, anti-Giα, and anti-Gβ antibodies. Crude membrane (5 μg protein/well) was separated by SDS-PAGE and used for Western blotting

proteins, respectively, as well as those in normal cells (Fig. 11.4). The Gβ antibody also recognized a 35-kDa protein in both cells.

Next, an experiment was carried out to examine whether GSS directly influences on adenylyl cyclase activity in follicle cells treated with CaFSW. On the basis of previous observations (Mita and Nagahama 1991), crude membrane preparations of follicle cells were incubated with GSS (20 nM) in the absence or presence of GTP (0.1 mM). Without GTP, GSS hardly stimulated adenylyl cyclase regardless of CaFSW treatment (Fig. 11.5a). The addition of GTP (0.1 mM) markedly enhanced the GSS-stimulated adenylyl cyclase activity in follicle cells without CaFSW treatment. Adenylyl cyclase activity in CaFSW-treated follicle cells was also stimulated by GSS in the presence of GTP, but the activation was a slight. In contrast, nonhydrolyzable GTP analogues, such as GTP-γS and GppNHp, activated adenylyl cyclase in follicle cells regardless of CaFSW treatment (Fig. 11.5b). These findings suggest that signal transduction associated with GSS in CaFSW-treated follicle cells does not transmit well from GSS receptor to Gsα protein. It is considered that the lack of response to GSS in CaFSW-treated follicle cells occurs because of a failure of signal transduction.

11.4   Cell Extracts from Follicle Cells Treated with CaFSW

Because the NBS for GSS declined in follicle cells after treatment with CaFSW (Table 11.1), it may be possible that the cell extract contains a receptor or binding protein of GSS. After follicle cells were suspended in CaFSW, supernatant obtained

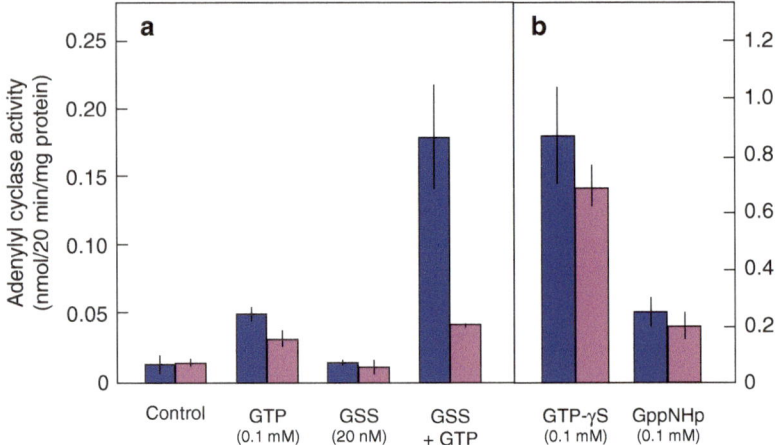

Fig. 11.5 Effect of GSS (**a**) and guanosine triphosphate (GTP) analogues (**b**) on adenylyl cyclase activity in starfish ovarian follicle cells. After treatment of follicle cells with either ASW (*blue*) or CaFSW (*purple*), their membrane preparations were used for adenylyl cyclase assay as described previously (Mita and Nagahama 1991). Values shown are means for duplicate determination

Fig. 11.6 Effect of cell extract on GSS-induced 1-MeAde production by starfish ovarian follicle cells. Isolated follicle cells in normal condition were incubated with 10 nM GSS in the presence of various concentrations of cell extract for 2 h at 20 °C. *Symbols* and *bars* represent means for three independent samples and SEM, respectively

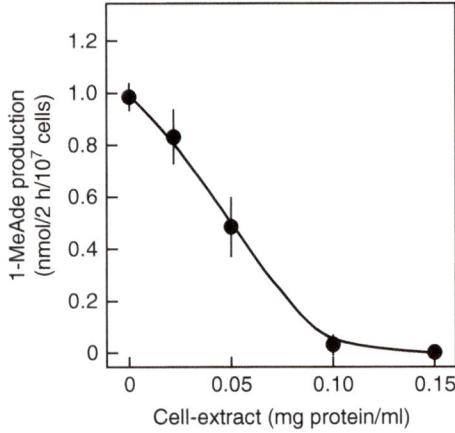

by centrifugation at 500 g for 15 min at 4 °C was dialyzed and lyophilized (Fig. 11.1). The lyophilized sample as a cell extract was dissolved in ASW and used for the experiment. When follicle cells were incubated in ASW containing 10 nM GSS in the presence or absence of cell extract at various concentrations, 1-MeAde production decreased in a dose-dependent manner of cell extract (Fig. 11.6). This result suggests that the cell extract contained some substance that disturbed GSS action on 1-MeAde production. It may be assumed that the cell extract includes a binding protein for GSS such as a component of the GSS receptor. Further studies on GSS receptors in follicle cells should provide useful insights into the hormonal regulation of starfish reproduction.

11.5 Conclusion

1. Brief treatment of follicle cells of the starfish *Asterina pectinifera* with CaFSW deprived the follicle cells of 1-MeAde and cAMP production upon stimulation by GSS.
2. Loss of capacity in CaFSW-treated follicle cells was irreversible.
3. GSS did not stimulate adenylyl cyclase in CaFSW-treated follicle cells.
4. Affinity and number of receptors in CaFSW-treated follicle cells were relatively inferior to those in intact cells.
5. Both Gsα and Giα were detected immunologically in membranes of CaFSW-treated follicle cells as well as those of nontreated cells.
6. It is possible that signal transduction for GSS in CaFSW-treated follicle cells does not flow well from GSS receptors to G proteins.

References

Kanatani H (1985) Oocyte growth and maturation in starfish. In: Metz CB, Monroy A (eds) Biology of fertilization, vol 1. Academic, New York, pp 119–140

Kanatani H, Shirai H (1970) Mechanism of starfish spawning. III. Properties and action of meiosis-inducing substance produced in gonad under influence of gonad-stimulating substance. Dev Growth Differ 12:119–140

Kanatani H, Shirai H, Nakanishi K, Kurokawa T (1969) Isolation and identification of meiosis-inducing substance in starfish, *Asterias amurensis*. Nature (Lond) 221:273–274

Mita M (1985) Effect of cysteine and its derivatives on 1-methyladenine production by starfish follicle cells. Dev Growth Differ 27:563–572

Mita M (1994) Effect of Ca²⁺-free seawater treatment on 1-methyladenine production in starfish ovarian follicle cells. Dev Growth Differ 36:389–395

Mita M, Nagahama Y (1991) Involvement of G-proteins and adenylate cyclase in the action of gonad-stimulating substance on starfish ovarian follicle cells. Dev Biol 144:262–268

Mita M, Ueta N, Nagahama Y (1987) Regulatory functions of cyclic adenosine 3′,5′-monophosphate in 1-methyladenine production by starfish follicle cells. Biochem Biophys Res Commun 147(1): 8–12

Mita M, Yoshikuni M, Ohno K, Shibata Y, Paul-Prasanth B, Pichayawasin S, Isobe M, Nagahama Y (2009) A relaxin-like peptide purified from radial nerves induces oocyte maturation and ovulation in the starfish, *Asterina pectinifera*. Proc Natl Acad Sci USA 106:9507–9512

Mita M, Yamamoto K, Nagahama Y (2011) Interaction of relaxin-like gonad-stimulating substance with ovarian follicle cells of the starfish *Asterina pectinifera*. Zool Sci 28:764–769

Scatchard G (1949) The attraction of proteins for small molecules and ions. Ann N Y Acad Sci 51:660–676

Shirai H, Kanatani H (1972) 1-Methyladenosine ribohydrolase in the starfish ovary and its relation to oocyte maturation. Exp Cell Res 75:79–88

Chapter 12
Novel Isoform of Vitellogenin Expressed in Eggs Is a Binding Partner of the Sperm Proteases, HrProacrosin and HrSpermosin, in the Ascidian *Halocynthia roretzi*

Mari Akasaka, Koichi H. Kato, Ken Kitajima, and Hitoshi Sawada

Abstract Vitellogenin is a precursor of yolk protein that is necessary for embryonic development. Vitellogenin is a large multidomain protein consisting of a signal peptide, a heavy-chain lipovitellin, a phosvitin, a light-chain lipovitellin, a von Willebrand factor type D domain (vWF-D), and a C-terminal coding region (CT), which are processed to respective domains after uptake into oocytes. It is currently believed that only lipovitellin and phosvitin domains are necessary for nutrient supply to oocytes. Thus, molecular species of vitellogenin lacking these domains are not known. We recently found that two novel isoforms of vitellogenin, both of which possess vWF-D and CT domains but not a lipovitellin or phosvitin domain, are expressed in the gonad of the ascidian *Halocynthia roretzi*. In situ hybridization revealed that mRNAs of these proteins are specifically expressed in oocytes and test cells, accessory cells in the perivitelline space of ascidian eggs. Immunocytochemistry showed that these proteins are localized around the surface of test cells in immature oocytes. Immunoelectron microscopy revealed that vitellogenin associates with vesicles located beneath the vitelline coat (VC) before fertilization but that it dissociates from the VC after fertilization. These results, together with our previous

M. Akasaka
Division of Biological Sciences, Graduate School of Science, Nagoya University,
Furo-cho, Chikusa-ku, Nagoya 464-8602, Japan

K.H. Kato
Graduate School of Natural Sciences, Nagoya City University,
Mizuho-ku, Nagoya 467-8501, Japan

K. Kitajima
Bioscience and Biotechnology Center, Nagoya University, Nagoya 464-8601, Japan

H. Sawada (✉)
Sugashima Marine Biological Laboratory, Graduate School of Science,
Nagoya University, Sugashima, Toba 517-0004, Japan
e-mail: hsawada@bio.nagoya-u.ac.jp

H. Sawada et al. (eds.), *Sexual Reproduction in Animals and Plants*,
DOI 10.1007/978-4-431-54589-7_12, © The Author(s) 2014

results showing that vWF-D and CT domains are capable of binding to the two sperm proteases HrProacrosin and HrSpermosin led us to propose that novel isoforms of vitellogenin, which are expressed in oocytes and test cells and released to the perivitelline space during oocyte maturation, may participate in gamete interaction upon fertilization.

Keywords Ascidian • Binding partner • Sperm protease • Vitellogenin

12.1   Vitellogenin Is a Binding Partner of Sperm Proteases

To accomplish successful fertilization, sperm must bind to and penetrate through the extracellular glycoprotein matrix surrounding the egg, which is called the zona pellucida (ZP) in mammals and the vitelline coat (VC) in marine invertebrates (McRorie and Williams 1974; Wassarman 1987; Sawada 2002). In mammals, an acrosomal trypsin-like protease, acrosin [EC 3.4.21.10], had long been believed to be a lytic agent, lysin, which makes a small hole for sperm penetration through the ZP of the ovum (Müller-Esterl and Fritz 1981; Urch et al. 1985a, b). However, because mouse sperm lacking the *acrosin* gene can penetrate through the ZP, albeit with some delay (Baba et al. 1994), it is currently thought that acrosin is not essential for the penetration of sperm through the ZP but is involved, at least in part, in the dispersal of acrosomal contents during the acrosome reaction (Yamagata et al. 1998) and in the secondary binding of sperm to the ZP (Howes et al. 2001; Howes and Jones 2002). However, more detailed studies on the targets of sperm proteases, including acrosin, are necessary to elucidate the roles of sperm proteases in fertilization.

Sperm trypsin-like proteases have been believed to take some part in fertilization of the ascidian *Halocynthia roretzi* because sperm–egg interaction was inhibited by protease inhibitors (Hoshi et al. 1981; Lambert et al. 2002). Then, two trypsin-like proteases, HrProacrosin and HrSpermosin, were purified from *H. roretzi* sperm (Sawada et al. 1984a), and it was also revealed that these proteases play important roles in ascidian fertilization (Sawada et al. 1984b, 1996; Sawada and Someno 1996). Both these proteases possess potential regions for protein–protein interaction: two CUB domains in the C-terminus of HrProacrosin and a proline-rich region in the N-terminus of HrSpermosin. Previously, we explored the binding partners of these proteases using affinity beads immobilized with the CUB domain peptide or a proline-rich region synthesized in *Escherichia coli* and succeeded in isolating several VC proteins (Kodama et al. 2001, 2002). However, they were not identified because the N-terminal sequences of these proteins showed no significant homology to any protein.

Recently, we carried out 5′- and 3′-rapid amplification of cDNA ends (5′-RACE)-polymerase chain reaction (PCR) on the basis of these N-terminal sequences, showing that some of these VC proteins are identical to the C-terminal coding region (CT) and von Willebrand factor type D (vWF-D) domain, which is located at the C-terminal region of vitellogenin (Akasaka et al. 2010). The binding ability between the vitellogenin C-terminus and CUB domain of HrProacrosin was confirmed by an in vitro pulldown assay (Fig. 12.1a).

12.2   Novel Isoforms of Vitellogenin are Expressed in the Gonad

Vitellogenin is a major precursor of yolk protein, which is necessary for embryonic development (Wallace and Selman 1985), and a large multidomain molecule that consists of the following regions/domains from the N-terminus to the C-terminus: a signal peptide, lipovitellin-1 (heavy chain), phosvitin, lipovitellin-2 (light chain), von Willebrand factor type D (vWF-D), and C-terminal coding region (CT) (Finn 2007). Gene expression of vitellogenin is induced in the liver by an estrogen-like hormone, and the synthesized protein is then transferred to the ovary via the bloodstream and is taken up into immature oocytes by clathrin-mediated endocytosis (Wallace and Selman 1990). Although the gene encoding vitellogenin in *H. roretzi* is a single copy (Fig. 12.1b), several lengths of mRNAs were detected mainly in the hepatopancreas and only two shorter species were weakly detected in the gonad by Northern blot analysis (Fig. 12.1c). These results suggest that mRNAs of vitellogenin were expressed in a tissue-specific manner, and it is inferred that shorter isoforms may play a specific role in the gonad.

To clarify the domain composition of vitellogenin expressed in the gonad, we determined the full-length cDNA sequences of two isoforms of vitellogenin by 5′-RACE-PCR using an *H. roretzi* gonad cDNA library and named vitellogenin S1 and S2 (Akasaka et al. 2013). A BLAST search of the deduced amino-acid sequences of vitellogenin S1 and S2 against genomic data of *H. roretzi* showed that S1 and S2 are alternatively spliced isoforms from a gene model of vitellogenin named Hr.Aug-120507.S000436.g02527, and the difference between S1 and S2 appears to be caused by the presence or absence of three exons (personal communication) (Fig. 12.1d). Unexpectedly, neither of these contains lipovitellin and phosvitin, which are necessary for vitellogenin functioning as a nutrient source, but contain vWF-D and CT domains. Concerning the vWF-D and CT domains, the CGXC motif of vWF-D and polycysteine residues of CT appear to participate in the folding of vitellogenin by forming a disulfide linkage (Mayadas and Wagner 1992; Mouchel et al. 1996). To the best of our knowledge, there is no report about the occurrence of vitellogenin isoform consisting of only vWF-D and CT domains that lacks both lipovitellin and phosvitin.

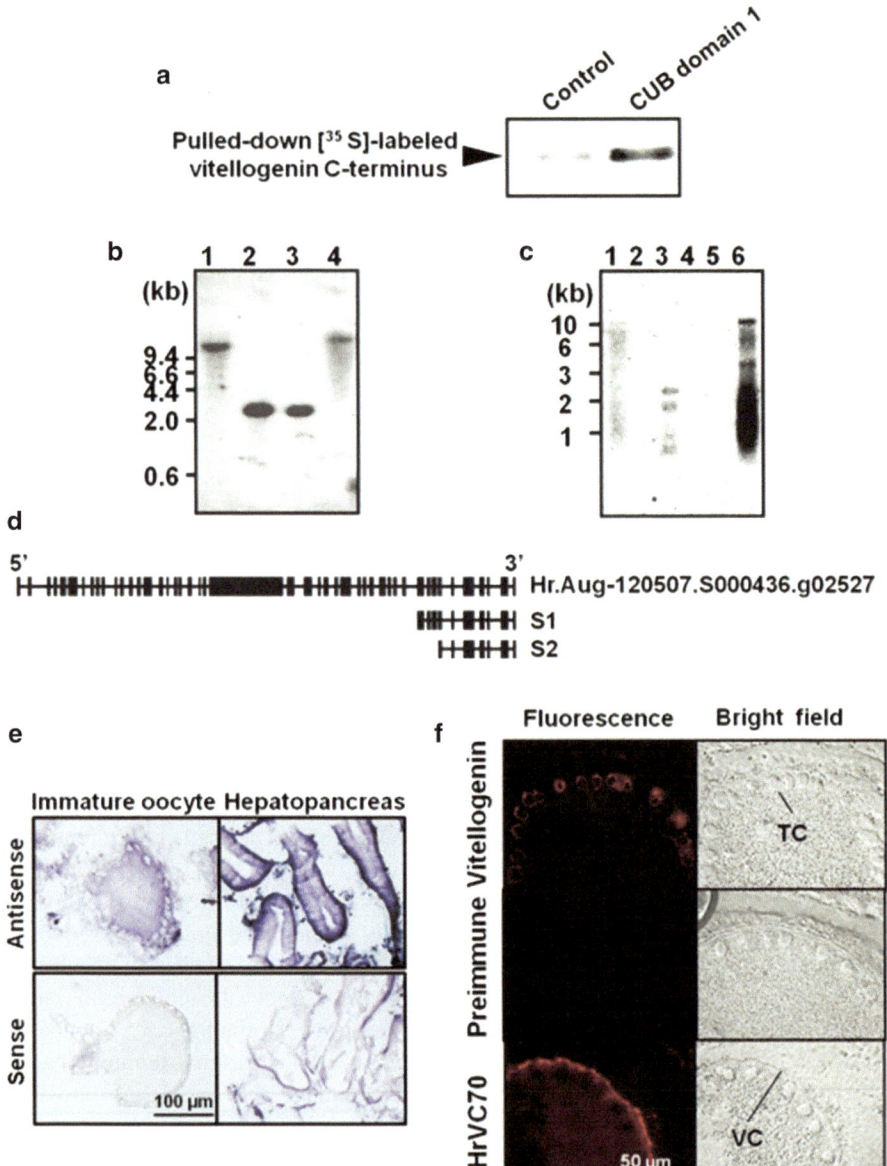

Fig. 12.1 (**a**) Pulldown assay of CUB domain 1 with vitellogenin C-terminus. The [^{35}S]-labeled vitellogenin C-terminus was pulled down with agarose beads immobilized with a recombinant CUB domain 1 protein. (**b**) Southern blotting. *Halocynthia roretzi* genome was digested with *Bam*HI (*lane 1*), *Eco*RI (*lane 2*), *Hind*III (*lane 3*), or *Pst*I (*lane 4*), and subjected to agarose gel electrophoresis. Southern blotting was carried out using a DIG-labeled probe corresponding to the vitellogenin C-terminus. (**c**) Northern blotting. Poly(A)$^+$ RNA obtained from each tissue was subjected to agarose gel electrophoresis followed by Northern blotting using a DIG-labeled probe corresponding to the vitellogenin C-terminus. *Lane 1*, muscle; *lane 2*, intestine; *lane 3*, gonad; *lane 4*, gill; *lane 5*, endostyle; *lane 6*, hepatopancreas. Note that 2.0- and 2.5-kb mRNA signals

12.3 Localization of Vitellogenin in Immature Oocytes

Although there are several reports showing the expression of vitellogenin in the gonad, there have been few experiments examining the detailed expression site of vitellogenin mRNA. To clarify the detailed expression site of vitellogenin, in situ hybridization against the sectioned gonad and hepatopancreas was carried out. As a positive control, the expression of vitellogenin mRNA in the hepatopancreas was confirmed, which coincided well with the expression pattern of vitellogenin mRNA in the hepatopancreas as revealed by Northern blot analysis (Fig. 12.1c). In immature oocytes, vitellogenin mRNA was detected in the cytosol (Fig. 12.1e): In particular, it appears to be concentrated on the surface of the test cells. Test cells are accessory cells in ascidian eggs that are embedded in immature oocytes and released to the perivitelline space during oocyte maturation. Several functions of test cells have been proposed from the results of ultrastructural observations; however, those functions are still controversial and poorly understood.

Because vitellogenin appears to be synthesized in oocytes and test cells, we carried out immunohistochemistry of vitellogenin in immature oocytes in the gonad using a mouse antiserum against a C-terminal region of vitellogenin. The results showed that vitellogenin appears to be specifically localized on the surface of test cells or the border between oocytes and test cells (Fig. 12.1f). As a marker for the VC, anti-HrVC70 antiserum was used to distinguish the outer structure of oocytes because HrVC70 is a major component of the VC in *H. roretzi* (Sawada et al. 2002). It was clarified that vitellogenin exists in neither the VC nor yolk but specifically on the surface of test cells or the intermembrane space between test cells and immature oocytes in the gonad. These results indicate that vitellogenin S1 and S2 are expressed and synthesized in immature oocytes, and probably also in test cells at the stage of immature oocytes, and are localized at the surface or intermembrane space between immature oocytes and test cells.

Fig. 12.1 (continued) were detected in the gonad, and stronger and longer signals were detected in hepatopancreas. (**d**) Putative gene model of vitellogenin that resulted from the search for the *H. roretzi* genome database. Exon and intron are represented by a *square* and a *bar*, respectively. Based on the gene model, S1 and S2 are predicted to be composed of ten exons and seven exons, respectively. (**e**) In situ hybridization of vitellogenin. An antisense probe of vitellogenin C-terminus was hybridized with sections of immature oocytes and hepatopancreas. Note that test cells, which are accessory cells of ascidian eggs located in the perivitelline space in mature eggs, are located in the periphery in immature oocytes. Sense probe was also hybridized with immature oocytes and hepatopancreas as a control. A 1-ng probe sample was hybridized in all specimens. (**f**) Immunohistochemisty of vitellogenin in immature oocytes. A specimen of the section of gonad eggs was treated with anti-vitellogenin C-terminus primary antiserum (*upper panels*), preimmune serum (*middle panels*), and anti-HrVC70 (*lower panels*). Vitellogenin was shown located on the surface of test cells. Anti-HrVC70 was used to distinguish the VC because HrVC70 is a major component of the VC. Test cell and vitelline coat are indicated as *TC* and *VC*, respectively. *Right* bright-field images, *left* fluorescent images in each panel

12.4   Localization of Vitellogenin in Mature Eggs

It is known that test cells move out from the egg periphery to the perivitelline space
during ovulation (Tucker 1942). To investigate the localization of vitellogenin in
terms of the movement of test cells, whole-mount immunohistochemistry was first
carried out in mature eggs (unfertilized eggs) and the fertilized eggs using antiserum
against vitellogenin C-terminus. The results showed that vitellogenin exists in the
VC of unfertilized eggs, but not in the VC of fertilized eggs (Fig. 12.2a). Then,
detailed observation on the localization of vitellogenin in the perivitelline space was
attempted, but nonspecific autofluorescence within the egg was too strong to observe
fluorescence in the perivitelline space (data not shown). Therefore, observation by
immunoelectron microscopy was performed using the fixed unfertilized and fertilized eggs. As a result, gold particles detecting vitellogenin were observed in vesicles in
the perivitelline space in both unfertilized and fertilized eggs, and the vesicles containing vitellogenin adhered to the VC of unfertilized eggs and dissociated from
the VC after fertilization (Fig. 12.2b). It is suggested that vitellogenin appears to
be incorporated into vesicles and secreted to the perivitelline space on ovulation or
oocyte maturation, most of which is located beneath the VC in unfertilized eggs.

12.5   Future Perspective

Our results shown here indicate that the novel isoform of vitellogenin is synthesized
in test cells as well as in immature oocytes and is released to the perivitelline
space, associating with vesicles. Although the biological roles of the novel type of
vitellogenin remain to be elucidated, it should be emphasized that the expression,
localization, and behavior of novel truncated vitellogenin isoforms during oocyte
maturation and fertilization are new findings in this study. Because vitellogenin S1
and S2 contain vWF-D and CT domains, which we previously identified as binding
proteins of the sperm proteases HrProacrosin and HrSpermosin, these novel vitellogenin species may play an important role in gamete interaction, in particular in the
process of sperm passage through the perivitelline space or gamete fusion, although
it remains to be determined whether these vitellogenin species are exposed to the
surface of the small vesicles. Further studies are necessary to elucidate the biological
roles of these novel vitellogenin species during ascidian fertilization.

Interestingly, neither of the novel vitellogenin species possesses a typical signal
sequence. Therefore, it could be presumed that these proteins synthesized in oocytes
and test cells are incorporated into vesicles and accumulated in the boundary
between immature oocytes and test cells. During oocyte maturation, these vesicles
may be secreted into the perivitelline space and mostly locate beneath the VC. The
detailed transport of vitellogenin from test cells to vesicles in mature eggs is still
unclear. Further observation of vitellogenin in the perivitelline space of mature eggs
is necessary.

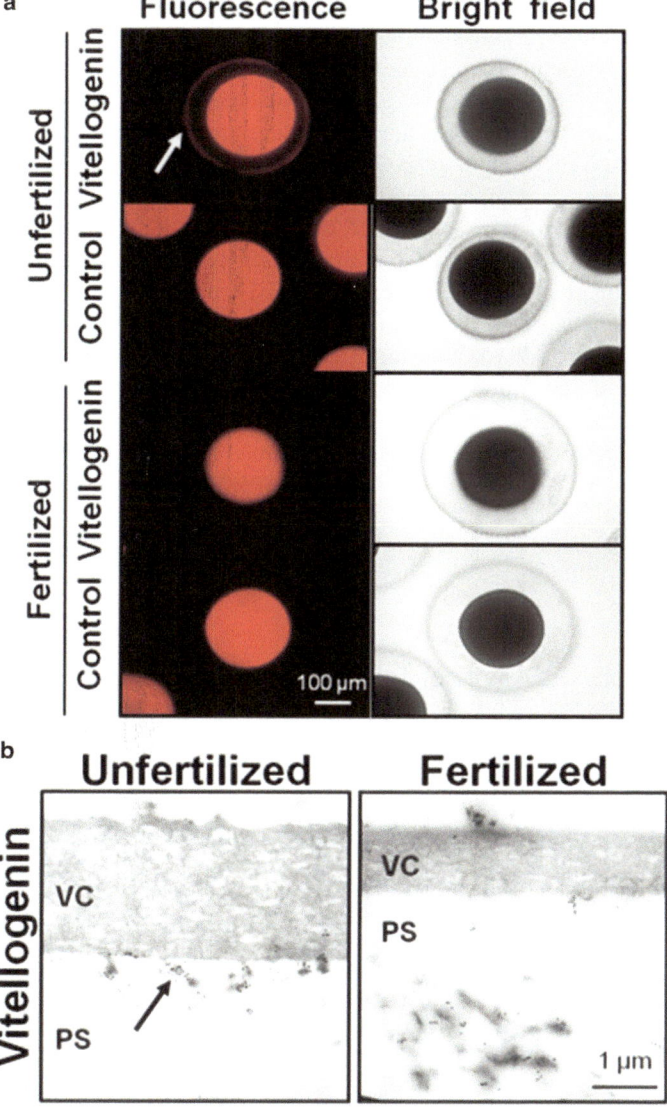

Fig. 12.2 (**a**) Fluorescence immunohistochemistry of vitellogenin C-terminus on unfertilized and fertilized eggs. Both unfertilized and fertilized eggs were treated with anti-vitellogenin C-terminus antiserum and control preimmune serum. Note that vitellogenin is present on the VC of unfertilized eggs (*arrow*) and disappears in fertilized eggs. *Right* bright-field images, *left* fluorescent images in each panel. (**b**) Immunoelectron micrographs of the perivitelline spaces of unfertilized and fertilized eggs. Unfertilized and fertilized eggs were sectioned and treated with anti-vitellogenin C-terminus antiserum and visualized by using a 10-nm gold particle-conjugated secondary antibody. In the unfertilized egg (*left*), gold particles are mainly observed in vesicles (*arrow*), which adhere to the inner side of the VC. After fertilization, vesicles with vitellogenin detach from the VC and localize in the perivitelline space. The VC becomes thinner after fertilization. Vitelline coat and perivitelline space are indicated as *VC* and *PS*, respectively

In connection with these vesicles, it is notable that CD9, which is a tetra-spanning membrane protein expressed in mouse oocytes that is essential for membrane fusion with sperm, was recently found to be secreted to the perivitelline space with the aid of small vesicles secreted from oocytes called "exosomes," and it was revealed that association of exosomes and sperm in the perivitelline space is essential for sperm–oocyte membrane fusion (Miyado et al. 2008). Moreover, the incorporation of vitellogenin into exosomes has been reported in the fruit fly and in humans (Brasset et al. 2006; Von Wald et al. 2010). These results led us to speculate that vitellogenin-associating vesicles may behave as an exosome in the mouse egg and play some important roles in gamete fusion or other fertilization processes. Further studies are necessary to elucidate the functions of the novel isoforms of vitellogenin in ascidian fertilization. Possible involvement of vitellogenin in gamete fusion should also be considered by analogy of the aforementioned recent discovery in CD9.

References

Akasaka M, Harada Y, Sawada H (2010) Vitellogenin C-terminal fragments participate in fertilization as egg-coat binding partners of sperm trypsin-like proteases in the ascidian *Halocynthia roretzi*. Biochem Biophys Res Commun 392(4):479–484

Akasaka M, Kato KH, Kitajima K, Sawada H (2013) Identification of novel isoforms of vitellogenin expressed in ascidian eggs. J Exp Zool B Mol Dev Evol 320(2):118–128

Baba T, Azuma S, Kashiwabara S, Toyoda Y (1994) Sperm from mice carrying a targeted mutation of the acrosin gene can penetrate the oocyte zona pellucida and effect fertilization. J Biol Chem 269(59):31845–31849

Brasset E, Taddei A, Arnaud F, Faye B, Fausto A, Mazzini M, Giorgi F, Vaury C (2006) Viral particles of the endogenous retrovirus ZAM from *Drosophila melanogaster* use a pre-existing endosome/exosome pathway for transfer to the oocyte. Retrovirology 3:26

Finn RN (2007) Vertebrate yolk complexes and the functional implications of phosvitins and other subdomains in vitellogenins. Biol Reprod 76(6):926–935

Hoshi M, Numakunai T, Sawada H (1981) Evidence for participation of sperm proteinases in fertilization of the solitary ascidian, *Halocynthia roretzi*: effects of protease inhibitors. Dev Biol 86(1):117–121

Howes L, Jones R (2002) Interactions between zona pellucida glycoproteins and sperm proacrosin/acrosin during fertilization. J Reprod Immunol 53(1-2):181–192

Howes E, Pascall JC, Engel W, Jones R (2001) Interactions between mouse ZP2 glycoprotein and proacrosin; a mechanism for secondary binding of sperm to the zona pellucida during fertilization. J Cell Sci 114(Pt 22):4127–4136

Kodama E, Baba T, Yokosawa H, Sawada H (2001) cDNA cloning and functional analysis of ascidian sperm proacrosin. J Biol Chem 276(27):24594–24600

Kodama E, Baba T, Kohno N, Satoh S, Yokosawa H, Sawada H (2002) Spermosin, a trypsin-like protease from ascidian sperm: cDNA cloning, protein structures and functional analysis. Eur J Biochem 269(2):657–663

Lambert CC, Someno T, Sawada H (2002) Sperm surface proteases in ascidian fertilization. J Exp Zool 292(1):88–95

Mayadas TN, Wagner DD (1992) Vicinal cysteines in the prosequence play a role in von Willebrand factor multimer assembly. Proc Natl Acad Sci USA 89(8):3531–3535

McRorie RA, Williams WL (1974) Biochemistry of mammalian fertilization. Annu Rev Biochem 43:777–803

Miyado K, Yoshida K, Yamagata K, Sakakibara K, Okabe M, Wang X, Miyamoto K, Akutsu H, Kondo T, Takahashi Y, Ban T, Ito C, Toshimori K, Nakamura A, Ito M, Miyado M, Mekada E, Umezawa A (2008) The fusing ability of sperm is bestowed by CD9-containing vesicles released from eggs in mice. Proc Natl Acad Sci USA 105(35):12921–12926

Mouchel N, Trichet V, Betz A, Le Pennec JP, Wolff J (1996) Characterization of vitellogenin from rainbow trout (*Oncorhynchus mykiss*). Gene (Amst) 174(1):59–64

Müller-Esterl W, Fritz H (1981) Sperm acrosin. Methods Enzymol 80(pt C):621–632

Sawada H (2002) Ascidian sperm lysin system. Zool Sci 19(2):139–151

Sawada H, Someno T (1996) Substrate specificity of ascidian sperm trypsin-like proteases, spermosin and acrosin. Mol Reprod Dev 45(2):240–243

Sawada H, Yokosawa H, Ishii S (1984a) Purification and characterization of two types of trypsin-like enzymes from sperm of the ascidian (Prochordata) *Halocynthia roretzi*. Evidence for the presence of spermosin, a novel acrosin-like enzyme. J Biol Chem 259(5):2900–2904

Sawada H, Yokosawa H, Someno T, Saino T, Ishii S (1984b) Evidence for the participation of two sperm proteases, spermosin and acrosin, in fertilization of the ascidian, *Halocynthia roretzi*: inhibitory effects of leupeptin analogs on enzyme activities and fertilization. Dev Biol 105(1):246–249

Sawada H, Iwasaki K, Kihara-Negishi F, Ariga H, Yokosawa H (1996) Localization, expression, and the role in fertilization of spermosin, an ascidian sperm trypsin-like protease. Biochem Biophys Res Commun 222(2):499–504

Sawada H, Sakai N, Abe Y, Tanaka E, Takahashi Y, Fujino J, Kodama E, Takizawa S, Yokosawa H (2002) Extracellular ubiquitination and proteasome-mediated degradation of the ascidian sperm receptor. Proc Natl Acad Sci USA 99(3):1223–1228

Tucker GH (1942) The histology of the gonads and development of the egg envelopes of an ascidian (*Styela plicata* Lesueur). J Morphol 70(1):81–113

Urch UA, Wardrip NJ, Hedrick JL (1985a) Limited and specific proteolysis of the zona pellucida by acrosin. J Exp Zool 233(3):479–483

Urch UA, Wardrip NJ, Hedrick JL (1985b) Proteolysis of the zona pellucida by acrosin: the nature of the hydrolysis products. J Exp Zool 236(2):239–243

Von Wald T, Monisova Y, Hacker M, Yoo S, Penzias A, Reindollar R, Usheva A (2010) Age-related variations in follicular apolipoproteins may influence human oocyte maturation and fertility potential. Fertil Steril 93(7):2354–2361

Wallace RA, Selman K (1985) Major protein changes during vitellogenesis and maturation of *Fundulus* oocytes. Dev Biol 110(2):492–498

Wallace RA, Selman K (1990) Ultrastructural aspects of oogenesis and oocyte growth in fish and amphibians. J Electron Microsc Tech 16(3):175–201

Wassarman PM (1987) Early events in mammalian fertilization. Annu Rev Cell Biol 3:109–142

Yamagata K, Murayama K, Okabe M, Toshimori K, Nakanishi T, Kashiwabara S, Baba T (1998) Acrosin accelerates the dispersal of sperm acrosomal proteins during acrosome reaction. J Biol Chem 273(17):10470–10474

Chapter 13
Actin Cytoskeleton and Fertilization in Starfish Eggs

Luigia Santella, Nunzia Limatola, and Jong Tai Chun

Abstract Starfish oocytes provide optimal opportunities to study meiotic progression and fertilization in vitro. A large and synchronized population of oocytes in the gonad can be induced to undergo maturation by addition of the hormone 1-methyladenine (1-MA). Successful monospermic fertilization is normally achieved when the eggs are in the interval between breakdown of the large nucleus (germinal vesicle) and extrusion of the first polar body. Insemination outside this temporal window frame gives rise to polyspermy. Although immature oocytes may become polyspermic because they are unable to form the fertilization envelope, which has been intuitively believed to serve as a mechanical block to polyspermy, overly mature eggs usually incorporate supernumerary sperm despite full elevation of the fertilization envelope. During the course of meiotic progression of the oocytes and of egg activation at fertilization, the cortical actin cytoskeleton undergoes dynamic changes. In this review, we discuss the role of the actin cytoskeleton during the meiotic maturation and at fertilization, focusing on its modulatory effects on intracellular Ca^{2+} signaling, cortical granule exocytosis, and sperm incorporation.

Keywords Actin • Ca^{2+} signaling • Ectoplasm • Polyspermy • Sea urchin • Starfish

13.1 Introduction

Oocytes of marine animals are an exceptional model system for the study of fertilization and embryonic development. Although ovulation in mammals only releases a few oocytes, starfish and sea urchin provide many eggs that are highly accessible. Fertilization and subsequent development of echinoderms can be easily observed with

L. Santella (✉) • N. Limatola • J.T. Chun
Laboratory of Cellular and Developmental Biology, Stazione Zoologica Anton Dohrn,
Villa Comunale 1, 80121 Naples, Italy
e-mail: santella@szn.it

H. Sawada et al. (eds.), *Sexual Reproduction in Animals and Plants*,
DOI 10.1007/978-4-431-54589-7_13, © The Author(s) 2014

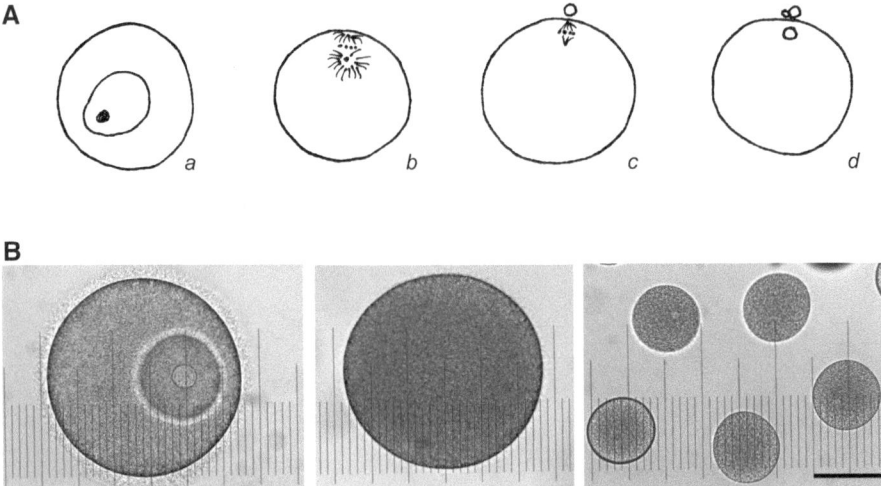

Fig. 13.1 Maturation stage and fertilization of starfish and sea urchin eggs. (**A**) Schematic diagrams of meiotic maturation stages of starfish oocytes: prophase I with large nucleus (4n) and nucleolus (*a*), metaphase I (*b*), metaphase II with the first polar body (2n) extruded (*c*), and completion of meiosis with female pronucleus (n) (*d*). Starfish and sea urchin eggs are normally fertilized at stages *b* and *d*, respectively. (Modified from Just 1939). (**B**) Micrograph of starfish and sea urchin eggs on the same scale. *Bar* 100 μm. *Left*, immature oocyte of *Astropecten aranciacus* at the germinal vesicle (GV) stage; *middle*, the same oocyte after germinal vesicle breakdown (GVDB); *right*, postmeiotic eggs of sea urchin (*Paracentrotus lividus*)

minimal experimental requirement, as they reproduce by "external fertilization" in seawater. At the peak of the breeding season, sea urchin ovaries are filled with eggs that have completed meiosis and contain haploid DNA (Vacquier 2011). Thus, a substantial population (<20 %) of the oocytes at the GV (germinal vesicle; i.e., the nucleus) stage can be obtained only before the peak of breeding season. At variance with the sea urchin, starfish oocytes in the female gonad at the peak of the breeding season are fully grown, yet they are arrested at prophase I of meiosis. Only upon hormonal stimulation (methyladenine, 1-MA) from the adhering follicle cells, the oocytes re-enter the meiotic cycle and are spawned into seawater (Kanatani 1973; Meijer and Guerrier 1984). Thus, even if belonging to the same phylum, sea urchin and starfish display a fundamental difference in the way in which their eggs are fertilized. Sea urchin eggs are fertilized after the completion of meiosis (haploid stage, n), but starfish eggs are normally fertilized before the completion of the first meiotic division (Fig. 13.1A). In starfish, the period between germinal vesicle breakdown (GVBD) and the extrusion of the first polar body (2n) represents the optimal period for successful fertilization. Indeed, insemination of the starfish oocytes before or after this period results in a high rate of polyspermic fertilization. As an experimental model, starfish provides some advantages over sea urchin eggs in that (a) meiotic maturation can be induced in vitro, and (b) the large cell dimensions (Fig. 13.1B) facilitate microinjection and fluorescence imaging (Santella et al. 2008, 2012; Santella and Chun 2011). Starfish oocytes thus provide an optimal opportunity to examine the cytoplasmic changes that occur during meiotic maturation and fertilization.

13.2 Cytoplasmic Changes During Meiotic Maturation of Oocytes

13.2.1 Morphological Transition

During meiotic maturation, oocytes display changes in the nucleus, where chromosome reshuffling and dissolution of the nuclear membranes take place, but also morphological alteration of their surface. Before and after the maturation process, a striking structural reorganization of the cortex occurs in sea urchin eggs (Runnström 1963; Franklin 1963; Lönning 1967; Longo 1978). Transmission electron microscopy (TEM) has revealed that numerous long microvilli that are present on the surface of immature oocytes nearly disappear in mature eggs (Dale and Santella 1985). These microvilli are filled with actin filaments (Tilney and Jaffe 1980). The cortex of starfish oocytes also undergoes mechanical and morphological changes during meiotic maturation. The microvilli become shorter, and the cortical granules are translocated and positioned perpendicularly to the plasma membrane (Hirai and Shida 1979; Longo et al. 1995; Santella et al. 1999). Change of the actin cytoskeleton is often shown by the transient formation of a F-actin spike a few minutes after 1-MA addition (Schroeder 1981; Schroeder and Stricker 1983), whereas the rigidity of the cell and its F-actin content decrease at the time of GVDB (Shôji et al. 1978; Heil-Chapdelaine and Otto 1996). Thus, it appears that the actin cytoskeleton is regulated differently, presumably playing different roles, in different subcellular domains. Because the cell nucleus is enriched with actin molecules that normally do not form filaments (Clark and Merriam 1977), the role of the "spilling" actin at the time of GVDB, when the nucleoplasm inevitably intermixes with cytoplasm, is of obvious interest.

13.2.2 Signaling Pathways to Meiotic Maturation

The starfish oocyte maturation hormone (1-MA) operates through a receptor on the plasma membrane because microinjected 1-MA does not induce oocyte maturation (Kanatani and Hiramoto 1970; Shida and Hirai 1978). The identity of this receptor has not been determined as yet, but it is known that its activation leads to the release of G$\beta\gamma$ proteins (Jaffe et al. 1993; Chiba et al. 1995), which in turn activate a phosphatidylinositol-3-kinase (PI3K)-producing phosphatidylinositol-3,4,5-trisphosphate (PIP3) from phosphatidylinositol-4,5-bisphoshate (PIP2).

An important target of the PI3K signaling pathway is the Akt/PKB kinase, which directly downregulates Myt1 and upregulates Cdc25, leading to the activation of the universal cell-cycle regulator, the cyclin B-Cdc2 complex (Masui 2001; Prigent and Hunt 2004; Kishimoto 2011).

13.2.3 Intracellular Ca^{2+} Increase During Meiotic Maturation

Artificial elevation of intracellular Ca^{2+} level can induce nuclear maturation of starfish oocytes, suggesting that Ca^{2+} is the functional downstream effector of 1-MA (Moreau et al. 1978). Nonetheless, the maturation of starfish oocytes can proceed without detectable Ca^{2+} transients or in the presence of Ca^{2+} chelators (Witchel and Steinhardt 1990), suggesting that the breakdown of the nuclear envelope and meiotic resumption may not be under the control of Ca^{2+}. However, direct microinjection of the Ca^{2+} chelator BAPTA into the GV blocked the disassembly of the nuclear envelope (Santella and Kyozuka 1994; Santella et al. 1998), suggesting that Ca^{2+} signal in the GV subcellular domain may be important. A more recent investigation has shown that the 1-MA-triggered Ca^{2+} wave always initiates in the vegetal hemisphere of the starfish oocyte and propagates to the animal pole through the superficial domains of the cortex, raising the question of the involvement of the Ca^{2+}-releasing pathway linked to the 1,4,5-inositol trisphosphate ($InsP_3$) receptors, as the latter are known to be more sensitive in the animal pole of the egg (Lim et al. 2003; Kyozuka et al. 2008). Because 1-MA can induce the Ca^{2+} response in the absence of external Ca^{2+}, the involvement of nicotinic acid adenine dinucleotide phosphate (NAADP) in the process also becomes questionable, as NAADP is known to act mainly by inducing Ca^{2+} influx in starfish oocytes.

13.2.4 Sensitization of the Ca^{2+}-Releasing Mechanisms

Starfish oocytes undergoing meiotic maturation become progressively more sensitive to $InsP_3$. Thus, mature eggs respond to the same dose of $InsP_3$ with a much higher level of Ca^{2+} release (Chiba et al. 1990; Lim et al. 2003). This response might be partly explained by the observation that the endoplasmic reticulum (ER), which constitutes the major intracellular Ca^{2+} store of the oocyte, undergoes extensive restructuring during meiotic maturation (Stricker 2006). The morphological changes of the ER parallel the alteration of the actin cytoskeleton, which readily responds to 1-MA and forms thick bundles that align perpendicularly to the egg surface 50 min after the addition of the activating hormone (Santella et al. 2008; Santella and Chun 2011). Because the sensitization of the $InsP_3$-dependent Ca^{2+}-releasing mechanism induced by 1-MA is blocked by agents promoting actin depolymerization, that is, latrunculin-A (LAT-A), it appears that changes of the actin cytoskeleton have a role in optimizing the Ca^{2+}-releasing mechanism in maturing oocytes (Lim et al. 2002). The actin cytoskeleton may contribute to the process, either by assisting the mobilization of Ca^{2+} from the ER or by changing the microdomain environment of the ion channels. We have provided support for this idea by demonstrating that the magnitude, kinetics, or onset of Ca^{2+} signaling in response to the fertilizing sperm, or to Ca^{2+}-linked second messengers such as $InsP_3$ and cyclic ADP ribose (cADPr), are significantly modified by interfering with the dynamic changes of the actin cytoskeleton: that is, with the actin-binding protein cofilin (Nusco et al. 2006), LAT-A, with Jasplakinolide, heparin (Puppo et al. 2008), or the PIP2-sequestering domain of PLC-δ1 (Chun et al. 2010). Interestingly, the exposure of the mature eggs of some starfish species, such as *Astropecten aranciacus*, to these treatments triggers an auto-catalyzing Ca^{2+} wave in the absence of fertilizing sperms or of Ca^{2+}-mobilizing second messengers. The Ca^{2+} wave induced by LAT-A originates largely from the intracellular stores, and its spatiotemporal pattern in Ca^{2+}-free seawater is highly reminiscent of the fertilization Ca^{2+} wave induced by sperm, although it requires a latent period of 6–10 min (Lim et al. 2002). This effect was not observed in the GV-stage oocytes, suggesting that the threshold for the triggering of the spontaneous Ca^{2+} release is much lower in mature eggs, and that the trigger itself is dependent on the actin cytoskeleton. Alternatively, Ca^{2+} may be released during the process of actin treadmilling (Lange and Gartzke 2006), and it may simply reflect a slower turnover rate of the actin filaments in immature oocytes. In other words, F-actin may contribute to intracellular Ca^{2+} homeostasis and ion flux either by modulating the ion channel activities, or by a more direct mechanism in which the polymerization and depolymerization cycle of actin filaments in the local environment may sequester and release cytosolic Ca^{2+} (Lange and Gartzke 2006; Chun and Santella 2009).

13.2.5 Changes of the Electrical Property of the Plasma Membrane During Meiotic Maturation

The aforementioned changes of cytoplasm are accompanied by decrease of oocyte stiffness within 7–8 min after 1-MA addition and by the decrease in K^+ conductance that leads to depolarization of the membrane potential (Miyazaki et al. 1975; Miyazaki and Hirai 1979). An abrupt switching of the membrane potential from the initial level of −70 to −90 mV to a new stable state of −10 to −20 mV takes place concurrently with, or shortly before, GVBD in the maturing oocytes of *A. aranciacus* (Dale et al. 1979). Thus, the resting potential of mature eggs is far less negative than that of the immature oocytes. However, the membrane potential can be restored to the original level of the GV-stage oocytes as the eggs become overripe in seawater (Miyazaki et al. 1975; Miyazaki and Hirai 1979).

13.3   Signals of Fertilization and Egg Activation

13.3.1   Generation and Propagation of the Intracellular Ca²⁺ Wave

Mature eggs are metabolically repressed, but interaction with the fertilizing sperm triggers a series of events that are collectively termed "egg activation." The two earliest detectable changes at fertilization are rapid depolarization of the egg plasma membrane and Ca^{2+} influx, which are followed by the intracellular Ca^{2+} wave propagating from the site of successful sperm–egg interaction (Steinhardt et al. 1971; Shen and Buck 1993). It has been shown that intracellular Ca^{2+} increase at fertilization can be recapitulated by the combined effect of Ca^{2+}-mobilizing second messengers, namely, $InsP_3$, NAADP, and cADPr. $InsP_3$ is generated by the hydrolysis of phosphatidylinositol-4,5-bisphosphate (PIP2) in the plasma membrane by phospholipase C. Several isozymes that catalyze this process have been identified, and fertilization is thought to activate at least one of them. In analogy to the $InsP_3$ receptor (Streb et al. 1983; Furuichi et al. 1989), cADPr may bind to the cADPr-sensitive ryanodine receptor complex and release Ca^{2+} from the internal stores (Carafoli et al. 2001; Lee 2002; Santella and Chun 2013). The Ca^{2+}-releasing activities of both $InsP_3$ and ryanodine receptors are facilitated by Ca^{2+} itself through a "Ca^{2+}-induced Ca^{2+} release" (CICR) mechanism, whereas the activity of the channels responding to NAADP does not display direct CICR properties and may involve a distinct intracellular (or extracellular) Ca^{2+} store (Santella et al. 2000; Moccia et al. 2003; Galione and Churchill 2002). At variance with the findings in the sea urchin eggs where NAADP releases Ca^{2+} from lysosome-like acidic stores (Churchill et al. 2002), the Ca^{2+} increase induced by microinjected NAADP in starfish eggs appear to require external Ca^{2+} and is mediated by the activation of ion channels on the plasma membrane (Runft et al. 2002; Santella et al. 2004). Thus, these three second messengers may contribute to distinct aspects of Ca^{2+} signaling in fertilized eggs of starfish. NAADP may trigger an initial Ca^{2+} increase in the egg cortex whereas $InsP_3$ contributes to egg-wide propagation of the Ca^{2+} waves (Lim et al. 2001; Santella et al. 2004; Santella and Chun 2013).

13.3.2   Morphological Changes of the Egg Cortex During Fertilization

Within a few minutes after fertilization, the actin bundles in the subplasmalemmal zones begin to be translocated centripetally in concert with the internalization of the sperm (Terasaki 1996; Puppo et al. 2008; Vasilev et al. 2012). In addition, the sperm that makes the initial contact with the egg induces the formation of the fertilization cone which is filled with actin filaments (Tilney and Jaffe 1980). Microinjection of starfish eggs with heparin not only inhibits Ca^{2+} signaling but also induces hyperpolymerization of subplasmalemmal actin (Puppo et al. 2008).

These eggs with perturbed actin cytoskeleton exhibit multiple fertilization cones but fail to form the functional tethering that leads to sperm incorporation. Similarly, eggs whose actin meshwork at the plasma membrane is perturbed by ionomycin fail to display the centripetal movement of the F-actin fibers at fertilization and often fail to incorporate sperm (Vasilev et al. 2012).

13.3.3 Changes of the Electrical Property of the Plasma Membrane at Fertilization during Meiotic Maturation

Fertilization also induces rapid changes of the membrane potentials in eggs. After successful binding of sperm, sea urchin eggs either fire a rapid action potential to a positive value of about +20 mV or display a step depolarization (DeFelice and Dale 1979); this is followed by a slower depolarization that initiates simultaneously with the cortical granule exocytosis and elevation of the fertilization envelope. The membrane potential of the fertilized sea urchin eggs remains at positive values for a few minutes but gradually returns to the negative value.

13.4 Block to Polyspermy

Although attracting multiple sperm, the eggs normally incorporate a single sperm in vitro. The elevation of the thick fertilization envelope in fertilized eggs of echinoderm led to the idea that such a structure may serve as a mechanical block to polyspermy. However, this process is rather slow; thus, the block to polyspermy may require an additional faster mechanism. A very rapid event that occurs before elevation of the vitelline envelope in fertilized eggs is of special interest. Just (1939) was among the first to describe the possibility of the fast block to polyspermy. He stated, "one can follow the *wave in the ectoplasm* which begins at the point of sperm-entry and sweeps over the egg," and emphasized that "with membrane-separation, the eggs undergo some change; and it is *this change*—not its result, membrane separation— which constitutes the block to the entrance of additional spermatozoa." He then added that "this block, which is more subtle than the mechanical obstacle interposed by the presence of a separated membrane, is established *before* membrane-separation occurs." Indeed, in one of the first experiments to study the nature of the block to polyspermy, Rothschild and Swann (1951, 1952) demonstrated that sea urchin eggs establish monospermic fertilization as early as 2 s after insemination, suggesting that a fast mechanism must be established on the egg surface to reduce the probability of successful refertilization.

As to the physicochemical event underlying the fast block to polyspermy, several hypotheses have been set forth, including those based on the early increase of hydrogen peroxide and protease that would inhibit the binding of supernumerary sperm (Vacquier et al. 1972). A hypothesis that drew special attention proposed that the fast partial block to polyspermy is electrically mediated by the rapidly changing membrane potential of the egg at fertilization, which renders the egg unreceptive to supernumerary sperm (Jaffe 1976). In support of the view that the establishment of monospermy is linked to the amplitude of the membrane depolarization at fertilization, sea urchin eggs artificially arrested at positive potentials were shown to be prevented from being fertilized by sperm. The "electrical block" hypothesis has been somewhat controversial (Dale and Monroy 1981; Dale and DeFelice 2010). Its main shortcoming, admitted also by its supporters, is that the electrical block is not absolute and can be overridden by high concentration of sperm. Thus, it has been generally agreed that insemination with excess sperm causes polyspermic fertilization in the eggs of many marine animals. However, Just (1939) had already contradicted this idea by showing that eggs of *Arbacia* were never polyspermic even "if a thick sperm-suspension be added to the eggs as early as one second after the first insemination." He added, "I have also made the initial insemination with the heaviest sperm-concentration procurable, i.e., 'dry' sperm as it exudes from the male, and have obtained only mono-spermic fertilization." He showed that polyspermy in *Arbacia* may occur when the ectoplasm of the fertilizable eggs was injured and thus slowed in reacting to the stimulus of the first sperm. Such injury was very apparent in the eggs with extremely heavy insemination. This observation and the failure of positively held membrane potentials to preclude fertilization of the eggs by excess sperm (Jaffe 1976) imply that an additional factor besides membrane potential may ensure the integrity of eggs and thus guide monospermy. Dale (1985) reported that the sperm receptivity of sea urchin oocytes and eggs was independent of their resting potentials when their jelly layers were kept intact, suggesting that the properties of the egg surface may confer fine regulation of monospermic fertilization.

13.5   Meiotic Stages of Oocytes and Polyspermy

The idea that polyspermy is linked to the regional anatomy of egg surface and cortex can be examined in starfish oocytes at different meiotic maturation stages. As mentioned, the morphology of the starfish oocyte changes with meiotic progression, as judged by the structure of microvilli and cortical actin cytoskeleton. Meiotically immature oocytes are prone to polyspermy upon insemination and fail to elevate the fertilization envelope (Fig. 13.2). They lack an effective fast or slow block to polyspermy (Santella et al. 2012). The progress of cytoplasmic maturation in sea urchin oocytes is also accompanied by structural modifications of the surface layer in a direction that favors monospermic fertilization (Runnström and Monné 1945; Runnström 1963). However, most starfish eggs overmatured beyond the extrusion of second polar bodies display polyspermy at fertilization despite their ability to elevate a normal fertilization envelope (Fujimori and Hirai 1979). The explanation for the polyspermy of these overmature eggs might be twofold. Firstly, the tendency to produce polyspermy may be linked to the electrical properties of the plasma membrane. For instance, the resting potential of the mature eggs of *A. pectinifera* at

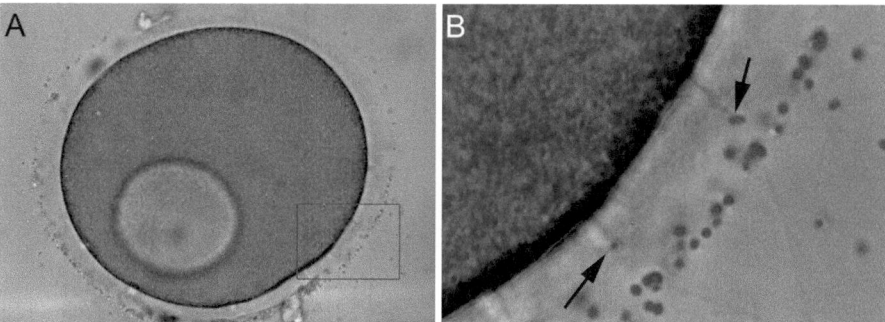

Fig. 13.2 Fertilization of the immature oocytes of *Astropecten aranciacus* at the GV stage. (**A**) Micrograph of fertilized oocyte 5 min after insemination. Fertilization at this stage leads to formation of multiple fertilization cones, which are indicative of polyspermy. Because of the lack of cortical granules exocytosis, there is no formation of the fertilization envelope. (**B**) Enlarged view of the region of interest marked with a *blue box* in **A**. Fertilizing sperm were stained with Hoechst 33342; sperm making successful contacts are marked with *arrows*

the optimum period is much less negative (-30 mV) than those of immature (-70 mV) and overmature eggs (-55 mV) because the membrane permeability to K$^+$ decreases during GVBD but is restored during overmaturation (Miyazaki et al. 1975). The peak of the activation potential in the overmature eggs at fertilization (-4 mV) is also not as high as in the mature eggs at the optimum period ($+12$ mV), rendering them less effective in preventing the entry of supernumerary sperm in the light of the "electrical block" hypothesis (Miyazaki and Hirai 1979; Dale et al. 1979; Moody and Bosma 1985). Second, the increased tendency to polyspermy outside the aforementioned "optimum period" might be caused by the lack of the natural physiological propensity of the egg cytoplasm. Classical studies had indicated that the dispersal of nuclear contents into the cytoplasm and cortex is a prerequisite for cytoplasmic maturation (Delage 1901; Chambers 1921). After the GVDB, a signal or a diffusible substance from the GV may interact with the egg cortex to induce the structural modifications that are essential for egg activation and monospermic sperm incorporation. Indeed, at variance with previous findings (Hirai et al. 1971), the InsP$_3$-dependent Ca^{2+} response was much slower, with no cortical granules exocytosis in the mature eggs of *A. pectinifera* whose GV had been removed before the addition of 1-MA, whereas Ca^{2+} signaling and elevation of the vitelline layer in response to NAADP and fertilizing sperm were not much affected by enucleation (Lim et al. 2001). On the other hand, eggs matured cortically with low doses of 1-MA displayed an increased rate of polyspermy (Hirohashi et al. 2008). These eggs underwent no GVDB but displayed full elevation of the fertilization envelope. Thus, it is conceivable that either underexposure or overexposure to the nucleus-borne signals during maturation and overmaturation leads to subtle differences in the structure and function of the egg *ectoplasm*, which may disarray the eggs toward polyspermy.

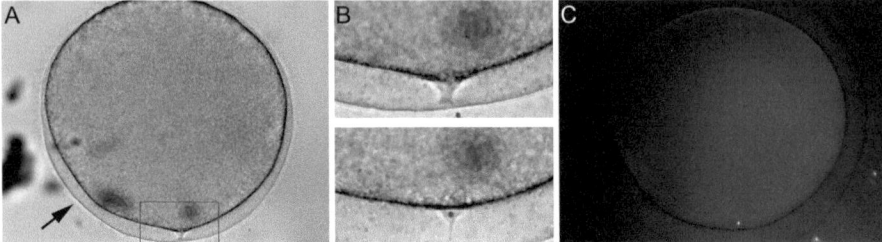

Fig. 13.3 Fertilization of *Astropecten aranciacus* eggs during optimal period. (**A**) Micrograph of fertilized egg 4 min after insemination. Fertilization at interval between GVDB and extrusion of the first polar body leads to monospermy with formation of one fertilization cone and coordinated elevation of the fertilization. (**B**) Enlarged views of the sperm entry site. A Hoechst 33342-stained sperm is tethered to the fertilization cone (*top panel*) before being engulfed (*bottom panel*) 3 min later. (**C**) Visualization of the single Hoechst 33342-stained sperm incorporated into the egg (10 min after insemination)

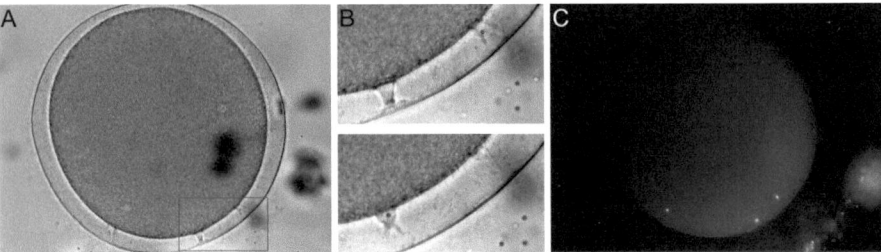

Fig. 13.4 Fertilization of overmature eggs of *Astropecten aranciacus*. Eggs were fertilized 4 h after 1-methyladenine (1-MA) treatment. (**A**) Micrograph of fertilized egg 4 min after insemination. Although the fertilization envelopes are fully elevated, multiple fertilization cones are formed in most cases. (**B**) Enlarged view of sperm entry site. Hoechst 33342-stained sperm (*top panel*) are engulfed, as the body of the fertilization cone is detached from the elevating fertilization envelopes 1 min later (*bottom panel*). (**C**) Visualization of the three Hoechst 33342-stained sperm inside the fertilized egg (10 min after insemination)

The translocation of the cortical granules and vesicles that takes place during the meiotic maturation of oocytes may contribute to shielding the eggs from the attack of supernumerary sperm (Longo et al. 1995). At fertilization, these granules and vesicles undergo exocytosis and their extrusion gives rise to full elevation of fertilization envelope (Fig. 13.3). However, the finding that polyspermy can take place even with full elevation of the fertilization envelope in overmature eggs (Fig. 13.4) indicates that the mechanical block by the fertilization envelope is no guarantee for monospermic fertilization. The fast mechanism preventing the entry of supernumerary sperm may have become disrupted in these eggs. As mentioned, this could be caused by the altered electrical property of the plasma membrane, or may be attributed to the overall failure of the cytoplasmic mechanism in the egg surface that controls sperm interaction and incorporation.

13.6   Role of the Actin Cytoskeleton

One of the most evident morphological consequences of meiotic maturation and overmaturation in starfish eggs is the change of the actin cytoskeleton. Although the signaling pathways leading to it are beyond the scope of this review, the relevance of the actin cytoskeleton to the control of sperm interaction and incorporation may be worth mentioning. Egg microvilli may be involved in deciding sperm receptivity, and their number on the surface of aged mammalian oocytes is severely reduced (Santella et al. 1985). In starfish, a decline in the proper functional interactions between sperm and aged eggs has been suggested (Chambers 1921; Just 1939). Our recent studies on starfish eggs have indicated that the alteration of the structural organization in subplasmalemmal actin filaments renders the egg surface less reactive to the first sperm and more receptive to supernumerary sperm (Puppo et al. 2008; Chun et al. 2010; Vasilev et al. 2012). Eggs with hyperpolymerization of the subplasmalemmal actin displayed the formation of numerous irregularly shaped fertilization cones (thus, polyspermic), yet also displayed unfunctional tethering of the sperm, suggesting a more direct role of actin in sperm interaction and incorporation (Puppo et al. 2008). As mentioned, actin has a modulatory role on the intracellular Ca^{2+} release and influx (Lim et al. 2002; Nusco et al. 2006; Kyozuka et al. 2008; Chun et al. 2010), but it must be emphasized that actin also is a critical factor in controlling exocytosis of the cortical granules. Despite the early claim that intracellular Ca^{2+} elevation is solely important for exocytosis (Steinhardt and Epel 1974; Vacquier 1975; Whitaker and Baker 1983), we have shown that, when the actin cytoskeleton in the cortex is perturbed, intracellular Ca^{2+} increase does not induce discharge of the cortical granules (Kyozuka et al. 2008; Puppo et al. 2008; Chun et al. 2010).

13.7   Concluding Remarks

Although belonging to the same phylum, sea urchin and starfish have evolved to use a fundamentally different temporal scheme at fertilization. Although fertilization of sea urchin awaits the completion of the egg meiosis, natural fertilization of starfish anticipates the meiotic program. To ensure monospermy, the starfish eggs must be fertilized when they are at the incomplete stage of meiosis; that is, during the interval between the GVDB and the extrusion of the first polar body. The reason why starfish evolved to adopt this seemingly premature fertilization scheme has not been clarified, but it may have been inevitable because the same hormone 1-MA not only stimulates oocyte maturation but also promotes spawning of mature gametes (Meijer and Guerrier 1984). Anyhow, the capability of controlling meiotic maturation with 1-MA in vitro makes starfish an excellent model system in the study of the relationship among the cellular events that take place during meiotic maturation, fertilization, and egg activation: that is, intracellular Ca^{2+} signaling, dynamic changes of the actin cytoskeleton, and the modulation of ion channel activities.

The rapid rearrangement of the actin cytoskeleton during meiotic maturation and fertilization may be a well-calculated reconfiguration of the cell structure that is linked to the fine regulation of Ca^{2+} signaling, vesicle exocytosis, and sperm incorporation. On the other hand, although the fertilization envelope of the echinoderm egg may provide mechanical protection of the early embryo, the other functions that have been proposed may require careful examination in future studies. The fully elevated fertilization membrane was not sufficient to ensure monospermic fertilization (Fig. 13.4) nor was it a chemical barrier to protect the zygote or early embryo (Vasilev et al. 2012). In view of the relative ease of applying imaging and electrophysiological methods following microinjection, starfish oocytes may provide opportunities to study the reciprocal regulation of cytoskeletal changes and ion channel activities.

References

Carafoli E, Santella L, Branca D, Brini M (2001) Generation, control, and processing of cellular calcium signals. Crit Rev Biochem Mol Biol 36(2):107–260

Chambers R (1921) Microdissection studies. III. Some problems in the maturation and fertilization of the echinoderm egg. Biol Bull 41(6):318–350

Chiba K, Kado RT, Jaffe LA (1990) Development of calcium release mechanisms during starfish oocyte maturation. Dev Biol 140(2):300–306

Chiba K, Longo FJ, Kontani K, Katada T, Hoshi M (1995) A periodic network of G protein beta gamma subunit coexisting with cytokeratin filament in starfish oocytes. Dev Biol 169(2): 415–420

Chun JT, Santella L (2009) Roles of the actin-binding proteins in intracellular Ca^{2+} signalling. Acta Physiol (Oxf) 195(1):61–70

Chun JT, Puppo A, Vasilev F, Gragnaniello G, Garante E, Santella L (2010) The biphasic increase of PIP2 in the fertilized eggs of starfish: new roles in actin polymerization and Ca^{2+} signaling. PLoS One 5(11):e14100

Churchill GC, Okada Y, Thomas JM, Genazzani AA, Patel S, Galione A (2002) NAADP mobilizes Ca^{2+} from reserve granules, lysosome-related organelles, in sea urchin eggs. Cell 111(5):703–708

Clark TG, Merriam RW (1977) Diffusible and bound actin nuclei of *Xenopus laevis* oocytes. Cell 12(4):883–891

Dale B (1985) Sperm receptivity in sea urchin oocytes and eggs. J Exp Biol 118:85–97

Dale B, DeFelice L (2010) Polyspermy prevention: facts and artifacts? J Assist Reprod Genet 28(3):199–207

Dale B, Monroy A (1981) How is polyspermy prevented? Gamete Res 4:151–169

Dale B, Santella L (1985) Sperm–oocyte interaction in the sea-urchin. J Cell Sci 74:153–167

Dale B, de Santis A, Hoshi M (1979) Membrane response to 1-methyladenine requires the presence of the nucleus. Nature (Lond) 282(5734):89–90

DeFelice LJ, Dale B (1979) Voltage response to fertilization and polyspermy in sea urchin eggs and oocytes. Dev Biol 72(2):327–341

Delage Y (1901) Etudes experimentales sur la maturation cytoplasmique et sur la parthenogenese artificielle chez les echinodermes. Arch Zool Exp Gen 9(3):284–336

Franklin L (1963) Morphology of gamete membrane fusion and of sperm entry into oocytes of the sea urchin. J Cell Biol 25(2):81–100

Fujimori T, Hirai S (1979) Differences in starfish oocyte susceptibility to polyspermy during the course of maturation. Biol Bull 157(2):249–257

Furuichi T, Yoshikawa S, Miyawaki A, Wada K, Maeda N, Mikoshiba K (1989) Primary structure and functional expression of the inositol 1,4,5-trisphosphate-binding protein P400. Nature (Lond) 342(6245):32–38

Galione A, Churchill GC (2002) Interactions between calcium release pathways: multiple messengers and multiple stores. Cell Calcium 32(5–6):343–354

Heil-Chapdelaine RA, Otto JJ (1996) Characterization of changes in F-actin during maturation of starfish oocytes. Dev Biol 177(1):204–216

Hirai S, Shida H (1979) Shortening of microvilli during the maturation of starfish oocyte from which vitelline coat was removed. Bull Mar Biol St Asamushi Tohoku Univ 16:161–167

Hirai S, Kubota J, Kanatani H (1971) Induction of cytoplasmic maturation by 1-methyladenine in starfish oocytes after removal of the germinal vesicle. Exp Cell Res 68(1):137–143

Hirohashi N, Harada K, Chiba K (2008) Hormone-induced cortical maturation ensures the slow block to polyspermy and does not couple with meiotic maturation in starfish. Dev Biol 318(1): 194–202

Jaffe LA (1976) Fast block to polyspermy in sea urchin eggs is electrically mediated. Nature (Lond) 261(5555):68–71

Jaffe LA, Gallo CJ, Lee RH, Ho YK, Jones TL (1993) Oocyte maturation in starfish is mediated by the beta gamma-subunit complex of a G-protein. J Cell Biol 121(4):775–783

Just EE (1939) The biology of the cell surface. P. Blakiston's son & Co., Inc. Philadelphia, USA

Kanatani H (1973) Maturation-inducing substance in starfishes. Int Rev Cytol 35(1):253–298

Kanatani H, Hiramoto Y (1970) Site of action of 1-methyladenine in inducing oocyte maturation in starfish. Exp Cell Res 61(2):280–284

Kishimoto T (2011) A primer on meiotic resumption in starfish oocytes: the proposed signaling pathway triggered by maturation-inducing hormone. Mol Reprod Dev 78(10–11):704–707

Kyozuka K, Chun JT, Puppo A, Gragnaniello G, Garante E, Santella L (2008) Actin cytoskeleton modulates calcium signaling during maturation of starfish oocytes. Dev Biol 320(2):426–435

Lange K, Gartzke J (2006) F-actin-based Ca signaling-a critical comparison with the current concept of Ca signaling. J Cell Physiol 209(2):270–287

Lee HC (ed) (2002) Cyclic ADP-ribose and NAADP. Structures, metabolism and functions. Kluwer Academic, Dordrecht

Lim D, Kyozuka K, Gragnaniello G, Carafoli E, Santella L (2001) NAADP$^+$ initiates the Ca^{2+} response during fertilization of starfish oocytes. FASEB J 15(12):2257–2267

Lim D, Lange K, Santella L (2002) Activation of oocytes by latrunculin A. FASEB J 16(9): 1050–1056

Lim D, Ercolano E, Kyozuka K, Nusco GA, Moccia F, Lange K, Santella L (2003) The M-phase-promoting factor modulates the sensitivity of the Ca^{2+} stores to inositol 1,4,5-trisphosphate via the actin cytoskeleton. J Biol Chem 278(43):42505–42514

Longo FJ (1978) Insemination of immature sea urchin (*Arbacia punctulata*) eggs. Dev Biol 62(2):271–291

Longo FJ, Woerner M, Chiba K, Hoshi M (1995) Cortical changes in starfish (*Asterina pectinifera*) oocytes during 1-methyladenine-induced maturation and fertilisation/activation. Zygote 3(3): 225–239

Lönning S (1967) Studies of the ultrastructure of sea urchin eggs and the changes induced at insemination. Sarsia 30(1):31–48

Masui Y (2001) From oocyte maturation to the in vitro cell cycle: the history of discoveries of maturation-promoting factor (MPF) and cytostatic factor (CSF). Differentiation 69(1):1–17

Meijer L, Guerrier P (1984) Maturation and fertilization in starfish oocytes. Int Rev Cytol 86(1): 129–196

Miyazaki S, Hirai S (1979) Fast polyspermy block and activation potential. Correlated changes during oocyte maturation of a starfish. Dev Biol 70(2):327–340

Miyazaki SI, Ohmori H, Sasaki S (1975) Potassium rectifications of the starfish oocyte membrane and their changes during oocyte maturation. J Physiol (Lond) 246(1):55–78

Moccia F, Lim D, Nusco GA, Ercolano E, Santella L (2003) NAADP activates a Ca^{2+} current that is dependent on F-actin cytoskeleton. FASEB J 17(13):1907–1909

Moody WJ, Bosma MM (1985) Hormone-induced loss of surface membrane during maturation of starfish oocytes: differential effects on potassium and calcium channels. Dev Biol 112(2): 396–404

Moreau M, Guerrier P, Doree M, Ashley CC (1978) Hormone-induced release of intracellular Ca^{2+} triggers meiosis in starfish oocytes. Nature (Lond) 272(5650):251–253

Nusco GA, Chun JT, Ercolano E, Lim D, Gragnaniello G, Kyozuka K, Santella L (2006) Modulation of calcium signalling by the actin-binding protein cofilin. Biochem Biophys Res Commun 348(1):109–114

Prigent C, Hunt T (2004) Oocyte maturation and cell cycle control: a farewell symposium for Pr. Marcel Dorée. Biol Cell 96(3):181–185

Puppo A, Chun JT, Gragnaniello G, Garante E, Santella L (2008) Alteration of the cortical actin cytoskeleton deregulates Ca^{2+} signaling, monospermic fertilization, and sperm entry. PLoS One 3(10):e3588

Rothschild L, Swann MM (1951) The fertilization reaction in the sea urchin. The probability of a successful sperm–egg collision. J Exp Biol 28(3):403–416

Rothschild L, Swann MM (1952) The fertilization reaction in the sea urchin. The block to polyspermy. J Exp Biol 29(3):469–483

Runft LL, Jaffe LA, Mehlmann LM (2002) Egg activation at fertilization: where it all begins. Dev Biol 245(2):237–254

Runnström J (1963) Sperm-induced protrusions in sea urchin oocytes: a study of phase separation and mixing in living cytoplasm. Dev Biol 7(1):38–50

Runnström J, Monné L (1945) On some properties of the surface layers of immature and mature sea urchin eggs, especially the changes accompanying nuclear and cytoplasmic maturation. Arkiv für Zoologie 36A(18):1–26

Santella L, Chun JT (2011) Actin, more than just a housekeeping protein at the scene of fertilization. Sci China Life Sci 54(8):733–743

Santella L, Chun JT (2013) Calcium signaling by cyclic ADP-ribose and NAADP. In: Lennarz WJ Lane MD (eds.) The Encyclopedia of Biological Chemistry 3:331–336. Academic Press, Waltham, MA, USA

Santella L, Kyozuka K (1994) Reinitiation of meiosis in starfish oocytes requires an increase in nuclear Ca^{2+}. Biochem Biophys Res Commun 203(1):674–680

Santella L, Alikani M, Talansky BE, Cohen J, Dale B (1985) Is the human oocyte plasma membrane polarized? Hum Reprod 7(7):999–1003

Santella L, De Riso L, Gragnaniello G, Kyozuka K (1998) Separate activation of the cytoplasmic and nuclear calcium pools in maturing starfish oocytes. Biochem Biophys Res Commun 252(1):1–4

Santella L, De Riso L, Gragnaniello G, Kyozuka K (1999) Cortical granule translocation during maturation of starfish oocytes requires cytoskeletal rearrangement triggered by InsP3-mediated Ca^{2+} release. Exp Cell Res 248(2):567–574

Santella L, Kyozuka K, Genazzani AA, De Riso L, Carafoli E (2000) Nicotinic acid adenine dinucleotide phosphate-induced Ca^{2+} release. Interactions among distinct Ca^{2+} mobilizing mechanisms in starfish oocytes. J Biol Chem 275(12):8301–8306

Santella L, Lim D, Moccia F (2004) Calcium and fertilization: the beginning of life. Trends Biochem Sci 29(8):400–408

Santella L, Puppo A, Chun JT (2008) The role of the actin cytoskeleton in calcium signaling in starfish oocytes. Int J Dev Biol 52(5–6):571–584

Santella L, Vasilev F, Chun JT (2012) Fertilization in echinoderms. Biochem Biophys Res Commun 425(3):588–594

Schroeder TE (1981) Microfilament-mediated surface change in starfish oocytes in response to 1-methyladenine: implications for identifying the pathway and receptor sites for maturation-inducing hormones. J Cell Biol 90(2):362–371

Schroeder TE, Stricker SA (1983) Morphological changes during maturation of starfish oocytes: surface ultrastructure and cortical actin. Dev Biol 98(2):373–384

Shen SS, Buck WR (1993) Sources of calcium in sea urchin eggs during the fertilization response. Dev Biol 157(1):157–169

Shida H, Hirai S (1978) Site of 1-methyladenine receptors in the maturation of starfish oocytes. Dev Growth Differ 20:205–211

Shôji Y, Hamaguchi MS, Hiramoto Y (1978) Mechanical properties of the endoplasm in starfish oocytes. Exp Cell Res 117(1):79–87

Steinhardt RA, Epel D (1974) Activation of sea-urchin eggs by a calcium ionophore. Proc Natl Acad Sci USA 71(5):1915–1919

Steinhardt RA, Lundin L, Mazia D (1971) Bioelectric responses of the echinoderm egg to fertilization. Proc Natl Acad Sci USA 68(10):2426–2430

Streb H, Irvine RF, Berridge MJ, Schulz I (1983) Release of Ca^{2+} from a nonmitochondrial intracellular store in pancreatic acinar cells by inositol-1,4,5-trisphosphate. Nature (Lond) 306(5938):67–69

Stricker SA (2006) Structural reorganizations of the endoplasmic reticulum during egg maturation and fertilization. Semin Cell Dev Biol 17(2):303–313

Terasaki M (1996) Actin filament translocations in sea urchin eggs. Cell Motil Cytoskeleton 34(1): 48–56

Tilney LG, Jaffe LA (1980) Actin, microvilli, and the fertilization cone of sea urchin eggs. J Cell Biol 87(3):771–782

Vacquier VD (1975) The isolation of intact cortical granules from sea urchin eggs: calcium ions trigger granule discharge. Dev Biol 43(1):62–74

Vacquier VD (2011) Laboratory on sea urchin fertilization. Mol Reprod Dev 78(8):553–564

Vacquier VD, Epel D, Douglas LA (1972) Sea urchin eggs release protease activity at fertilization. Nature (Lond) 237(5349):34–36

Vasilev F, Chun JT, Gragnaniello G, Garante E, Santella L (2012) Effects of ionomycin on egg activation and early development in starfish. PLoS One 7(6):e39231

Whitaker MJ, Baker PF (1983) Calcium-dependent exocytosis in an in vitro secretory granule plasma membrane preparation from sea urchin eggs and the effects of some inhibitors of cytoskeletal function. Proc R Soc Lond B Biol Sci 218(1213):397–413

Witchel HJ, Steinhardt RA (1990) 1-Methyladenine can consistently induce a fura-detectable transient calcium increase which is neither necessary nor sufficient for maturation in oocytes of the starfish *Asterina miniata*. Dev Biol 141(2):393–398

Chapter 14
Focused Proteomics on Egg Membrane Microdomains to Elucidate the Cellular and Molecular Mechanisms of Fertilization in the African Clawed Frog *Xenopus laevis*

Ken-ichi Sato, A.K.M. Mahbub Hasan, and Takashi W. Ijiri

Abstract Involvement of protein tyrosine kinase-dependent signal transduction (PTK signaling) in fertilization was initially demonstrated by studies using sea invertebrates: namely, an increase of tyrosine phosphorylation in egg or embryo proteins is shown to occur within minutes after gamete interaction. Among vertebrate species so far studied are fish, frog, and some mammalian species in which the importance of PTK signaling for fertilization or activation of development has been shown. In this review chapter, we summarize our experimental data that explore the role played by the tyrosine kinase Src in fertilization of the African clawed frog *Xenopus laevis*. In addition, we introduce our recent approaches that focus on the structure and function of egg membrane microdomains (MDs), where the Src PTK signaling machinery is organized. Finally, we propose a hypothesis that gamete membrane interaction at fertilization is accompanied by mutual signaling cross-talk between egg and sperm using the egg MDs as scaffolds and discuss the versatility of our hypothesis in general understanding of the sexual reproduction mechanism.

Keywords Cross-talk between gametes • In vitro reconstitution • Membrane microdomains • Signal transduction • Tyrosine phosphorylation

K. Sato (✉) • T.W. Ijiri
Laboratory of Cell Signaling and Development, Department of Molecular Biosciences,
Faculty of Life Sciences, Kyoto Sangyo University, Kyoto 603-8555, Japan
e-mail: kksato@cc.kyoto-su.ac.jp

A.K.M. Mahbub Hasan
Laboratory of Cell Signaling and Development, Department of Molecular Biosciences,
Faculty of Life Sciences, Kyoto Sangyo University, Kyoto 603-8555, Japan

Laboratory of Gene Biology, Department of Biochemistry and Molecular Biology,
University of Dhaka, Dhaka 1000, Bangladesh

H. Sawada et al. (eds.), *Sexual Reproduction in Animals and Plants*,
DOI 10.1007/978-4-431-54589-7_14, © The Author(s) 2014

14.1 Src PTK Signaling and Fertilization

Protein tyrosine phosphorylation was initially discovered as an intracellular phenomenon to be associated with malignancy of cancer cells (Hunter 2009). The discovery was made in the process of research on the molecular function of a gene product of Rous sarcoma virus, that is, v-Src, the memorably first example of so-called cancer gene or oncogene. Such a breakthrough finding, in conjunction with another epoch-making discovery of the first example of cellular cancer gene or proto-oncogene, c-Src (hereafter Src), has led many researchers to study the physiological and pathological importance of protein tyrosine phosphorylation catalyzed by Src and other protein tyrosine kinases (PTKs) (e.g., Abelson kinase, epidermal growth factor receptor/kinase, insulin receptor/kinase) (Abram and Courtneidge 2000; Hunter and Cooper 1985; Jove and Hanafusa 1987; Thomas and Brugge 1997). Now, it is well established that a variety of cellular functions involve this kind of posttranslational modification of proteins. Under this background, fertilization is one of the earliest and pioneering as well as contemporary subjects in the PTK research field (Kinsey 2013; Sato et al. 2004; Sato 2008).

Dasgupta and Garbers (1983) published the first report on protein tyrosine phosphorylation in the fertilization study: they demonstrated that PTK activity toward synthetic peptide substrates is present in unfertilized sea urchin eggs and continues to increase during early embryogenesis. Given that early embryos of sea urchin, similar to rapidly proliferating cancer cells, undergo several cycles of very fast cell division, that is, cleavage, it seems to be natural that they employ high PTK activity. Further detailed studies by Kinsey's group and some other researchers (including us) using sea urchins and other animal species, however, highlighted facts of specific importance in this cellular system, that PTK activity is rapidly and transiently activated in response to fertilization, or more precisely, gamete interaction (Abassi and Foltz 1994; Ciapa and Epel 1991; Kamel et al. 1986; Ribot et al. 1984; Sato et al. 1996; Wu and Kinsey 2000), and that the activated PTK may be responsible for sperm-induced increase(s) in calcium concentrations within the fertilized eggs, whose occurrence is believed to be indispensable for the subsequent initiation of embryonic development, in other words, "egg activation" (Giusti et al. 1999; Kinsey and Shen 2000; Runft and Jaffe 2000; Sato et al. 2000).

As described, compelling evidence suggests that egg-associated PTK signaling serves an important role in sperm-induced egg activation of nonmammalian species. As well appreciated in somatic cell systems, PTK signaling usually requires the binding of extracellular ligands to their cell-surface receptors. Therefore, it has been thought that gamete interaction at fertilization may act as a ligand-like signal to stimulate an egg-surface receptor so that the intracellular PTK signaling is triggered. On the other hand, eggs of mammalian species (e.g., mouse) and some other nonmammalian species (e.g., bird, newt) seem to employ sperm-derived factor(s), which would be incorporated into the egg cytoplasm, for the sperm-induced calcium responses: molecules identified so far include phospholipase Cζ (Swann and Lai 2013) and citrate synthase (Iwao 2012). This observation may reflect that gamete membrane interaction-mediated and gamete membrane fusion-mediated egg activation systems employ their specific molecular machinery and, perhaps more importantly, some other differences of fundamental importance that are found between these species: aquatic or terrestrial life, and external or internal fertilization.

In this review article, we explain why we analyze egg membrane microdomains (MDs) and summarize briefly our achievements, with a special focus on the finding of the egg MDs as structural and functional platform for sperm-induced Src PTK signaling. Then, we introduce our current research projects, evaluating the hypothesis that Src and the other MD-associated molecules constitute a signaling network for successful gamete interaction and activation of development. About 10 years ago, we published one review article in the journal *Proteomics* on our focused proteomics project on *Xenopus* egg MDs (Sato et al. 2002a). Therefore, a part of this review article can be considered as updated information for the MDs projects.

14.2   Characterization of Src as a Mediator of Gamete Interaction and Egg Activation

It was 1996 when we published the first paper on the *Xenopus* egg Src (hereafter xSrc) and its possible involvement in fertilization signaling (Sato et al. 1996). In that paper, we showed data about chromatographic fractionation of membrane-associated proteins and in vitro protein kinase assay, by which we could detect an elevation of the activity of xSrc in response to fertilization. Our further studies demonstrated that pharmacological (e.g., use of inhibitors; Sato et al. 1999, 2000, 2001) or molecular biological inhibition of xSrc (i.e., expression of kinase-negative mutant of xSrc; manuscript in preparation) impairs the ability of *Xenopus* eggs to undergo calcium reactions and egg activation in response to sperm (Sato et al. 1999) and that *Xenopus* eggs can be activated in a Src-dependent manner by artificial egg activators that interacts with the egg surface (i.e., RGD peptide and cathepsin B; Sato et al. 1999; Mahbub Hasan et al. 2005) and by hydrogen peroxide that may directly activate xSrc (Sato et al. 2001). These results suggest that xSrc acts between gamete interaction/fusion and an increase in intracellular calcium concentration at fertilization (Fig. 14.1). In support of this, phospholipase Cγ, whose activation leads to the hydrolysis of phosphatidylinositol-4,5-bisphosphates and production of the intracellular second messengers diacylglycerol and inositol-1,4,5-trisphosphate (direct activator for intracellular calcium release from endoplasmic reticulum), was shown to be a substrate of the activated xSrc (Sato et al. 2000, 2001, 2003).

Other targets of the activated xSrc include Shc (Aoto et al. 1999), hnRNP K (Iwasaki et al. 2008), lipovitellin 2 (Kushima et al. 2011), and pp40, whose identity is not yet demonstrated (manuscript in preparation) (Fig. 14.1). All these proteins and PLCγ are, however, cytoplasmic proteins resembling xSrc, so that they could not be directly involved in the gamete interaction and subsequent Src activation. Given that xSrc seems to be involved in the upstream signaling for egg activation and perhaps other cellular functions for embryonic development (e.g., translational control of maternal mRNAs), our goal has shifted to understand the mechanisms of gamete interaction and subsequent xSrc activation. Under these circumstances, we became interested in analyzing the egg plasma membrane, or more specifically,

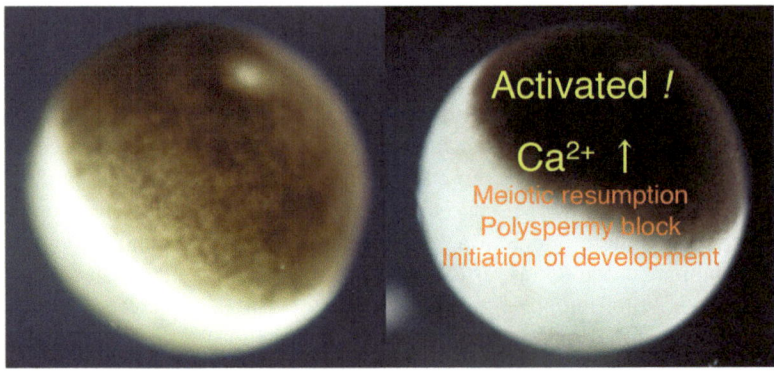

Unfertilized egg $\longrightarrow$ Fertilized egg

Tyrosine kinase-dependent signal transduction

Fig. 14.1 Molecular mechanisms of fertilization and activation of development in *Xenopus laevis*. Shown are photographs for a *X. laevis* egg before (*left*) and after (*right*) fertilization. Note that cortical contraction in the pigmented area of the egg (animal hemisphere) is evident after fertilization. Successful gamete membrane interaction and fusion leads to a transient increase in intracellular Ca^{2+} concentration in the egg, by which a series of events, collectively called "egg activation," takes place. The egg activation events include meiotic resumption, block to polyspermy, nuclear fusion, and initiation of embryonic development. Studies from our research group and others have demonstrated that tyrosine kinase-dependent signal transduction serves a pivotal role in connecting gamete membrane interaction and Ca^{2+}-dependent egg activation

membrane microdomains and their associated molecules. In the following sections, we introduce membrane microdomains and describe our achievements were obtained by the study on this subject.

14.3 Focused Proteomics on *Xenopus* Egg MDs: Achievements and Problems

14.3.1 Rationale to Study MDs for Exploring the Mechanism of Fertilization

A growing body of knowledge indicates that cellular plasma membranes consist of mixtures of heterogeneously organized substructures, whose specific identities depend on the composition of lipids, proteins, and their associated carbohydrates (Brown and London 1998; Simons and Ikonen 1997; Simons and Sampaio 2011). These membrane substructures are also called as membrane subdomains or microdomains: we prefer to use the term "microdomains" (MDs) because it would reflect the real scale of these membrane substructures (i.e., diameters of submicrometers or a few micrometers) (Pike 2006).

Experimentally, conventional biochemical fractionation methods are used to isolate MDs; hallmarks for obtaining MDs are such criteria as insolubility under certain cell extraction methods (e.g., resistance to detergent extraction) and tendency to float under ultracentrifugation (i.e., low density). The resulting low-density and detergent-insoluble membranes (LD-DIMs) are often regarded as lipid/membrane "rafts" and are enriched in cholesterol and some specific subsets of components (e.g., sphingolipids, gangliosides, signaling proteins). Several lines of evidence indicate that MDs constitute a platform/scaffold for capturing/sensing extracellular signals (e.g., growth factors, environmental stimuli), as well as for transmitting the signals into the cytoplasm and nucleus, by which a variety of cellular functions are exerted (Simons and Ikonen 1997; Simons and Toomre 2000).

In the fertilization research field, Kitajima and colleagues demonstrated first the presence of the sea urchin sperm MDs and their possible involvement in fertilization (Ohta et al. 1999). Publication of this leading report and our data showing that *Xenopus* egg fertilization involves Src PTK signaling (see foregoing) led us to examine the presence and physiological importance of MDs in *Xenopus* eggs. For a more detailed introduction to MDs, please also refer to our recent publication (Mahbub Hasan et al. 2011).

14.3.2 Xenopus *Egg MDs Projects: Achievements and Problems*

14.3.2.1 Discovery of Egg MDs as an Important Resource for Fertilization Study

In 2002, we published the first report on *Xenopus* egg MDs (Sato et al. 2002b), in which we demonstrated the following: (1) extraction of unfertilized eggs in the presence of Triton X-100 and subsequent ultracentrifugation under stepwise gradients of sucrose concentration yield LD-DIMs fractions (see above) that are enriched in cholesterol, the GM1 ganglioside, and most importantly for us, xSrc; (2) the LD-DIMs fractions that are prepared from fertilized eggs contain at least three tyrosine-phosphorylated proteins, two of which are xSrc and PLCγ; (3) a similar and more augmented pattern of protein tyrosine phosphorylation is seen in the LD-DIMs that are prepared from hydrogen peroxide-activated eggs; and (4) in the LD-DIMs of eggs that are activated by the calcium ionophore A23187, no increase in protein tyrosine phosphorylation is observed. These results suggest that the LD-DIMs fractions contain MDs, in which sperm-induced Src PTK signaling is operating (Fig. 14.1).

The functional importance of MDs in fertilization is suggested by the studies using methyl-β-cyclodextrin (MβCD), a drug that causes disruption of the cholesterol-dependent membrane structures. We found that unfertilized *Xenopus* eggs that are preincubated with this substance fail to undergo tyrosine phosphorylation of the LD-DIMs-associated proteins, intracellular calcium release, and other egg activation events (Sato et al. 2002b). Such inhibitory effect of MβCD could be canceled by the addition of excess amounts of cholesterol; therefore, it was not simply caused by

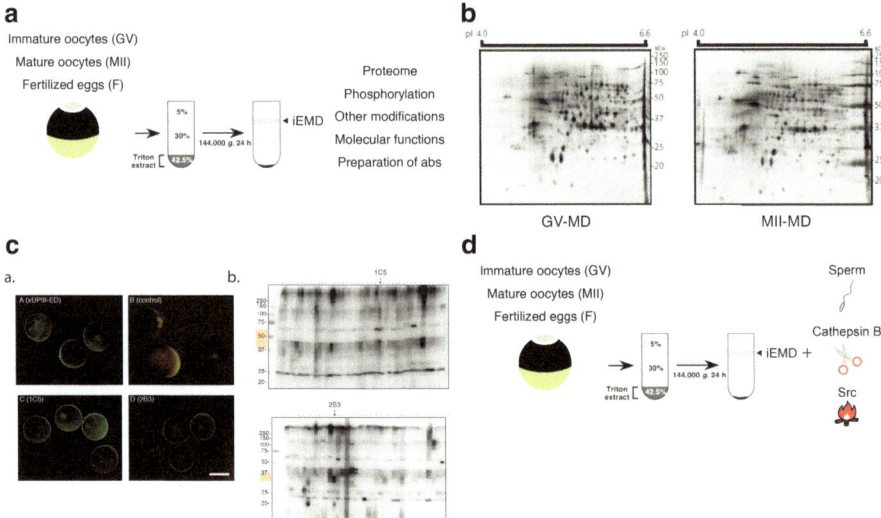

Fig. 14.2 *Xenopus* egg microdomains (MDs) project. (**a**) Our ongoing snapshot-type MDs project includes conventional, differential proteome analysis (b shows example images), analysis of post-translational modifications such as phosphorylation, molecular targeting analysis, and preparation and characterization of monoclonal antibody library (c shows example images). These experiments involve the use of MDs that are isolated from fully grown, immature oocytes at the germinal vesicle stage (GV oocytes), mature and unfertilized oocytes/eggs at the second meiotic metaphase (MII oocytes), and fertilized embryos, all of which are called iEMD (isolated egg MDs). (**b**) Proteome patterns for MDs-associated proteins in GV oocytes (*left*) and MII oocytes (*right*), both of which are separated by two-dimensional gel electrophoresis (first dimension, isoelectric focusing at the pI range of 4.0–6.6; second dimension, SDS-polyacrylamide gel electrophoresis) and visualized by silver staining. (**c**) In panels shown in *a*, indirect immunofluorescent experiments are performed to verify the binding of monoclonal antibodies to the surface of unfertilized *Xenopus* eggs. In *b*, the binding of the antibody to egg proteins is examined by immunoblotting. (**d**) Our ongoing MDs project also includes in vitro reconstitution of signal transduction events associated with fertilization. In this project, iEMD are subjected to in vitro treatment with sperm, cathepsin B, Src, or some others, and analyzed for their biochemical (e.g., tyrosine phosphorylation of MDs-associated proteins) as well as cell biological responses (e.g., ability of the sperm to fertilize eggs)

toxicity of the substance. Moreover, we found that the addition of sperm to the LD-DIMs, which were isolated from unfertilized eggs, caused an increase in tyrosine phosphorylation of proteins that are present in the LD-DIMs. These results argue the idea that LD-DIMs contain MDs, in which both the receptor for sperm and the signaling machinery for sperm-induced Src activation are pre-organized.

The aforementioned data have led us to consider two major directions of the MDs project (Sato et al. 2002a) (Fig. 14.2). The first subproject is to identify novel fertilization-related components by characterization of MDs-associated molecules. A major achievement in this subproject is the identification of a type I transmembrane protein uroplakin III (UPIII) that is thought to be involved in gamete interaction, which may involve the action of sperm-derived protease, and regulation of xSrc activity, in that UPIII may be involved in the negative regulation of xSrc in unfertilized eggs (Mahbub Hasan et al. 2005, 2007; Mammadova et al. 2009; Sakakibara et al. 2005a) (Fig. 14.1). The second subproject is to examine in vitro reconstitution of the fertilization signaling events with the use of isolated MDs. We have succeeded in reconstituting sperm-induced egg activation events such as tyrosine phosphorylation of xSrc and PLCγ, calcium responses, and resumption of the meiotic cell cycle by using isolated, unfertilized egg MDs and cytostatic factor-arrested unfertilized egg extracts (Sato et al. 2003, 2006). In the remaining part of this subsection, we describe achievements and current problems in these two subprojects.

14.3.2.2 Characterization of UPIII as a Novel Component of Fertilization

UPIII was originally identified by mass spectrometric analysis of a prominently tyrosine-phosphorylated 30-kDa protein that is present in the LD-DIMs fractions of fertilized eggs (for more detail on UPIII and other UP family proteins, see Mahbub Hasan et al. 2011). Coexpression of UPIII and xSrc in human embryonic kidney cells results in tyrosine phosphorylation of UPIII. These results suggest that the cytoplasmic part of UPIII acts as a target of xSrc in fertilized *Xenopus* eggs, although physiological relevance of the phosphorylation is unknown. It was also shown that treatment of *Xenopus* eggs with cathepsin B, a mimetic enzyme of the sperm-derived protease that can promote parthenogenetic activation of the eggs, causes partial degradation of UPIII and activation of xSrc. These two events were inhibited when eggs are pretreated with an antibody that is raised against the extracellular domain of UPIII. More importantly, the same antibody also effectively inhibits sperm-induced egg activation. These results suggest that the extracellular part of UPIII acts as a sperm receptor, by which it transmits the sperm signal into the egg cytoplasm (via xSrc activation) (Fig. 14.3). One potential mechanism of UPIII-xSrc connection is that, in unfertilized *Xenopus* eggs, the molecular complex consisting of UPIII and UPIb, a well-known binding partner of UPIII, suppresses the activity of xSrc, and that the proteolysis of UPIII at fertilization leads to the liberation of the activated xSrc. This idea is suggested by studies using the overexpression systems of culture cells (Mahbub Hasan et al. 2007).

14.3.2.3 In Vitro Reconstitution of Fertilization Signaling by Isolated MDs

Reportedly, cellular and molecular insights into the cell-cycle events (e.g., mitosis, meiotic resumption) have been well documented by studies using cell-free extracts that are prepared from unfertilized *Xenopus* eggs (Murray 1991). In particular, cytostatic factor-arrested unfertilized egg extracts (CSF extracts) are used to reconstitute meiosis and other cell-cycle events, as seen in *Xenopus* oocytes or embryos (Maresca and Heald 2006; Ohsumi et al. 2006). In this experimental

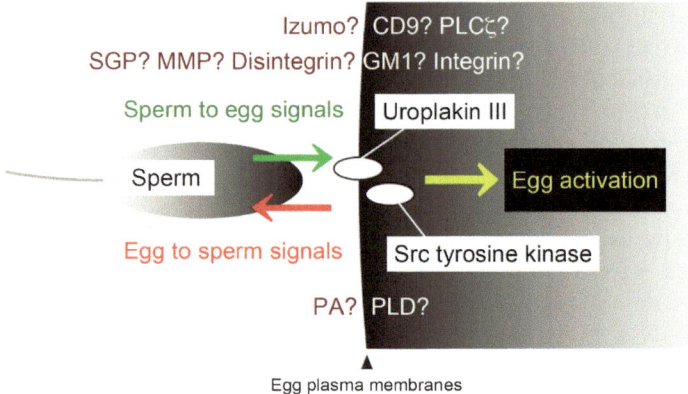

Fig. 14.3 Signaling cross-talk between sperm and egg MDs. A number of gamete membrane-associated components have been implicated in fertilization, some of which seem to be species specific whereas others are of universal importance in a wide variety of species. In *Xenopus laevis*, sperm-induced egg activation involves proteolytic cleavage of UPIII and activation of the tyrosine kinase xSrc, both of which are localized events in the egg MDs. Our recent data suggest that membrane interaction between egg and sperm via the egg MDs also involves modulation of fertilizing property in sperm. Such bidirectional signaling between egg and sperm may be crucial for completion of successful gamete interaction/fusion and initiation of embryonic development

platform, however, the only events to be evaluated are cytoplasmic and nuclear functions. Our idea in the second MDs subproject is to employ the isolated MDs as a resource to reconstitute plasma membrane-associated functions and their interactions with cytoplasmic environments at fertilization (Fig. 14.2). As described earlier, the addition of sperm to the isolated MDs causes an elevation in tyrosine phosphorylation of MDs-associated proteins. This phenomenon actually involves the activation of xSrc as well as phosphorylation of UPIII (Sato et al. 2003, 2006; unpublished results), both of which are the earliest events that are seen in fertilized *Xenopus* eggs. Therefore, we sought to combine the sperm-treated MDs with CSF extracts to reconstitute events of fertilization from the plasma membranes through the cytoplasm. As a result, we have found that the sperm-treated MDs can promote a transient calcium release, dephosphorylation of mitogen-activated protein kinase, and cell-cycle progression as judged by the morphological change in sperm-derived nuclei (Sato et al. 2003). In addition, the isolated MDs that are pretreated with hydrogen peroxide (Src PTK signaling is stimulated), but not those pretreated with A23187 (Src PTK signaling is not stimulated), are shown to promote the aforementioned events. These results are consistent with the idea that egg MDs can be used as an experimental platform to reconstitute membrane-associated, Src PTK-dependent signaling events at fertilization (Sato et al. 2003).

Given this background, there are two major problems to be solved in our MDs project: one is when and how sperm-dependent signaling functions of MDs are established; and the other is how egg MDs interact with sperm and what is the consequence of this interaction.

14.3.3   Ongoing Approaches to Explore the Physiological Functions of MDs

As already described, our MDs project has involved two major subprojects: snapshot analysis of MDs-associated proteins and in vitro reconstitution of MDs functions (Fig. 14.2), and these two subprojects have raised new problems and questions. To explore further these problems and questions, we are now undergoing a panel of experiments. In the following subsections, we describe the aims and current states of those ongoing MDs projects.

14.3.3.1   Evaluation of UPIII and MDs Functions in Immature Oocytes

In general, ovarian and immature oocytes cannot be fertilized because they are not yet fully competent to undergo sperm-induced developmental activation. It has long been appreciated that hormone-induced oocyte maturation is an event in that oocytes acquire fertilization competence by establishing intracellular conditions (e.g., meiotic cell-cycle stage, subcellular localizations of endoplasmic reticulum and other intracellular components, biochemical properties including the activity of certain protein kinases such as maturation-promoting factor). However, it seems that oocyte maturation also involves an alteration in the structure and function of the oocyte plasma membranes. To access this possibility, we are now analyzing expression, subcellular localization, and molecular interaction of UPIII in the course of oocyte growth (oogenesis) and progesterone-induced oocyte maturation. In addition, we are examining whether sperm-induced PTK signaling can work in the immature oocyte MD. Results so far obtained suggest that oocyte maturation leads to the exposure of the extracellular domain of UPIII on the oocyte surface and that, at a similar timing, MDs become fully competent for sperm-induced PTK signaling (manuscript in preparation).

14.3.3.2   Gain- and Loss-of-Function Experiments on xSrc and UPIII

We have prepared immature oocytes expressing xSrc of either kinase-active or kinase-negative mutants, and we are now examining in vitro maturation of the oocytes and their activation in response to artificial egg activators: the calcium ionophore A23187, cathepsin B, and hydrogen peroxide (manuscript in preparation). We have also constructed mutant UPIII, in which its possible target amino acids for tryptic protease in the extracellular domain are mutated to be proteolysis resistant. We are now characterizing molecular function of the mutant UPIII (named UPIII-RRAA mutant) by using expression systems with human embryonic kidney 293 cells. Results so far obtained demonstrate that the UPIII-RRAA mutant, as does the wild-type UPIII, localizes to plasma membranes/MDs and is capable of inactivating the coexpressed xSrc in 293 cells (manuscript in preparation). In addition, more importantly, it has been shown that the UPIII-RRAA is actually resistant to proteolysis by cathepsin B. We are now examining the sperm-induced PTK signaling function of MDs that are prepared from 293 cells expressing the UPIII of either the wild type or RRAA mutant. These experiments would not only evaluate the function of the UPIII-RRAA mutant but also suggest that the sperm receptor function can be reconstituted in MDs of the cultured cells.

14.3.3.3 Unbiased Approaches to Identify and Characterize Novel Components

Although the aforementioned two approaches focus on functions of certain well-characterized egg MD-associated proteins (i.e., xSrc and UPIII), the following two approaches are aimed to discover novel fertilization-related components that localize to the egg MDs. The first one is to generate a library of monoclonal antibodies in which plasma membrane-associated components are used as antigens. In *Xenopus*, some egg surface-interacting substances have been reported to act as an egg activator: examples include disintegrin peptides (Iwao and Fujimura 1996; Shilling et al. 1998) and cathepsin B (Mizote et al. 1999). As we mentioned earlier, an antibody that recognizes the extracellular domain of UPIII has been shown to inhibit normal fertilization (Sakakibara et al. 2005a). Therefore, we expect that an antibody(s) of the library, which can bind to the surface (in other words, some known or unknown components essential for fertilization) of unfertilized *Xenopus* eggs, would be able to activate or inhibit egg functions. We are now screening monoclonal antibody clones that satisfy the following criteria: (1) to bind to the egg surface in indirect immunofluorescent experiments, (2) to bind to protein bands in immunoblotting experiments, and (3) to inhibit normal fertilization or to activate the egg in the absence of sperm. The second approach is to identify the molecular basis of polarity in the egg surface. In *Xenopus*, sperm entry point is limited to somewhere in the pigmented area of the egg: the animal hemisphere. Our hypothesis is that this limitation is the result of uneven distribution of sperm receptor or its inhibitory machinery. To access this problem, we are now undergoing comparative analysis of the animal and vegetal hemisphere-associated proteins. Other kinds of differential proteome analysis have been done in our previous study, in that an antibody that recognizes fertilization-induced plasma membrane-associated antigens in *Xenopus* eggs is generated by subtractive immunization (Sakakibara et al. 2005b). Therefore, these two ongoing experiments would also be useful for further evaluation of such an "unbiased" approach to discover fertilization-related egg components.

14.3.3.4 Analysis of Signaling Cross-Talk Between MDs and Sperm or Egg Cytoplasm

In this ongoing project, we analyze the interactions of the egg MDs with sperm and the egg cytoplasm. Regarding the egg MD–sperm interactions, we have preliminary data showing that the isolated egg MDs do not inhibit normal fertilization when they are present in the insemination media, but rather recover the ability of sperm to fertilize the eggs that are pretreated with the anti-UPIII extracellular domain antibody (manuscript in preparation). In other words, it is suggested that the MD-treated sperm somehow overcome the absence of functional UPIII on the egg surface. In *Xenopus*, no evidence is available with regard to the presence of extracellular MDs- or other membrane-containing vesicles (such as exosomes in the mouse; Miyado et al. 2008; manuscript in preparation). Therefore, we assume that, exactly at the time of gamete membrane interaction and fusion, egg-"associated" MDs act not only as a platform for sperm-induced Src PTK signaling in fertilized eggs but also as an unknown signaling trigger for sperm, by which sperm may acquire the ability to fertilize eggs, in a UPIII-dependent manner (Fig. 14.3). We now assess this "bidirectional signaling between sperm and the egg MDs" hypothesis by analyzing what kind of cellular functions in sperm are modulated after their interactions with the isolated MDs, and what kind of molecule(s) in the sperm surface are involved in the interactions with the isolated MDs. One interesting thought is the involvement of PTK signaling in sperm, which has been well documented in several other aspects of sperm functions (Ijiri et al. 2012).

14.3.3.5   Analysis of Signaling Cross-Talk Between MDs and Egg Mitochondria

We are also interested in functional interactions between egg MDs and organelles such as mitochondria. This interest is based on the fact that serine/threonine-specific protein kinase Akt, whose activation is shown in fertilized *Xenopus* eggs (Mammadova et al. 2009), localizes predominantly to mitochondria and that two mitochondrion-associated proteins of ~40 kDa on SDS-polyacrylamide gel electrophoresis (named pp40) are identified as possible Src substrates at fertilization (manuscript in preparation). These two proteins are also present in MD fractions. In some cell systems, cell-surface receptor/kinases [e.g., epidermal growth factor receptor (EGFR)] and membrane-associated cytoplasmic protein kinases (e.g., Akt, Src) are shown to translocate or localize to mitochondria in response to cell stimuli (Demory et al. 2009; Hebert-Chatelain 2013). Therefore, it will be interesting to examine whether fertilization involves a similar signaling linkage between MDs and mitochondria by way of characterization of these novel Src substrates. Our interest in mitochondria also involves a possible relationship of their known functions and fertilization. In somatic cell systems, mitochondria serve a pivotal role in ATP production and survival control. Thus, this ongoing project will explore the presence of MDs–mitochondria interactions at fertilization and their physiological relevance to the metabolism and viability of developing embryos. In this connection, we have started to analyze quantitative and spatiotemporal regulation of ATP in oocytes and eggs during oocyte maturation and fertilization (Ijiri et al. 2014).

14.4   Summary and Perspectives

Our ongoing egg-MDs project, consisting of snapshot experiments and in vitro reconstitution experiments, molecular targeting approaches and unbiased proteomics approaches, and cell biological approaches involving sperm and egg mitochondria; aim to understand how structure and function of egg MDs are developed and how egg MDs and MD-associated molecules contribute to successful fertilization and subsequent embryogenesis. We believe that the knowledge obtained with this project in itself would contribute to general understanding of the fertilization system. In addition, fertilization system involves, with no exception, membrane interaction and fusion between female and male gametes, in spite of a countless variety of species-specific differences in the fertilization system such as molecules of fertilization (e.g. requirement of Src PTK signaling), systems for egg activation (e.g., physiological monospermy and polyspermy), extracellular environment for fertilization (e.g., internal and external fertilization), and even animal and nonanimal fertilization. Therefore, our approach would be useful, at least in part, for fertilization studies on a wide range of other organisms.

Acknowledgments This work is supported by Grants-in-Aid on Innovative Areas (22112522, 24112714) from the Ministry of Education, Culture, Sports, Science, and Technology, Japan to K.S.

References

Abassi YA, Foltz KR (1994) Tyrosine phosphorylation of the egg receptor for sperm at fertilization. Dev Biol 164(2):430–443

Abram CL, Courtneidge SA (2000) Src family tyrosine kinases and growth factor signaling. Exp Cell Res 254(1):1–13

Aoto M, Sato K, Takeba S, Horiuchi Y, Iwasaki T, Tokmakov AA, Fukami Y (1999) A 58-kDa Shc protein is present in *Xenopus* eggs and is phosphorylated on tyrosine residues upon egg activation. Biochem Biophys Res Commun 258(2):265–270

Brown DA, London E (1998) Functions of lipid rafts in biological membranes. Annu Rev Cell Dev Biol 14:111–136

Ciapa B, Epel D (1991) A rapid change in phosphorylation on tyrosine accompanies fertilization of sea urchin eggs. FEBS Lett 295(1–3):167–170

Dasgupta JD, Garbers DL (1983) Tyrosine protein kinase activity during embryogenesis. J Biol Chem 258(10):6174–6178

Demory ML, Boerner JL, Davidson R, Faust W, Miyake T, Lee I, Hüttemann M, Douglas R, Haddad G, Parsons SJ (2009) Epidermal growth factor receptor translocation to the mitochondria: regulation and effect. J Biol Chem 284(52):36592–36604

Giusti AF, Carroll DJ, Abassi YA, Terasaki M, Foltz KR, Jaffe LA (1999) Requirement of a Src family kinase for initiating calcium release at fertilization in starfish eggs. J Biol Chem 274(41):29318–29322

Hebert-Chatelain E (2013) Src kinases are important regulators of mitochondrial functions. Int J Biochem Cell Biol 45(1):90–98

Hunter T (2009) Tyrosine phosphorylation: thirty years and counting. Curr Opin Cell Biol 21(2):140–146

Hunter T, Cooper JA (1985) Protein-tyrosine kinases. Annu Rev Biochem 54:897–930

Ijiri TW, Mahbub Hasan AK, Sato K (2012) Protein-tyrosine kinase signaling in the biological functions associated with sperm. J Signal Transduct 2012:181560

Ijiri TW, Kishikawa J, Imamura H, Iwao Y, Yokoyama K, Sato K (2014) ATP imaging system in the *Xenopus laevis* oocyte. In: Sawada H, Inoue N, Iwano M (eds) Sexual reproduction in animals and plants. Springer, Heidelberg, pp 181–186

Iwao Y (2012) Egg activation in physiological polyspermy. Reproduction 144(1):11–22

Iwao Y, Fujimura T (1996) Activation of *Xenopus* eggs by RGD-containing peptides accompanied by intracellular Ca^{2+} release. Dev Biol 177(2):558–567

Iwasaki T, Koretomo Y, Fukuda T, Paronetto MP, Sette C, Fukami Y, Sato K (2008) Expression, phosphorylation, and mRNA-binding of heterogeneous nuclear ribonucleoprotein K in *Xenopus* oocytes, eggs, and early embryos. Dev Growth Differ 50(1):23–40

Jove R, Hanafusa H (1987) Cell transformation by the viral *src* oncogene. Annu Rev Cell Biol 3:31–56

Kamel C, Veno PA, Kinsey WH (1986) Quantitation of a *src*-like tyrosine protein kinase during fertilization of the sea urchin egg. Biochem Biophys Res Commun 138(1):349–355

Kinsey WH (2013) Intersecting roles of protein tyrosine kinase and calcium signaling during fertilization. Cell Calcium 53(1):32–40

Kinsey WH, Shen SS (2000) Role of the Fyn kinase in calcium release during fertilization of the sea urchin egg. Dev Biol 225(1):253–264

Kushima S, Mammadova G, Mahbub Hasan AK, Fukami Y, Sato K (2011) Characterization of lipovitellin 2 as a tyrosine-phosphorylated protein in oocytes, eggs and early embryos of *Xenopus laevis*. Zool Sci 28(8):550–559

Mahbub Hasan AK, Sato K, Sakakibara K, Ou Z, Iwasaki T, Ueda Y, Fukami Y (2005) Uroplakin III, a novel Src substrate in *Xenopus* egg rafts, is a target for sperm protease essential for fertilization. Dev Biol 286(2):483–492

Mahbub Hasan AK, Ou Z, Sakakibara K, Hirahara S, Iwasaki T, Sato K, Fukami Y (2007) Characterization of *Xenopus* egg membrane microdomains containing uroplakin Ib/III complex: roles of their molecular interactions for subcellular localization and signal transduction. Genes Cells 12(2):251–267

Mahbub Hasan AK, Fukami Y, Sato K (2011) Gamete membrane microdomains and their associated molecules in fertilization signaling. Mol Reprod Dev 78(10–11):814–830

Mammadova G, Iwasaki T, Tokmakov AA, Fukami Y, Sato K (2009) Evidence that phosphatidylinositol 3-kinase is involved in sperm-induced tyrosine kinase signaling in *Xenopus* egg fertilization. BMC Dev Biol 9:68

Maresca TJ, Heald R (2006) Methods for studying spindle assembly and chromosome condensation in *Xenopus* egg extracts. Methods Mol Biol 322:459–474

Miyado K, Yoshida K, Yamagata K, Sakakibara K, Okabe M, Wang X, Miyamoto K, Akutsu H, Kondo T, Takahashi Y, Ban T, Ito C, Toshimori K, Nakamura A, Ito M, Miyado M, Mekada E, Umezawa A (2008) The fusing ability of sperm is bestowed by CD9-containing vesicles released from eggs in mice. Proc Natl Acad Sci USA 105(35):12921–12926

Mizote A, Okamoto S, Iwao Y (1999) Activation of *Xenopus* eggs by proteases: possible involvement of a sperm protease in fertilization. Dev Biol 208(1):79–92

Murray AW (1991) Cell cycle extracts. Methods Cell Biol 36:581–605

Ohsumi K, Yamamoto TM, Iwabuchi M (2006) Oocyte extracts for the study of meiotic M-M transition. Methods Mol Biol 322:445–458

Ohta K, Sato C, Matsuda T, Toriyama M, Lennarz WJ, Kitajima K (1999) Isolation and characterization of low density detergent-insoluble membrane (LD-DIM) fraction from sea urchin sperm. Biochem Biophys Res Commun 258(3):616–623

Pike LJ (2006) Rafts defined: a report on the Keystone Symposium on Lipid Rafts and Cell Function. J Lipid Res 47(7):1597–1598

Ribot HD Jr, Eisenman EA, Kinsey WH (1984) Fertilization results in increased tyrosine phosphorylation of egg proteins. J Biol Chem 259(8):5333–5338

Runft LL, Jaffe LA (2000) Sperm extract injection into ascidian eggs signals Ca^{2+} release by the same pathway as fertilization. Development (Camb) 127(15):3227–3236

Sakakibara K, Sato K, Iwasaki T, Kitamura K, Fukami Y (2005a) Generation of an antibody specific to *Xenopus* fertilized eggs by subtractive immunization. Genes Cells 10(4):345–356

Sakakibara K, Sato K, Yoshino K, Oshiro N, Hirahara S, Mahbub Hasan AK, Iwasaki T, Ueda Y, Iwao Y, Yonezawa K, Fukami Y (2005b) Molecular identification and characterization of *Xenopus* egg uroplakin III, an egg raft-associated transmembrane protein that is tyrosine-phosphorylated upon fertilization. J Biol Chem 280(15):15029–15037

Sato K (2008) Signal transduction of fertilization in frog eggs and anti-apoptotic mechanism in human cancer cells: common and specific functions of membrane microdomains. Open Biochem J 2:49–59

Sato K, Aoto M, Mori K, Akasofu S, Tokmakov AA, Sahara S, Fukami Y (1996) Purification and characterization of a Src-related p57 protein-tyrosine kinase from *Xenopus* oocytes. Isolation of an inactive form of the enzyme and its activation and translocation upon fertilization. J Biol Chem 271(22):13250–13257

Sato K, Iwao Y, Fujimura T, Tamaki I, Ogawa K, Iwasaki T, Tokmakov AA, Hatano O, Fukami Y (1999) Evidence for the involvement of a Src-related tyrosine kinase in *Xenopus* egg activation. Dev Biol 209(2):308–320

Sato K, Tokmakov AA, Iwasaki T, Fukami Y (2000) Tyrosine kinase-dependent activation of phospholipase Cγ is required for calcium transient in *Xenopus* egg fertilization. Dev Biol 224(2):453–469

Sato K, Ogawa K, Tokmakov AA, Iwasaki T, Fukami Y (2001) Hydrogen peroxide induces Src family tyrosine kinase-dependent activation of *Xenopus* eggs. Dev Growth Differ 43(1):55–72

Sato K, Iwasaki T, Sakakibara K, Itakura S, Fukami Y (2002a) Towards the molecular dissection of fertilization signaling: Our functional genomic/proteomic strategies. Proteomics 2(9):1079–1089

Sato K, Iwasaki T, Ogawa K, Konishi M, Tokmakov AA, Fukami Y (2002b) Low density detergent-insoluble membrane of *Xenopus* eggs: subcellular microdomain for tyrosine kinase signaling in fertilization. Development (Camb) 129(4):885–896

Sato K, Tokmakov AA, He CL, Kurokawa M, Iwasaki T, Shirouzu M, Fissore RA, Yokoyama S, Fukami Y (2003) Reconstitution of Src-dependent phospholipase Cγ phosphorylation and transient calcium release by using membrane rafts and cell-free extracts from *Xenopus* eggs. J Biol Chem 278(40):38413–38420

Sato K, Iwasaki T, Hirahara S, Nishihira Y, Fukami Y (2004) Molecular dissection of egg fertilization signaling with the aid of tyrosine kinase-specific inhibitor and activator strategies. Biochim Biophys Acta 1697(1–2):103–121

Sato K, Yoshino K, Tokmakov AA, Iwasaki T, Yonezawa K, Fukami Y (2006) Studying fertilization in cell-free extracts: focusing on membrane/lipid raft functions and proteomics. Methods Mol Biol 322:395–411

Shilling FM, Magie CR, Nuccitelli R (1998) Voltage-dependent activation of frog eggs by a sperm surface disintegrin peptide. Dev Biol 202(1):113–124

Simons K, Ikonen E (1997) Functional rafts in cell membranes. Nature (Lond) 387(6633):569–572

Simons K, Sampaio JL (2011) Membrane organization and lipid rafts. Cold Spring Harbor Perspect Biol 3(10):a004697

Simons K, Toomre D (2000) Lipid rafts and signal transduction. Nat Rev Mol Cell Biol 1(1):31–39, Erratum in Nat Rev Mol Cell Biol 2(3):216

Swann K, Lai FA (2013) PLCζ and the initiation of Ca^{2+} oscillations in fertilizing mammalian eggs. Cell Calcium 53(1):55–62

Thomas SM, Brugge JS (1997) Cellular functions regulated by Src family kinases. Annu Rev Cell Dev Biol 13:513–609

Wu W, Kinsey WH (2000) Fertilization triggers activation of Fyn kinase in the zebrafish egg. Int J Dev Biol 44(8):837–841

Chapter 15
Egg Activation in Polyspermy: Its Molecular Mechanisms and Evolution in Vertebrates

Yasuhiro Iwao

Abstract In amphibians, most urodeles (newts) exhibit polyspermy physiologically, but primitive urodeles (*Hynobius*) and anurans (frogs) exhibit monospermy. Several fertilizing sperm induce multiple small Ca^{2+} waves in the polyspermic egg, but a single large Ca^{2+} wave occurs in the monospermic egg. The Ca^{2+} waves in newt eggs are caused by a sperm-specific citrate synthase localized outside the mitochondria. The single Ca^{2+} wave at monospermy is necessary for eliciting a fast block to polyspermy, whereas the small multiple Ca^{2+} waves provide slower egg activation to permit the entry of several sperm at polyspermy. Physiological polyspermy seems to be evolved in association with the increase in size of eggs in urodeles, reptiles, and birds laying larger yolky eggs. The sperm factor (citrate synthase) operating in slower egg activation in polyspermic eggs is already prepared in the monospermic urodele *Hynobius*. We have focused on comparative studies in fertilization among amphibians to understand the role of egg activation in establishment of polyspermy with discussion of the evolution in vertebrates.

Keywords Amphibians • Ca^{2+} wave • Citrate synthase • IP3 receptor • Polyspermy block

Y. Iwao (✉)
Laboratory of Molecular Developmental Biology, Department of Applied Molecular Biosciences, Graduate School of Medicine, Yamaguchi University,
753-8512 Yamaguchi, Japan
e-mail: iwao@yamaguchi-u.ac.jp

H. Sawada et al. (eds.), *Sexual Reproduction in Animals and Plants*,
DOI 10.1007/978-4-431-54589-7_15, © The Author(s) 2014

15.1 Introduction

Fertilization is indispensable for sexual reproduction in animals. Both sperm and eggs are highly specialized to ensure development with the diploid genome. We have focused on the molecular mechanism of fertilization in amphibians, which are one of the best models for studying fertilization, in particular, for understanding the evolution in egg activation and polyspermy blocks in vertebrates. There are two different types of fertilization in vertebrates: monospermy, in which only one sperm penetrates into the egg, and physiological polyspermy, in which several sperm enter the egg at normal fertilization. Fertilization in the ancestral vertebrates seems to be monospermic, because not only deuterostome invertebrates such as sea urchins and ascidians, but also most fishes, including a primitive fish, the lamprey, exhibit monospermy (Iwao 2012). Amphibians consist of three groups: anurans (frogs and toads), urodeles (newts and salamanders), and caecilians (limbless amphibians) (Iwao 2000a). Although there is little information on fertilization of caecilians, most anurans exhibit external and monospermic fertilization (Table 15.1). Only one sperm is incorporated into the egg, and other sperm are prevented from entering the fertilized egg. Several blocks to polyspermy operate to exclude the extra sperm outside the egg plasma membrane. In contrast, most urodeles exhibit internal fertilization and the female stores the sperm in a spermatheca near the cloaca (Akiyama et al. 2011). The eggs are inseminated by a small number of sperm released from the spermatheca just before oviposition. Although several sperm enter a physiologically polyspermic urodele egg, development with the diploid genome is ensured by the

Table 15.1 Characteristics of fertilization in amphibians

Species	Mode of fertilization	Ca^{2+} wave	Positive fertilization potential	Fast polyspermy block	Sperm citrate synthase	Size of egg (diameter in mm)
Anurans						
Discoglossus pictus	External occasional polyspermy	Multiple[a]	+	–	ND	1.6
Xenopus laevis	External monospermy	Single	+	+	–	1.2
Bufo japonicus	External monospermy	Single[a]	+	+	ND	1.8
Urodeles						
Hynobius nebulosus	External monospermy	Single[a]	+	+	+	2.4
Andrias japonicus	External polyspermy	ND	ND	–	ND	5.0
Cynops pyrrhogaster	Internal polyspermy	Multiple	–	–	+	2.3

ND not determined
[a]Based on the transient opening Ca^{2+}-activated Cl^- channels

intracellular block to polyspermy in egg cytoplasm (Fankhauser 1948; Iwao and Elinson 1990; Iwao et al. 1993, 2002). Only one sperm nucleus forms a zygote nucleus with the egg nucleus, and the other extra sperm nuclei degenerate before cleavage. However, the most primitive *Hynobius* salamanders exhibit external fertilization and monospermy as with anurans (Iwao 1989). Thus, comparative studies in fertilization among amphibians will provide better understanding of the role of egg activation in establishment of polyspermy during vertebrate evolution.

15.2 Egg Activation at Physiologically Polyspermic Fertilization

The egg nucleus in unfertilized eggs of vertebrates is arrested at metaphase of the second meiotic division until the sperm breaks the attest at fertilization. Fertilization provides the sperm nucleus into the egg, as well as initiates its embryonic development, which is called egg activation. An increase in $[Ca^{2+}]_i$ in the egg cytoplasm induced by the fertilizing sperm is essential for egg activation in vertebrate fertilization (Iwao 2000b). In monospermic anurans, the fertilizing sperm induces a transient and single $[Ca^{2+}]_i$ increase that propagates from a sperm entry site toward the opposite site on the whole egg surface as a $[Ca^{2+}]_i$ wave (Fontanilla and Nuccitelli 1998). The $[Ca^{2+}]_i$ increase causes the opening of Cl^- channels on the egg plasma membrane to produce a positive fertilization potential within 1 s (Kline and Nuccitelli 1985), which prevents the entry of another sperm as a fast, but temporal, block to polyspermy (Cross and Elinson 1980; Iwao 1989; Iwao and Jaffe 1989). The Ca^{2+} wave then induces exocytosis of cortical granules, resulting in transformation of the vitelline coat into a fertilization coat (Hedrick 2008). The fertilization coat prevents extra sperm from reaching the fertilized egg, as a slow and permanent block to polyspermy. In the monospermic salamander *Hynobius nebulosus*, the eggs exhibit a large positive fertilization potential mediated by opening of Ca^{2+}-activated Cl^- channels (Iwao 1989). The fast and transient opening of Cl^- channels indicates a single Ca^{2+} wave at fertilization. Polyspermy is also prevented by a positive fertilization potential without formation of a fertilization coat in monospermic salamanders. Thus, the fast generation of a single Ca^{2+} wave soon after the entry of the first sperm is important for the accomplishment of the fast polyspermy block in monospermic species.

In contrast, in physiologically polyspermic newts, small and multiple increases of $[Ca^{2+}]_i$ occur at fertilization (Grandin and Charbonneau 1992; Yamamoto et al. 1999; Harada et al. 2011). The $[Ca^{2+}]_i$ increase at the sperm entry site propagates as a Ca^{2+} wave (Harada et al. 2011). The small Ca^{2+} wave is induced by each fertilizing sperm in the polyspermic egg. Because 2 to 20 sperm enter an egg at normal newt fertilization (Iwao et al. 1985), multiple Ca^{2+} waves induce $[Ca^{2+}]_i$ increase, lasting 30–40 min after the first sperm entry (Harada et al. 2011). The $[Ca^{2+}]_i$ increase induces no change (Charbonneau et al. 1983) or a very small hyperpolarization in response to each sperm entry (Iwao 1985) in the egg membrane potential. Sperm entry into newt eggs is not sensitive to the positive membrane potential (Iwao and Jaffe 1989). No cortical granule is observed in urodele eggs, indicating lack of fertilization envelope formation (Iwao 2000a).

15.3 The Signaling Mechanism of $[Ca^{2+}]_i$ Increase Induced by the Fertilizing Sperm

The Ca^{2+} increase at fertilization is caused by the release of Ca^{2+} ions from the endoplasmic reticulum (ER), a major intracellular Ca^{2+} store in the egg cytoplasm (Fig. 15.1). Inositol 1,4,5-trisphosphate (IP3) generated by phospholipase C (PLC) opens Ca^{2+} channels of IP3 receptors on the ER. However, the mechanism for induction of $[Ca^{2+}]_i$ increase by the fertilizing sperm is quite different between monospermic and physiologically polyspermic eggs. In monospermic anurans, a sperm agonist (ligand) probably binds an egg receptor at contact between the sperm and egg membranes, and then a signal for stimulating IP3 production is transmitted into the egg cytoplasm. Indeed, *Xenopus* eggs are activated by external treatment with tryptic sperm protease (Iwao et al. 1994; Mizote et al. 1999), which can hydrolyze one of the candidates of egg receptors, uroplakin III (UP III), on the egg plasma

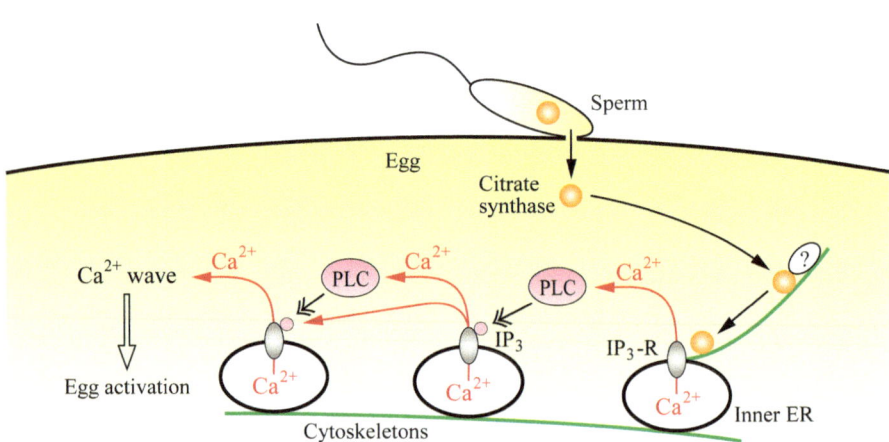

Fig. 15.1 Model of signaling for Ca^{2+} increase in a physiologically polyspermic newt egg. The sperm-specific citrate synthase is introduced from the sperm cytoplasm into the egg cytoplasm after sperm–egg fusion. Citrate synthase, in association with cytoskeletons, sensitizes the inositol-1,4,5-trisphosphate (IP3) receptor on the inner endoplasmic reticulum (*ER*) to release Ca^{2+} ions. The local Ca^{2+} increase propagates through the inner ER with cytoskeletons as a Ca^{2+} wave by the activation of phospholipase C (*PLC*) to produce IP3 or stimulation of IP3 receptors

membrane (Sakakibara et al. 2005; Sato et al. 2003). The cleavage of UP III induces activation of Src kinase and PLCγ to produce IP3 in egg cytoplasm (Mahbub Hasan et al. 2005, 2007; Ijiri et al. 2012). In addition, the $[Ca^{2+}]_i$ increase in *Xenopus* eggs is induced by external treatment of the egg with RGD-containing peptides (Iwao and Fujimura 1996) or KTE-containing peptides (Shilling et al. 1998), which binds integrins on the plasma membrane accompanied by activation of Src kinase (Sato et al. 1999). RGDS peptide also activates the eggs of monosperrmic *Hynobius* salamanders (Iwao, unpublished observations, 2012). Although the precise interaction between those molecules remains to be investigated, the initial Ca^{2+} release induced at the sperm entry site is propagated through further activation of PLCγ or direct sensitization of IP3 receptors on ER abundant in egg cortex, resulting in the formation of a single Ca^{2+} wave.

In polyspermic newt eggs, the signal for egg activation is provided from sperm cytoplasm after sperm and egg fusion (Fig. 15.1). Injection of an extract containing newt sperm cytoplasm into unfertilized eggs induces egg activation accompanied by a Ca^{2+} wave (Yamamoto et al. 2001; Harada et al. 2007, 2011). A sperm-specific form of citrate synthase is purified from the sperm extract as one of the major components of the sperm factor for egg activation (Harada et al. 2007). A large amount of citrate synthase is localized in the neck to the midpiece of newt sperm (Fig. 15.2A), but a smaller amount is also distributed under the plasma membrane around the nucleus (Fig. 15.2B). Injection of not only purified citrate synthase protein, but also mRNA of citrate synthase, induces egg activation with a Ca^{2+} increase (Harada et al. 2007). A single newt sperm contains about 2 pg citrate synthase, but injection of sperm cytoplasm equivalent to one sperm activates about 20 % of the eggs, indicating that the entry of at least two sperm is necessary for activating the egg. This estimation corresponds well to the observation that a small Ca^{2+} wave is induced by each sperm entry in the polyspermic newt egg (Harada et al. 2011). How does the sperm-specific citrate synthase induce the Ca^{2+} wave in the egg cytoplasm? In some cases, the Ca^{2+} wave is preceded by a small spike-like Ca^{2+} increase (Harada et al. 2011). The sperm tryptic protease seems to be involved in the small and nonpropagative Ca^{2+} increase, but this is insufficient for inducing the Ca^{2+} increase to cause egg activation, probably because of the lack of cortical ER in newt eggs. The inner ER forms a larger complex with some cytoskeletons and is required to trigger a Ca^{2+} wave by the sperm factor (Harada et al. 2011). Egg activation not only by injection of the sperm factor, but also by fertilizing sperm, is probably mediated by the enzymatic activity of sperm citrate synthase (Harada et al. 2011). Reactive substrates of citrate synthase, acetyl CoA and oxaloacetate, induce Ca^{2+} increase to cause egg activation, but citrate does not. The reverse reaction might occur in egg cytoplasm containing a large amount of citrate, and acetyl CoA might then sensitize IP3 receptors to release Ca^{2+}. Further investigation is, however, necessary for determining the exact changes of those substances at fertilization. Furthermore, it is possible that citrate synthase interacts with other molecules such as cytoskeletons (see Fig. 15.1; Iwao and Masui 1995). Investigations into the role of microtubules and microfilaments are important for clarifying the Ca^{2+}-signaling cascade by the sperm factor.

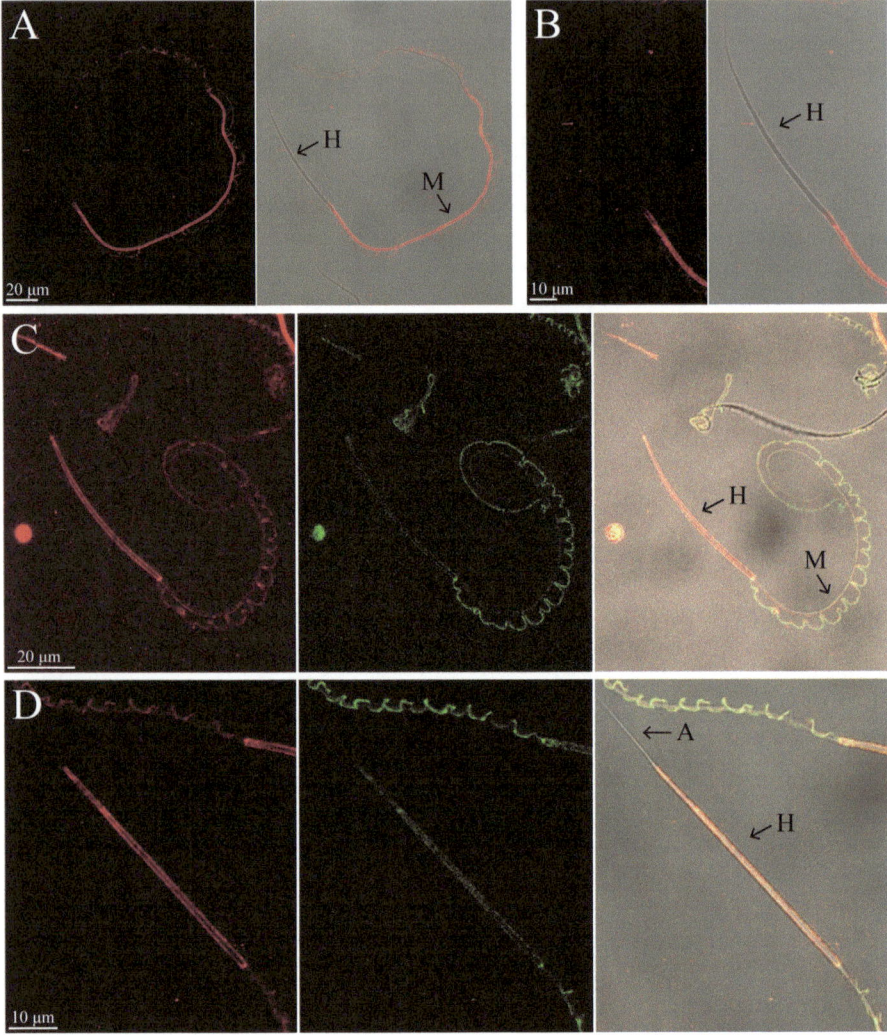

Fig. 15.2 (**A**), (**B**) Newt *Cynops pyrrhogaster* sperm show localization of citrate synthase (*red*) on *left* and merge with the differential interference contrast (DIC) image on *right*. (**C**), (**D**) Salamander *Hynobius nebulosus* sperm show citrate synthase (*red*) on *left*, α-tubulin (*green*) in *middle*, and merge with DIC image on *right*. *A* acrosome, *H* head region, *M* midpiece

15.4 Evolution of a Sperm Factor in Vertebrate Fertilization

It is worth discussing the species specificity of sperm factors to understand the evolution of egg activation in vertebrates. Although it is reported that extract of the monospermic *Xenopus* sperm induces Ca²⁺ oscillation when injected into mouse eggs (Dong et al. 2000), and the injection of several sperm into a *Xenopus* egg causes egg activation (Aarabi et al. 2010), no activity to activate *Xenopus* eggs is detected in homologous sperm extract (Harada et al. 2011). *Xenopus* eggs do not respond to the newt sperm factor, and no citrate synthase is detected in *Xenopus* sperm (Table 15.1). Not only polyspermic newt sperm, but also monospermic *Hynobius* sperm, contain a large amount of citrate synthase under the plasma membrane in the head region, except for the acrosomal region (Fig. 15.2C). Citrate synthase is distributed in close association with microtubules (Fig. 15.2D). A large amount of sperm citrate synthase is observed in mammalian mouse sperm, but not in fish carp sperm (Iwao and Harada, unpublished observations, 2011). Thus, the extramitochondrial localization of citrate synthase in the sperm appears to be acquired in the transition between monospermy and physiological polyspermy in urodele amphibians.

Taken together, the large and single Ca^{2+} wave induced by the first sperm entry is necessary for ensuring monospermy to elicit the positive fertilization potential mediated by Ca^{2+}-activated Cl^- channels in monospermic vertebrates, such as lampreys (Kobayashi et al. 1994), frogs, and *Hynobius* salamanders (Table 15.1). In the bony fishes, a single Ca^{2+} wave is induced by a fertilizing sperm (Gilkey et al. 1978; Webb and Miller 2013), but monospermy is ensured by a micropyle (canal) on the hard chorion, through which only one sperm approaches the egg (Iwamatsu 2000). Thus, the single Ca^{2+} wave at egg activation is characteristic of monospermic vertebrates (Iwao 2012). In this connection, it is interesting to know the Ca^{2+} increase at the physiological polyspermy of large eggs in sharks and chimera (Hart 1990). In contrast, multiple Ca^{2+} waves are necessary for egg activation in physiological polyspermy because a single newt sperm does not have a sufficient amount of sperm factor to induce egg activation and multiple Ca^{2+} increases are necessary for complete activation of the large eggs (Iwao 2012). Some transitional characteristics are, however, observed in occasionally polyspermic eggs of the frog *Discoglossus picutus* with multiple Ca^{2+} increases (Talevi 1989), or in external and polyspermic fertilization in the Japanese giant salamander *Andrias japonicus* (Table 15.1) (Iwao 2000a). Physiological polyspermy probably appeared in species whose egg size was more than about 2 mm in diameter (Table 15.1). Reptiles and birds lay larger and yolky eggs, but their Ca^{2+} increase at polyspermy remains to be investigated. In primitive mammals, the monotrematous platypus laying big eggs exhibits physiological polyspermy (Gatenby and Hill 1924). Although a small and yolkless egg of the higher eutherian mouse exhibits monospermy, it elicits multiple Ca^{2+} increase to ensure sufficient egg activation (Ozil 1990; Ducibella et al. 2002). Sperm-specific PLCζ is known as a potent sperm factor for egg activation in mammals (Saunders et al. 2002; Kouchi et al. 2004) and birds (Mizushima et al. 2009). In mammalian egg activation, the role of sperm citrate synthase remains unknown.

15.5 Perspective

Thus, comparative studies in fertilization among vertebrates provide better understanding of the role of egg activation in establishment of polyspermy during evolution. Because egg activation by sperm citrate synthase is tightly linked to slow egg activation in physiological polyspermy, investigations in polyspermic birds and reptiles are important to clarify the evolution of egg activation in vertebrates. It is also interesting to know the mechanisms of egg activation in bony fishes that exhibit monospermy but lack the fast electrical block to polyspermy. In addition, investigations in invertebrates, such as ascidians and sea urchins, may provide us with the ancestral and universal mechanisms of egg activation during the evolution of animal reproduction.

Acknowledgments We thank Tomoyo Ueno for her help on preparing the manuscript. This work was supported in part by a Grant-in-Aid for Scientific Research on Innovative Areas from MEXT (22112518, 24112712).

References

Aarabi M, Qin Z, Xu W, Mewburn J, Oko R (2010) Sperm-borne protein, PAWP, initiates zygotic development in *Xenopus laevis* by eliciting intracellular calcium release. Mol Reprod Dev 77(3):249–256

Akiyama S, Iwao Y, Miura I (2011) Evidence for true fall-mating in Japanese newt *Cynops pyrrhogaster*. Zool Sci 28(10):758–763

Charbonneau M, Moreau M, Picheral B, Vilain JP, Guerrier P (1983) Fertilization of amphibian eggs: a comparison of electrical responses between anurans and urodeles. Dev Biol 98(2):304–318

Cross NL, Elinson RP (1980) A fast block to polyspermy in frogs mediated by changes in the membrane potential. Dev Biol 75(1):187–198

Dong JB, Tang TS, Sun FZ (2000) *Xenopus* and chicken sperm contain a cytosolic soluble protein factor which can trigger calcium oscillations in mouse eggs. Biochem Biophys Res Commun 268(3):947–951

Ducibella T, Huneau D, Angelichio E, Xu Z, Schultz RM, Kopf GS, Fissore R, Madoux S, Ozil JP (2002) Egg-to-embryo transition is driven by differential responses to Ca^{2+} oscillation number. Dev Biol 250(2):280–291

Fankhauser G (1948) The organization of the amphibian egg during fertilization and cleavage. Ann N Y Acad Sci 49 (Art 5):684–708

Fontanilla RA, Nuccitelli R (1998) Characterization of the sperm-induced calcium wave in *Xenopus* eggs using confocal microscopy. Biophys J 75(4):2079–2087

Gatenby JB, Hill JP (1924) On an ovum of *Ornithorhynchus* exhibiting polar bodies and polyspermy. J Cell Sci s2-68 (270):229–238

Gilkey JC, Jaffe LF, Ridgway EB, Reynolds GT (1978) A free calcium wave traverses the activating egg of the medaka, *Oryzias latipes*. J Cell Biol 76(2):448–466

Grandin N, Charbonneau M (1992) Intracellular free Ca^{2+} changes during physiological polyspermy in amphibian eggs. Development (Camb) 114(3):617–624

Harada Y, Matsumoto T, Hirahara S, Nakashima A, Ueno S, Oda S, Miyazaki S, Iwao Y (2007) Characterization of a sperm factor for egg activation at fertilization of the newt *Cynops pyrrhogaster*. Dev Biol 306(2):797–808

Harada Y, Kawazoe M, Eto Y, Ueno S, Iwao Y (2011) The Ca^{2+} increase by the sperm factor in physiologically polyspermic newt fertilization: its signaling mechanism in egg cytoplasm and the species-specificity. Dev Biol 351(2):266–276

Hart NH (1990) Fertilization in teleost fishes: mechanisms of sperm–egg interactions. Int Rev Cytol 121:1–66

Hedrick JL (2008) Anuran and pig egg zona pellucida glycoproteins in fertilization and early development. Int J Dev Biol 52(5–6):683–701

Ijiri TW, Mahbub Hasan AK, Sato K (2012) Protein-tyrosine kinase signaling in the biological functions associated with sperm. J Signal Transduct 2012:181560

Iwamatsu T (ed) (2000) Fertilization in fishes. Fertilization in Protozoa and metazoan animals. Springer, Berlin

Iwao Y (1985) The membrane potential changes of amphibian eggs during species- and cross-fertilization. Dev Biol 111(1):26–34

Iwao Y (1989) An electrically mediated block to polyspermy in the primitive urodele *Hynobius nebulosus* and phylogenetic comparison with other amphibians. Dev Biol 134(2):438–445

Iwao Y (ed) (2000a) Fertilization in amphibians. Fertilization in Protozoa and metazoan animals. Springer, Berlin

Iwao Y (2000b) Mechanisms of egg activation and polyspermy block in amphibians and comparative aspects with fertilization in other vertebrates. Zool Sci 17(6):699–709

Iwao Y (2012) Egg activation in physiological polyspermy. Reproduction 144(1):11–22

Iwao Y, Elinson RP (1990) Control of sperm nuclear behavior in physiologically polyspermic newt eggs: possible involvement of MPF. Dev Biol 142(2):301–312

Iwao Y, Fujimura T (1996) Activation of *Xenopus* eggs by RGD-containing peptides accompanied by intracellular Ca^{2+} release. Dev Biol 177(2):558–567

Iwao Y, Jaffe LA (1989) Evidence that the voltage-dependent component in the fertilization process is contributed by the sperm. Dev Biol 134(2):446–451

Iwao Y, Masui Y (1995) Activation of newt eggs in the absence of Ca^{2+} activity by treatment with cycloheximide or D_2O. Dev Growth Differ 37(6):641–651

Iwao Y, Yamasaki H, Katagiri C (1985) Experiments pertaining to the suppression of accessory sperm in fertilized newt eggs. Dev Growth Differ 27(3):323–331

Iwao Y, Sakamoto N, Takahara K, Yamashita M, Nagahama Y (1993) The egg nucleus regulates the behavior of sperm nuclei as well as cycling of MPF in physiologically polyspermic newt eggs. Dev Biol 160(1):15–27

Iwao Y, Miki A, Kobayashi M, Onitake K (1994) Activation of *Xenopus* eggs by an extract of *cynops* sperm. Dev Growth Differ 36(5):469–479

Iwao Y, Murakawa T, Yamaguchi J, Yamashita M (2002) Localization of γ-tubulin and cyclin B during early cleavage in physiologically polyspermic newt eggs. Dev Growth Differ 44(6):489–499

Kline D, Nuccitelli R (1985) The wave of activation current in the *Xenopus* egg. Dev Biol 111(2):471–487

Kobayashi W, Baba Y, Shimozawa T, Yamamoto TS (1994) The fertilization potential provides a fast block to polyspermy in lamprey eggs. Dev Biol 161(2):552–562

Kouchi Z, Fukami K, Shikano T, Oda S, Nakamura Y, Takenawa T, Miyazaki S (2004) Recombinant phospholipase Cζ has high Ca^{2+} sensitivity and induces Ca^{2+} oscillations in mouse eggs. J Biol Chem 279(11):10408–10412

Mahbub Hasan AK, Sato K, Sakakibara K, Ou Z, Iwasaki T, Ueda Y, Fukami Y (2005) Uroplakin III, a novel Src substrate in *Xenopus* egg rafts, is a target for sperm protease essential for fertilization. Dev Biol 286(2):483–492

Mahbub Hasan AK, Ou Z, Sakakibara K, Hirahara S, Iwasaki T, Sato K, Fukami Y (2007) Characterization of *Xenopus* egg membrane microdomains containing uroplakin Ib/III complex: roles of their molecular interactions for subcellular localization and signal transduction. Genes Cells 12(2):251–267

Mizote A, Okamoto S, Iwao Y (1999) Activation of *Xenopus* eggs by proteases: possible involvement of a sperm protease in fertilization. Dev Biol 208(1):79–92

Mizushima S, Takagi S, Ono T, Atsumi Y, Tsukada A, Saito N, Shimada K (2009) Phospholipase Cζ mRNA expression and its potency during spermatogenesis for activation of quail oocyte as a sperm factor. Mol Reprod Dev 76(12):1200–1207

Ozil JP (1990) The parthenogenetic development of rabbit oocytes after repetitive pulsatile electrical stimulation. Development (Camb) 109(1):117–127

Sakakibara K, Sato K, Yoshino K, Oshiro N, Hirahara S, Mahbub Hasan AK, Iwasaki T, Ueda Y, Iwao Y, Yonezawa K, Fukami Y (2005) Molecular identification and characterization of *Xenopus* egg uroplakin III, an egg raft-associated transmembrane protein that is tyrosine-phosphorylated upon fertilization. J Biol Chem 280(15):15029–15037

Sato K, Iwao Y, Fujimura T, Tamaki I, Ogawa K, Iwasaki T, Tokmakov AA, Hatano O, Fukami Y (1999) Evidence for the involvement of a Src-related tyrosine kinase in *Xenopus* egg activation. Dev Biol 209(2):308–320

Sato K, Tokmakov AA, He CL, Kurokawa M, Iwasaki T, Shirouzu M, Fissore RA, Yokoyama S, Fukami Y (2003) Reconstitution of Src-dependent phospholipase Cγ phosphorylation and transient calcium release by using membrane rafts and cell-free extracts from *Xenopus* eggs. J Biol Chem 278(40):38413–38420

Saunders CM, Larman MG, Parrington J, Cox LJ, Royse J, Blayney LM, Swann K, Lai FA (2002) PLCζ: a sperm-specific trigger of Ca²⁺ oscillations in eggs and embryo development. Development (Camb) 129(15):3533–3544

Shilling FM, Magie CR, Nuccitelli R (1998) Voltage-dependent activation of frog eggs by a sperm surface disintegrin peptide. Dev Biol 202(1):113–124

Talevi R (1989) Polyspermic eggs in the anuran *Discoglossus pictus* develop normally. Development (Camb) 105(2):343–349

Webb SE, Miller AL (2013) Ca²⁺ signaling during activation and fertilization in the eggs of teleost fish. Cell Calcium 53(1):24–31

Yamamoto S, Yamashita M, Iwao Y (1999) Rise of intracellular Ca²⁺ level causes the decrease of cyclin B1 and Mos in the newt eggs at fertilization. Mol Reprod Dev 53(3):341–349

Yamamoto S, Kubota HY, Yoshimoto Y, Iwao Y (2001) Injection of a sperm extract triggers egg activation in the newt *Cynops pyrrhogaster*. Dev Biol 230(1):89–99

Chapter 16
ATP Imaging in *Xenopus laevis* Oocytes

Takashi W. Ijiri, Jun-ichi Kishikawa, Hiromi Imamura,
Yasuhiro Iwao, Ken Yokoyama, and Ken-ichi Sato

Abstract Adenosine 5′-triphosphate (ATP) is a major energy currency for various chemical reactions in living cells. In spite of its significant roles, the local ATP time-course in *Xenopus* oocyte/egg has not been studied fully yet. Therefore, our goal is to understand the molecular mechanisms of maturation and fertilization by analyzing ATP in oocytes and eggs. A fluorescence resonance energy transfer (FRET)-based ATP indicator, named ATeam, has been developed and enabled ATP imaging. We are applying this latest experimental technique to observe local ATP in *Xenopus* oocytes and eggs during maturation and fertilization. First, we used full-grown oocytes to set up the experimental conditions. To observe ATeam fluorescence in *Xenopus* oocytes, the translucent oocytes were prepared after injecting ATeam protein into the vegetal hemisphere. Then they were observed under fluorescence microscopy, and FRET was measured by analyzing their images using software. ATeam displayed strong FRET with low background. More importantly, FRET signal of ATeam consistently increased in response to the addition of ATP, suggesting that ATeam works in the translucent oocytes. Using this *Xenopus* ATP imaging system, ATP distribution is investigated not only in oocytes during maturation but also in eggs during fertilization.

Keywords ATP imaging • Maturation • Oocyte • *Xenopus laevis*

T.W. Ijiri • J. Kishikawa • K. Yokoyama (✉) • K. Sato (✉)
Department of Molecular Biosciences, Faculty of Life Sciences, Kyoto Sangyo
University, Kyoto 603-8555, Japan
e-mail: yokoken@cc.kyoto-su.ac.jp; kksato@cc.kyoto-su.ac.jp

H. Imamura
The Hakubi Center for Advanced Research, Kyoto University, Kyoto 606-8302, Japan

Y. Iwao
Department of Applied Molecular Biosciences, Graduate School of Medicine,
Yamaguchi University, Yamaguchi 753-8512, Japan

H. Sawada et al. (eds.), *Sexual Reproduction in Animals and Plants*,
DOI 10.1007/978-4-431-54589-7_16, © The Author(s) 2014

16.1   Introduction

Adenosine 5′-triphosphate (ATP) works as intracellular energy transfer for many reactions of the living body in a wide variety of organisms. For all its importance, ATP distribution in the cells could not be investigated because methods for local ATP detection were lacking. However, Imamura et al. (2009) developed an ATP indicator, named ATeam, that consists of CFP and YFP fused with the ε-subunit of FoF_1-ATPase. The ε-subunit changes its conformation into a folded form upon ATP binding, resulting in increase of YFP/CFP ratio under higher ATP concentration. In other words, measurement of its fluorescence resonance energy transfer (FRET) efficiency enables ATP imaging. [For the details of ATeam, please refer the original publication (Imamura et al. 2009).] The number of reports has increased lately about successful ATP observations using ATeam in worms (Kishikawa et al. 2012), in the hepatitis C virus-replicating cells (Ando et al. 2012), in plant cells (Hatsugai et al. 2012), and in flies and worms (Tsuyama et al. 2013) since the first report in HeLa cells (Imamura et al. 2009).

The study of ATP metabolism in the oocyte/egg has been performed mainly in mammals. It was suggested that oxidative phosphorylation in mitochondria is activated during maturation (Brinster 1971; Eppig 1976) and fertilization (Dumollard et al. 2004). In *Xenopus* oocytes, vast numbers of mitochondria are produced suddenly near the nucleus at early oogenesis; they then localize to the vegetable pole side in the progress of oogenesis. To understand how ATP metabolism occurs in the *Xenopus* oocyte or egg, observation of local ATP is needed. Therefore, we started ATP imaging in *Xenopus* oocytes/eggs using ATeam. In this chapter, we show results to set up the ATP imaging system in *Xenopus* oocytes were successful and discuss its future implications.

16.2   Methodology

16.2.1   Purification of ATeam Protein

ATeam protein was prepared according to the procedures by Imamura et al. (2009). First, *Escherichia coli* transformed by ATeam expression plasmid was cultured in LB medium overnight, then the collected bacteria were dissolved in buffer A [100 mM Na_3PO_4 (pH 8.0), 200 mM NaCl, and 10 mM imidazole]. After treatment by a sonicator, they were applied to a Ni-NTA column and eluted by buffer A with 200 mM imidazole. Furthermore, the concentrated ATeam fraction was purified by gel filtration and replaced in 20 mM Tris-HCl (pH 8.0) and 150 mM NaCl. After adding glycerol to purified ATeam protein, it was frozen rapidly in liquid nitrogen and stored at −80 °C until use. Protein concentration was determined by the absorption at 515 nm of YFP.

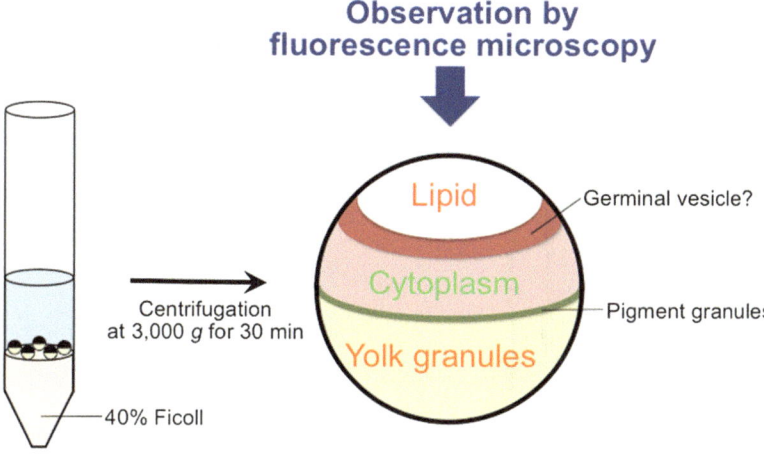

Fig. 16.1 Schematic diagram of the method for translucent oocytes. *Left*: Oocytes were centrifuged at 3,000 *g* for 30 min on a 40 % Ficoll cushion. *Right*: The oocyte stratified into lipid, a semitransparent layer, clear cytoplasm, pigment granules, and yolk granules from the centripetal to centrifugal side after this treatment

16.2.2 Preparation of the Translucent **Xenopus** Oocytes

Xenopus oocytes and eggs contain pigment and yolk granules that interrupt the observation of fluorescence under microscopy. Therefore, the preparation of translucent oocytes or eggs is needed to observe the FRET signal of ATeam. In this report, we prepared the translucent oocytes using the methods by Iwao et al. (2005). Ovaries were dissected out from adult frogs (*Xenopus laevis*) and treated with 1 mg/ml collagenase. After washing, stage VI full-grown oocytes were stored in modified birth solution [MBS: 88 mM NaCl, 1 mM KCl, 2.4 mM NaHCO$_3$, 10 mM HEPES, 0.82 mM MgSO$_4$, 0.33 mM Ca(NO$_3$)$_2$, 0.41 mM CaCl$_2$, pH 7.6] with 0.01 % chloramphenicol. In advance, ATeam protein (23 or 46 nl, corresponding to 0.1 or 0.2 pmol, respectively) was injected into vegetal hemisphere of the oocytes. The oocytes were put onto 40 % Ficoll in MBS, then centrifuged at 3,000 *g* for 30 min (Fig. 16.1). After this treatment, lipids form a cap-like structure on the top of the oocytes; germinal vesicle may surround this lipid layer, and translucent cytoplasm occupies the remaining area in the animal hemisphere, while pigment granules set on the equator and yolk granules consists in whole vegetal hemisphere (Fig. 16.1). These translucent oocytes were put in MBS until observation.

16.2.3 Observation Under Microscopy and Image Analysis

The translucent oocytes were observed under fluorescence microscopy (Zeiss, Axioplan2) with the filters D480/30 for CFP and D535/40 for YFP (Photometrics, DualView2). The fluorescence images were taken by a digital camera (Hamamatsu,

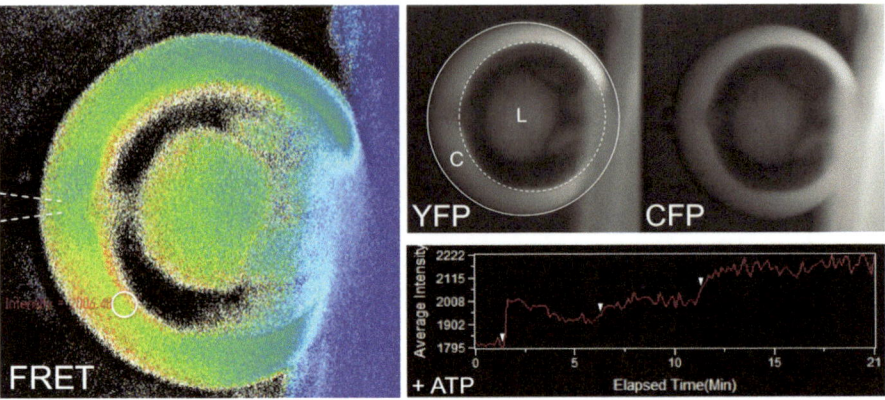

Fig. 16.2 Monitoring of ATeam in *Xenopus* oocytes. FRET signal of ATeam increased after each of three injections of ATP. *Left*: The *white circle* in the FRET image is the spot where average intensity was measured for the line graph. The two *white dotted lines* indicate a glass needle injected into translucent cytoplasm. *Right upper*: Each image of ATeam YFP and CFP. *L* lipid layer, *C* translucent cytoplasm. *Right bottom*: *White heads* in the line graph show the times when ATP was injected. *Note*: Viewing angle by fluorescence microscopy was from animal pole to vegetal pole. The diameter of the *Xenopus* oocyte is 1.2–1.3 mm

ORCA-Flash2.8) every 10 s. The fluorescence intensity was measured and then FRET signal of ATeam (YFP/CFP) was calculated using MetaMorph software (Molecular Devices). The results were shown as average intensity (Fig. 16.2), which is 1,000 fold of the FRET value. For the example in Fig. 16.2, 0.2 pmol ATeam protein was injected into an oocyte before treatment for translucent oocytes, then 9.2 nl 16.5 mM ATP solution was injected three times into an oocyte through a glass needle, which was inserted into the translucent cytoplasm beforehand.

16.3   Injected ATeam Protein Works in *Xenopus* Oocytes

Our preliminary trial using robust ATeam protein resulted in bright fluorescence over all the translucent cytoplasm of unfertilized egg without any background (Ijiri, Kishikawa, Imamura, Iwao, Yokoyama, and Sato, unpublished results). Therefore, the next trial should be performed with minimal ATeam amount to detect accurate ATP distribution in the *Xenopus* oocyte or egg. This time we used full-grown oocytes for long-term observation because they are easily handed compared to an unfertilized egg; that is, the oocytes have moderate strength and are not activated by stimuli such as injection. After several trials, we decided that 0.1–0.2 pmol of protein injection per oocyte is suitable for the *Xenopus* ATP imaging system. In this study, we used modified ATeam, which is optimized for the lower temperature of most model animals (20–25 °C) (Tsuyama et al. 2013). Under these experimental considerations, our strategy worked well, and either 0.1 or 0.2 pmol of ATeam protein also produced similar results. First, we could observe a strong FRET signal of ATeam with low background over all the translucent cytoplasm in the edge of an oocyte, although a lipid layer blocked in the central part (Fig. 16.2). Then, the FRET did not decrease markedly during the observation time: at least 1 h (data not shown). The most important thing to note here is that ATeam reacted with the increase of ATP concentration during observation. In this test to evaluate the imaging system, ATP was injected into an oocyte three times between moderate intervals, resulting in constant increase of FRET after each of the three injections (Fig. 16.2). These data suggest that ATeam works accurately in *Xenopus* translucent oocytes.

16.4   Conclusions and Future Directions

Here, we have succeeded in observing the FRET signal of ATeam using translucent oocytes in *Xenopus*. In most cases, ATeam mRNA was injected into the subject of the research, whereas ATeam protein was used in our *Xenopus* system. As far as we know, this is the first report about ATP imaging in living *Xenopus* oocytes. In our methodology, oocytes have to be treated by centrifugation on Ficoll cushion to prepare the translucent cytoplasm; however, this treatment does not disrupt the cleavage of embryos, at least until blastula stage (Iwao et al. 2005). Therefore, for further investigation, we will observe the time-course of local ATP distribution in the oocytes after adding progesterone to detect the difference during oocyte maturation. We will also apply this imaging system with unfertilized eggs to observe local ATP in *Xenopus* eggs at fertilization. To understand the biological significance of ATP distribution in oocytes and eggs, it is needed to confirm whether ATP localizes with mitochondria using a mitochondrion-selective proves. It is also important to examine the correlation between ATP and Ca^{2+} with calcium indicators during maturation as well as during fertilization. Another approach will use inhibitors for glycolysis, 2-deoxyglucose, and for oxidative phosphorylation, KCN and oligomycin A, to define whether ATP is produced only by oxidative phosphorylation in oocytes during maturation and in eggs during fertilization. Our future knowledge from this study can contribute to understanding the fundamental mechanisms of ATP metabolism in vertebrate oocytes and eggs.

References

Ando T, Imamura H, Suzuki R et al (2012) Visualization and measurement of ATP levels in living cells replicating hepatitis C virus genome RNA. PLoS Pathog 8(3):e1002561

Brinster RL (1971) Oxidation of pyruvate and glucose by oocytes of the mouse and rhesus monkey. J Reprod Fertil 24(2):187–191

Dumollard R, Marangos P, Fitzharris G et al (2004) Sperm-triggered [Ca^{2+}] oscillations and Ca^{2+} homeostasis in the mouse egg have an absolute requirement for mitochondrial ATP production. Development (Camb) 131(13):3057–3067

Eppig JJ (1976) Analysis of mouse oogenesis in vitro. Oocyte isolation and the utilization of exogenous energy sources by growing oocytes. J Exp Zool 198(3):375–382

Hatsugai N, Perez Koldenkova V, Imamura H et al (2012) Changes in cytosolic ATP levels and intracellular morphology during bacteria-induced hypersensitive cell death as revealed by real-time fluorescence microscopy imaging. Plant Cell Physiol 53(10):1768–1775

Imamura H, Nhat KP, Togawa H et al (2009) Visualization of ATP levels inside single living cells with fluorescence resonance energy transfer-based genetically encoded indicators. Proc Natl Acad Sci USA 106(37):15651–15656

Iwao Y, Uchida Y, Ueno S et al (2005) Midblastula transition (MBT) of the cell cycles in the yolk and pigment granule-free translucent blastomeres obtained from centrifuged *Xenopus* embryos. Dev Growth Differ 47(5):283–294

Kishikawa J, Fujikawa M, Imamura H et al (2012) MRT letter: expression of ATP sensor protein in *Caenorhabditis elegans*. Microsc Res Tech 75(1):15–19

Tsuyama T, Kishikawa JI, Han YW et al (2013) In vivo fluorescent ATP imaging of *Drosophila melanogaster* and *Caenorhabditis elegans* by using a genetically encoded fluorescent ATP biosensor optimized for low temperatures. Anal Chem 85(16): 7889–7896

Chapter 17
Mitochondrial Activation and Nitric Oxide (NO) Release at Fertilization in Echinoderm Eggs

Tatsuma Mohri and Keiichiro Kyozuka

Abstract In the cell, mitochondria are important organelles to supply energy for essential cell activities. It is unclear when and how mitochondrial activity increases on fertilization. We refer to mitochondrial activation here as mitochondrial inner-membrane hyperpolarization. We measured fluorescence changes in mitochondrial inner-membrane potential ($\Delta\Psi_m$) during fertilization in eggs of the echinoderm *Hemicentrotus pulcherrimus* and the sand dollar *Clypeaster japonicus* using a mitochondrial dye, MitoTracker Red CMXR (MTR). Increase in fluorescence was detected after insemination in most eggs, which indicates hyperpolarization in $\Delta\Psi_m$. To obtain the relationship between changes in intracellular Ca^{2+} ($[Ca^{2+}]_i$) and $\Delta\Psi_m$, simultaneous measurements of $[Ca^{2+}]_i$ and $\Delta\Psi_m$ were carried out using MTR and Ca Green dextran. Results showed that an increase in $\Delta\Psi_m$ occurred just after the increase in $[Ca^{2+}]_i$. Moreover, microinjection of 1 mM BAPTA suppressed the $\Delta\Psi_m$. These results suggest that the mitochondrial activation is dependent on $[Ca^{2+}]_i$ increase. However, nitric oxide (NO) release (ΔNO) at fertilization in sea urchin eggs is a well-known phenomenon, although its function is not clear. Our previous study demonstrated that ΔNO at fertilization occurred at a peak of $[Ca^{2+}]_i$ and was inhibited by cyanide, a mitochondrial inhibitor, suggesting that ΔNO is related to mitochondrial activation. Simultaneous measurements of $[Ca^{2+}]_i$ and $\Delta\Psi_m$ revealed that the timing of the increase of $\Delta\Psi_m$ was a little earlier than that of ΔNO. We conclude both ΔNO and changes in $\Delta\Psi_m$ are downstream phenomena caused by sperm-induced $[Ca^{2+}]_i$ increase. We hypothesize that mitochondrial activation is involved in ΔNO. Therefore, we are investigating the issue further.

T. Mohri (✉)
National Institutes of Natural Sciences, National Institute
for Physiological Sciences, Okazaki 444-8787, Japan
e-mail: tsmohri@nips.ac.jp

K. Kyozuka
Research Center for Marine Biology, Asamushi, Graduate School of Life Science,
Tohoku University, Asamushi Aomori 039-3501, Japan

H. Sawada et al. (eds.), *Sexual Reproduction in Animals and Plants*,
DOI 10.1007/978-4-431-54589-7_17, © The Author(s) 2014

Keywords Changes in mitochondrial inner-membrane potential ($\Delta\Psi_m$) • Intracellular Ca^{2+} ($[Ca^{2+}]_i$) • Nitric oxide (NO) release (ΔNO)

17.1   Introduction

Fertilization is thought to be one of the most remarkable phenomena, in which a large amount of energy is required for a zygote to increase diverse cellular activities leading to structural and qualitative changes for development from a dormant unfertilized state. At fertilization, mitochondria could be significant in increasing various cellular activities, including supplying the energy. Generally, mitochondria have been recognized as multifunctional organelles playing crucial roles in cellular control. Their roles are to provide the main cellular energy in the form of adenosine triphosphate (ATP), to generate and regulate reactive oxygen species, to buffer cytosolic Ca^{2+}, and to regulate apoptosis through a mitochondrial permeability transition pore (Wallace et al. 2010). Therefore, it is assumed that mitochondrial activity should increase accompanying the sequential phenomena, called "egg activation," that comprise changes in intracellular concentration of Ca^{2+} ($[Ca^{2+}]_i$), pH, nitric oxide (NO) release (ΔNO), other second-messenger signals, and cytoplasmic changes, leading to activation of many important enzymes and factors for development. In sea urchins, early studies on respiration demonstrated that a burst of respiration started immediately after fertilization and reached the peak level within 3 min (Foerder et al. 1978; Fujiwara and Yasumasu 1997; Warburg 1908). One third of the respiration at fertilization should be responsible for mitochondrial respiration. However, when and how the mitochondria really activate remains unclear. According to the chemiosmotic theory, measuring mitochondrial inner-membrane potential ($\Delta\Psi_m$) has been used as an index of mitochondrial activity because $\Delta\Psi_m$ is utilized to provide energy for ATP production (Solani et al. 2007). Mitochondrial activation is believed to be by changes in hyperpolarization of $\Delta\Psi_m$ (Fujihara et al. 2007). On the other hand, ΔNO at fertilization has been reported in sea urchin eggs, although its main function is still obscure (Leckie et al. 2003; Mohri et al. 2008). We found in our previous study that mitochondrial activity has some relationship to ΔNO because mitochondrial inhibitors such as CN^- and NaN_3 inhibited ΔNO, and an egg stratified by centrifugation showed colocalization by double-staining with the mitochondrial dye MitoRed and the NO indicator DAF-FM DA (Mohri et al. 2008). Therefore, we have investigated the precise relationship between $\Delta\Psi_m$ and ΔNO.

Here, we report whether mitochondrial activity truly increases after insemination by measuring changes in fluorescence of a mitochondrial inner-membrane potential sensitive dye as $\Delta\Psi_m$ at fertilization using MitoTracker Red CMXR (MTR). For the relationship between ΔNO and $\Delta\Psi_m$, we show the effect of mitochondrial inhibitors carbonyl cyanide-*4*-(trifluoromethoxy)phenylhydrazone (FCCP) and CN^- on ΔNO. We also report the relationship between $[Ca^{2+}]_i$ and $\Delta\Psi_m$ by simultaneous measurements and show the $[Ca^{2+}]_i$ dependency of $\Delta\Psi_m$ by injection of a Ca^{2+} chelator, 1,2-bis(o-aminophenoxy)ethane-*N*,*N*,*N'*,*N'*-tetraacetic acid (BAPTA), because $[Ca^{2+}]_i$ increase has been reported to induce mitochondrial activation in various types of cells such as heart, muscle, liver, kidney, pancreas, and nervous system (Griffiths and Rutter 2009; Gunter and Sheu 2009).

17.2 Materials and Methods

17.2.1 Gametes

Eggs of the sea urchin *Hemicentrotus pulcherrimus* and the sand dollar *Clypeaster japonicus* were used for experiments. Eggs were obtained by injecting 0.2–0.4 ml 0.1 mM acetylcholine into the coelomic cavity. Whole testes removed from males were stored dry at 4 °C. Gametes were used within 10 h after shedding, only when more than 95 % of the eggs elevated simultaneously with symmetrical fertilization envelopes (FE) and cleaved regularly through the four-cell stage. Experiments were carried out at 20–21 °C for *H. pulcherrimus* and 24 °C for *C. japonicus*, respectively, using experimental seawater (ESW) [natural seawater (NSW) passed through a membrane filter (0.22 μm), buffered with 10 mM tris(hydroxymethyl)-ethylaminopropane sulfonic acid (TAPS), and adjusted to pH 8.2 with NaOH]. The NSW used was collected from surface water off the coast just outside Asamushi Marine Biological Station.

17.2.2 Measurements of $\Delta\Psi_m$, ΔNO, and $[Ca^{2+}]_i$

$\Delta\Psi_m$ was measured as fluorescence changes in eggs loaded with a mitochondrial dye, MitoTracker Red CMXR (MTR; Life Technologies, USA). Eggs were immersed for 10 min in 1,000 nM MTR dissolved in ESW. Nitric oxide release (ΔNO) measurement was carried out in eggs loaded with 12.5–50 μM DAF-FM DA (DAF) for 15 min or 2.5 μM DAR-4M AM (DAR) for 60 min, respectively. For measurement of intracellular Ca^{2+} ($[Ca^{2+}]_i$), Ca green dextran (CGD) was microinjected into eggs by a slightly modified Hiramoto's method (Hiramoto 1974). Fluorescence images were detected using a high-speed fluorescence imaging system (HSFIS) (AquaCosmos/Ratio; Hamamatsu Photonics, Hamamatsu, Japan) attached to an inverted microscope (TE-300; Nikon, Tokyo, Japan). Some experiments were carried out using a confocal microscope (A1R; Nikon). In the HSFIS, excitation light at 490 nm and 560 nm was obtained with a grating monochrometer (POLYCHROME-2; T.I.L.L Photonics, Martinsried, Germany), a dual dichroic beam splitter (FF01-512/630-25; Semrock, Rochester, NY, USA), and a dual emission filter at 512/630 nm were used for a measurement of CGD, DAF, DAR, and MTR fluorescence. Fluorescence images of eggs loaded with those dyes were acquired every 2 and 5 s using a cooled CCD camera (C6790; Hamamatsu Photonics). Processing was performed with Image J (a public domain image processing software). Changes in fluorescence for both $\Delta\Psi_m$ and $[Ca^{2+}]_i$ were expressed on a graph in terms of ratio values, $R(F/F_0)$, where F is a background-subtracted fluorescent image of the egg during fertilization and F_0 a background-subtracted fluorescent image just before insemination. Therefore, the initial value of R is 1 and fractional increase is $R - 1$. ΔNO is expressed with displacement values as dF $(F - F_0)$ in a graph because fluorescence of DAF is accumulative and we found it much more suitable to express ΔNO as dF rather than a ratio.

A final concentration of 0.1–2 mM NaCN and 10–20 µM carbonyl cyanide-4-(trifluoromethoxy)phenylhydrazone (FCCP) in ESW was used for experiments. In some experiments, bright-field images were simultaneously monitored with fluorescence measurement using red light at >640 nm by another dichroic beam splitter (Chroma Tech, Bellows Falls, VT, USA) and by another CCD camera (C2400-77; Hamamatsu Photonics) to obtain morphological changes in fertilizing eggs. The images were recorded on a DVD recorder (RDR-HX10; Sony, Tokyo, Japan).

17.2.3 Experimental Procedure on the Microscopes

The experimental chamber was made from a 35 × 10 mm plastic culture dish lid (Falcon #1008; Becton Dickinson Japan, Tokyo, Japan) in which a hole of 20–22 mm in diameter was cut out in the bottom and cemented by a circular coverglass (24–26 mm in diameter) using dental wax. Poly-L-lysine at the concentration of 0.005 % was used to facilitate adhesion of eggs. The chamber was mounted on the stage of the inverted microscope. About five to ten de-jellied eggs loaded with MTR were pipetted into the chamber filled with 3 or 4 ml ESW using a volume-controllable glass pipette connected to a mouthpiece with a silicone tube. For simultaneous measurements of $[Ca^{2+}]_i$ and $\Delta\Psi_m$, eggs loaded with MTR were set in the chamber and then microinjected with CGD.

Insemination was carried out by gently pipetting 30 µl of diluted suspension of sperm (1–2 µl of dry sperm/10 ml ESW) into the bath ~1 cm away from the experimental egg after starting simultaneous measurements of fluorescence images and bright-field images. Elevation of the fertilization envelope (FE) and other morphological changes were monitored by bright-field images. After the experiment, egg development was observed at least up to the two-cell stage. Occasionally, development was observed until a higher stage or hatching stage.

17.3   Results and Discussion

17.3.1   Mitochondrial Activation (Inner-Membrane Hyperpolarization) at Fertilization

Eggs of *H. pulcherrimus* loaded with 50 nM MTR were inseminated and changes in fluorescence reflecting $\Delta\Psi_m$ were measured. At 50–100 s after insemination, an increase in fluorescence was observed (Fig. 17.1A). To examine whether the increase is caused by mitochondrial inner-membrane hyperpolarization, we first

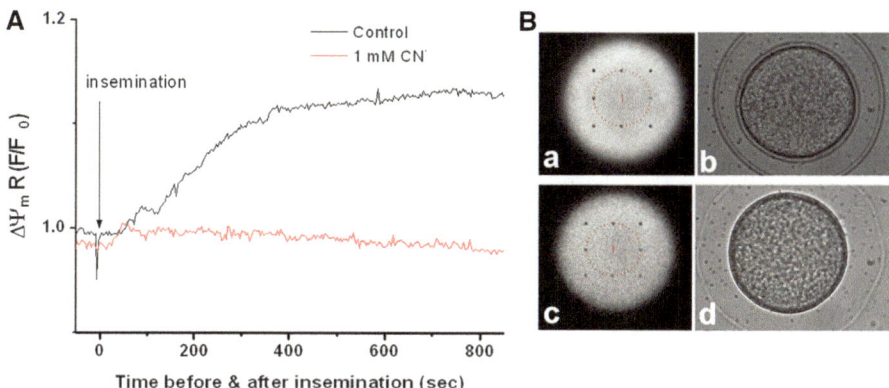

Fig. 17.1 Mitochondrial activation takes place at fertilization. (**A**) *Black* and *red lines* present typical patterns of mitochondrial inner-membrane potential ($\Delta\Psi_m$) during fertilization in a control egg and an egg in the presence of 1 mM NaCN, respectively. (**B**) Corresponding images of fluorescence (*a* and *c*) and bright field after each experiment (*b* and *d*). *a* and *b* are a control egg, *c* and *d* are in the presence of NaCN; they correspond to the *black* and *red lines* in a, respectively. *Red circle* indicates region of interest (ROI) where fluorescence intensities is acquired. *Note*: Fertilization envelope (FE) elevated in both cases. In *d*, FE looks pale and higher

used MTR in concentrations ranging from 1 to 1,000 nM. In most concentrations we tested, the changes in fluorescence were positive (Fig. 17.2). The positive changes indicate hyperpolarization unless the dye, MTR, aggregates within mitochondria and quenches fluorescence. To clarify whether the fluorescence is quenched at the concentration of the dye used, we next applied an uncoupler, FCCP, that dissipates the mitochondrial inner-membrane potential. When, after insemination, 10–20 µM FCCP was applied to eggs showing increase in $\Delta\Psi_m$, the fluorescence signal immediately decreased and then increased but eventually gradually decreased (Fig. 17.3a). When FCCP was added to unfertilized eggs, the fluorescence signal dropped promptly and then continuously decreased. More than 20 min later, those eggs were inseminated. A small increase appeared but eventually showed a gradual decrease in fluorescence (Fig. 17.3b). These results suggest the increase in fluorescence was caused by the hyperpolarization of $\Delta\Psi_m$. All these data suggest that mitochondrial activation takes place at fertilization in sea urchin eggs.

17.3.2 Inhibition of Mitochondrial Activation ($\Delta\Psi_m$) by CN- or FCCP

CN^- is a well-known inhibitor of mitochondrial respiratory complex IV (Fujiwara et al. 2000). Figure 17.4b demonstrates that ΔNO was inhibited by CN^- as previously reported (Mohri et al. 2008). Because CN^- inhibited ΔNO, we also examined whether CN^- affects $\Delta\Psi_m$. We measured $\Delta\Psi_m$ in eggs in the presence of CN^- when

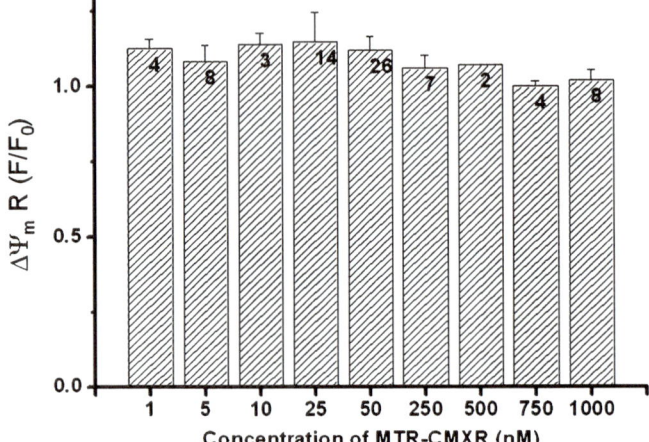

Fig. 17.2 Comparison of $\Delta\Psi_m$ between different concentrations of MitoTracker Red CMXR (MTR). Values of $\Delta\Psi_m$ were measured at 600 s after insemination and expressed as ratios (mean ± SD). The *numbers* below the *x*-axis represent concentrations of MTR used (in nM). *Number in each column* indicates the number of eggs examined

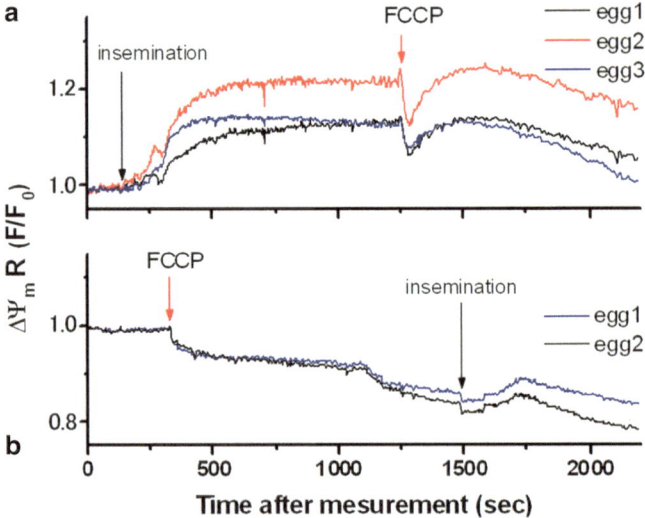

Fig. 17.3 Inhibition of $\Delta\Psi_m$ by carbonyl cyanide-*4*-(trifluoromethoxy)phenylhydrazone (FCCP). (**a**) Effect of FCCP addition on $\Delta\Psi_m$ in eggs after fertilization: 20 μM FCCP (final concentration) was added in three fertilized *Hemicentrotus pulcherrimus* eggs in the same chamber. (**b**) Effect of FCCP addition on $\Delta\Psi_m$ in unfertilized eggs: 20 μM FCCP was added in two unfertilized eggs in the same chamber. Insemination was carried out 20 min after addition of FCCP

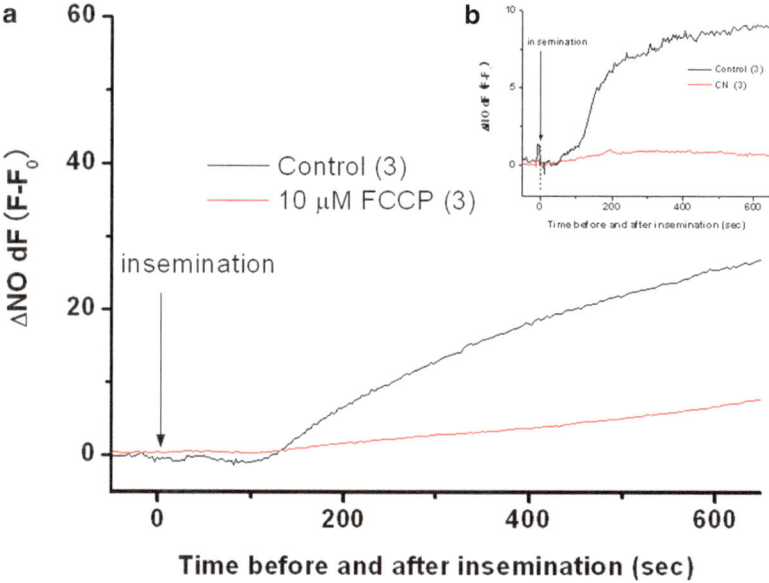

Fig. 17.4 (**a**) Suppression of ΔNO by FCCP. ΔNO was measured in the presence of 10 μM FCCP. ΔNOs in control (3) and eggs (3) treated with FCCP were compared. (**b**) Inhibition of ΔNO by CN⁻. ΔNO was measured using DAR. *Black* and *red lines* indicate average patterns of ΔNO during fertilization in three control eggs and three eggs in the presence of 1 mM NaCN, respectively. *Numbers* in parentheses indicate the mean number of eggs

inseminated. Eggs loaded with MTR were immersed in 1–2 mM NaCN dissolved in ESW for more than 10 min. $\Delta\Psi_m$ was almost completely inhibited, although very pale, and high FE always appeared (Fig.17.1 A, B-b). Similarly, eggs loaded with MTR were inseminated in the presence of 10 μM FCCP after 10-min pretreatment. None of the eggs showed a significant increase in $\Delta\Psi_m$ during fertilization, similar to a late part of the graph after insemination in Fig. 17.3b. Also, ΔNO was measured in the presence of 10 μM FCCP. Compared to the control, ΔNO was significantly suppressed (Fig. 17.4a). Results demonstrated that inhibition of mitochondrial activation by CN⁻ and FCCP induces inhibition of ΔNO. Taken together, the results suggest ΔNO is largely involved in mitochondrial activation.

17.3.3   Timing of $\Delta\Psi_m$ and ΔNO

In a single measurement of $\Delta\Psi_m$ or ΔNO, there is uncertainty on the timing of the phenomenon because it is difficult to know the exact time of sperm–egg fusion. Most of the uncertainty is in estimating the time between addition of sperm and cellular fertilization. In addition, there is human error between observation of the egg and recording the clock time. In echinoderm eggs, changes in $[Ca^{2+}]_i$ always start

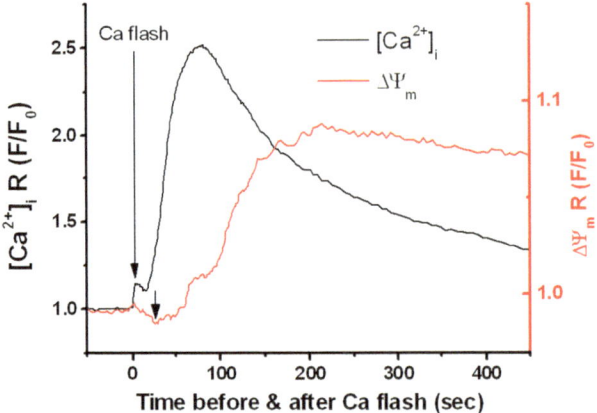

Fig. 17.5 Simultaneous measurements of $[Ca^{2+}]_i$ and $\Delta\Psi_m$. $[Ca^{2+}]_i$ and $\Delta\Psi_m$ were simultaneously measured using CGD and MTR. Time 0 was standardized by the occurrence of a "Ca flash," indicating Ca^{2+} influx from extracellular media through voltage-dependent Ca^{2+} channels in the egg plasma membrane. A *black short arrow* indicates a point at which $\Delta\Psi_m$ started to increase just after increasing $[Ca^{2+}]_i$ at fertilization. A small decreasing part in the beginning of $\Delta\Psi_m$ is possibly an artifact. Details are described in the text

with a small rise in $[Ca^{2+}]_i$ called a "Ca flash or cortical flash" (that is induced by Ca^{2+} influx from the whole surface of the egg through voltage-dependent Ca^{2+} channels). The Ca flash can be a very good indication of the time point of fertilization. Therefore, to examine more precisely the spatiotemporal relationship between $[Ca^{2+}]_i$ and $\Delta\Psi_m$ during fertilization, simultaneous measurements of $[Ca^{2+}]_i$ and $\Delta\Psi_m$ were carried out. The patterns of changes in $[Ca^{2+}]_i$ and $\Delta\Psi_m$ are shown in Fig. 17.5. Several seconds after the occurrence of the Ca flash $[Ca^{2+}]_i$ began to increase, reached a peak at 80 s, and gradually decreased to one fourth of the peak in 400 s after the Ca flash. On the other hand, changes in $\Delta\Psi_m$ showed a fraction of decrease several seconds after the Ca flash and then became an significant increase associated with $[Ca^{2+}]_i$ rise (Fig. 17.1A). The rise in $\Delta\Psi_m$ started at 25 s after the Ca flash, during a period of increasing $[Ca^{2+}]_i$ (small black arrow in Fig. 17.5). Mean value of the increase in $\Delta\Psi_m$ in the simultaneous measurements of $\Delta\Psi_m$ and $[Ca^{2+}]_i$ experiments was 29 ± 8 s (mean $\pm$ SD, $n = 6$). A small fraction of decrease was sometimes observed that may contain an artifact from fluorescence changes by cytoplasmic deformation at fertilization. Similarly, sequentially changing images of $[Ca^{2+}]_i$ and $\Delta\Psi_m$ are displayed in Fig. 17.6 where the Ca flash can be seen in the image at 0 in Fig. 17.6a. In separate experiments, $\Delta\Psi_m$ was measured in eggs sandwiched with a coverglass to reduce cytoplasmic deformation using a confocal microscope. $\Delta\Psi_m$ increased at 40 ± 11 s ($n = 3$) after insemination without a fraction of decrease (data not shown). The time of increase in $\Delta\Psi_m$ was also close to the value already mentioned using simultaneous measurements of $\Delta\Psi_m$ and $[Ca^{2+}]_i$.

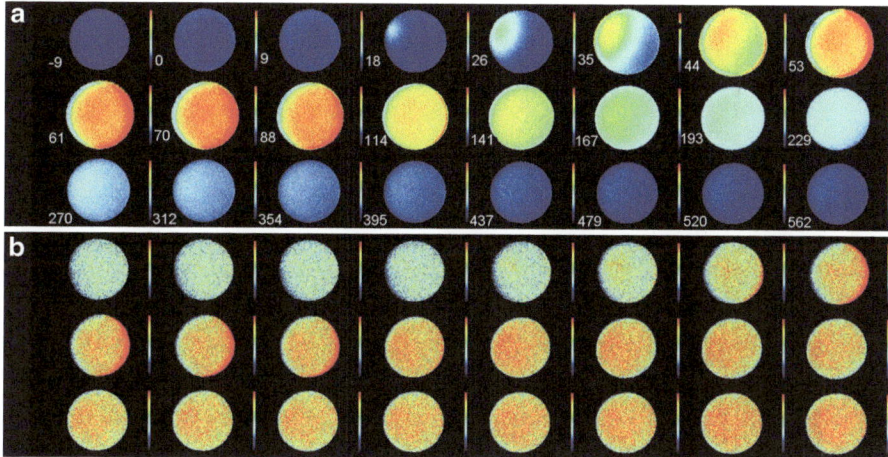

Fig. 17.6 Sequential images of $[Ca^{2+}]_i$ (**a**) and corresponding $\Delta\Psi_m$ (**b**) before and after Ca flash. *Number in the left lower corner* of each image indicates seconds (s) before and after Ca flash. The $[Ca^{2+}]_i$ images are always 270 ms ahead of the corresponding $\Delta\Psi_m$ images. Egg of images shown in Fig. 17.6 was different from that in Fig. 17.5

The rise time of ΔNO, 53.5 s, has been previously reported around a peak of $[Ca^{2+}]_i$ by the method of simultaneous measurement of ΔNO and the activation current (AC) in a voltage-clamped egg because a peak of AC that corresponds to a peak of $[Ca^{2+}]_i$ (Mohri et al. 1995; Mohri et al. 2008).

Taken together, all these results revealed that $\Delta\Psi_m$ is slightly earlier than ΔNO, suggesting that ΔNO is largely involved in mitochondrial activation.

17.3.4 $[Ca^{2+}]_i$ Dependency of $\Delta\Psi_m$

To confirm whether $\Delta\Psi_m$ depends on changes in $[Ca^{2+}]_i$, an egg microinjected with BAPTA at 1 mM within the egg and the control egg were set in the same field of the chamber and inseminated. The control egg showed normal $\Delta\Psi_m$ whereas the egg injected with 1 mM BAPTA showed no significant increase in $\Delta\Psi_m$ (Fig. 17.7). The egg injected with BAPTA showed no elevation of the fertilization envelope, indicating that sperm-induced changes in $[Ca^{2+}]_i$ were significantly suppressed. Sperm entries were confirmed by sperm aster formation under bright-field observation after the experiment. Thus, we concluded the increase in $\Delta\Psi_m$ requires, to some extent, an increase in $[Ca^{2+}]_i$ induced by sperm. It is consistent with earlier research that mitochondrial activation is induced by $[Ca^{2+}]_i$ increase in many other cell systems (Griffiths and Rutter 2009).

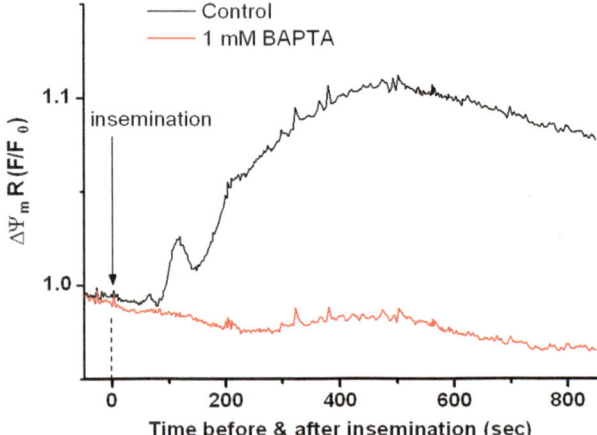

Fig. 17.7 $\Delta\Psi_m$ of control egg and an egg microinjected with 1 mM BAPTA. $\Delta\Psi_m$s were measured in a control egg (*black trace*) and an egg (*red trace*) injected with 1 mM BAPTA (final concentration in the egg) in the same chamber set in the stage of the microscope. Declining trace in control during 120–160 s was an artifact caused by fluorescence change by deformation of the egg while FE was forming

17.4   Conclusion

In the present study, we found that mitochondrial activation indicating hyperpolarization of $\Delta\Psi_m$ occurred downstream to the increase of $[Ca^{2+}]_i$ during fertilization, and that it started after the $[Ca^{2+}]_i$ rise in sea urchin and sand dollar eggs. We confirmed $[Ca^{2+}]_i$ dependency of $\Delta\Psi_m$ by an experiment in which an injection of BAPTA inhibited $\Delta\Psi_m$. Therefore, $\Delta\Psi_m$ was apparently $[Ca^{2+}]_i$ dependent, similar to ΔNO, as previously reported (Leckie et al. 2003; Mohri et al. 2008). Mitochondrial inhibitors CN^- and FCCP inhibited both ΔNO and $\Delta\Psi_m$, suggesting that mitochondria participate in ΔNO and may well regulate ΔNO. Furthermore, simultaneous measurements of $[Ca^{2+}]_i$ and $\Delta\Psi_m$ revealed mitochondrial activation took place slightly earlier than ΔNO, suggesting that mitochondria may be responsible for ΔNO. ΔNO has generally been considered to be produced by cytosolic NO synthase (NOS) (Thaler and Epel 2003). Our data suggest that mitochondria may have their own separate generation system of NO such as mitochondrial NOS, mtNOS (Finocchietto et al. 2009). The present data suggest mitochondrial activity unmistakably participates in ΔNO. Therefore, we are now continuing this investigation.

References

Finocchietto PV, Franco MC, Holod S, Gonzalez AS, Converso DP, Antico Arciuch VG, Serra MP, Poderoso JJ, Carreras MC (2009) Mitochondrial nitric oxide synthase: a masterpiece of metabolic adaptation, cell growth, transformation, and death. Exp Biol Med (Maywood) 234(9):1020–1028

Foerder CA, Klebanoff SJ, Shapiro BM (1978) Hydrogen peroxide production, chemiluminescence, and the respiratory burst of fertilization: interrelated events in early sea urchin development. Proc Natl Acad Sci USA 75(7):3183–3187

Fujihara T, Nakagawa-Izumi A, Ozawa T, Numata O (2007) High-molecular-weight polyphenols from oolong tea and black tea: purification, some properties, and role in increasing mitochondrial membrane potential. Biosci Biotechnol Biochem 71(3):711–719

Fujiwara A, Kamata Y, Asami K, Yasumasu I (2000) Relationship between ATP level and respiratory rate in sea urchin embryos. Dev Growth Differ 42(2):155–165

Fujiwara A, Yasumasu I (1997) Does the respiratory rate in sea urchin embryos increase during early development without proliferation of mitochondria? Dev Growth Differ 39(2):179–189

Griffiths EJ, Rutter GA (2009) Mitochondrial calcium as a key regulator of mitochondrial ATP production in mammalian cells. Biochim Biophys Acta 1787(11):1324–1333

Gunter TE, Sheu SS (2009) Characteristics and possible functions of mitochondrial Ca^{2+} transport mechanisms. Biochim Biophys Acta 1787(11):1291–1308

Hiramoto Y (1974) A method of microinjection. Exp Cell Res 87(2):403–406

Leckie C, Empson R, Becchetti A, Thomas J, Galione A, Whitaker M (2003) The NO pathway acts late during the fertilization response in sea urchin eggs. J Biol Chem 278(14):12247–12254

Mohri T, Ivonnet PI, Chambers EL (1995) Effect on sperm-induced activation current and increase of cytosolic Ca^{2+} by agents that modify the mobilization of [Ca^{2+}]$_i$. I. Heparin and pentosan polysulfate. Dev Biol 172(1):139–157

Mohri T, Sokabe M, Kyozuka K (2008) Nitric oxide (NO) increase at fertilization in sea urchin eggs upregulates fertilization envelope hardening. Dev Biol 322(2):251–262

Solani G, Sgarbi G, Lenaz G, Baracca A (2007) Evaluating mitochondrial membrane potential in cells. Biosci Rep 27:11–21

Thaler CD, Epel D (2003) Nitric oxide in oocyte maturation, ovulation, fertilization, cleavage and implantation: a little dab'll do ya. Curr Pharm Des 9(5):399–409

Wallace DC, Fan W, Procaccio V (2010) Mitochondrial energetics and therapeutics. Annu Rev Pathol 5:297–348

Warburg O (1908) Beobachtungen uber die oxydationsprozesse im seeigelei. Hoppe-Seyler's Z Physiol Chem 66:1–16

Chapter 18
Functional Roles of *spe* Genes in the Male Germline During Reproduction of *Caenorhabditis elegans*

Hitoshi Nishimura, Tatsuya Tajima, Skye Comstra, and Steven W. L'Hernault

Abstract The nematode *Caenorhabditis elegans* is an excellent model animal to study various biological phenomena, including reproduction. In this chapter, we focus on functional roles of spermatogenesis- or sperm-defective (*spe*) genes in the *C. elegans* male germline during reproduction. So far, approximately 190 mutants of *C. elegans* that are defective in male germline functions have been isolated, and many of them carry mutated alleles for one of the perhaps 60 *spe* genes. Most *spe* genes exhibit male germline-specific expression and play roles during spermatogenesis (spermatid production), spermiogenesis (spermatid activation into sperm), or fertilization. For example, *spe-8* class genes are indispensable for hermaphrodite-dependent spermiogenesis. If either of the *spe-8* class genes is aberrant, spermatids from mutant hermaphrodites, but not from males, arrest at an intermediate stage during spermiogenesis. In contrast, fertilization requires *spe-9* class genes. Hermaphrodites and males of *spe-9* class mutants produce otherwise normal sperm that are incapable of fertilizing oocytes. Because *C. elegans* oocytes have no egg coats, *spe-9* class genes are probably required for sperm to bind to and/or fuse with the oocyte plasma membrane. Intriguingly, several *spe* genes are likely to be orthologues of mammalian genes, suggesting that *C. elegans* and mammals share some common steps during male germline functions at the molecular level.

H. Nishimura (✉)
Department of Life Science, Setsunan University, 17-8 Ikeda-Nakamachi, Neyagawa, Osaka 572-8508, Japan

Department of Biology, Emory University, 1510 Clifton Road NE, Atlanta, GA 30322, USA
e-mail: nishimura@lif.setsunan.ac.jp

T. Tajima
Department of Life Science, Setsunan University, 17-8 Ikeda-Nakamachi, Neyagawa, Osaka 572-8508, Japan

S. Comstra • S.W. L'Hernault
Department of Biology, Emory University, 1510 Clifton Road NE, Atlanta, GA 30322, USA

H. Sawada et al. (eds.), *Sexual Reproduction in Animals and Plants*,
DOI 10.1007/978-4-431-54589-7_18, © The Author(s) 2014

Keywords *Caenorhabditis elegans* • Fertilization • Male germline • *spe* genes
• Spermiogenesis

18.1   Overview of *Caenorhabditis elegans* Reproduction

In many life science research fields, the nematode *Caenorhabditis elegans* is utilized as a model animal. The advantage in using this worm is that we can observe
biological phenomena in vivo and easily manipulate its genes, about 40 % of which
are shared with those of other animal species (*C. elegans* Sequencing Consortium
1998). In particular, the many *C. elegans* mutants available are so powerful in investigating reproduction because sperm and oocytes are haploid, cell types that are too
specialized to examine gamete-specific gene functions in vitro.

C. elegans somatic cells have five pairs of autosomes (I, II, III, IV, V) and one
pair of sex chromosomes (XX or XO). Therefore, there are two sexes in *C. elegans*,
an XX hermaphrodite and an XO male (Fig. 18.1).

In a hermaphrodite, spermatogenesis occurs in both arms of the U-shaped gonad
during the fourth larval (L4) stage. As L4 hermaphrodites become young adults,
spermatogenesis completely switches to oogenesis. Hence, adult hermaphrodites
are somatically females, despite the presence of self-sperm in the spermatheca.

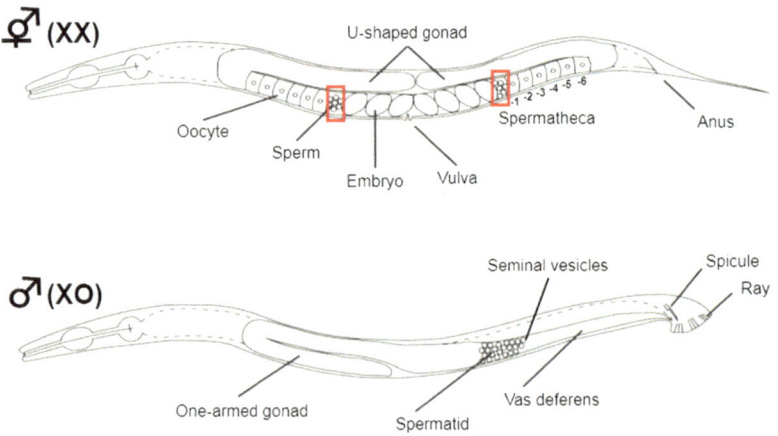

Fig. 18.1 Hermaphrodite and male of the nematode *Caenorhabditis elegans*. *Top:* An adult hermaphrodite (sex genotype, XX) has a U-shaped gonad that exhibits mirror image symmetry around
the single vulva. Location of each spermatheca is outlined by *squares*. *Numbers* below the gonad
indicate positions of oocytes according to their developmental stages. *Bottom:* An adult male (XO)
has a one-armed gonad. Spermatids accumulate in the single vas deferens until they are ejaculated
during mating to a hermaphrodite. (These figures were taken from the article by Nishimura and
L'Hernault 2010 with the publisher's permission)

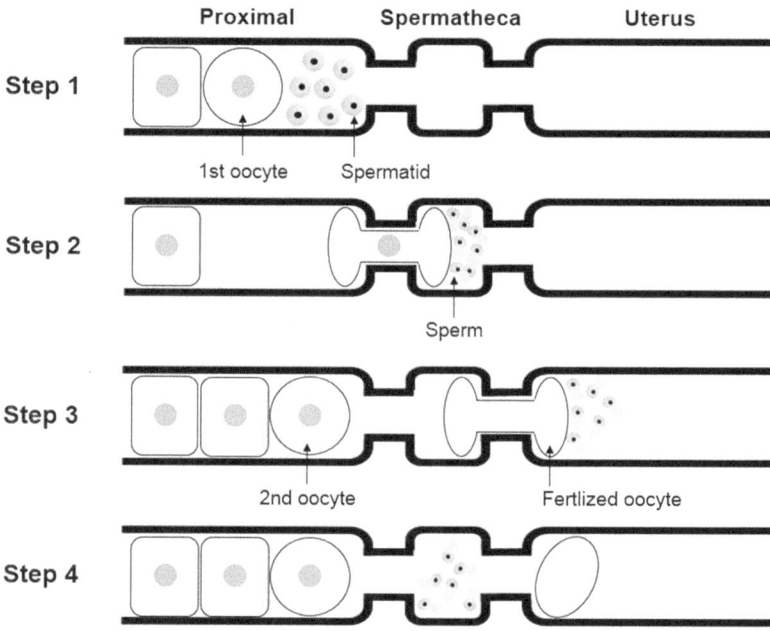

Fig. 18.2 Ovulation–fertilization cycle in the gonad of *C. elegans* hermaphrodite. Each figure shows an expanded area of the hermaphroditic gonad, including the proximal gonad, spermatheca, and uterus. For details, see Sect. 18.1

Oogenesis is ongoing from the distal to the proximal direction in the adult worm gonads, so that the fertilization-ready oocyte resides at the most proximal region (−1 position in Fig. 18.1). Figure 18.2 shows a scheme of the ovulation–fertilization cycle. Before the first ovulation, self-spermatids that had been produced during the L4 stage are localized in the proximal gonads of adult hermaphrodites (Step 1). When the first −1 oocyte is ovulated, it pushes the previously produced spermatids out of the proximal gonad into the spermatheca (Step 2). Those spermatids rapidly transform into sperm (self-sperm) in the spermatheca, and one of them fertilizes the first oocyte. Then, together with the remaining sperm, the fertilized oocyte moves into the uterus for the onset of embryogenesis (Step 3). The uterine sperm crawl back into the spermatheca and compete again for the next fertilization (Step 4). An adult hermaphrodite carries approximately 300 self-sperm in the spermatheca and eventually produces about 300 self-progeny. Thus, although the self-sperm are pushed into the uterus at every ovulation, most are capable of crawling back into the spermatheca so that nearly all the sperm can be consumed by fertilization. When self-fertilization occurs in wild-type hermaphrodites, 99.9 % of self-progeny are XX hermaphrodites.

In a male (Fig. 18.1), spermatogenesis first occurs at the L4 stage and lasts throughout the entire life of the animal. Wild-type male spermatids are barely activated into sperm in the gonad. During mating to a hermaphrodite, male spermatids are ejaculated into the uterus through the vulva and then activated into sperm. The male sperm subsequently crawl into the spermatheca to fertilize oocytes. If self-sperm already exist in the spermatheca, male sperm have an advantage to fertilize oocytes, probably because of the larger size and faster crawling velocity of male-derived sperm than those of self-sperm. Consequently, after mating, fertilization mostly occurs between oocytes and male sperm, rather than between oocytes and self-sperm. This observation also indicates that outcrossing of an XX hermaphrodite and an XO male results in about half the progeny being males, in contrast to self-fertilization.

18.2 Male Germline Functions in *C. elegans*

18.2.1 Spermatogenesis

As shown in Fig. 18.3, during *C. elegans* reproduction, the male germline is involved in spermatogenesis, spermiogenesis, and fertilization (Singson et al. 2008; L'Hernault 2009; Nishimura and L'Hernault 2010). Spermatogenesis is a process that produces four haploid spermatids from one diploid spermatocyte via meiosis (Fig. 18.3a). A primary spermatocyte is first divided into two secondary

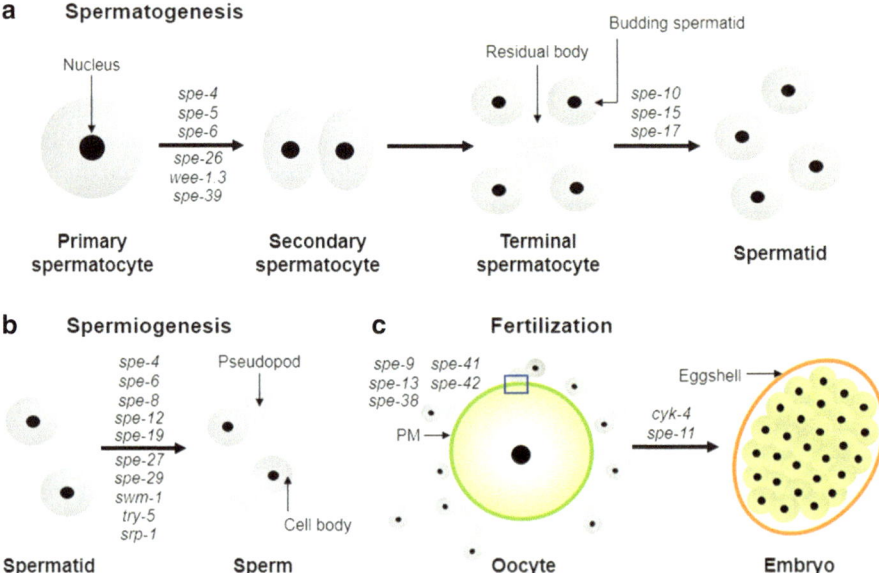

Fig. 18.3 *C. elegans spe* genes acting during male germline functions. A part of the *spe* genes that are involved in either of spermatogenesis (**a**), spermiogenesis (**b**), or fertilization (**c**) are shown in these figures. In **c**, a *square* shows that a sperm contacts the oocyte plasma membrane through its pseudopod. (These figures were taken from the article by Nishimura and L'Hernault 2010 with the publisher's permission)

spermatocytes during meiosis I. Then, each secondary spermatocyte undergoes meiosis II to generate two haploid spermatids, which have budded from an acellular residual body. The second cell division is asymmetrical, and the residual body receives many organelles and cytoplasmic proteins. One of the cytological features of mature spermatids is that this cell type possesses Golgi-derived, secretory membranous organelles (MOs) in the cytoplasm.

In contrast to mammals, *C. elegans* spermatogenesis is readily reproducible in vitro; spermatocytes that are released from dissected males can differentiate into spermatids in a simple, chemically defined medium (L'Hernault 2009; Nishimura and L'Hernault 2010). Because this in vitro spermatogenesis can be completed within approximately 90 min, use of *C. elegans* provides a significant advantage in studying spermatogenesis.

18.2.2 Spermiogenesis

During *C. elegans* spermiogenesis, a sessile round spermatid transforms into a motile, amoeboid sperm (Fig. 18.3b). Because no ribosomes are present in spermatids, spermiogenesis does not entail de novo protein synthesis. For crawling, each *C. elegans* sperm has a single pseudopod extending from its cell body, instead of a flagellum. The pseudopod also acts as a binding/fusion site with the oocyte plasma membrane (Fig. 18.3c). Nematode sperm, differing from other amoeboid-like cell types, utilize the major sperm protein (MSP) as their cytoskeletal protein.

The MOs, which are present in the cytoplasm of a spermatid, have a secretory function during spermiogenesis. The MOs fuse with the spermatid plasma membrane and release their contents extracellularly. Moreover, some sperm proteins that play essential roles in fertilization translocate from the MOs to the pseudopod surface (see Sect. 18.3.3). In this aspect, the MO might be analogous to the acrosome of flagellated sperm.

The seminal fluid contains the serine protease TRY-5, and this protein probably activates male-derived spermatids in vivo (Smith and Stanfield 2011) (see Sect. 18.3.2.2). TRY-5 might also activate hermaphrodite-derived spermatids, but hermaphrodites likely have a unknown, physiological activator(s) besides TRY-5. In vitro, spermatids can be activated into sperm by treatment with either the cationic ionophore Monensin, the weak base triethanolamine (TEA), the phosphoinositide-3-kinase inhibitor Wortmannin, or the bacterial serine protease mixture Pronase. Sperm generated by in vitro activation with TEA, but not Pronase, are competent to fertilize oocytes after the sperm are artificially inseminated into hermaphrodites.

18.2.3 Fertilization

C. elegans fertilization (Fig. 18.3c) is distinguished from mammalian fertilization in several ways. First, mammalian sperm undergo the acrosome reaction, which is essentially required before sperm–oocyte fusion. In contrast, *C. elegans* sperm lack an acrosome, probably because of the absence of any substantial egg coat (ex. the zona pellucida) surrounding *C. elegans* oocytes. Second, acrosome-reacted sperm of mammals bind to the oocyte plasma membrane at the equatorial region of the sperm head, whereas *C. elegans* sperm probably first contact the oocyte plasma membrane via the pseudopod (Fig. 18.3c), which functionally corresponds to the flagella of mammalian sperm. Third, in contrast to mammals, a system of in vitro fertilization (IVF) is not yet available for *C. elegans*.

On the other hand, the reproductive tract of *C. elegans* adult hermaphrodites is functionally analogous to that of mammalian females; the proximal gonad, spermatheca, and uterus of adult hermaphrodites play similar spatial roles to those of the ovary, oviduct, and uterus of mammals. As described in Sect. 18.1, adult hermaphrodites are somatically females and no longer produce self-sperm. Hence, *C. elegans* might provide valuable insights into certain aspects of in vivo fertilization that are analogous to mammalian fertilization.

18.3   *spe* Genes Acting During the *C. elegans* Male Germline Functions

18.3.1   What Are spe Genes?

Mutants lacking either type of *spe* (spermatogenesis- or sperm-defective) genes produce spermatocytes, spermatids, or sperm of which the functions are aberrant during spermatogenesis, spermiogenesis, and fertilization (Singson et al. 2008; L'Hernault 2009; Nishimura and L'Hernault 2010) (Fig. 18.3). *spe* mutant hermaphrodites produce very few progeny and instead lay unfertilized oocytes. However, mating to wild-type males allows *spe* mutant hermaphrodites to produce outcross progeny. Thus, sperm, but not oocytes, are functionally defective in *spe* mutants. In other words, wild-type sperm are necessary and sufficient to rescue the self-sterility of mutant hermaphrodites. Using these criteria, mutants that define about 60 *spe* genes have been isolated after treatment of hermaphrodites with chemical mutagens.

Most *spe* genes are expressed specifically or predominantly in the *C. elegans* male germline as expected, although some genes that play important roles in male germline functions are also expressed in other tissues. Among *spe* genes so far identified, *spe-8* class and *spe-9* class genes have been intensively studied; *spe-8* class genes (*spe-8*, *spe-12*, *spe-19*, *spe-27*, and *spe-29*) act in hermaphrodite-dependent spermiogenesis (see Sect. 18.3.2.1 and Table 18.1), whereas fertilization requires *spe-9* class genes (*spe-9*, *spe-13*, *spe-38*, *spe-41/trp-3*, and *spe-42*) (see Sect. 18.3.3 and Table 18.2).

Table 18.1 *spe* genes involved in spermiogenesis

Gene	Chr.	Predicted protein	Protein localization
spe-4	I	Seven-pass transmembrane aspartyl protease of the presenilin family (465 aa)	MOs of spermatid Cell body of sperm
spe-6	III	Ser/Thr kinase of the casein kinase 1 family (379 aa)	ND
spe-8	I	Nonreceptor type of tyrosine kinase with the SH2 domain (512 aa)	ND
spe-12	I	Transmembrane protein (255 aa)	ND
spe-19	V	Transmembrane protein with 11 potential phosphorylation sites in the cytoplasmic tail (300 aa)	ND
spe-27	IV	Soluble protein (131 aa)	ND
spe-29	IV	Transmembrane protein (66 aa)	ND
try-5	V	Soluble trypsin-like protease (327 aa)	Seminal fluid
swm-1	V	Soluble protein with two trypsin inhibitor-like domains (86 and 135 aa)	Likely seminal fluid
srp-1	V	Ovalbumin-like protein of the Serpin (serine protease inhibitor) family (366 aa)	MOs of spermatid (Ce and As) Cell body and pseudopod of sperm (As)[a]

Chr. chromosome, *MO* membranous organelles, *aa* amino acid, *ND* not determined, *Ce Caenorhabditis elegans, As Ascaris suum*
[a]The SRP-1 localization in *C. elegans* sperm is not clear

Table 18.2 *spe* genes involved in fertilization

Gene	Chr.	Predicted protein	Protein localization
spe-9	I	Transmembrane protein with ten EGF-like domains, similar to Delta, Serrate, and Jagged proteins (381 and/or 661 aa)	MOs of spermatid Pseudopod of sperm
spe-13	I	ND	ND
spe-38	I	Four-pass transmembrane protein (179 aa)	MOs of spermatid Pseudopod of sperm
spe-41 (*trp-3*)	III	Ca^{2+}-permeable cation channel of the TRPC family (854 aa)	MOs of spermatid Cell body and pseudopod of sperm
spe-42	V	Six-pass transmembrane protein with the DC-STAMP and RING finger domains (774 aa)	ND

Chr. chromosome, *aa* amino acid, *ND* not determined

18.3.2   spe *Genes Involved in Spermiogenesis*

18.3.2.1   *spe-8* Class-Dependent Spermiogenesis

Mutants lacking either of the *spe-8* class genes all exhibit nearly the same phenotype: hermaphrodites are self-sterile and males are cross-fertile [*spe-8* (L'Hernault et al. 1988; Shakes and Ward 1989), *spe-12* (L'Hernault et al. 1988; Shakes and Ward 1989; Nance et al. 1999), *spe-19* (Geldziler et al. 2005), *spe-27* (Minniti et al. 1996), and *spe-29* (Nance et al. 2000)]. These mutant hermaphrodites, but not males, produce spermatids that do not undergo spermiogenesis in vivo.

Figure 18.4 shows one of the predicted models for *C. elegans* spermiogenesis (Nishimura and L'Hernault 2010). Spermatids derived from both hermaphrodites and males of *spe-8* class mutants exhibit normal cytology before and after in vitro activation with Monensin or TEA (L'Hernault 2009; Nishimura and L'Hernault 2010) (see Sect. 18.2.2). However, by in vitro treatment with Pronase, *spe-8* class mutant spermatids from either sex arrest at an intermediate stage; spiky projections are extended from mutant spermatids, but they never transform into pseudopods. Hence, *spe-8* class mutant spermatids are likely to have defects in the spermiogenesis pathway that is affected by Pronase. These data suggest that both hermaphrodite- and male-derived spermatids contain two pathways for activation into sperm, which are dependent on or independent of *spe-8* class genes. The *spe-8* class-dependent pathway seems to act during spermiogenesis in hermaphrodites, whereas males probably utilize the *spe-8* class-independent pathway for spermatid activation. Among *spe-8* class genes, *spe-12* might be required for both the spermiogenesis pathways because *spe-12* mutant males show partially defective spermatid activation.

spe-6 and *spe-4* are required in spermatocytes for proper complex formation of the MO and the fibrous body (FB), which is an assembly of MSP bundles (L'Hernault 2009; Nishimura and L'Hernault 2010). Intriguingly, both these two genes also act in spermiogenesis; there are non-null, suppressor alleles of *spe-6* (Muhlrad and Ward 2002) and *spe-4* (Gosney et al. 2008) that rescue the self-sterility of *spe-8* class mutant hermaphrodites. Thereby, SPE-6 and SPE-4 (Table 18.1), which are a casein kinase 1-like Ser/Thr kinase and a presenilin 1-like aspartyl protease, respectively, appear to be downstream of the SPE-8 class proteins (Fig. 18.4). At present, how SPE-6 and SPE-4 function during spermiogenesis is controversial. One of the possible interpretations (Fig. 18.4) is that SPE-6 is active in resting spermatids to phosphorylate SPE-4, which contains predicted phosphorylation sites by casein kinase 1: this might disturb SPE-4 to appropriately cleave a membranous protein substrate(s), resulting in blocking spermatid activation. During spermiogenesis, SPE-6 activity is conversely reduced, and dephosphorylated SPE-4 becomes capable of proteolytically splitting its substrate(s).

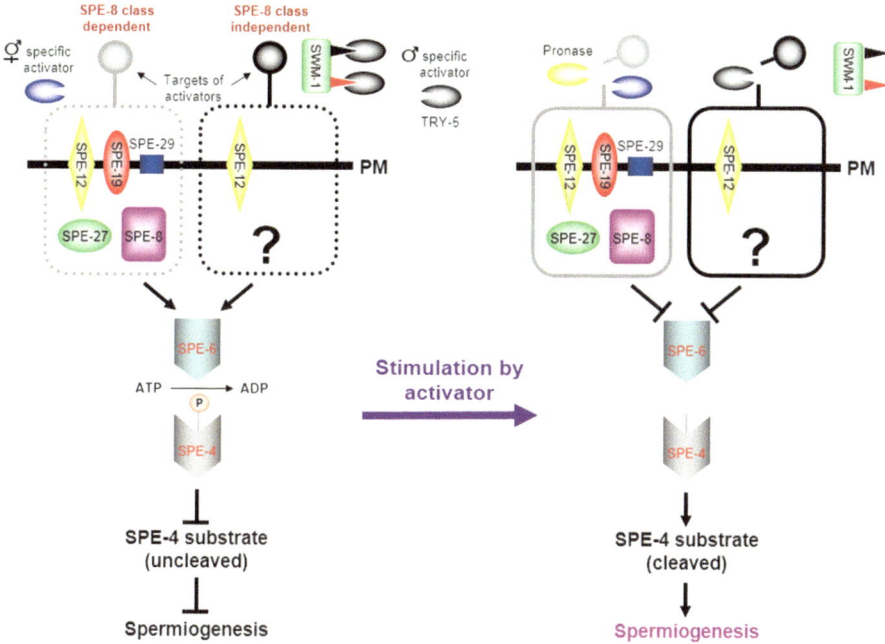

Fig. 18.4 Two predicted pathways for *C. elegans* spermiogenesis. To regulate spermiogenesis, there are likely two distinct pathways in all wild-type spermatids from hermaphrodites and males: one pathway is *spe-8* class dependent and another is *spe-8* class independent. Sex-specific activators probably determine which pathway is utilized. The *spe-8* class-dependent pathway appears to be stimulated by an unknown hermaphrodite-derived activator(s) or the serine protease mixture Pronase. For males, the serine protease TRY-5 presumably activates spermatids via the *spe-8* class-independent pathway, and its protease activity is blocked by the trypsin inhibitor-like protein SWM-1. To initiate spermiogenesis, these activators might be required to cleave a certain cell-surface protein(s). SPE-6, a casein kinase 1-like Ser/Thr kinase, is downstream of SPE-8 class proteins and is one of the common points between these two pathways. This kinase perhaps phosphorylates (shown as "P" in a ball) the presenilin 1-like aspartyl protease SPE-4, because SPE-4 has predicted phosphorylation sites by casein kinase 1. Then, by the phosphorylation, SPE-4 may become incapable of splitting a membranous protein substrate(s), leading to blocking spermiogenesis. On the other hand, SPE-6 activity presumably requires to be reduced so that dephosphorylated SPE-4 can cleave its substrate(s) for onset of spermiogenesis. Note that other interpretations for how SPE-6 and SPE-4 act downstream of SPE-8 class proteins are likely available. In this figure, the active and inactive status of each pathway are shown by *solid* and *broken* lines, respectively. *Thick black arrows* represent positive regulation, whereas negative regulation is expressed by *T-shaped lines*. *PM* plasma membrane. (These figures were taken from the article by Nishimura and L'Hernault 2010 with the publisher's permission)

18.3.2.2   *spe-8* Class-Independent Spermiogenesis

As described in Sect. 18.2.2, *try-5* encodes a soluble serine protease that is a component of the seminal fluids (Smith and Stanfield 2011) (Table 18.1). Hermaphrodites and males of *try-5* mutants are both fertile. However, the seminal fluids of *try-5* mutant males are incompetent to activate spermatids in vivo, unlike those of wild-type males. Moreover, males carrying double mutations in *try-5* and *spe-27* or *spe-29* are sterile, whereas male fertility is not affected by any single mutation in *spe-27*, *spe-29*, or *try-5*. Therefore, it is postulated from these findings that (1) TRY-5 protease probably activates male-derived spermatids via the *spe-8* class-independent pathway, and (2) male-derived spermatids can respond to a hermaphrodite-produced activator(s) that stimulates the *spe-8* class-dependent pathway.

swm-1 is also implicated in male fertility (Stanfield and Villeneuve 2006; Smith and Stanfield 2011). In *swm-1* mutant males, spermatids are ectopically activated into sperm within the male gonads, and these mutant sperm are not transferred normally into hermaphrodites during mating. Because *swm-1* mutant sperm are capable of fertilizing oocytes after artificial insemination of the mutant sperm into wild-type hermaphrodites, the sterility of *swm-1* mutant males is probably caused by this transfer defect. The *swm-1* gene encodes a soluble protein with two trypsin inhibitor-like domains (Table 18.1), both of which seem to react with TRY-5 and/or another serine protease(s) that is related to TRY-5 (ex. activator or substrate of TRY-5) (Smith and Stanfield 2011). Thus, SWM-1 is very likely to block the premature spermatid activation by TRY-5 within the seminal vesicles. After ejaculation, TRY-5 or its related serine protease(s) probably become free from SWM-1 in the uterus, and activation of male spermatids immediately occurs.

Recently, it was found that male spermatids of *Ascaris suum* (As) secrete the Serpin (serine protease inhibitor) family protein As_SRP-1, which seems to block the activity of As_TRY-5 (*Ascaris* orthologue of TRY-5) (Zhao et al. 2012). Therefore, *C. elegans* SPR-1 (Table 18.1), in addition to SWM-1, presumably regulates the TRY-5 protease activity to prevent male spermatids from ectopic activation in the male reproductive tract.

18.3.3   spe *Genes Involved in Fertilization*

As one of the *spe-9* class genes is disturbed, hermaphrodites and males produce otherwise normal sperm that cannot fertilize oocytes [*spe-9* (L'Hernault et al. 1988; Singson et al. 1998; Zannoni et al. 2003; Putiri et al. 2004), *spe-13* (L'Hernault et al. 1988; Putiri et al. 2004), *spe-38* (Chatterjee et al. 2005), *spe-41/trp-3* (Xu and Sternberg 2003), and *spe-42* (Kroft et al. 2005)]. There are at least two interpretations for this phenotype. First, *spe-9* class mutant sperm bind to the oocyte plasma membrane with very low affinities and thus mutant sperm are easily detached from the oocyte plasma membrane or cannot proceed to the next step, such as sperm–oocyte fusion. Second, *spe-9* class mutant sperm normally bind to the oocyte plasma membrane, but a defect in fusion occurs.

At any rate, it is very likely that SPE-9 class proteins (Fig. 18.5 and Table 18.2) play critical roles during sperm–oocyte interactions (binding or fusion).

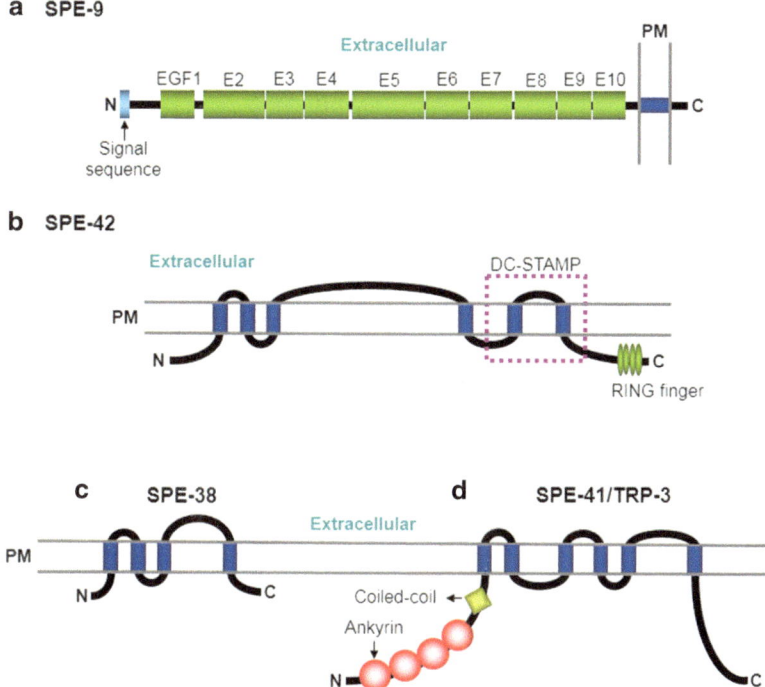

Fig. 18.5 SPE-9 class proteins presumably required for sperm–oocyte interactions. (**a**) SPE-9. *E*, EGF-like domain. (**b**) SPE-42. The region corresponding to the DC-STAMP domain is indicated by a *broken line*. (**c**) SPE-38. (**d**) SPE-41/TRP-3. In these figures, *cylinders* shown in the plasma membrane (*PM*) represent the transmembrane domains. *N* N-terminus, *C* C-terminus. (These figures were taken from the article by Nishimura and L'Hernault 2010 with the publisher's permission)

18.3.3.1 *spe-9*

SPE-9 is a single-pass transmembrane protein containing ten epidermal growth factor (EGF)-like domains (Fig. 18.5a). The overall domain structure of SPE-9 is similar to those of ligands for the Notch/LIN-12/GLP-1 family: Delta, Serrate, and Jagged1. Moreover, as these Notch family ligand proteins, the EGF-like domains of SPE-9 are essential for its function in *C. elegans* fertilization (Putiri et al. 2004). Therefore, SPE-9 might be a ligand for a putative sperm receptor(s) on the oocyte surface, although SPE-9 has no DSL domain, which characterizes the Delta/Serrate/LAG-2 family (Cordle et al. 2008). This hypothesis does not conflict with the fact that SPE-9 is localized on the surface of pseudopods, through which sperm bind/fuse with oocytes.

18.3.3.2 *spe-42*

SPE-42 is a six-pass transmembrane protein (Fig. 18.5b), but its localization is yet unclear. This protein contains two functional domains, the DC-STAMP (dendritic cell-specific transmembrane protein) and the C4C4-type RING finger domains (Fig. 18.5b). Mammalian DC-STAMPs are known to mediate cell–cell fusion of osteoclasts and foreign-body giant cells. Moreover, the *Drosophila* gene *sneaky*, presumably the orthologue of *spe-42*, is involved in sperm plasma membrane breakdown after the fly sperm enter the oocytes (Wilson et al. 2006). Thus, SPE-42 might act in sperm–oocyte fusion via the DC-STAMP domain.

The C-terminal RING finger domain of SPE-42 is essentially required for *C. elegans* fertilization (Wilson et al. 2011). The C4C4-type RING finger domains generally act in protein–protein interactions, suggesting that this domain of SPE-42 could be a binding site with another sperm protein(s). It is also possible that the SPE-42 RING finger domain might have a ubiquitin E3 ligase activity, similar to the human RING finger protein CNOT4, although most RING finger E3 ligases are a C3HC4 type. Consequently, SPE-42 possibly catalyzes the ubiquitination of a sperm protein(s) that acts during sperm–oocyte interactions to regulate the localization or function of that protein(s) (Nishimura and L'Hernault 2010).

18.3.3.3 *spe-38*

This gene encodes a four-pass transmembrane protein with no other significant domains (Fig. 18.5c). SPE-38 is located within the MOs in spermatids, but it appears on the cell surface after MO fusion with the spermatid plasma membrane. SPE-38 is probably involved in sperm–oocyte interactions because it localizes to pseudopods as does SPE-9. Recently, this protein was found to associate with SPE-41/TRP-3 to regulate the SPE-41/TRP-3 localization (Singaravelu et al. 2012).

18.3.3.4 *spe-41/trp-3*

SPE-41/TRP-3 belongs to the TRPC [transient receptor potential (TRP)-canonical] superfamily of cation channels (Fig. 18.5d). Indeed, SPE-41/TRP-3 was demonstrated to act as a calcium channel in sperm, but not in spermatids. This finding is in good agreement with the SPE-41/TRP-3 localization; SPE-41/TRP-3 is intracellularly localized in the MOs of spermatids, and during spermiogenesis, this protein translocates onto the surface of both the pseudopod and the cell body. Again, the pseudopod is the place where sperm–oocyte interactions are thought to occur. On the *C. elegans* sperm surface, SPE-41/TRP-3 probably forms a homo- or heterotetramer through the ankyrin and coiled-coil domains, as do other members of the TRP family. Sperm–oocyte binding might produce a signal to open the SPE-41/TRP-3 channel, and the ensuing calcium influx would trigger gamete fusion.

Moreover, we have recently found that a male germline-specific gene encoding an immunoglobulin (Ig)-like transmembrane protein is essentially required for *C. elegans* fertilization, similar to the mouse *Izumo1* gene (Inoue et al. 2005). This observation indicates that the *C. elegans* Ig-like gene belongs to the *spe-9* class, and functional studies of this gene are currently ongoing in our laboratories.

18.4 Perspectives

We see two major reasons why *C. elegans* is useful for the study of reproductive biology. First, as already described, numerous mutants that are defective in reproduction can be easily created. These mutants are powerful tools to identify *C. elegans* genes that are indispensable for reproduction. For example, one area that is especially important and poorly understood in any species is the mechanism of sperm–oocyte binding and fusion. *C. elegans spe-9* class mutants, in which otherwise normal sperm fail to properly bind to or fuse with the oocyte plasma membrane, are the largest collection of gamete interaction-defective mutants in any organism.

Second, *C. elegans* is optically transparent, which allows in vivo analysis of reproduction. Because small chemicals, peptides, and antibodies can be introduced into worms by soaking, feeding, or microinjection, this animal can be used as a kind of "in vivo test tube," which is a unique property in reproductive research. These experiments might result in development and evaluation of drugs for infertile or contraceptive therapies.

A current problem of the *C. elegans* reproductive research is that an IVF system is not available (see Sect. 18.2.3). Development of this system enables us to determine the precise point at which mutants affect fertilization. Moreover, an IVF assay would be useful to examine candidate drugs for infertile or contraceptive therapy in addition to testing them by in vivo *C. elegans* fertilization.

Acknowledgments These studies were supported by grants from the National Science Foundation (to S.W.L., IOB-0544180) and National Institute of Health (to S.W.L., GM082932), USA, and by Grant-in-aid for Scientific Research on Innovative Areas from the MEXT (to H.N., 24112716), Japan. We thank all past and current members of the L'Hernault lab and the Nishimura lab for valuable discussions and suggestions.

References

C. elegans Sequencing Consortium (1998) Genome sequence of the nematode *C. elegans*: a platform for investigating biology. Science 282:2012–2018. doi:10.1126/science.282.5396. 2012

Chatterjee I, Richmond A, Putiri E, Shakes DC, Singson A (2005) The *Caenorhabditis elegans spe-38* gene encodes a novel four-pass integral membrane protein required for sperm function at fertilization. Development (Camb) 132:2795–2808

Cordle J, Johnson S, Tay JZ, Roversi P, Wilkin MB, de Madrid BH, Shimizu H, Jensen S, Whiteman P, Jin B, Redfield C, Baron M, Lea SM, Handford PA (2008) A conserved face of the Jagged/Serrate DSL domain is involved in Notch *trans*-activation and *cis*-inhibition. Nat Struct Mol Biol 15:849–857. doi:10.1038/nsmb.1457

Geldziler B, Chatterjee I, Singson A (2005) The genetic and molecular analysis of *spe-19*, a gene required for sperm activation in *Caenorhabditis elegans*. Dev Biol 283:424–436

Gosney R, Liau WS, LaMunyon CW (2008) A novel function for the presenilin family member *spe-4*: inhibition of spermatid activation in *Caenorhabditis elegans*. BMC Dev Biol 8:44. doi:10.1186/1471-213X-8-44

Inoue N, Ikawa M, Isotani A, Okabe M (2005) The immunoglobulin superfamily protein Izumo is required for sperm to fuse with eggs. Nature (Lond) 434:234–238

Kroft TL, Gleason EJ, L'Hernault SW (2005) The *spe-42* gene is required for sperm-egg interactions during C. elegans fertilization and encodes a sperm-specific transmembrane protein. Dev Biol 286:169–181

L'Hernault SW (2009) The genetics and cell biology of spermatogenesis in the nematode *C. elegans*. Mol Cell Endocrinol 306:59–65. doi:0.1016/j.mce.2009.01.008

L'Hernault SW, Shakes DC, Ward S (1988) Developmental genetics of chromosome I spermatogenesis-defective mutants in the nematode *Caenorhabditis elegans*. Genetics 120: 435–452

Minniti AN, Sadler C, Ward S (1996) Genetic and molecular analysis of *spe-27*, a gene required for spermiogenesis in *Caenorhabditis elegans* hermaphrodites. Genetics 143:213–223

Muhlrad PJ, Ward S (2002) Spermiogenesis initiation in *Caenorhabditis elegans* involves a casein kinase 1 encoded by the *spe-6* gene. Genetics 161:143–155

Nance J, Minniti AN, Sadler C, Ward S (1999) *spe-12* encodes a sperm cell surface protein that promotes spermiogenesis in *Caenorhabditis elegans*. Genetics 152:209–220

Nance J, Davis EB, Ward S (2000) *spe-29* encodes a small predicted membrane protein required for the initiation of sperm activation in *Caenorhabditis elegans*. Genetics 156:1623–1633

Nishimura H, L'Hernault SW (2010) Spermatogenesis-defective (*spe*) mutants of the nematode *Caenorhabditis elegans* provide clues to solve the puzzle of male germline functions during reproduction. Dev Dyn 239:1502–1514. doi:10.1002/dvdy.22271

Putiri E, Zannoni S, Kadandale P, Singson A (2004) Functional domains and temperature-sensitive mutations in SPE-9, an EGF repeat-containing protein required for fertility in Caenorhabditis elegans. Dev Biol 272:448–459

Shakes DC, Ward S (1989) Initiation of spermiogenesis in *C. elegans*: a pharmacological and genetic analysis. Dev Biol 134:189–200

Singaravelu G, Chatterjee I, Rahimi S, Druzhinina MK, Kang L, Xu XZ (2012) Singson A (2012) The sperm surface localization of the TRP-3/SPE-41 Ca^{2+}-permeable channel depends on SPE-38 function in *Caenorhabditis elegans*. Dev Biol 365:376–383. doi:10.1016/j.ydbio.2012.02.037

Singson A, Mercer KB, L'Hernault SW (1998) The *C. elegans spe-9* gene encodes a sperm transmembrane protein that contains EGF-like repeats and is required for fertilization. Cell 93:71–79

Singson A, Hang JS, Parry JM (2008) Genes required for the common miracle of fertilization in *Caenorhabditis elegans*. Int J Dev Biol 52:647–656. doi:10.1387/ijdb.072512as

Smith JR, Stanfield GM (2011) TRY-5 is a sperm-activating protease in *Caenorhabditis elegans* seminal fluid. PLoS Genet 7:e1002375. doi:10.1371/journal.pgen.1002375

Stanfield GM, Villeneuve AM (2006) Regulation of sperm activation by SWM-1 is required for reproductive success of *C. elegans* males. Curr Biol 16:252–263

Wilson KL, Fitch KR, Bafus BT, Wakimoto BT (2006) Sperm plasma membrane breakdown during *Drosophila* fertilization requires *sneaky*, an acrosomal membrane protein. Development (Camb) 133:4871–4879

Wilson LD, Sackett JM, Mieczkowski BD, Richie AL, Thoemke K, Rumbley JN, Kroft TL (2011) Fertilization in *C. elegans* requires an intact C-terminal RING finger in sperm protein SPE-42. BMC Dev Biol 11:10

Xu XZ, Sternberg PW (2003) A *C. elegans* sperm TRP protein required for sperm-egg interactions during fertilization. Cell 114:285–297

Zannoni S, L'Hernault SW, Singson AW (2003) Dynamic localization of SPE-9 in sperm: a protein required for sperm-oocyte interactions in *Caenorhabditis elegans*. BMC Dev Biol 3:10

Zhao Y, Sun W, Zhang P, Chi H, Zhang MJ, Song CQ, Ma X, Shang Y, Wang B, Hu Y, Hao Z, Hühmer AF, Meng F, L'Hernault SW, He SM, Dong MQ, Miao L (2012) Nematode sperm maturation triggered by protease involves sperm-secreted serine protease inhibitor (serpin). Proc Natl Acad Sci USA 109:1542–1547. doi:10.1073/pnas.1109912109

Chapter 19
Origin of Female/Male Gender as Deduced by the Mating-Type Loci of the Colonial Volvocalean Greens

Hisayoshi Nozaki

Abstract Colonial Volvocales (green algae) are a model lineage for the study of the evolution of sexual reproduction because isogamy, anisogamy, and oogamy are recognized within the closely related group, and several mating type (sex)-specific genes were identified in the closely related unicellular *Chlamydomonas reinhardtii* during the past century. In 2006, we first identified a sex-specific gene within the colonial Volvocales using the anisogamous colonial volvocalean alga *Pleodorina starrii*, namely, a male-specific gene called "*OTOKOGI*," which is a homologue of the *minus* mating type-determining gene *MID* of the isogamous *C. reinhardtii*. Thus, it was speculated that the derived or *minus* mating type of *C. reinhardtii* is homologous to the male in the anisogamous/oogamous members of the colonial Volvocales. The discovery of the male-specific gene facilitated comparative studies of the mating-type locus (*MT*) (primitive sex chromosomal region) because it must be localized in *MT*. Recently, our international research group determined the genome sequence of *MT* in the oogamous *Volvox carteri*. *V. carteri MT* shows remarkable expansion and divergence relative to that from *C. reinhardtii*. Five new female-limited "*HIBOTAN*" genes and ten male-limited genes (including "*OTOKOGI*") were identified in *V. carteri MT*. These observations suggest that the origins of femaleness and maleness are principally affected by the evolution of *MT*, which has undergone a remarkable expansion and gain of new male- and female-limited genes. Our recent results regarding the evolution of the volvocalean *MT* gene *MAT3/RB* are also discussed in relationship to the evolution of male–female sexual dimorphism.

Keywords Gametes • Male–female sexual dimorphism • Mating-type locus • Sex-specific gene • Volvocales

H. Nozaki (✉)
Department of Biological Sciences, Graduate School of Science,
University of Tokyo, Hongo, Bunkyo-ku, Tokyo 113-0033, Japan
e-mail: nozaki@biol.s.u-tokyo.ac.jp

H. Sawada et al. (eds.), *Sexual Reproduction in Animals and Plants*,
DOI 10.1007/978-4-431-54589-7_19, © The Author(s) 2014

19.1   Introduction

There is a great deal of interest in sexual reproduction because we are the progeny of our parents who produced sperm and immotile eggs (oogamy), which differ markedly in size and motility. In contrast, primitive organisms sometimes exhibit isogamy (equally sized gametes); in this case, we cannot distinguish between males and females. Therefore, evolutionary transition from isogamy to anisogamy (small and large motile gametes) to oogamy (Bold and Wynne 1985), or the origin of male and female sexes, is one of the most interesting issues in biological research. Although such male–female sexual dimorphism has arisen repeatedly in the evolution of eukaryotes (Kirk 2006), no molecular genetic evidence had been reported for the evolutionary link between male and female and mating types of isogamy until our recent study (Nozaki et al. 2006a), possibly because animals and land plants have no extant isogamous relatives (Rokas et al. 2005; Laurin-Lemay et al. 2012).

Colonial volvocalean algae represent a useful model lineage for the study of such an evolutionary link because isogamy, anisogamy, and oogamy can be recognized within the closely related organisms (Nozaki et al. 2000; Herron et al. 2009), and several mating type-specific genes had been identified in the related unicellular model species, *Chlamydomonas reinhardtii* (Ferris and Goodenough 1994, 1997). This review focuses on studies of sexual reproduction of the colonial Volvocales as well as the related *C. reinhardtii*.

19.2   Sexual Reproduction in the Colonial Volvocales

Figure 19.1 presents a summary of the vegetative organization and modes of sexual reproduction in volvocine algae. It is interesting that, in this lineage, the primitive colonial organization also has a primitive type of sexual reproduction (Nozaki and Ito 1994).

In the colonial volvocalean *Gonium pectorale*, vegetative spheroids are flattened with 8 or 16 cells, and sexual reproduction is isogamous with no differentiation between female and male gametes. Both the two conjugating isogametes bear a mating papilla, which is called the bilateral mating papilla (Nozaki 1984, 1996; Mogi et al. 2012) (Figs. 19.1, 19.2). Plasmogamy is initiated by the union of the tips of the mating papillae from mating-type *plus* and *minus* gametes.

In *Eudorina elegans*, vegetative spheroids are spheroidal with 16 or 32 cells, and sexual reproduction is anisogamous with flagellate female and male gametes. In sexual reproduction, bundles of male gametes called "sperm packets" are formed. Sperm packets swim to a female spheroid and dissociate into individual male gametes that penetrate the female for gametic union (Goldstein 1964; Nozaki 1983, 1996) (Fig. 19.3).

Volvox carteri is oogamous and has more than 500 cells in a spheroid. When inducing sexual reproduction, dwarf male and special female sexual spheroids are formed. The male spheroid produces sperm packets that swim to the female spheroid, which contains a number of eggs (Nozaki 1988) (Fig. 19.4). *V. carteri* shows prominent male–female differentiation (Kirk 1998; Nozaki 1996).

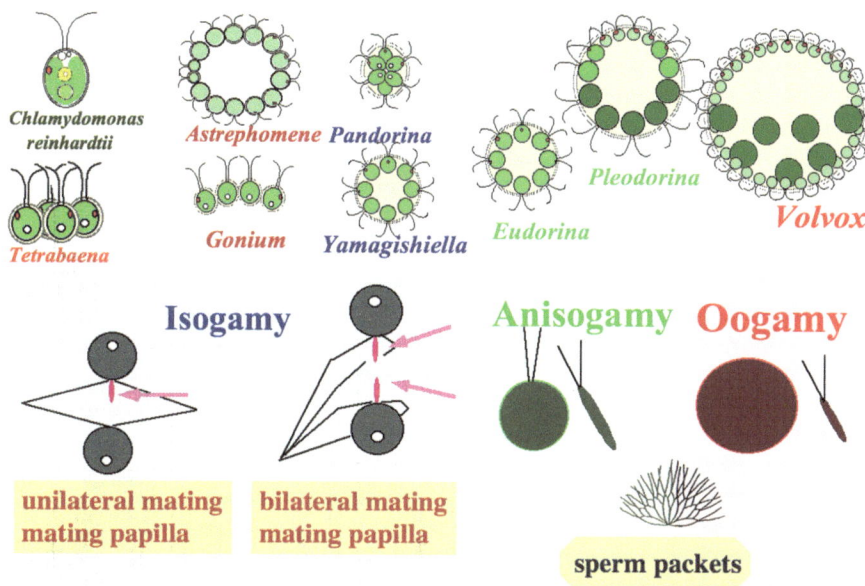

Fig. 19.1 Diagrams of vegetative organization and sexual reproduction in various members of the colonial Volvocales and *Chlamydomonas reinhardtii*. (Based on Nozaki and Ito 1994)

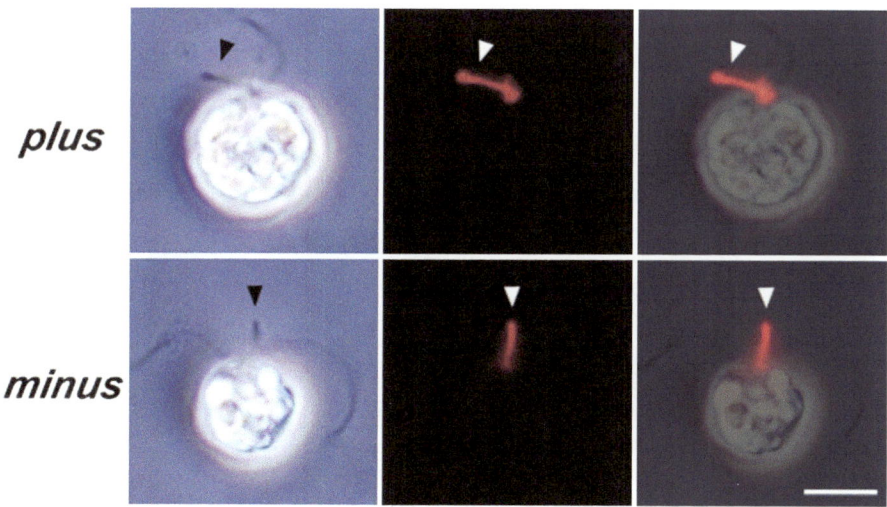

Fig. 19.2 Light microscopy of *Gonium pectorale* isogametes induced in separate mating type plus (*plus*) and mating type minus (*minus*) cultures. All adjacent panels show the same cell. *Bar* 10 μm. Phase-contrast (*left*) and immunofluorescence (actin) (*middle*) images are merged in the *right* panels (merge). Note naked gametes of both mating types bearing a tubular mating structure (*arrowheads*). Note the accumulation of actin (*red*) in the mating structure. (From Mogi et al. 2012; reproduced by permission of John Wiley and Sons, License No. 3079280348215)

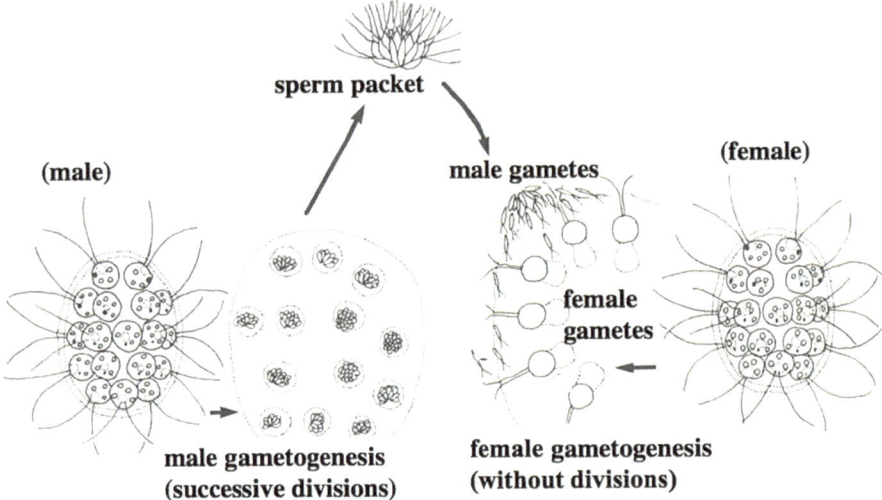

Fig. 19.3 Diagrams of anisogamous sexual reproduction in *Eudorina elegans*. (Based on Nozaki 1983)

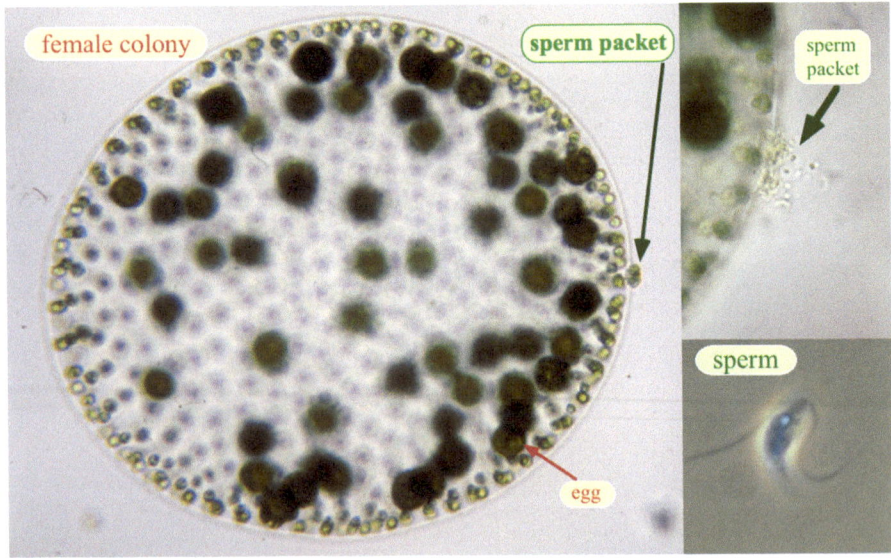

Fig. 19.4 Oogamous sexual reproduction of the multicellular green alga *Volvox carteri* f. *kawasakiensis*. Female spheroids and male spheroids with sperm packets develop in female and male strains, respectively, when sexually induced. The sperm packet attaching to the female spheroid containing many eggs (*left panel*). Sperm packet (*arrow*) dissociates into individual sperm that penetrate the female spheroid (*right upper panel*). Sperm (*right lower panel*). (Original photographs of the strains used by Nozaki 1988)

19.3 Determining the Evolutionary Process of Sexual Reproduction and Tracing Gender-Specific Genes of the Colonial Volvocales

To determine the evolutionary process of the sexual reproduction characteristics in colonial Volvocales, detailed and robust phylogenetic relationships of the members are needed. Therefore, we carried out cladistic analysis of morphological characteristics of vegetative and reproductive phases (Nozaki and Ito 1994; Nozaki et al. 1996) and multigene phylogeny of plastid protein-coding genes (Nozaki et al. 2000; Nozaki 2003). Based on our phylogenetic analyses of morphological and multigene sequence data, a possible evolutionary scenario of sexual reproduction could be deduced. In relationship to the increase in spheroid cell number, female and male genders evolved with the formation of sperm packets from isogamy with bilateral mating papillae (Fig. 19.5). Therefore, it was of interest to determine the genetic basis underlying this evolution of sex. However, mating type-specific genes had not

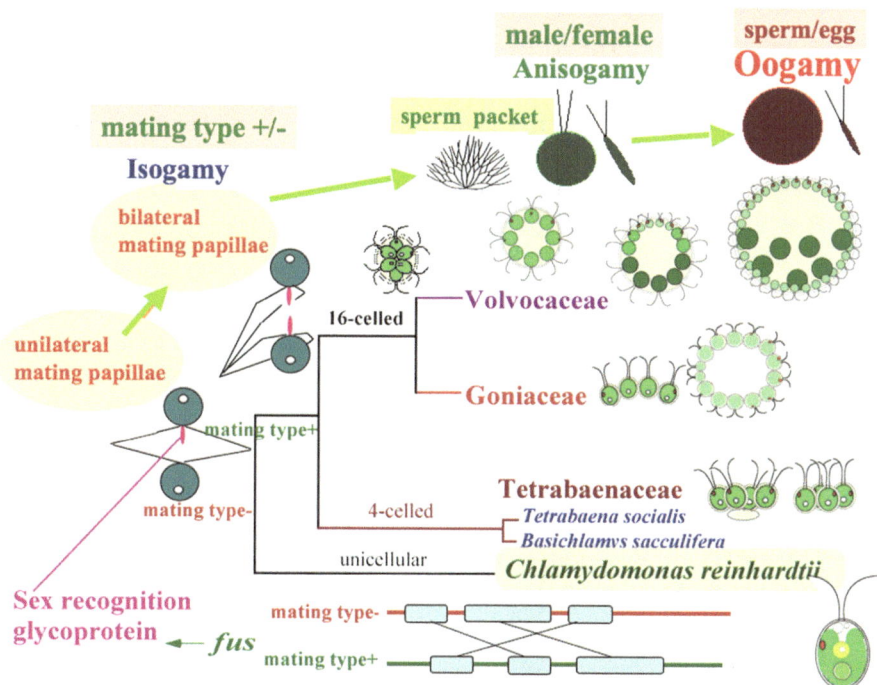

Fig. 19.5 Schematic representation of phylogenetic relationships within the colonial Volvocales as inferred from morphological and multigene sequence data (Nozaki and Ito 1994; Nozaki et al. 2000; Nozaki 2003). A possible evolutionary scenario of sexual reproduction is also shown

been identified within colonial Volvocales, despite studies of the mating type-specific genes, including the sex-determining minus dominance (*MID*) gene in the close relative *C. reinhardtii* (Ferris and Goodenough 1994, 1997).

There are two possible explanations for difficulties in identifying mating type-specific genes from colonial Volvocales. The first is the rapid evolution of the sex-related genes. Ferris et al. (1997) could not isolate *MID* homologues from *G. pectorale* and *V. carteri* based on Southern blotting, and they discussed the rapid evolution of sex-related genes. The second is loss of fertility or sexual activity during long-term maintenance of culture strains (Coleman 1975). Strains obtained from the culture collections that have been maintained in living culture for more than 10 years cannot generally be induced to reproduce sexually to study sexual reproduction, especially in anisogamous and oogamous species. Therefore, we used newly designed degenerate polymerase chain reaction (PCR) primers and new strains of the anisogamous volvocalean *Pleodorina starrii* to identify gender-specific genes. *P. starrii* was described by Nozaki et al. (2006a) based on samples collected in Japanese lakes in 2000–2001, and it is heterothallic with sperm packet formation occurring even within the male strain (Nozaki et al. 2006a, b).

19.4   Problems Regarding the Evolutionary Link Between Isogametic Mating Types and Male–Female Differentiation

In volvocine algae, sex is determined by a single mating-type locus (*MT*) with two haplotypes that specify sexual differentiation. Although *MT* segregates as a single Mendelian trait, it is a complex genomic region containing both shared and sex-limited genes that are rearranged with respect to each other and do not undergo meiotic recombination (Umen 2011). Sexual reproduction of the unicellular volvocalean *C. reinhardtii* is isogamic, with *plus* and *minus* mating types. The *minus* mating type of this species represents a "dominant sex" because occasionally produced diploid cells exhibit the *minus* mating phenotype (Ebersold 1967); when the *minus* mating type-determining gene *MID* is lacking, the phenotype of the sex changes from *minus* to *plus*, forming a fertilization tubule (Goodenough et al. 1982). Therefore, mating type *minus* in *C. reinhardtii* can be considered a "derived type of sex." However, the evolutionary link between isogamous mating type *plus/minus* and female–male differentiation cannot be deduced, even based on comparison of modes of uniparental inheritance of the organelle genomes. In *C. reinhardtii*, the plastid genome is inherited from the *plus* mating type parent, whereas the mitochondrial genome is inherited from the *minus* parent (Boynton et al. 1987). In contrast, both the plastid and mitochondrial genomes are inherited from the female parent in the oogamous alga *V. carteri* (Adams et al. 1990). Thus, homologues of some *C. reinhardtii* mating type-specific genes were needed to be identified in the anisogamous or oogamous members of colonial Volvocales.

19.5   Discovery of the Male-Specific Gene "*OTOKOGI*"

The male-specific gene isolated from *P. starrii* is a homologue of the *MID* gene of *C. reinhardtii* (Nozaki et al. 2006a). The *MID* gene is located in the rearranged (R) domain of the *minus MT* or primitive sex chromosome in *C. reinhardtii* (Ferris and Goodenough 1994). The R domain in the sex chromosome is different between *plus* and *minus* mating types. To isolate the *MID* gene in *P. starrii*, new degenerate primers were designed, and RNA was extracted from *P. starrii* male cultures that produced abundant male gametes in nitrogen-deficient medium (Nozaki et al. 2006a).

The new male-specific gene (*PlestMID*) is called "*OTOKOGI*," meaning "manliness" or "chivalry" in Japanese (Nozaki et al. 2006a; Nozaki 2008). Figure 19.6 shows the amino acid alignment of the *Pleodorina MID* gene, i.e., "*OTOKOGI*," and two *Chlamydomonas MID* genes. These three genes have essentially the same molecular structure with an RWP-RK domain (Fig. 19.6). Southern blotting analysis indicated that both "*OTOKOGI*" and its pseudogene are present in the nuclear genome of the male strain but lacking in the female strain (Nozaki et al. 2006a). Genomic PCR of six *Pleodorina* strains also demonstrated that "*OTOKOGI*" and its pseudogene are present only in male strains (Nozaki et al. 2006a). Using antibody staining, "OTOKOGI" protein was shown to be accumulated in the male gametic nuclei (Nozaki et al. 2006a) (Fig. 19.7). Therefore, the *Pleodorina* "*OTOKOGI*" gene may participate in differentiation of male gametes, and thus the dominant or *minus* mating type of the isogamous *C. reinhardtii* is homologous to the male in the anisogamous volvocalean *Pleodorina*. In Volvocales, the derived type of sex evolved to male, whereas the basic type of sex evolved to female.

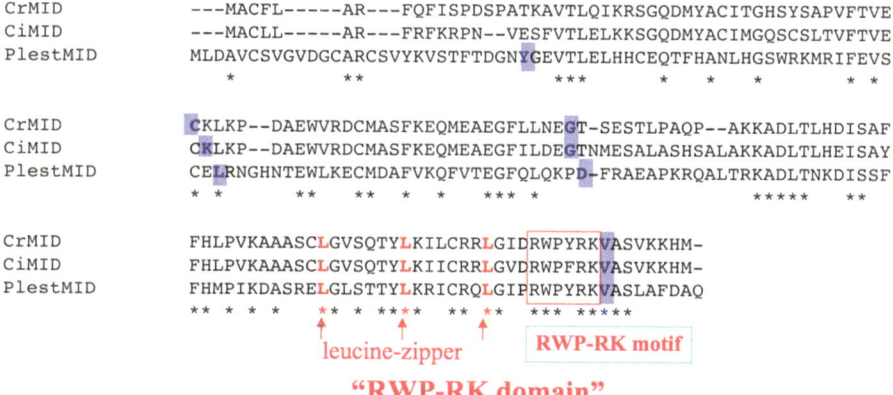

Fig. 19.6 Alignment of the sequences of male-specific "OTOKOGI" and two *Chlamydomonas* MID proteins (CrMID and CiMID). *Blue* shading indicates intron locations. (From Nozaki et al. 2006a; reproduced by permission of Elsevier Ltd., License No. 3079280808611)

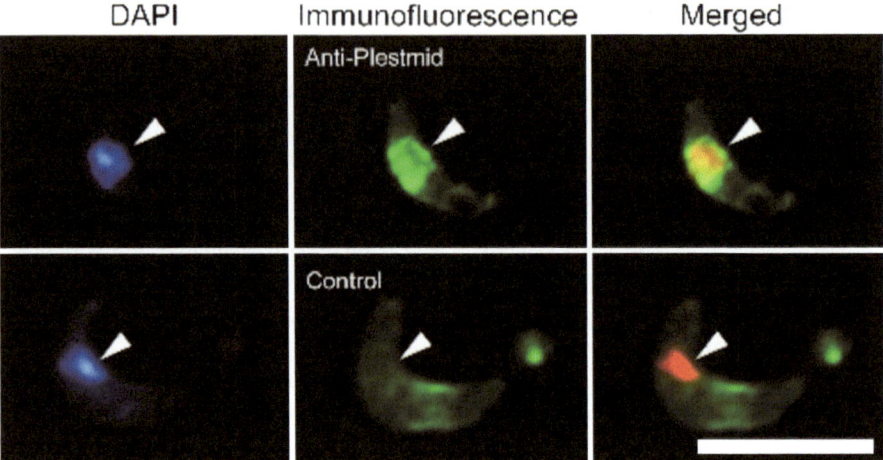

Fig. 19.7 Visualization of male-specific "OTOKOGI" protein in male gamete after release from sperm packets. "OTOKOGI" expression is obvious in nucleus (*arrowheads* in DAPI images). Specimens were double stained with 4′,6-diamidino-2-phenylindole (DAPI) and anti-"OTOKOGI" antibodies. All panels show identical cells. DAPI (pseudo-colored) and immunofluorescence images are merged (*right panels*). *Bar* 5 μm. (From Nozaki et al. 2006a,b; reproduced by permission of Elsevier Ltd., License No. 3079280808611)

19.6 Female-Limited "*HIBOTAN*" Genes Resolved in the *Volvox carteri* Genome

Even though the evolutionary link between male–female differentiation and isogametic mating types was determined (Nozaki et al. 2006a), genes contributing to the evolution of femaleness and maleness from isogametic mating types remained unclear. In colonial Volvocales, the female sex evolved from the original type of sex (*plus* mating type) in isogamy (Nozaki et al. 2006a), and female-specific genes are lacking or rare in higher animals and land plants. Thus, the female-specific gene "*HIBOTAN*" may be lacking in the evolution of femaleness of colonial Volvocales. Genomic data regarding *MT* or the sex chromosomal region of the anisogamous or oogamous members were needed.

In 2010, we published a paper on the genome analyses of the *MT* of *V. carteri* that exhibits the production of eggs and sperm (Ferris et al. 2010). This publication was based on co-research between the United States (USA) and Japan, which originally started in 2005 when Dr. Patrick Ferris came to Japan and became a guest researcher at the Graduate School of Science, University of Tokyo. The Japanese team contributed to cloning of the *V. carteri* "*OTOKOGI*" gene and construction of a male *V. carteri* BAC library.

The *V. carteri MT* shows remarkable expansion and divergence relative to that from *C. reinhardtii*, which has equal-sized gametes, and is about five times larger and contains more genes than the *C. reinhardtii MT*. Our transcriptome analysis

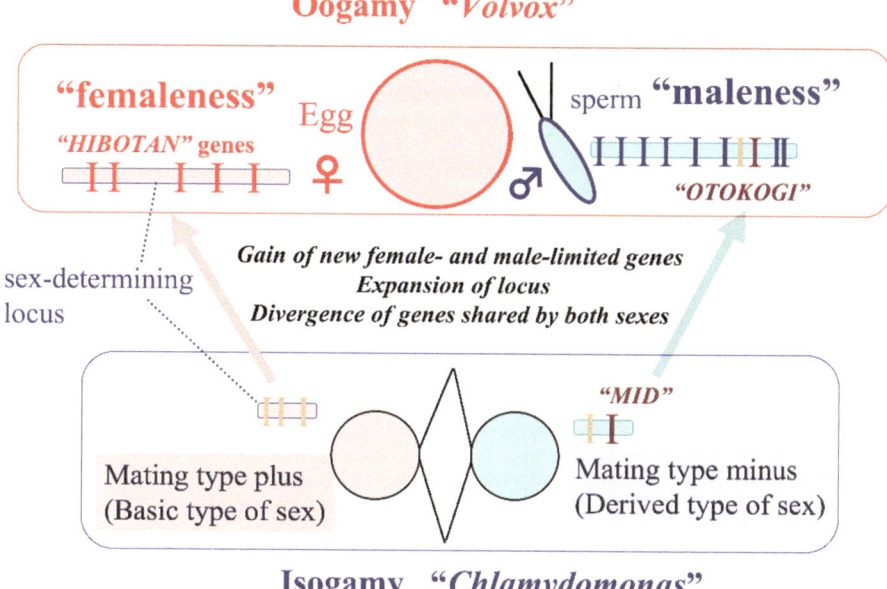

Fig. 19.8 Diagram of evolution of oogamy from isogamy and mating type (sex-determining) loci in the volvocalean algae. (Based on Nozaki et al. 2006a and Ferris et al. 2010)

using next-generation sequencing identified five new female-limited "*HIBOTAN*" genes and ten male-limited genes, including "*OTOKOGI*," in the *MT* locus of *V. carteri* (Fig. 19.8). None of the five "*HIBOTAN*" genes was identified in the *C. reinhardtii MT* (Ferris et al. 2010). Thus, the origins of femaleness and maleness are principally affected by the evolution of the *MT* or sex chromosome that has undergone remarkable expansion and gain of new male- and female-limited genes. However, our study (Ferris et al. 2010) was based on the comparison of only two evolutionary extremes, the isogamous unicellular *Chlamydomonas* and the ooga-mous multicellular *Volvox* (Fig. 19.5). Thus, genome information on the *MT* from intermediate colonial Volvocales is needed to determine the evolutionary signifi-cance of expansion of the *MT* as well as the male- and female-limited genes newly identified in *V. carteri* (Charlesworth and Charlesworth 2010).

19.7   Origin of the Gender-Based Divergence of the Mating Locus Gene *MAT3* from Oogamous *Volvox carteri*

Our recent comparative study of *MT* from oogamous *V. carteri*, with *MT* from an isogamous species, *C. reinhardtii*, revealed major differences in the size and sex-based differentiation of *MT* genes (Ferris et al. 2010). *V. carteri MT* was found to show a much higher degree of sex-based differentiation in its shared genes (those

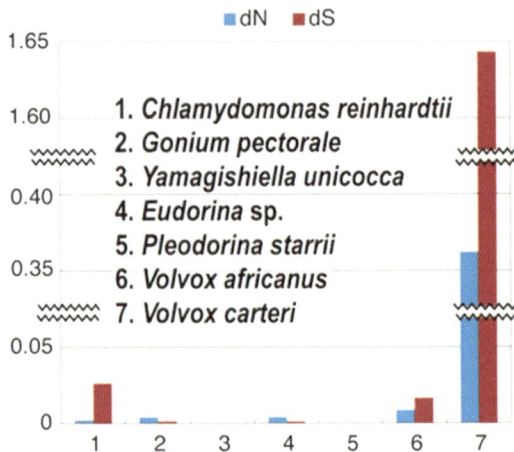

Fig. 19.9 Gender-based divergence of mating-type locus *MAT3* genes from the colonial Volvocales and *Chlamydomonas reinhardtii*. Bar graph depicts dN (number of substitutions per nonsynonymous site) and dS (number of substitutions per synonymous site) between *MAT3* alleles from each of the two mating types or sexes. (From Hiraide et al. 2013; reproduced by permission of Oxford University Press, License Number 3079271441882)

with an allele in both mating haplotypes or sexes). Shared genes that have become masculinized and feminized in sequence or expression, as occurred in *V. carteri*, are candidates for contributing to male–female sexual dimorphism. However, there had been no investigations of *MT* genes in the context of the isogamy to anisogamy/oogamy transition in volvocine algae other than the previously described comparison of *C. reinhardtii* and *V. carteri* (Ferris et al. 2010; Charlesworth and Charlesworth 2010).

One candidate regulator of gamete size is the mating locus gene *MAT3*, which encodes a homologue of the retinoblastoma (RB) tumor suppressor protein. In *C. reinhardtii*, *MAT3* is closely linked to *MT* and regulates cell size and cell-cycle progression (Umen and Goodenough 2001; Fang et al. 2006). In contrast to *C. reinhardtii*, in which the *minus* and *plus MAT3* alleles are nearly identical and function interchangeably (Umen and Goodenough 2001; Merchant et al. 2007), high degrees of male–female sequence differentiation and sex-regulated alternative splicing were observed for *V. carteri MAT3* (Ferris et al. 2010). These observations led Ferris et al. (2010) to suggest that *MAT3* homologues may be related to control of gamete size in colonial volvocine algae, as predicted earlier by the gamete size regulator recruitment model for the evolution of anisogamy/oogamy from isogamous mating types (Charlesworth 1978). Thus, Hiraide et al. (2013) sequenced the full-length coding regions of *MAT3* from *plus* and *minus* mating types of isogamous *Gonium pectorale* and *Yamagishiella unicocca*, from males and females of anisogamous *Eudorina* sp. and *Pleodorina starrii*, and from males and females of oogamous *Volvox africanus*. In contrast to *V. carteri*, *MAT3* homologues from the five colonial species examined had almost identical nucleotide sequences between the two sexes (Fig. 19.9). Our phylogenetic analysis of *MAT3* sequences suggested that the extensive *MAT3*

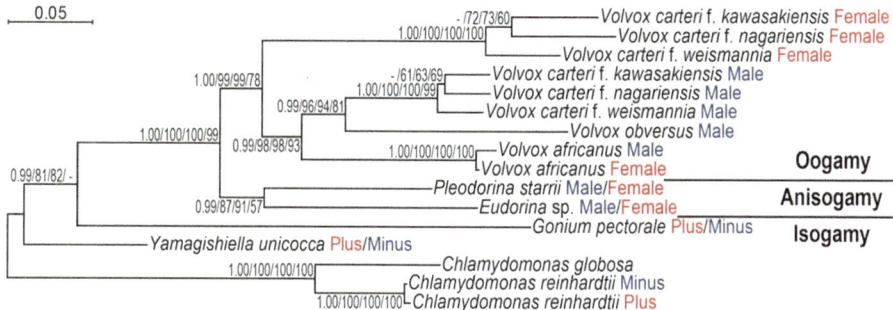

Fig. 19.10 Phylogeny of deduced amino acids of mating-type locus *MAT3* genes from colonial Volvocales and *Chlamydomonas*. Branch labels indicate, from *left* to *right*: posterior probabilities (≥0.90) from Bayesian inference/bootstrap values (≥50 %) obtained using 1,000 replicates with RAxML/Bootstrap values (≥50 %) using 1,000 replicates with maximum parsimony. (From Hiraide et al. 2013; reproduced by permission of Oxford University Press, License Number 3079271441882)

divergence in the *V. carteri* lineage may have occurred recently in the ancestor of the three *V. carteri* forms after their divergence from the anisogamous lineage containing *P. starrii* and *Eudorina* sp. (Fig. 19.10). These observations suggest the roles of genetic determinants other than or in addition to *MAT3* in the evolution of anisogamy in colonial volvocalean algae.

19.8   Conclusions

The considerable expansion of oogamous *V. carteri MT* is based mostly on the increase in the noncoding DNA region, but *V. carteri MT* contains more coding regions or genes than *C. reinhardtii MT* (Ferris et al. 2010). Similar expansion of noncoding DNA regions was recently recognized in the mitochondrial and plastid genomes from colonial Volvocales (Smith and Lee 2009, 2010; Hamaji et al. 2013; Smith et al. 2013). Smith et al. (2013) demonstrated that ratios of noncoding DNA regions in the mitochondrial and plastid genomes increase in relationship to the increase in spheroid cell number (from unicellular *Chlamydomonas* to *Gonium* to *Pleodorina* to *Volvox*) (see Fig. 19.1). This increase may be explained by the mutational hazard hypothesis (Lynch 2007), which argues that genome expansion is a product of a low effective population size (Ne), which results in increased random genetic drift, or a low mutation rate (μ), which reduces the burden of harboring excess DNA. Therefore, a gradual increase in *MT* size as a function of the increase in spheroid cell number, which may result in a decrease in Ne, may also be considered within colonial Volvocales. However, the origins and evolution of the many gender-specific genes found in *V. carteri MT* remain unexplained on the basis of the present limited data from colonial volvocalean *MT*.

Acknowledgments This work was supported by the MEXT/JSPS KAKENHI (grant numbers 24112707 and 24247042 to H.N.).

References

Adams CR, Stamer KA, Miller JK et al (1990) Patterns of organellar and nuclear inheritance among progeny of two geographically isolated strains of *Volvox carteri*. Curr Genet 18:141–153

Bold HC, Wynne MJ (1985) Introduction to the algae: structure and reproduction, 2nd edn. Prentice-Hall, Englewood Cliffs

Boynton JE, Harris EH, Burkhart BD et al (1987) Transmission of mitochondrial and chloroplast genomes in crosses of *Chlamydomonas*. Proc Natl Acad Sci USA 84:2391–2395

Charlesworth B (1978) The population genetics of anisogamy. J Theor Biol 73:347–357

Charlesworth D, Charlesworth B (2010) Evolutionary biology: the origins of two sexes. Curr Biol 20:R519–R521

Coleman AW (1975) Long-term maintenance of fertile algal clones: experience with *Pandorina* (Chlorophyceae). J Phycol 11:282–286

Ebersold WT (1967) *Chlamydomonas reinhardi*: heterozygous diploid strains. Science 157:447–449

Fang S-C, de los Reyes C, Umen JG (2006) Cell size checkpoint control by the retinoblastoma tumor suppressor pathway. PLoS Genet 2:e167

Ferris PJ, Goodenough UW (1994) The mating-type locus of *Chlamydomonas reinhardtii* contains highly rearranged DNA sequences. Cell 76:1135–1145

Ferris PJ, Goodenough UW (1997) Mating type in *Chlamydomonas* is specified by *Mid*, the minus-dominance gene. Genetics 146:859–869

Ferris PJ, Pavlovic G, Fabry S et al (1997) Rapid evolution of sex-related genes in *Chlamydomonas*. Proc Natl Acad Sci USA 94:8634–8639

Ferris P, Olson BJSC, De Hoff PL et al (2010) Evolution of an expanded sex-determining locus in *Volvox*. Science 328:351–354

Goldstein M (1964) Speciation and mating behavior in *Eudorina*. J Protozool 11:317–344

Goodenough UW, Detmers PA, Hwang C (1982) Activation for cell fusion in *Chlamydomonas*: analysis of wild-type gametes and nonfusing mutants. J Cell Biol 92:378–386

Hamaji H, Smith DR, Noguchi H et al (2013) Mitochondrial and plastid genomes of the colonial green alga *Gonium pectorale* give insights into the origins of organelle DNA architecture within the Volvocales. PLoS One 8:e57177

Herron MD, Hackett JD, Aylward FO et al (2009) Triassic origin and early radiation of multicellular volvocine algae. Proc Natl Acad Sci USA 106:3254–3258

Hiraide R, Kawai-Toyooka H, Hamaji T et al (2013) The evolution of male–female sexual dimorphism predates the gender-based divergence of the mating locus gene *MAT3/RB*. Mol Biol Evol 30:1038–1040

Kirk DK (1998) *Volvox*: molecular genetic origins of multicellularity and cellular differentiation. Cambridge University Press, Cambridge

Kirk DL (2006) Oogamy: inventing the sexes. Curr Biol 16:R1028–R1030

Laurin-Lemay S, Brinkmann H, Philippe H (2012) Origin of land plants revisited in the light of sequence contamination and missing data. Curr Biol 22:R593–R594. doi:10.1016/j.cub.2012.06.013

Lynch M (2007) The origins of genome architecture. Sinauer, Sunderland

Merchant SS, Prochnik SE, Vallon O et al (2007) The *Chlamydomonas* genome reveals the evolution of key animal and plant functions. Science 318:245–250

Mogi Y, Hamaji T, Suzuki M et al (2012) Evidence for tubular mating structures induced in each mating type of heterothallic *Gonium pectorale* (Volvocales, Chlorophyta). J Phycol 48:670–674

Nozaki H (1983) Sexual reproduction in *Eudorina elegans* (Chlorophyta, Volvocales). Bot Mag (Tokyo) 96:103–110

Nozaki H (1984) Newly found facets in the asexual and sexual reproduction in *Gonium pectorale* (Chlorophyta, Volvocales). Jpn J Phycol 32:130–133

Nozaki H (1988) Morphology, sexual reproduction and taxonomy of *Volvox carteri* f. *kawasakiensis* f. nov. (Chlorophyta) from Japan. Phycologia 27:209–220

Nozaki H (1996) Morphology and evolution of sexual reproduction in the Volvocaceae (Chlorophyta). J Plant Res 109:353–361

Nozaki H (2003) Origin and evolution of the genera *Pleodorina* and *Volvox* (Volvocales). Biologia 58(4):425–431

Nozaki H (2008) A new male-specific gene "*OTOKOGI*" from *Pleodorina starrii* (Volvocaceae, Chlorophyta) unveiling an origin of male and female. Biologia 63:772–777

Nozaki H, Ito M (1994) Phylogenetic relationships within the colonial Volvocales (Chlorophyta) inferred from cladistic analysis based on morphological data. J Phycol 30:353–365

Nozaki H, Ito M, Watanabe MM et al (1996) Ultrastructure of the vegetative colonies and systematic position of *Basichlamys* (Volvocales, Chlorophyta). Eur J Phycol 31:67–72

Nozaki H, Misawa K, Kajita T et al (2000) Origin and evolution of the colonial Volvocales (Chlorophyceae) as inferred from multiple, chloroplast gene sequences. Mol Phylogenet Evol 17:256–268

Nozaki H, Mori T, Misumi O et al (2006a) Males evolved from the dominant isogametic mating type. Curr Biol 16:R1018–R1020

Nozaki H, Ott FD, Coleman AW (2006b) Morphology, molecular phylogeny and taxonomy of two new species of *Pleodorina* (Volvoceae, Chlorophyceae). J Phycol 42:1072–1080

Rokas A, Krüger D, Carroll SB (2005) Animal evolution and the molecular signature of radiations compressed in time. Science 310:1933–1938

Smith DR, Lee RW (2009) The mitochondrial and plastid genomes of *Volvox carteri*: bloated molecules rich in repetitive DNA. BMC Genomics 10:132

Smith DR, Lee RW (2010) Low nucleotide diversity for the expanded organelle and nuclear genomes of *Volvox carteri* supports the mutational-hazard hypothesis. Mol Biol Evol 27:2244–2256

Smith DR, Hamaji T, Olson BJSC et al (2013) Organelle genome complexity scales positively with organism size in volvocine green algae. Mol Biol Evol 30(4):793–797

Umen JG (2011) Evolution of sex and mating loci: an expanded view from volvocine algae. Curr Opin Microbiol 14:634–641

Umen JG, Goodenough UW (2001) Control of cell division by a retinoblastoma protein homolog in *Chlamydomonas*. Genes Dev 15:1652–1661

Part III
Allorecognition in Male–Female Interaction

Chapter 20
Allorecognition and Lysin Systems During Ascidian Fertilization

Hitoshi Sawada, Kazunori Yamamoto, Kei Otsuka, Takako Saito, Akira Yamaguchi, Masako Mino, Mari Akasaka, Yoshito Harada, and Lixy Yamada

Abstract Ascidians (primitive chordates) are hermaphroditic animals that release sperm and eggs almost simultaneously, but several species, including *Halocynthia roretzi* and *Ciona intestinalis*, show strict self-sterility. In *H. roretzi*, a 70-kDa vitelline coat (VC) protein consisting of 12 EGF-like repeats (HrVC70) appears to be a promising candidate for the self/non-self-recognition (or allorecognition) system during gamete interaction. After sperm recognizes the VC as non-self, the sperm extracellular ubiquitin-proteasome system appears to degrade HrVC70, allowing sperm to penetrate through the VC with the aid of sperm trypsin-like proteases.

In *C. intestinalis*, egg-side highly polymorphic fibrinogen-like ligands on the VC (v-Themis-A and v-Themis-B) and cognate sperm-side hypervariable region-containing polysystin-1-like receptors (s-Themis-A and s-Themis-B) seem to be responsible for allorecognition in gamete interaction. Recently, we noticed that a novel pair of v-Themis-B2 and s-Themis-B2 and an acid-extractable VC protein called Ci-v-Themis-like may take part in gamete interaction or allorecognition. When sperm recognizes the VC as self, the sperm undergoes a drastic Ca^{2+} influx, which is one of the major intracellular self-recognition responses within sperm, resulting in sperm detachment from the VC or in sperm becoming quiescent. These allorecognition systems and self-recognition responses within sperm are very similar to the self-incompatibility system in flowering plants.

Keywords Allorecognition • Ascidian • Lysin • Protease • Proteasome • Self-incompatibility • Sperm

H. Sawada (✉) • K. Yamamoto • K. Otsuka • T. Saito • A. Yamaguchi
M. Mino • M. Akasaka • Y. Harada • L. Yamada
Sugashima Marine Biological Laboratory, Graduate School of Science,
Nagoya University, Sugashima, Toba 517-0004, Japan
e-mail: hsawada@bio.nagoya-u.ac.jp

H. Sawada et al. (eds.), *Sexual Reproduction in Animals and Plants*,
DOI 10.1007/978-4-431-54589-7_20, © The Author(s) 2014

20.1   Introduction

Sexual reproduction is an excellent reproductive strategy to elicit genetic diversity in the next generation, but most flowering plants and several animals, including ascidians, are hermaphrodites. Therefore, it seems beneficial for these hermaphroditic organisms to acquire a self-sterility or self-incompatibility system to avoid self-fertilization.

Ascidians (tunicates) are hermaphroditic marine invertebrates (primitive chordates), but several species, including *Halocynthia roretzi* and *Ciona intestinalis*, that release sperm and eggs almost simultaneously show strict self-sterility. In animal fertilization, it is indispensable for sperm to penetrate through the proteinaceous egg investment called the vitelline coat (VC) in marine invertebrates and zona pellucida (ZP) in mammals (McRorie and Williams 1974; Morton 1977; Sawada 2002). Because VC-free eggs are self-fertile in ascidians, it is thought that a self/non-self-recognition system, which is also referred to as an allorecognition system, is involved in the interaction between sperm and the VC of eggs. Therefore, sperm lysin, a lytic agent that makes a small hole in the VC for sperm passage, must be activated or exposed to the sperm surface after sperm recognizes the VC as non-self (Sawada 2002).

To investigate the allorecognition and lysin systems during ascidian fertilization, we have been mainly using two solitary ascidian species, *H. roretzi* and *C. intestinalis*, because readily fertilizable sperm and eggs are easily obtained. *H. roretzi* is particularly useful for biochemical studies because this species is cultured in Japan for human consumption and also large quantities of gametes can be easily collected (Sawada 2002). On the other hand, *C. intestinalis* is very useful for genetic analysis and molecular biological approaches because a genome database is available and also genetic analysis can be easily carried out using adults that are sexually matured within 3 months. In this chapter, we summarize the sperm proteases, including the ubiquitin-proteasome system (UPS), that are involved in ascidian fertilization as lysins and also the allorecognition or self-incompatibility system functioning in fertilization of *H. roretzi* and *C. intestinalis*.

20.2   Allorecognition and Lysin Systems in *H. roretzi*

20.2.1   Allorecognition in **H. roretzi**

It is known that self-sterile *H. roretzi* eggs become self-fertile when the eggs are treated with acid (pH 2–3) for 1 min and also that immature and VC-free eggs are self-fertile (Fuke 1983; Fuke and Numakunai 1996). We therefore speculated that a certain allorecognition factor is attached to the VC during oocyte maturation and that the putative factor may be detached from the VC or denatured by a weak acid. To test this possibility, VCs were isolated from immature and mature eggs and

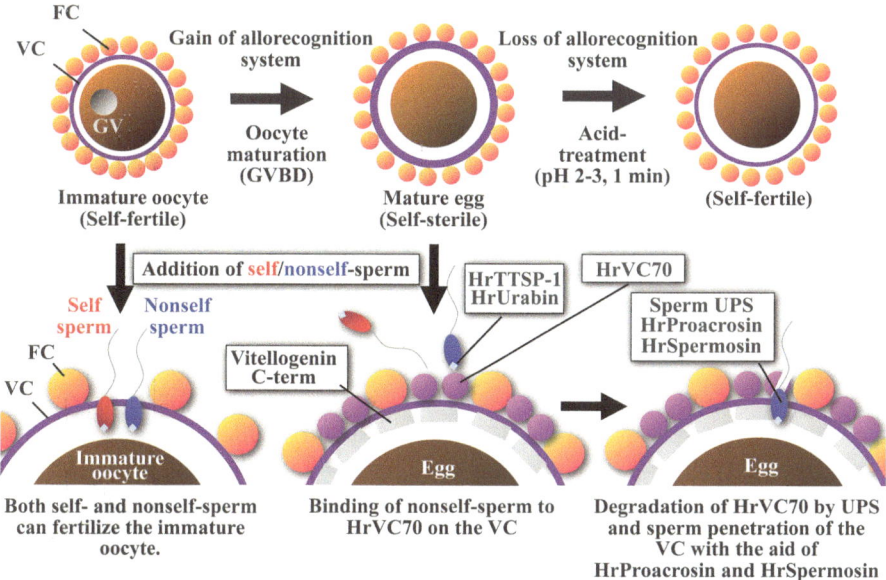

Fig. 20.1 Working hypothesis for the possible roles of sperm proteases as a lysin and allorecognition system in *Halocynthia roretzi*. In *H. roretzi*, HrVC70, consisting of 12 EGF-like repeats with a high degree of polymorphism among individuals, attaches to the vitelline coat (*VC*) during oocyte maturation, resulting in prevention of self-fertilization. The sperm-side binding partners of HrVC70 appear to be HrTTSP-1 and HrUrabin, and if the sperm recognizes the VC as non-self, sperm UPS may be activated, enabling sperm to penetrate through the VC. HrVC70 has a nature to be extracted by weak acid, which allows self-sperm to penetrate through the VC. During sperm passage through the VC, sperm trypsin-like proteases HrProacrosin and HrSpermosin also play some important roles by enabling movable binding to the VC components, which are C-terminal fragments of vitellogenin

subjected to sodium dodecyl sulfate-polyacrylamide gel electrophoresis (SDS-PAGE). We found that the amount of the 70-kDa main component, HrVC70, of mature eggs is markedly larger than that of immature oocytes, suggesting that HrVC70 is attached to the VC during oocyte maturation. In addition we noticed that HrVC70 is easily solubilized and extracted from the isolated VC by 1–10 mM HCl and that sperm are capable of binding to HrVC70-immobilized agarose beads. We also found that non-self sperm rather than self-sperm efficiently bound to HrVC70 and that HrVC70 isolated from non-self eggs more efficiently inhibited the fertilization than did that from self-eggs (Sawada et al. 2004). From these results, together with the fact that HrVC70 shows a high degree of polymorphism among individuals and that even a single amino-amino acid substitution in EGF-like repeat regions in Notch protein is sufficient to cause Notch-signaling diseases (Artavanis-Tsakonas et al. 1995), we concluded that HrVC70 is a promising candidate for allorecognition in fertilization of *H. roretzi*. Although it is still unclear whether the amino-acid substitution in HrVC70 is actually responsible for allorecognition during gamete interaction in *H. roretzi*, all the biochemical data so far obtained support the idea that HrVC70 is a key protein involved in allorecognition (Fig. 20.1).

As sperm-side binding partners of HrVC70, HrTTSP-1 (type II transmembrane serine protease) and *u*nique *RA*FT-derived *bin*ding (HrUrabin), partner for HrVC70 (a GPI-anchored CRISP family protein) have been identified by yeast two-hybrid screening (Harada and Sawada 2007) and far Western blot analysis, respectively (Urayama et al. 2008). HrTTSP1 has an estimated molecular mass of 337 kDa and contains 23 CCP/SCP/Sushi domains, 3 ricin B domains, and 1 CUB domain in its extracellular region. Although HrTTSP-1 contains several potentially interesting domains, its biological function has not been studied in detail. In contrast, HrUrabin appears to play a key role in fertilization because an anti-HrUrabin antibody can inhibit fertilization and also binding of allo-recognizable sperm to HrVC70 agarose beads. However, HrUrabin had little polymorphism among individuals and showed no difference in its binding ability to HrVC70 from self-eggs and non-self eggs. Therefore, it is thought that HrUrabin is unable to directly distinguish self- and non-self-HrVC70 but participates in the process of allorecognition in a broad sense, because the antibody against HrUrabin potently inhibited binding of allo-recognizable sperm to HrVC70 (Urayama et al. 2008) (Fig. 20.1).

There is an orthologue of HrVC120, a precursor of HrVC70, in another ascidian species, *Halocynthia aurantium*, which is a close relative species of *H. roretzi* inhabiting the northern part of Japan. The mature protein HaVC80 consists of 13 epidermal growth factor (EGF)-like repeats, 1 repeat longer than HrVC70, and is derived from the precursor protein HaVC130 consisting of 14 EGF-like repeats and a C-terminal ZP domain (Ban et al. 2005). HrVC120 is very similar to HaVC130 (83.4 % identity based on their amino-acid sequences), and the 8th EGF domain of *HrVC120* gene appears to have been duplicated during evolution. HaVC80 is also highly polymorphic among individuals in restricted regions between the first and second Cys residues, the third and fourth Cys residues, and the EGF-domain-connected regions, where similar polymorphisms are observed in HrVC70 (Ban et al. 2005). Further studies are needed to elucidate the binding partners and roles of HaVC80 in fertilization.

20.2.2 *Lysin in* H. roretzi

Hoshi et al. (1981) first reported that trypsin-like protease(s) and chymotrypsin-like protease(s) are indispensable for sperm penetration through the VC of eggs in *H. roretzi* by examining the effects of various protease inhibitors on fertilization of intact and VC-free eggs (Hoshi et al. 1981; Hoshi 1985). Two trypsin-like proteases, HrAcrosin and HrSpermosin, were then purified from *H. roretzi* sperm using Boc-Val-Pro-Arg-MCA, the strongest inhibitor of fertilization among the substrates tested (Sawada et al. 1982, 1984a). Although it was suggested that both these proteases participate in fertilization by comparing the effects of various leupeptin analogues (peptidyl-rgininal) on fertilization and enzymatic activities (Sawada et al. 1984b; Sawada and Someno 1996) and also by examining the inhibitory ability of anti-spermosin antibody on fertilization (Sawada et al. 1996), the purified enzymes showed little degrading activity toward VC proteins (Sawada et al., unpublished data).

Both HrProacrosin (precursor of HrAcrosin) and HrSpermosin possess several candidate regions for protein–protein interaction, that is, two CUB domains in the C-terminus of HrProacrosin and a Pro-rich region in the N-terminus of the light chain of HrSpermosin (Kodama et al. 2001, 2002). Two VC proteins (25- and 30-kDa VC proteins) were identified as binding proteins to these proteases (Akasaka et al. 2010). By cDNA sequencing, it was revealed that the 25- and 30-kDa proteins correspond to the C-terminal region of high molecular mass vitellogenin, which belongs to a family of lipid transfer proteins (Akasaka et al. 2010, 2013). We propose that sperm binds to the C-terminal fragments of vitellogenin located on the VC, the binding being mediated by the sperm-side HrProacrosin CUB domain and HrSpermosin Pro-rich region, and then sperm proteases degrade these VC proteins or the precursor regions, enabling sperm to detach from and penetrate the VC. These sequential actions may explain the phenomena of sperm binding to and penetrating through the VC (Fig. 20.1) (working hypothesis: for details, see another chapter by Akasaka et al.).

Because the purified preparations of ascidian sperm trypsin-like proteases were unable to efficiently degrade the VC, we focused on a chymotrypsin-like protease as a potential VC lysin. As Suc-Leu-Leu-Val-Tyr-MCA was the strongest inhibitor of *H. roretzi* fertilization among the peptide substrates tested (Sawada et al. 1983), Suc-Leu-Leu-Val-Tyr-MCA-hydrolyzing protease was purified and identified as 20S and 26S (or 26S-like) proteasomes (Saito et al. 1993). The UPS is one of the most important intracellular protein-degradation systems in eukaryotic cells (Hershko and Ciechanover 1998; Finley 2009; Tanaka 2009). In this system, intracellular short-lived and aberrant proteins are tagged with ubiquitin by sequential actions of ubiquitin-activating enzyme E1, ubiquitin-conjugating enzyme E2 and ubiquitin ligase E3 and then degraded by the 26S proteasome in an ATP-dependent manner. The 26S proteasome is made up of the 20S proteasome, a barrel-shaped protease complex consisting of four stacked heptameric rings $\alpha_7\beta_7\beta_7\alpha_7$, and the 19S regulatory particle (19S RP)/PA700, consisting of 19 subunits, including 6 ATPase subunits and a ubiquitin-recognizing subunit S5a. The 20S proteasome has three protease activities, that is, caspase-like (β_1), trypsin-like (β_2), and chymotrypsin-like (β_5) activities (Tanaka 2009).

The 26S proteasome-containing fraction partially purified from activated sperm showed a weak VC-degrading activity in *H. roretzi* (Saitoh et al. 1993). In addition, HrVC70 was ubiquitinated and degraded by the purified sperm 26S proteasome in the presence of ATP and ubiquitin (Sawada et al. 2002a). The extracellular UPS appears to play a key role in ascidian fertilization for the following main reasons. First, *H. roretzi* fertilization was inhibited by proteasome inhibitors, including MG115 and MG132, and also by anti-proteasome antibody and anti-multi-ubiquitin chain-specific monoclonal antibody FK2 (Sawada et al. 2002a, b). Second, Suc-Leu-Leu-Val-Tyr-MCA-hydrolyzing proteasome activity, which was specifically inhibited by MG115, was detected in the sperm head region under an epifluorescence microscope when activated by alkaline seawater (Sawada et al. 2002b). Third, sperm proteasomes, as well as HrVC70-ubiquitinating enzyme, ATP, and ubiquitin, were partially released from sperm when activated by alkaline seawater (Sakai et al. 2003; Sawada et al., unpublished data). Fourth, HrVC70 on the VC appears to be ubiquitinated upon insemination on the basis of results of Western blotting and immunocytochemistry using the monoclonal antibody FK2 (Sawada et al. 2002a; Sakai et al. 2003).

HrVC70-ubiquitinating enzyme has been purified from sperm exudate, a fraction released from activated sperm, by DEAE-cellulose chromatography, ubiquitin-agarose chromatography, and 10–40 % glycerol density gradient centrifugation (Sakai et al. 2003). The molecular size of the enzyme was estimated to be approximately 700 kDa by glycerol density gradient centrifugation (Sakai et al. 2003). The purified enzyme exhibited activity in artificial seawater and required a high concentration (~10 mM) of Ca^{2+} for its activity. These enzymatic features also support our idea that the purified enzyme functions extracellularly in seawater. Furthermore, apyrase, which depletes ATP and inhibits HrVC70-ubiquitinating activity, inhibited fertilization when added to the surrounding seawater. These results indicate that a novel extracellular 700-kDa HrVC70-ubiquitinating enzyme complex plays a pivotal role in ubiquitination of HrVC70. Although there are two Lys residues in HrVC70, Lys234 and Lys636, only Lys234 was identified to be ubiquitinated by a ubiquitin-conjugation assay using several site-directed Lys-to-Arg mutant recombinant proteins of HrVC70 (Sawada et al. 2005). Because it is widely believed that only one molecular species of E1 is committed to every ubiquitination reaction, our findings imply the existence of a novel E1-containing complex, functioning extracellularly during fertilization.

20.3   Allorecognition and Lysin Systems in *C. intestinalis*

20.3.1   Proposed Hypotheses of Allorecognition in C. intestinalis

Several candidate molecules involved in the self-incompatibility (SI) system in *C. intestinalis* have been proposed. Kawamura and colleagues found that an acid extract of the VC has the ability to inhibit the binding of non-self sperm, but not self-sperm, to the VC (Kawamura et al. 1991). They partially purified several factors responsible for this activity and revealed that there are a non-allo-recognizable glucose-enriched inhibitor of gamete-binding and Glu/Gln-enriched peptide modulators, which serve as acceptors of non-self sperm, and that certain combinations of these factors show specific inhibitory ability toward the binding of non-self sperm (Kawamura et al. 1991; Harada and Sawada 2008). However, further detailed studies have not been carried out. De Santis and colleagues showed that SI becomes effective several hours after germinal vesicle breakdown (De Santis and Pinto 1991). Because removal of follicle cells prevents the onset of self-sterility, they proposed that follicle cells release a certain self-sterility factor(s) that binds to the VC (De Santis and Pinto 1991). By analogy to the mammalian cellular immune system, they proposed that peptides produced by proteasome-mediated proteolysis are loaded onto Cihsp70, which is a molecular chaperone assumed to be an ancestor protein of major histo-compatibility complex (MHC) class I and II molecules in lower vertebrates or invertebrates, and delivered to the surface of the VC (Marino et al. 1998, 1999). They also showed that the proteasome inhibitor clasto-lactacystin beta-lacton and anti-HSP70 antibodies prevented the onset of self-sterility (Marino et al. 1999). From these results, they speculated that Cihsp70 and a self-peptide produced by proteasomal degradation might be involved in the SI system of *C. intestinalis*, which might share the origin of the vertebrate immune system. However, the antigenic peptide fragments on Cihsp70 on the VC have not yet been identified. On the other hand, Khalturin, Bosch, and colleagues performed PCR-based subtraction experiments and compared gonad cDNAs between genetically unrelated individuals. They identified several candidate genes that are expressed in developing oocytes or/and follicle cells and are polymorphic among individuals, including CiS7 (EGF-like repeat-containing gene), vCRL1 [Sushi (or SCR)-domain-containing gene], and multiple homologues of HrVC70 (EGF-like repeat- and ZP-domain-containing genes) (Khalturin et al. 2005; Kürn et al. 2007a, b). However, they recently reported that vCRL1 genes are not related to the SI system in fertilization of *C. intestinalis,* but the s- and v-Themis system, which we reported previously (Harada et al. 2008; described below), plays a key role in SI (Sommer et al. 2012). In addition, several molecular interactions appear to be involved in gamete interaction, although the involvement in allorecognition is not known. For example, it has been reported that the terminal fucose residue on VC glycoproteins (Rosati and De Santis 1980) and sperm-side fucosidase may make an enzyme–substrate complex, allowing interaction between sperm and the VC of eggs (Hoshi 1986; Matsumoto et al. 2002). On the other hand, we reported that sperm-side CiUrabin, a GPI-anchored CRISP family protein located at the sperm head and tail, is capable of binding to CiVC57, an EGF-like-repeat-containing major glycoprotein on the VC, which can support the interaction between sperm and the VC of eggs (Yamaguchi et al. 2011; Yamada et al. 2009).

20.3.2 Two or Three Pairs of s-Themis and v-Themis, the Key Molecules in Allorecognition in C. intestinalis

In the early part of the twentieth century, Morgan published several papers on the SI system of an ascidian, *C. intestinalis* (Morgan 1911, 1939, 1942, 1944). He reported that the VC is a barrier against self-fertilization and that the SI system is abolished by treatment of eggs with acidic seawater or protease (Morgan 1939). By acid-induced self-fertilization, he raised many selfed F_1 siblings and examined cross-fertility and cross-sterility among them (Morgan 1942, 1944). Cross-sterility is rarely observed in wild populations, but cross-sterile combinations are sometimes observed in selfed or experimentally cross-fertilized siblings, suggesting that self-sterility is genetically governed. The self-fertilized individuals gave a considerable number of cross-sterile combinations, among which he recognized two types of cross-sterility: bi-directional and one-way (Morgan 1942, 1944). Morgan proposed a "haploid sperm hypothesis" to explain the occurrence of one-way cross-sterility (Morgan 1942, 1944), where SI specificity is determined by haploid expression in sperm and diploid expression in eggs. According to his hypothesis, a parent heterozygous at the SI locus (A/a) produces two populations of sperm (A-expressing and a-expressing sperm), either of which can fertilize both types of homozygous eggs (A/A and a/a eggs). In contrast, sperm (A-expressing and a-expressing sperm) from two types of homozygotes (A/A and a/a individuals) are sterile to heterozygous eggs (A/a eggs), because heterozygous eggs (or VC) express both types of female SI gene products, either of which must be recognized by sperm as self. Thus, once a one-way cross-sterile pair of individuals was found, an egg-donating individual should be a heterozygote at a SI-responsible locus, whereas a sperm-donating individual should be a homozygote.

Based on these criteria, Harada et al. (2008) searched for a candidate SI locus by determining the DNA sequence at about 70 genetic markers in 14 chromosomes, and they noticed that two loci, loci A and B, located in chromosomes 2q and 7q, respectively, are involved in SI in this species (Harada et al. 2008) (Figs. 20.2, 20.3). Among the proteins encoded in locus A, a fibrinogen-C-terminus-like protein, referred to as v-Themis-A, with a high degree of polymorphism among individuals was detected on the VC by proteome analysis. On the other hand, of genes encoded in locus A, four genes were found to be expressed in the testis, among which a PKD-1-like protein, referred to as s-Themis-A, was identified as a candidate sperm-side receptor protein with a hypervariable region in its N-terminal region (Themis is a Greek goddess who is the embodiment of divine order, law, and custom and prohibits incest.). Although there is no overall synteny between loci A and B, a similar gene pair of proteins (v-Themis-B and s-Themis-B) was identified in locus B (Harada et al. 2008). Interestingly, *v-Themis-A/B* genes were located in the first intron of *s-Themis-A/B* genes, respectively, in the opposite direction in both cases (Fig. 20.2a). These features indicate a tight linkage between *s-Themis* and *v-Themis* genes, not allowing the segregation of putative binding partners. Based on results of genetic analysis, it has been proposed that when sperm-side s-Themis-A and s-Themis-B interact with the same allelic v-Themis-A and v-Themis-B, respectively, on the VC, sperm must recognize the VC as self, resulting in blocking fertilization.

Polymorphisms among individuals of *s/v-Themis-A* and *s/v-Themis-B* genes were investigated: molecular phylogenetic analysis indicated coevolution between *s-Themis-A/B* and *v-Themis-A/B* (unpublished data).

Within locus B, the genome sequence data showed one additional pair of *s-Themis* and *v-Themis* genes. We first thought that this gene model is caused by a possible miss-assembly. To clarify this point, we determined the DNA sequence in this region after cloning from the genome derived from one individual. The data confirmed the occurrence of a novel pair of *s/v-Themis* genes, which we tentatively call *s-Themis-B2* and *v-Themis-B2*. The sequence of this gene pair is almost identical to that of the pair of *s/v-Themis-B* except for the *v-Themis-B2* region and a hypervariable region of *s-Themis-B* (Fig. 20.2c). In addition, *v-Themis-B2* was detected in the VC by proteomic analysis (Yamada et al., unpublished data). Furthermore, cross-fertility/sterility experiments among individuals, whose *s/v-Themis-A/B/B2*

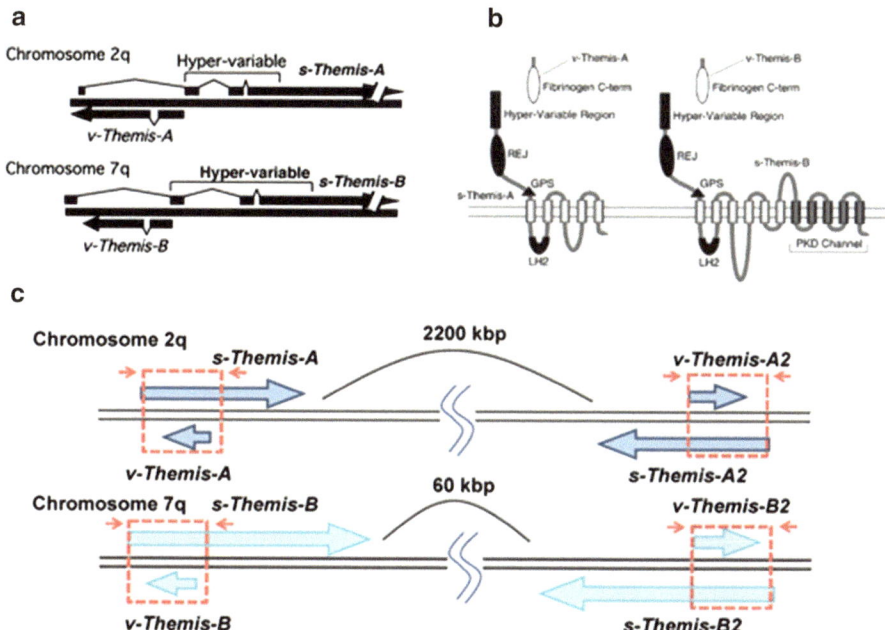

Fig. 20.2 Genes and proteins involved in the self-incompatibility (SI) system in *Ciona intestinalis*. **a** The pair of *s-Themis-A* and *v-Themis-A* located in chromosome 2q and the pair of *s-Themis-B* and *v-Themis-B* in chromosome 7q play a pivotal role in SI of *C. intestinalis*. **b** s-Themis-A and s-Themis-B possess a hypervariable region in their N-termini and have 5 and 11 transmembrane domains, respectively, in their C-termini. Five C-terminal transmembrane domains of s-Themis-B (and s-Theims-B2) showed homology to a cation channel. **c** A novel pair called *s-Themis-B2* and *v-Themis-B2*, which resides 60 kbp apart from s/v-*Themis-B* loci, may be involved in this SI system. A newly found pair of *v-Themis-A2* and *s-Themis-A2*, which resides 2.2 Mbp apart from *s/v-Themis-A* loci, appears to be a pseudo-gene pair. *Arrows* indicate the amplified regions to identify the allelic variety (haplotypes). Note that s-Themis-B and s-Themis-B2 contain a cation-channel domain in their C-terminal regions, which may be responsible for ionic flow during SI response. It is also known that sperm undergoes drastic Ca^{2+} influx when attached to the self-VC, a self-recognition signal in *C. intestinalis* (see also Fig. 20.3)

genes were checked by direct sequencing, showed that not only *s/v-Themis-A* and *s/v-Themis-B* but also *s/v-Themis-B2* plays a key role in the SI system (Yamada et al., unpublished data). By close inspection of the DNA sequence around the *s/v-Themis-A* region, we noticed one additional gene pair similar to *s/v-Themis-A*, called *s/v-Themis-A2*. However, our genetic analysis revealed that this gene pair appears to be a pseudo-gene pair (Yamamoto et al., unpublished data).

Sperm behavior and intracellular Ca^{2+} concentration ($[Ca^{2+}]_i$) in response to self/non-self recognition were also investigated. We found that sperm motility markedly decreased within 5 min after attachment to the VC of self-eggs but not after attachment to the VC of non-self eggs and that the apparent decrease in sperm motility was suppressed in low-Ca^{2+} seawater (Saito et al. 2012). It was also revealed that sperm detach from the self-VC or stop motility within 5 min after binding to the self-VC. As s-Themis-B contains a cation-channel domain in its C terminus, we monitored

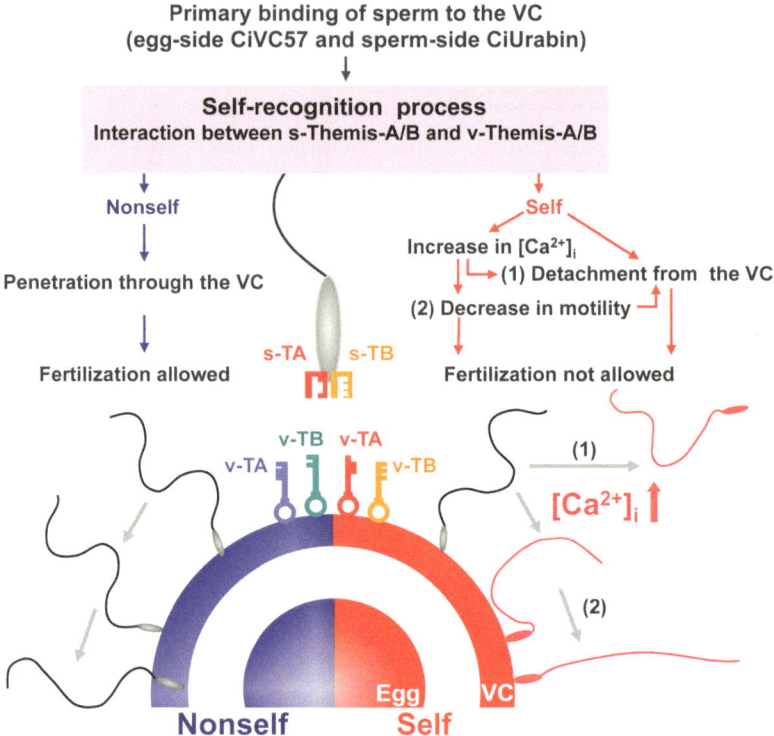

Fig. 20.3 Working hypothesis of self-incompatibility in *Ciona intestinalis*. We propose that sperm increase [Ca²⁺]ᵢ and detach from the VC when both s-Themis-A/B (s-TA and S-TB; "keyholes") on the sperm surface recognize respective v-Themis-A/B (v-TA and v-TB; "keys") on the VC as self. Sperm remaining on the self-VC change their waveform and motility. Non-self sperm remain on the VC and penetrate through the VC to fertilize the egg. Although only s/v-Themis-A and s/v-Themis-B are depicted in this figure, s/v-Themis-B2 may also participate in this SI system

sperm $[Ca^{2+}]_i$ by real-time $[Ca^{2+}]_i$ imaging using Fluo-8H-AM, a cell-permeable Ca^{2+} indicator. Interestingly, we found that sperm $[Ca^{2+}]_i$ rapidly and dramatically increased and was maintained at a high level in the head and flagellar regions when sperm interacted with the self-VC but not when the sperm interacted with the non-self VC (Saito et al. 2012). The increase in $[Ca^{2+}]_i$ was also suppressed by low-Ca^{2+} seawater (Saito et al. 2012). These results indicate that the sperm self-recognition signal triggers $[Ca^{2+}]_i$ increase or Ca^{2+} influx, which induces an SI response to reject self-fertilization.

As described in the preceding section, it has been reported that a non-self sperm-recognizing factor was identified in an acid extract of the VC in *C. intestinalis*. However, v-Themis-A and -B were hardly solubilized from the VC by acid treatment. In contrast, a novel factor, called Ci-v-Themis-like, which has the same molecular architecture, consisting of a coiled-coil domain and C-terminal fibrinogen-like domain, as that of v-Themis-A and -B except for having no apparent polymorphism, is a major acid-extractable VC protein. Ci-v-Themis-like appears to be an ancestral protein of v-Themis-A/B based on results of molecular phylogenetic analysis (Otsuka et al. 2013). Although there is no direct evidence indicating the participation of this protein in SI, our preliminary data showed interaction with v-Themis-A proteins, suggesting that the interaction between Ci-v-Themis-like on the VC and its sperm-side binding partners participates in gamete interaction to support the Themis-mediated allorecognition system. Further studies on the role of Ci-v-Themis-like are now in progress in our laboratory.

20.3.3 Lysin in C. intestinalis

To elucidate the VC lysin in *C. intestinalis*, effects of protease inhibitors on fertilization were examined (Pinto et al. 1990; Hoshi 1985). In contrast to *H. roretzi*, chymostatin, but not leupeptin, showed a strong inhibitory effect on fertilization. Then, a chymotrypsin-like protease was purified from *C. intestinalis* sperm (Marino et al. 1992). The purified preparation of a 24-kDa protease had a weak activity to impair the VC on the basis of electron microscopic observation. Later, it was found that proteasome inhibitors, rather than chymostatin, potently inhibited the fertilization (Sawada et al. 1998). By analogy to the UPS, which functions as a lysin in *H. roretzi*, the UPS may play a key role in fertilization, probably as a lysin, also in *C. intestinalis* (Sawada et al. 1998).

20.4 Future Perspective

As discussed here, allorecognition or self-incompatibility systems in ascidians were seen to be very similar to the SI system in flowering plants. In flowering plants, SI-responsible proteins are different in different families (Table 20.1). In Brassicaceae, both genes of pollen-side SP11/SCP and pistil-side SRK (S-receptor kinase) are highly polymorphic and tightly linked (Takayama and Isogai 2005; Iwano and Takayama 2012). A drastic increase in intracellular Ca^{2+} within pollen is known in Papaveraceae, resulting in caspase-like protease-mediated cell death, which appears to be similar to the SI response in *C. intestinalis*. From this aspect, sexual reproductive strategies might be much more common between animals and plants than previously thought, although the SI-responsible proteins themselves are considerably diverged.

Acknowledgments This study was supported in part by Grant-in-Aids for Scientific Research on Innovative Areas from MEXT, Japan to H.S. (21112001, 21112002) and to L.Y. (22112511). We are grateful to the staff of the Research Center for Marine Biology Asamushi, Graduate School of Science, Tohoku University. We also thank Drs. Kazuo Inaba and Kogiku Shiba of Shiomoda Marine Research Center, the University of Tsukuba, for their collaboration in intracellular Ca^{2+} imaging under a fluorescent microscope.

Table 20.1 Self-incompatibility system in plants and animals

Family	Female determinant	Male determinant
Flowering plants		
Brassicaceae	SRK	SP11/SCR
Solanaceae, Rosaceae	S-RNase	SLF/SBP
Papaveraceae	PrsS	PrpS
Ascidians		
Cionidae		
(*Ciona intestinalis*)	s-Themis-A, -B, -B2	v-Themis-A, -B, -B2
(*Ciona intestinalis*)	Not identified	(Gln-enriched VC peptides, Cihsp70, vCRL1 etc.)
Pyuridae		
(*Halocynthia roretzi*)	HrVC70	HrTTSP-1[a], HrUrabin[a]
(*Halocynthia aurantium*)	HaVC80	Not identified

For details, see reviews (Takayama and Isogai 2005; Iwano and Takayama 2012; Sawada 2002; Harada and Sawada 2008)
[a]These proteins are potential candidates, but it is not known whether these proteins are directly involved in SI

References

Akasaka M, Harada Y, Sawada H (2010) Vitellogenin C-terminal fragments participate in fertilization as egg-coat binding partners of sperm trypsin-like proteases in the ascidian *Halocynthia roretzi*. Biochem Biophys Res Commun 392:479–484

Akasaka M, Kato KH, Kitajima K, Sawada H (2013) Identification of novel isoforms of vitellogenin expressed in ascidian eggs. J Exp Zool B Mol Dev Evol 320:118–128

Artavanis-Tsakonas S, Matsumoto K, Fortini ME (1995) Notch signaling. Science 268:225–232

Ban S, Harada Y, Yokosawa H, Sawada H (2005) Highly polymorphic vitelline-coat protein HaVC80 from the ascidian, *Halocynthia aurantium*: structural analysis and involvement in self/nonself recognition during fertilization. Dev Biol 286:440–451

De Santis R, Pinto MR (1991) Gamete self-discrimination in ascidians: a role for the follicle cells. Mol Reprod Dev 29:47–50

Finley D (2009) Recognition and processing of ubiquitin-protein conjugates by the proteasome. Annu Rev Biochem 78:477–513

Fuke TM (1983) Self and nonself recognition between gametes of the ascidian, *Halocynthia roretzi*. Roux's Arch Dev Biol 192:347–352

Fuke M, Numakunai M (1996) Establishment of self-sterility of eggs in the ovary of the solitary ascidian, *Halocynthia roretzi*. Roux's Arch Dev Biol 205:391–400

Harada Y, Sawada H (2007) Proteins interacting with the ascidian vitelline-coat sperm receptor HrVC70 as revealed by yeast two-hybrid screening. Mol Reprod Dev 74:1178–1187

Harada Y, Sawada H (2008) Allorecognition mechanisms during ascidian fertilization. Int J Dev Biol 52:637–645

Harada Y, Takagaki Y, Sugnagawa M, Saito T, Yamada L, Taniguchi H, Shobuchi E, Sawada H (2008) Mechanism of self-sterility in a hermaphroditic chordate. Science 320:548–550

Hershko A, Ciechanover A (1998) The ubiquitin system. Annu Rev Biochem 67:425–479

Hoshi M (1985) Lysin. In: Metz CB (ed) Biology of fertilization, vol 2. Academic, New York

Hoshi M (1986) Sperm glycosidase as a plausible mediator of sperm binding to the vitelline envelope in ascidians. Adv Exp Med Biol 207:251–260

Hoshi M, Numakunai T, Sawada H (1981) Evidence for participation of sperm proteinases in fertilization of the solitary ascidian, *Halocynthia roretzi:* effects of protease inhibitors. Dev Biol
86:117–121

Iwano M, Takayama S (2012) Self/non-self discrimination in angiosperm self-incompatibility.
Curr Opin Plant Biol 15:78–83

Kawamura K, Nomura M, Kameda T, Shimamoto H, Nakauchi M (1991) Self-nonself recognition
activity extracted from self-sterile eggs of the ascidian, *Ciona intestinalis.* Dev Growth Differ
33:139–148

Khalturin K, Kurn U, Pinnow N, Bosch TC (2005) Towards a molecular code for individuality in
the absence of MHC: screening for individually variable genes in the urochordate *Ciona intestinalis.* Dev Comp Immunol 29:759–773

Kodama E, Baba T, Yokosawa H, Sawada H (2001) cDNA cloning and functional analysis of
ascidian sperm proacrosin. J Biol Chem 276:24594–24600

Kodama E, Baba T, Kohno N, Satoh S, Yokosawa H, Sawada H (2002) Spermosin, a trypsin-like
protease from ascidian sperm: cDNA cloning, protein structures and functional analysis. Eur J
Biochem 269:657–663

Kürn U, Sommer F, Bosch TC, Khalturin K (2007a) In the urochordate *Ciona intestinalis* zona
pellucida domain proteins vary among individuals. Dev Comp Immunol 31:1242–1254

Kürn U, Sommer F, Hemmrich G, Bosch TC, Khalturin K (2007b) Allorecognition in urochordates: identification of a highly variable complement receptor-like protein expressed in follicle
cells of Ciona. Dev Comp Immunol 31:360–371

Marino R, De Santis R, Hirohashi N, Hoshi M, Pinto MR, Usui N (1992) Purification and characterization of a vitelline coat lysin from Ciona intestinalis. Mol Reprod Dev 32:383–388

Marino R, Pinto MR, Cotelli F, Lamia CL, De Santis R (1998) The hsp70 protein is involved in the
acquisition of gamete self-sterility in the ascidian *Ciona intestinalis.* Development (Camb)
125:899–907

Marino R, De Santis R, Giuliano P, Pinto MR (1999) Follicle cell proteasome activity and acid
extract from the egg vitelline coat prompt the onset of self-sterility in *Ciona intestinalis*
oocytes. Proc Natl Acad Sci USA 96:9633–9636

Matsumoto M, Hirata J, Hirohashi N, Hoshi M (2002) Sperm–egg binding mediated by sperm
α-L-fucosidase in the ascidian, *Halocynthia roretzi.* Zool Sci 19:43–48

McRorie RA, Williams WL (1974) Biochemistry of mammalian fertilization. Annu Rev Biochem
43:777–803

Morgan TH (1911) Cross- and self-fertilization in *Ciona intestinalis.* Roux Arch Entwicklungsmech
30:206–235

Morgan TH (1939) The genetic and the physiological problems of self-sterility in *Ciona.* III.
Induced self-fertilization. J Exp Zool 80:19–54

Morgan TH (1942) The genetic and the physiological problems of self-sterility in *Ciona.* V. The
genetic problem. J Exp Zool 90:199–228

Morgan TH (1944) The genetic and the physiological problems of self-sterility in *Ciona.* VI.
Theoretical discussion of genetic data. J Exp Zool 95:37–59

Morton DB (1977) The occurrence and function of proteolytic enzymes in the reproductive tract
and of mammals. In: Barret AJ (ed) Proteinases in mammalian cells and tissues. North-Holland,
New York, pp 450–500

Otsuka K, Yamada L, Sawada H (2013) cDNA cloning, localization and candidate binding partners
of acid-extractable vitelline-coat protein Ci-v-Themis-like in the ascidian *Ciona intestinalis.*
Mol Reprod Dev 80:840–848

Pinto MR, Hoshi M, Marino R, Amoroso A, De Santis R (1990) Chymotrypsin-like enzymes are
involved in sperm penetration through the vitelline coat of *Ciona intetstinalis.* Mol Reprod Dev
26:319–323

Rosati F, De Santis R (1980) Role of the surface carbohydrates in sperm–egg interaction in *Ciona intestinalis*. Nature (Lond) 283:762–764

Saito T, Shiba K, Inaba K, Yamada L, Sawada H (2012) Self-incompatibility response induced by calcium increase in sperm of the ascidian *Ciona intestinalis*. Proc Natl Acad Sci USA 109:4158–4162

Saitoh Y, Sawada H, Yokosawa H (1993) High-molecular-weight protease complex (proteasome) of sperm of the asicidan, *Halocynthia roretzi*: isolation, characterization, and physiological roles in fertilization. Dev Biol 158:238–244

Sakai N, Sawada H, Yokosawa H (2003) Extracellular ubiquitin system implicated in fertilization of the ascidian, *Halocynthia roretzi*: isolation and characterization. Dev Biol 264:299–307

Sawada H (2002) Ascidian sperm lysin system. Zool Sci 19:139–151

Sawada H, Someno T (1996) Substrate specificity of ascidian sperm trypsin-like proteases, spermosin and acrosin. Mol Reprod Dev 45:240–243

Sawada H, Yokosawa H, Hoshi M, Ishii S (1982) Evidence for acrosin-like enzyme in sperm extract and its involvement in fertilization of the ascidian, *Halocynthia roretzi*. Gamete Res 5:291–301

Sawada H, Yokosawa H, Hoshi M, Ishii S (1983) Ascidian sperm chymotrypsin-like enzyme: participation in fertilization. Experientia (Basel) 39:377–378

Sawada H, Yokosawa H, Ishii S (1984a) Purification and characterization of two types of trypsin-like enzymes from sperm of the ascidian (Prochordata) *Halocynthia roretzi*. Evidence for the presence of spermosin, a novel acrosin-like enzyme. J Biol Chem 259:2900–2904

Sawada H, Yokosawa H, Someno T, Saino T, Ishii S (1984b) Evidence for the participation of two sperm proteases, spermosin and acrosin, in fertilization of the ascidian, *Halocynthia roretzi*: inhibitory effects of leupeptin analogs on enzyme activities and fertilization. Dev Biol 105:246–249

Sawada H, Iwasaki K, Kihara-Negishi F, Ariga H, Yokosawa H (1996) Localization, expression, and the role in fertilization of spermosin, an ascidian sperm trypsin-like protease. Biochem Biophys Res Commun 222:499–504

Sawada H, Pinto MR, De Santis R (1998) Participation of sperm proteasome in fertilization of the phlebobranch ascidian *Ciona intestinalis*. Mol Reprod Dev 50:493–498

Sawada H, Sakai N, Abe Y, Tanaka E, Takahashi Y, Fujino J, Kodama E, Takizawa S, Yokosawa H (2002a) Extracellular ubiquitination and proteasome-mediated degradation of the ascidian sperm receptor. Proc Natl Acad Sci USA 99:1223–1228

Sawada H, Takahashi Y, Fujino J, Flores SY, Yokosawa H (2002b) Localization and roles in fertilization of sperm proteasome in the ascidian *Halocynthia roretzi*. Mol Reprod Dev 62:271–276

Sawada H, Tanaka E, Ban E, Yamasaki C, Fujino J, Ooura K, Abe Y, Matsumoto K, Yokosawa H (2004) Self/nonself recognition in ascidian fertilization: vitelline coat protein HrVC70 is a candidate allorecognition molecule. Proc Natl Acad Sci USA 101:15615–15620

Sawada H, Akasaka M, Yokota N, Sakai N (2005) Modification of ascidian fertilization related gamete proteins by ubiquitination, proteolysis, and glycosylation. In: Tokumoto T (ed) New impact on protein modifications in the regulation of reproductive system. Research Signpost, Kerala, pp 61–81

Sommer F, Awazu S, Anton-Erxleben F, Jian D, Klimovich AV, Samoilovich MP, Stow Y, Krüss M, Gelhaus C, Kürn U, Bosch TC, Khalturin K (2012) Blood system formation in the urochordate *Ciona intestinalis* requires the variable receptor vCRL1. Mol Biol Evol 29:3081–3093

Takayama S, Isogai A (2005) Self-incompatibility in plants. Annu Rev Plant Biol 56:467–489

Tanaka K (2009) The proteasome: overview of structure and functions. Proc Jpn Acad Sci B 85:12–36

Urayama S, Harada Y, Nakagawa Y, Ban S, Akasaka M, Kawasaki N, Sawada H (2008) Ascidian sperm glycosylphosphatidylinositol-anchored CRISP-like protein as a binding partner for an allorecognizable sperm receptor on the vitelline coat. J Biol Chem 283:21725–21733

Yamada L, Saito T, Taniguchi H, Sawada H, Harada Y (2009) Comprehensive egg coat proteome of the ascidian *Ciona intestinalis* reveals gamete recognition molecules involved in self-sterility. J Biol Chem 284:9402–9410

Yamaguchi A, Saito T, Yamada L, Taniguchi H, Harada Y, Sawada H (2011) Identification and localization of the sperm CRISP family protein CiUrabin involved in gamete interaction in the ascidian *Ciona intestinalis*. Mol Reprod Dev 78:488–497

Chapter 21
Self-Incompatibility in the Brassicaceae

Megumi Iwano, Kanae Ito, Hiroko Shimosato-Asano,
Kok-Song Lai, and Seiji Takayama

Abstract Self-incompatibility (SI) in angiosperms prevents inbreeding and promotes outcrossing to generate genetic diversity. SI in the Brassicaceae is controlled by the *S*-haplotype-specific interaction between pollen ligand (*S*-locus protein 11, SP11 or SCR) and its stigmatic receptor (*S*-receptor kinase, *SRK*). SP11/SCR binding to cognate SRK induces autophosphorylation of SRK, which triggers a signaling cascade leading to the rejection of self-pollen. However, the mechanism of self-pollen rejection downstream of this ligand–receptor interaction is unknown. Here, we generated self-incompatible *Arabidopsis thaliana* accession C24 for the forward-genetic approach and live-cell imaging of SI in the Brassicaceae. Furthermore, for reverse-genetic analysis, we extended the Arabidopsis Targeting Induced Local Lesions IN Genomes (TILLING) resources by developing a new population of ethyl methanesulfonate (EMS)-induced mutant lines in *A. thaliana* accession C24. We believe that the reverse-genetic approach is a useful tool for identifying genes that function in the SI signaling pathway of the Brassicaceae.

Keywords Arabidopsis • Brassicaceae • Self-incompatibility • TILLING

21.1 Introduction

Angiosperms have developed self-incompatibility (SI) as a genetic system to prevent inbreeding and thereby promote outcrossing to generate genetic diversity. SI is based on the self/non-self discrimination between male and female. In many angiosperms, SI is controlled by a single locus, designated *S*, with multiple haplotypes (de Nettancourt 2001). Each *S*-haplotype encodes both male-specificity and

M. Iwano (✉) • K. Ito • H. Shimosato-Asano • K.-S. Lai • S. Takayama
Graduate School of Biological Sciences, Nara Institute of Science and Technology,
8916-5 Takayama-cho, Ikoma, Nara 630-0192, Japan
e-mail: m-iwano@bs.naist.jp

H. Sawada et al. (eds.), *Sexual Reproduction in Animals and Plants*,
DOI 10.1007/978-4-431-54589-7_21, © The Author(s) 2014

female-specificity determinants (*S*-determinants), and self/non-self discrimination is accomplished by the *S*-haplotype-specific interaction between these *S*-determinants. Both the male and female determinants are polymorphic and are inherited as one segregating unit. The variants of this gene complex are called *S*-haplotypes. Self/non-self recognition operates at the level of protein–protein interactions between the two determinants, and an incompatible response occurs when both determinants originate from the same *S*-haplotype.

21.2 Self/Non-Self Recognition System in the Brassicaceae

In the Brassicaceae, the self/non-self discrimination between male and female occurs on papilla cells covering the stigma surface of the pistil. When cross-pollen lands on the papilla cell, the pollen hydrates and germinates. The pollen tube penetrates the surface of the papilla cell and enters the style, ultimately resulting in cross-fertilization. By contrast, when self-pollen lands on the papilla cell, pollen hydration and germination are inhibited (Fig. 21.1).

The female determinant is *S*-receptor kinase (SRK) (Takasaki et al. 2000). SRK consists of an SLG-like extracellular domain, a transmembrane domain, and an intracellular serine/threonine kinase domain. SRK spans the plasma membrane of the stigma papilla cell. The male determinant is *S*-locus protein 11 (SP11; also called *S*-locus cysteine-rich protein, SCR) (Schopfer et al. 1999; Takayama et al. 2000). SP11 is a small basic cysteine-rich protein that is predominantly expressed in the anther tapetum and accumulates in the pollen coat during pollen maturation (Iwano et al. 2003) (Fig. 21.2). Upon pollination, SP11/SCR penetrates the papilla cell wall and binds SRK in an *S*-haplotype-specific manner. This binding induces the autophosphorylation of SRK, triggering a signaling cascade that results in the rejection of self-pollen (Takayama et al. 2001; Takayama and Isogai 2005; Iwano and Takayama 2012) (Fig. 21.3).

The self-recognition, that is, the *S*-haplotype-specific interaction between SP11 and its cognate SRK, has been shown by a series of biochemical studies in *Brassica rapa*. A binding experiment using ^{125}I-labeled-S_8-SP11 suggested that it strongly

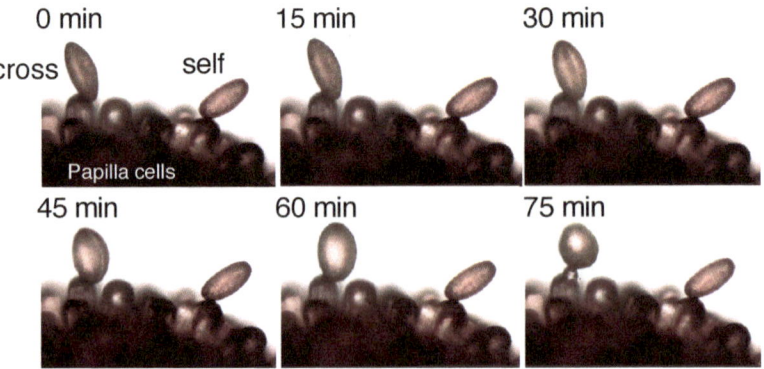

Fig. 21.1 Self- and cross-pollination in *Brassica rapa*

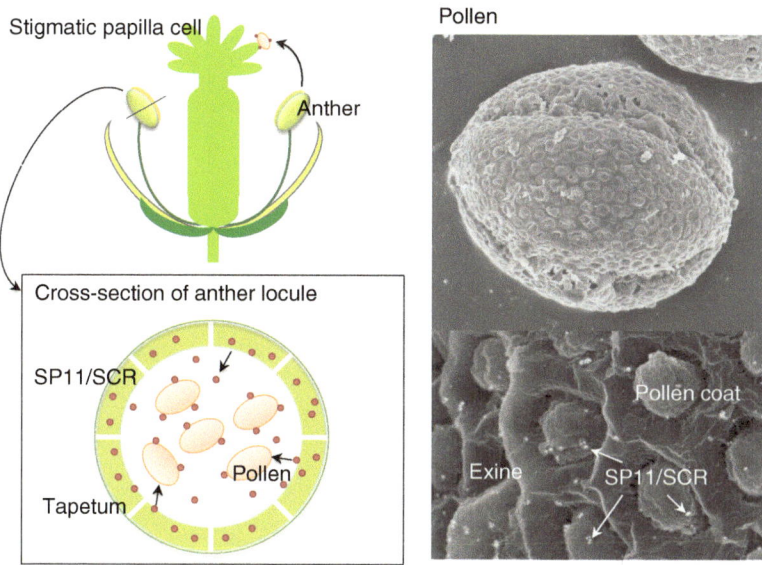

Fig. 21.2 Expression of SP11/SCR in the anther locule and localization of SP11/SCR on the pollen surface

binds to the stigmatic membrane of S_8-haplotype ($K_d = 0.7$ nM) but not of the S-haplotype (Takayama et al. 2001). Cross-linking and immunological analyses suggested that ^{125}I-labeled-S_8-SP11 directly binds to S_8-SRK and a 60-kDa protein in the stigmatic membrane of S_8-haplotype (Takayama et al. 2001). Affinity purification and LC-MS/MS analysis of SP11-binding stigmatic proteins have revealed that the 60-kDa protein is a truncated form of SRK (tSRK) containing the extracellular, transmembrane, and part of the intracellular juxtamembrane domains (Shimosato et al. 2007). Interestingly, an artificially expressed dimerized form of eSRK exhibited high-affinity binding to SP11. Another recent study suggested that two regions in the extracellular domain of SRK mediated the homo-dimerization of eSRK (Naithani et al. 2007). Taken together, these studies suggested that SRK on the stigmatic membrane is in an equilibrium between the inactive monomeric or dimeric low-affinity forms and the dimeric active high-affinity form, and that the SP11/SCR binding to its cognate SRK stabilizes its dimeric active form, which is expected to trigger the SI responses in the papilla cell (Shimosato et al. 2009) (Fig. 21.3).

21.3 SI Signaling Cascade Leading to Rejection of Self-Pollen

To date, the only candidates for signaling molecules acting downstream of SP11/SRK have been MLPK, the membrane-anchored M-locus protein kinase (Murase et al. 2004), and ARC1, an arm repeat-containing protein with E3 ubiquitin-ligase

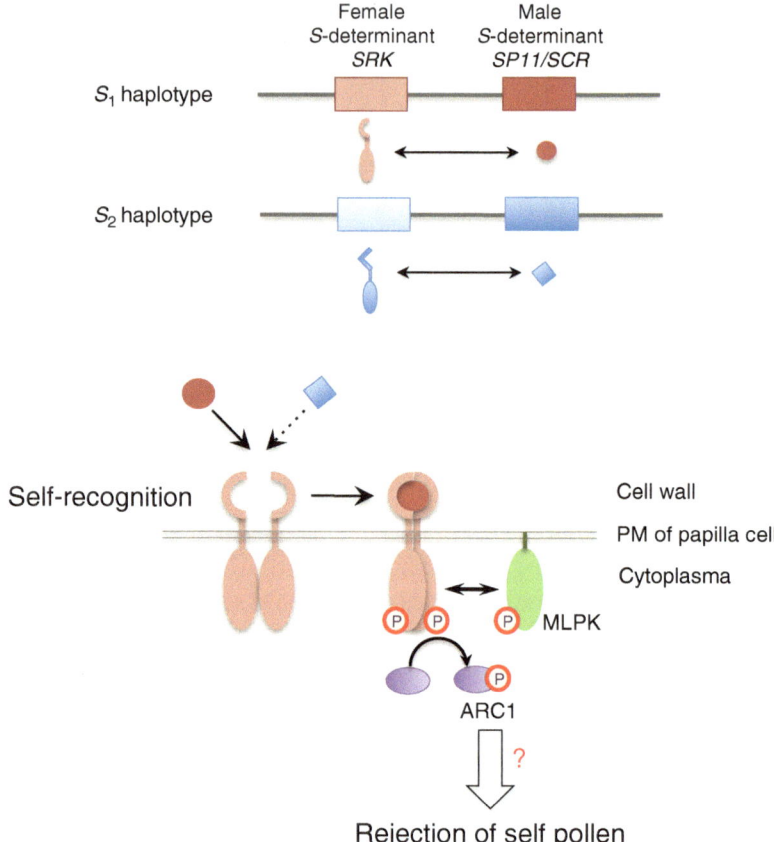

Fig. 21.3 Self-recognition self-incompatibility (SI) in Brassicaceae. The *S*-locus encodes female and male *S*-determinants, designated SRK and SP11 (or SCR), respectively. In self-pollination, the self (same *S*-haplotype)-specific SP11 binding to its cognate SRK stabilizes SRK in an active dimeric form on plasma membrane (PM), which triggers SI responses in the stigmatic papilla cell. MLPK and ARC1 were the only SP11/SRK downstream signaling molecular candidates

activity (Stone et al. 1999) (Fig. 21.3). MLPK was identified as a positive mediator of SI signaling in a genetic analysis of a self-compatible mutant of *B. rapa* var. Yellow Sarson (Murase et al. 2004). Upon self-pollination, MLPK is thought to interact directly with SRK to form an SRK-MLPK receptor complex on the plasma membrane and enhance SI signaling (Kakita et al. 2007). ARC1 is a potent positive mediator of this signal transduction pathway (Stone et al. 2003), and can be phosphorylated in vitro by both SRK and MLPK, suggesting that ARC1 is recruited to an SRK-MLPK complex at the plasma membrane (Samuel et al. 2008). ARC1 is predicted to promote self-pollen rejection in the self-incompatibility response by negatively regulating Exo70A1 and blocking the delivery of secretory vesicles to the pollen contact site (Samuel et al. 2009). However, the observation that suppression of *ARC1* expression results in incomplete breakdown of SI in both *Brassica napus* and *Arabidopsis lyrata* (Stone et al. 1999; Indriolo et al. 2012) might suggest the existence of another unknown signaling pathway acting downstream of SP11/ SRK.

Pollen hydration is the earliest step of cross-pollination. The regulation of water transport from a papilla cell to a pollen grain is one of the most important steps in the rejection of self-pollen. The lipid and proteinaceous components of the pollen coat are essential to pollen hydration during pollen-foot formation in *Brassica oleracea* (Elleman and Dickinson 1986). In *B. rapa*, monitoring transiently expressed GFP-mTalin, and rhodamine-phalloidin staining showed the concentration of actin bundles at the cross-pollen attachment site and actin reorganization at the self-pollen attachment site (Iwano et al. 2007). Additionally, the application of cross-pollen coat induces actin polymerization in the apical region of the papilla, whereas the application of self-pollen coat is associated with a decrease in actin filaments in the apical region. The actin-depolymerizing drug cytochalasin D (CD) significantly inhibited pollen hydration and germination during cross-pollination, further emphasizing a role for actin in these processes. Furthermore, electron tomography using ultrahigh-voltage electron microscopy revealed the close association of the actin cytoskeleton with an apical vacuole network. Self-pollination disrupted the vacuole network, whereas cross-pollination led to vacuolar rearrangements toward the site of pollen attachment. Taken together, these data suggested that self- and cross-pollination differentially affect the dynamics of the actin cytoskeleton, leading to changes in vacuolar structure that might be associated with hydration and germination (Iwano et al. 2007). In *B. napus* and *Arabidopsis thaliana*, immunostaining using anti-tubulin antibodies found that moderate changes in the microtubule network were observed after self-incompatible pollinations, but a more distinct localized breakdown of the microtubule network was observed during compatible pollinations (Samuel et al. 2011). Visualization over time of the morphological and physiological changes in the stigmatic papilla cell during self- and cross-pollination is a useful method for understanding SI. However, the application of live-cell imaging to *B. rapa* and *B. oleracea* has been difficult because of the low efficiency of transformation and the variability of pollination timing. To investigate the SI downstream signaling cascade leading to rejection of self-pollen in the Brassicaceae, generation of SI *Arabidopsis* was thought to be useful.

21.4 Generations of SI *Arabidopsis*

Arabidopsis thaliana is a popular model plant in the Brassicaceae. *A. thaliana* is the first plant to have its genome sequenced and is a popular tool for understanding the molecular biology of many plant traits. Its small size, rapid life cycle, and easy genetic transformation are also advantageous for research. In addition, this plant is well suited for live-cell imaging during pollination. However, *A. thaliana* is a self-compatible species; by contrast, *Arabidopsis lyrata* is a self-incompatible species in

genus *Arabidopsis*. Previously, a comparative analysis of the *S*-locus region of
A. lyrata and its homologous region in *A. thaliana* (Col-0) identified orthologues of
the *SRK* and *SCR* genes (Kusaba et al. 2001). However, none of the three candidate
SCR orthologues was predicted to encode full-length SCR proteins; therefore, they
were designated *ΨSCR1*, *ΨSCR2*, and *ΨSCR3*. The predicted *SRK* orthologue was
also thought to be inactive because it contains a premature stop codon. Thus, self-
compatibility in *A. thaliana* is associated with the inactivation of SI specificity
genes. Introduction of functional *SP11/SCR* and *SRK* gene pairs isolated from
A. lyrata into *A. thaliana* accession C24 conferred stable SI responses (Nasrallah
et al. 2004). In the resulting SI *Arabidopsis*, pollen hydration and germination were
arrested after self-pollination with SP11/SCR pollen, but normal pollen germination
and pollen tube penetration were observed after pollination with wild-type (WT)
pollen (SC). The establishment of the monitoring systems using this SI *Arabidopsis*
is a useful tool to visualize the physiological events during SI- and SC-pollination.

21.5   A TILLING Resource for Functional Genomics
in *Arabidopsis thaliana* Accession C24

Many reverse-genetic resources have been developed for functional genetic studies.
Because site-directed mutagenesis is not effective in plants, random mutagenesis
approaches, including insertional (Wisman et al. 1998; Alonso et al. 2003), chemical (McCallum et al. 2000), and fast neutron mutagenesis (Li et al. 2001), have been
used to establish reverse-genetic platforms. In *Arabidopsis*, insertional mutagenic
techniques using T-DNA or transposons have become popular tools for functional
genomics. However, insertional mutagenesis often leads to complete gene knock-
outs, making it difficult to associate nuanced phenotypes with essential genes
(Jander et al. 2002). Similarly, radiation mutagenesis, for example, fast neutrons,
often induces large genomic deletions that affect multiple genes (Li et al. 2001). By
contrast, classical chemical mutagenesis using a mutagen such as ethyl methanesulfonate (EMS) induces an array of interesting point mutations with different impacts
on gene function. Such allelic series are desirable because they generate a wide
repertoire of mutant phenotypes covering a range of severity, which provide more
insight into a gene's function. Moreover, individual plants carrying point mutations
can be identified easily through a powerful method called TILLING (Targeting
Induced Local Lesions IN Genomes).

TILLING is a reverse-genetic method that takes advantages of classical mutagenesis, sequence databases, and high-throughput PCR-based screening for point mutations in a targeted sequence (Henikoff et al. 2004). The key advantage of TILLING
over competing methods is that it can be applied to any plant species, regardless of
ploidy level, genome size, or genetic background (Kurowska et al. 2011). TILLING
extends genomic resources, particularly in organisms lacking reverse-genetic tools,
where mutants with a range of phenotypic severity are highly desirable. Since the
inception of TILLING, this method has been applied to various organisms including

Table 21.1 Ethyl methanesulfonate (EMS)-mutagenized TILLING resources in the Brassicaceae

Species	Ploidy level	Population size (line)	Mutation density (kb^{-1})	References
Arabidopsis thaliana (Col-1)	2*x*	3,072	1/300	Greene et al. (2003)
Arabidopsis thaliana (Ler)	2*x*	3,712	1/89	Martín et al. (2009)
Arabidopsis thaliana (C24)	2*x*	3,509	1/345	Lai et al. (2013)
Brassica oleracea	2*x*	2,263	1/447	Himelblau et al. (2009)
Brassica rapa	2*x*	9,216	1/60	Stephenson et al. (2010)

Cucumis melo L. (González et al. 2011), *Solanum lycopersium* (Minioa et al. 2010), *B. napus* (Wang et al. 2008; Harloff et al. 2012), *B. oleracea* (Himelblau et al. 2009), *B. rapa* (Stephenson et al. 2010), *Lotus japonicus* (Perry et al. 2009), *Zea mays* (Till et al. 2004), *Oryza sativa* (Till et al. 2007), *Drosophila* (Winkler et al. 2005), and zebrafish (Wienholds et al. 2003).

To date, *Arabidopsis* TILLING resources are only available in accessions Columbia (Col-0) (Greene et al. 2003) and Landsberg *erecta* (L*er*) (Martín et al. 2009). Reverse genetic tools for many commonly used *Arabidopsis* accessions are still limited, in particular accession C24, which is genetically distinct from accession Col-0 (Barth et al. 2002; Törjek et al. 2003). C24 is distinguished physiologically from other familial accessions in terms of tolerance to drought (Bechtold et al. 2010), ozone (Brosche et al. 2010), and frost (Rohde et al. 2004), and enhanced basal resistance to pathogens (Bechtold et al. 2010). The transgenic *A. thaliana* accession C24 also exhibited a robust and stable self-incompatible (SI) phenotype (Rea et al. 2010), which served as a good model for understanding SI signaling. In addition, a large portion of its genomic sequence was available (Schneeberger et al. 2011), making accession C24 an excellent alternative tool for plant research.

To take advantage of this tool, the Arabidopsis TILLING resources from the GAH molecules by developing a new population of EMS-induced mutant lines in *A. thaliana* accession C24 X (Lai et al. 2013). From approximately 8,000 *A. thaliana* C24 seeds treated with 25 mM EMS, 3,620 M1 seedlings were obtained, all of which were used to generate the M2 population. An M2 population with a total of 3,509 individual plants was successfully recovered for use in TILLING. This M2 population also contained 77 partial-seed set lines (semi-sterile) and 125 very low seed set lines (sterile). This population, including semi-sterile and sterile phenotypes, represents a valuable genetic resource for use in forward-genetic screens aimed at isolating novel genes affecting reproduction. Each M2 plant sampled for DNA used in TILLING was originally isolated from a distinct individual M1 plant to ensure independence of the mutations within the population. Ultimately, DNA from M2 plants and M3 seeds from 3,509 lines were stored for TILLING analysis (Lai et al. 2013). The TILLING collection represents the third TILLING resource reported for *A. thaliana* to date (Table 21.1). TILLING for selected genes from this new collection successfully identified allelic series of induced point mutations, including sense, missense, and nonsense mutations.

21.6 Conclusion

Self-recognition in the Brassicaceae is clearly mediated by a haplotype-specific interaction between pollen ligand (SP11/SCR) and its stigmatic receptor kinase (SRK). To clarify the downstream signaling pathway leading to self-pollen rejection, SI *Arabidopsis* was generated by the introduction of functional SP11/SCR and SRK gene pairs isolated from *A. lyrata* into *A. thaliana* accession C24 and conferred stable SI responses. The SI *Arabidopsis* is useful for the forward-genetic approach and live-cell imaging. In addition, TILLING resource was established in *A. thaliana* accession C24 for reverse genetic approach. The combination of forward- and reverse-genetic approaches with live-cell imaging will be a useful tool for identifying genes that function in the SI signaling pathway in the Brassicaceae.

References

Alonso JM, Stepanova AN, Leisse TJ et al (2003) Genome-wide insertional mutagenesis of *Arabidopsis thaliana*. Science 301:653–657

Barth S, Melchinger AE, Lubberstedt T (2002) Genetic diversity in *Arabidopsis thaliana* L. Heynh. investigated by cleaved amplified polymorphic sequences (CAPS) and inter-simple sequence repeat (ISSR) markers. Mol Ecol 11:495–505

Bechtold U, Lawson T, Mejia-Carranza J et al (2010) Constitutive salicylic acid defenses do not compromise seed yield, drought tolerance and water productivity in *Arabidopsis* accession C24. Plant Cell Environ 33:1959–1973

Brosche M, Merilo E, Mayer F et al (2010) Natural variation in ozone sensitivity among *Arabidopsis thaliana* accessions and its relation to stomatal conductance. Plant Cell Environ 33:914–925

Elleman CJ, Dickinson HG (1986) Pollen–stigma interactions in *Brassica*. IV. Structural reorganization in the pollen grains during hydration. J Cell Sci 80:141–157

González M, Xu M, Esteras C et al (2011) Towards a TILLING platform for functional genomics in Piel de Sapo melons. BMC Res Notes 4:289

Greene EA, Codomo CA, Taylor NE et al (2003) Spectrum of chemically induced mutations from a large-scale reverse genetic screen in *Arabidopsis*. Genetics 164:731–740

Harloff HJ, Lemcke S, Mittasch J et al (2012) A mutation screening platform for rapeseed (*Brassica napus* L.) and the detection of sinapine biosynthesis mutants. Theor Appl Genet 124:957–969

Henikoff S, Till BJ, Comai L (2004) TILLING: traditional mutagenesis meets functional genomics. Plant Physiol 135:630–636

Himelblau E, Gilchrist EJ, Buono K et al (2009) Forward and reverse genetics of rapid-recycling *Brassica oleracea*. Theor Appl Genet 118:953–961

Indriolo E, Tharmapalan P, Wright SI, Goring DR (2012) The *ARC1* E3 ligase gene is frequently deleted in self-compatible Brassicaceae species and has a conserved role in *Arabidopsis lyrata* self-pollen rejection. Plant Cell 24:4607–4620

Iwano M, Shiba H, Funato M et al (2003) Immunohistochemical studies on translocation of pollen *S*-haplotype determinant in self-incompatibility of *Brassica rapa*. Plant Cell Physiol 44:428–436

Iwano M, Shiba H, Matoba K et al (2007) Actin dynamics in papilla cells of *Brassica rapa* during self- and cross-pollination. Plant Physiol 144:72–81

Iwano M, Takayama S (2012) Self/non-self discrimination in angiosperm self-incompatibility. Curr Opin Plant Biol 15:78–83

Jander G, Norris SR, Rounsley SD et al (2002) *Arabidopsis* map-based cloning in the post-genome era. Plant Physiol 129:440–450

Kakita M, Murase K, Iwano M et al (2007) Two distinct forms of *M*-locus protein kinase localize to the plasma membrane and interact directly with *S*-locus receptor kinase to transduce self-incompatibility signaling in *Brassica rapa*. Plant Cell 19:3961–3973

Kurowska M, Daszkowska-Golec A, Gruszka D, Marzec M et al (2011) TILLING: a shortcut in functional genomics. J Appl Genet 52:371–390

Kusaba M, Dwyer K, Hendershot J et al (2001) Self-incompatibility in the genus *Arabidopsis*: characterization of the S locus in the outcrossing *A. lyrata* and its autogamous relative *A. thaliana*. Plant Cell 13:627–643

Lai KS, Kaothien-Nakayama P, Iwano M et al (2013) A TILLING resource for functional genomics in *Arabidopsis thaliana* accession C24. Genes Genet Syst 87:291–297

Li X, Song Y, Century K, Straight S et al (2001) A fast neutron deletion mutagenesis-based reverse genetics system for plants. Plant J 27:235–242

Martín B, Ramiro M, Martínez-Zapater JM, Alonso-Blanco C (2009) A high-density collection of EMS-induced mutations for TILLING in Landsberg *erecta* genetic background of *Arabidopsis*. BMC Plant Biol 9:147

Minioa S, Petrozza A, D'Onofrio O et al (2010) A new mutant genetic resource for tomato crop improvement by TILLING technology. BMC Res Notes 3:69

McCallum CM, Comai L, Greene EA, Henikoff S (2000) Targeted screening for induced mutations. Nat Biotechnol 18:455–457

Murase K, Shiba H, Iwano M et al (2004) A membrane-anchored protein kinase involved in *Brassica* self-incompatibility signaling. Science 303:1516–1519

Naithani S, Chookajorn T, Ripoll DR, Nasrallah JB (2007) Structural modules for receptor dimerization in the *S*-locus receptor kinase extracellular domain. Proc Natl Acad Sci USA 104:12211–12216

Nasrallah ME, Liu P, Sherman-Broyles S et al (2004) Natural variation in expression of self-incompatibility in *Arabidopsis thaliana*: implications for the evolution of selfing. Proc Natl Acad Sci USA 101:16070–16074

de Nettancourt D (2001) Incompatibility and incongruity in wild and cultivated plants, 2nd edn. Springer, Berlin

Perry J, Brachmann A, Welham T et al (2009) TILLING in *Lotus japonicus* identified large allelic series for symbiosis genes and revealed a bias in functionally defective ethyl methanesulfonate alleles toward glycine replacements. Plant Physiol 51:1281–1291

Rea AC, Liu P, Nasrallah JB (2010) A transgenic self-incompatible *Arabidopsis thaliana* model for evolutionary and mechanistic studies of crucifer self-incompatibility. J Exp Bot 61:1897–1906

Rohde P, Hincha DK, Heyer AG (2004) Heterosis in the freezing tolerance of crosses between two *Arabidopsis thaliana* accessions (Columbia-O and C24) that show differences in non-acclimated and acclimated freezing tolerance. Plant J 38:790–799

Samuel MA, Mudgil Y, Salt JN et al (2008) Interactions between the S-domain receptor kinases and AtPUB-ARM E3 ubiquitin ligases suggest a conserved signaling pathway in *Arabidopsis*. Plant Physiol 147:2084–2095

Samuel MA, Chong YT, Haasen KE et al (2009) Cellular pathways regulating responses to compatible and self-incompatible pollen in Brassica and *Arabidopsis* stigmas intersect at Exo70A1, a putative component of the exocyst complex. Plant Cell 21:2655–2671

Samuel MA, Tang W, Jamshed M et al (2011) Proteomic analysis of *Brassica* stigmatic proteins following the self-incompatibility reaction reveals a role for microtubule dynamics during pollen responses. Mol Cell Proteomics 10:M111.011338

Schneeberger K, Ossowski S, Ott F et al (2011) Reference-guided assembly of four diverse *Arabidopsis thaliana* genomes. Proc Natl Acad Sci USA 108:10249–10254

Schopfer CR, Nasrallah ME, Nasrallah JB (1999) The male determinant of self-incompatibility in *Brassica*. Science 286:1697–1700

Shimosato H, Yokota N, Shiba H et al (2007) Characterization of the SP11/SCR high-affinity binding site involved in self/nonself recognition in *Brassica* self-incompatibility. Plant Cell 19:107–117

Stephenson P, Baker D, Girin T et al (2010) A rich TILLING resource for studying gene function in *Brassica rapa*. BMC Plant Biol 10:62

Stone SL, Arnoldo M, Goring DR (1999) A breakdown of *Brassica* self-incompatibility in *ARC1* antisense transgenic plants. Science 286:1729–1731

Stone SL, Anderson EM, Mullen RT et al (2003) ARC1 is an E3 ubiquitin ligase and promotes the ubiquitination of proteins during the rejection of self-incompatible *Brassica pollen*. Plant Cell 15:885–898

Takasaki T, Hatakeyama K, Suzuki G et al (2000) The *S* receptor kinase determines self-incompatibility in *Brassica* stigma. Nature (Lond) 403:913–916

Takayama S, Isogai A (2005) Self-incompatibility in plants. Annu Rev Plant Biol 56:467–489

Takayama S, Shiba H, Iwano M et al (2000) The pollen determinant of self-incompatibility in *Brassica campestris*. Proc Natl Acad Sci USA 97:1920–1925

Takayama S, Shimosato H, Shiba H et al (2001) Direct ligand-receptor complex interaction controls *Brassica* self-incompatibility. Nature (Lond) 413:534–538

Till BJ, Reynolds SH, Weil C et al (2004) Discovery of induced point mutations in maize genes by TILLING. BMC Plant Biol 4:12

Till BJ, Cooper J, Tai TH et al (2007) Discovery of chemically induced mutations in rice by TILLING. BMC Plant Biol 7:19

Törjek O, Berger D, Meyer RC et al (2003) Establishment of a high-efficiency SNP-based framework marker set for *Arabidopsis*. Plant J 36:122–140

Wang N, Wang Y, Tian F et al (2008) A functional genomics resource for *Brassica napus*: development of an EMS mutagenized population and discovery of *FAE1* point mutations by TILLING. New Phytol 180:751–765

Wienholds E, van Eeden F, Kosters M et al (2003) Efficient target-selected mutagenesis in zebrafish. Genome Res 13:2700–2707

Winkler S, Schwabedissen A, Backasch D et al (2005) Target-selected mutant screen by TILLING in *Drosophila*. Genome Res 15:718–723

Wisman E, Hartmann U, Sagasser M et al (1998) Knock-out mutants from an En-1 mutagenized *Arabidopsis thaliana* population generate phenylpropanoid biosynthesis phenotypes. Proc Natl Acad Sci USA 95:12432–12437

Chapter 22
Signaling Events in Pollen Acceptance or Rejection in the *Arabidopsis* Species

Emily Indriolo, Darya Safavian, and Daphne R. Goring

Abstract The initial events of pollen–pistil interactions are fundamentally important in flowering plants because they influence successful fertilization. These early events include the recognition of pollen grains through signaling events in the pistil that will lead to the acceptance of a compatible pollen grain or the rejection of an incompatible pollen grain. There has been much research into this field in the Brassicaceae, as this family includes many agriculturally important crops such as canola, radish, turnip, and cabbage. However, this review focuses on what is known about the early pollen–pistil interactions in the experimentally tractable *Arabidopsis* genus, including *Arabidopsis thaliana* (a self-compatible species) and *Arabidopsis lyrata* (a self-incompatible species). Compatible pollinations are driven by the ability of the pistil to provide the resources for an acceptable pollen grain to hydrate, germinate, and fertilize the ovule. Self-incompatible species have a receptor–ligand signaling pathway that rejects self-pollen grains, preventing inbreeding and encouraging genetic diversity within the species. There is some overlap between these two pathways, and current research is looking for unknown elements and downstream events following the initial recognition of a pollen grain in *Arabidopsis*.

Keywords Compatible pollen response • E3 ubiquitin ligase • Exocyst • Self-incompatibility • Vesicle secretion

E. Indriolo • D. Safavian
Department of Cell and Systems Biology, University of Toronto,
Toronto, ON, Canada M5S 3B2

D.R. Goring (✉)
Department of Cell and Systems Biology, University of Toronto,
Toronto, ON, Canada M5S 3B2

Centre for the Analysis of Genome Evolution and Function, University of Toronto,
Toronto, ON, Canada M5S 3B2
e-mail: d.goring@utoronto.ca

H. Sawada et al. (eds.), *Sexual Reproduction in Animals and Plants*,
DOI 10.1007/978-4-431-54589-7_22, © The Author(s) 2014

22.1   Introduction

Throughout the course of land plant evolution, various mating strategies have occurred over time, with the most recent innovation being the development of flowers. With flowers, a successful fertilization is typically determined by interactions between the male pollen and the female pistil. Therefore, a great deal of research has focused on pollen–pistil interactions, and one targeted area has been uncovering the molecular and cellular mechanisms of early pollen–pistil interactions in species of the mustard family (Brassicaceae) such as the *Arabidopsis* and *Brassica* genera. In this family, selective cell–cell interaction events occur between the pollen grain and the surface of the pistil. These pollen–pistil interactions trigger an active recognition system to allow for the acceptance of compatible pollen and the subsequent successful fertilization or pollen rejection if the self-incompatibility system is activated. Much of our understanding of these early pollen–pistil interaction events has come from research on the *Brassica* genus (reviewed in Hiscock and Allen 2008; Chapman and Goring 2010; Iwano and Takayama 2012). This chapter largely focuses on reviewing the early stages following compatible and self-incompatible pollinations for the closely related Arabidopsis genus.

22.2   Early Compatible Pollen–Pistil Interactions in *Arabidopsis thaliana*

Pollen first comes in contact with the stigma at the top of the pistil, and the ability of the stigma to discriminate between compatible and other foreign pollen depends on whether the plant possesses wet- or dry-type stigmas. Wet stigmas have abundant surface secretions that indiscriminately capture pollen and allow germination to occur. In contrast, surface secretions are absent in dry stigmas, and as a result there is a much tighter regulation of these early pollen–stigma interaction stages (Dickinson 1995). *Arabidopsis* species possesses a dry stigma that is covered with papillae on the stigmatic surface. Once a compatible pollen grain comes in contact with a stigmatic papilla, the early stages of pollen capture and adhesion, pollen hydration, and germination are closely regulated (Elleman et al. 1992; Preuss et al. 1993; Kandasamy et al. 1994; Zinkl et al. 1999). The ability to conduct genetic screens with relative ease in *A. thaliana* has aided in identifying a number of factors that regulate pollen–stigma interactions (Preuss et al. 1993; Hulskamp et al. 1995; Nishikawa et al. 2005).

22.2.1   Pollen–Stigma Components for Compatible Pollen Acceptance

Following pollination, two components of the pollen grain come in contact with the stigmatic papilla: the exine, which is the sculptured outermost layer of the pollen grain, and the lipid- and protein-rich pollen coat that is present in the pockets of the exine (Elleman et al. 1992; Kandasamy et al. 1994). The initial binding of compatible *A. thaliana* pollen to the stigmatic papilla was determined to be mediated by the pollen exine, while the pollen coat was not required (Zinkl et al. 1999). Consistent with this, a genetic screen for mutants disrupted in this initial adhesion stage resulted in a number of mutants with exine defects (Nishikawa et al. 2005; Dobritsa et al. 2011). The pollen coat is then involved in further adhesion of the pollen grain to the stigmatic papilla. At this stage, the pollen coat flows out from the pollen grain toward the stigmatic papilla to mix with the lipidic and proteinaceous surface of the stigmatic papilla, creating a more robust connection at the pollen–stigma interface (Elleman et al. 1992; Preuss et al. 1993; Kandasamy et al. 1994; Zinkl et al. 1999). The result is an interface between the pollen grain and the papilla where signaling is hypothesized to occur, and water is transferred from the papilla to the pollen grain for hydration (Elleman and Dickinson 1990; Preuss et al. 1993).

Pollen hydration is necessary for the desiccated pollen grain to become metabolically active and proceed to the next stage of pollen germination and formation of a pollen tube, and components of the pollen–stigma interface are important to this process (Preuss et al. 1993). On the female side, the specific lipid content in the stigmatic cuticle was indirectly implicated in the control of pollen hydration through the study of the *A. thaliana fiddlehead* (*fdh*) mutant. Normally, pollen hydration and pollen tube growth are restricted to mature stigmas and cannot be supported on other tissues (Kandasamy et al. 1994; Ma et al. 2012). However, the *fdh* mutant allowed pollen hydration and pollen tube growth to occur on nonstigmatic surfaces such as the entire shoot epidermis (Lolle and Cheung 1993). Interestingly, *fdh* mutant leaves were found to have increased levels of long-chain lipids, and the *FDH* gene is predicted to encode an enzyme involved in lipid biosynthesis (Lolle et al. 1997; Yephremov et al. 1999; Pruitt et al. 2000).

The importance of pollen coat lipids in the hydration process was determined by the study of impaired pollen hydration in *A. thaliana eceriferum* (*cer*) mutants (Preuss et al. 1993; Hulskamp et al. 1995). The *cer* mutants have defects in the long-chain lipid synthesis, and there was a reduction or loss of pollen coat on the surface of the *cer* mutant pollen grains. As a result, these pollen grains failed to hydrate on the stigma, but this defect could be rescued by high environmental humidity where normal pollen hydration, pollen tube growth, and successful seed set were observed (Preuss et al. 1993; Hulskamp et al. 1995). Thus, work on the *cer* and *fdh* mutants suggests that long-chain lipids are required on both surfaces (i.e., the pollen and the stigma) to support pollen hydration.

Although the molecular mechanism of water transfer from the stigmatic papilla to the pollen grain has yet to be determined, changes to the impermeable pollen coat with the formation of the pollen–stigma interface are proposed to create a passageway for water to flow from the stigma to the pollen grain (Elleman and Dickinson 1986; Elleman et al. 1992; Murphy 2006). This may involve changes in the lipid properties of the pollen–stigma interface through the actions of lipid-binding oleosin-like proteins or lipases that have been identified in the *A. thaliana* pollen coat (Mayfield et al. 2001). For example, the pollen glycine-rich protein 17 (GRP17) contains an oleosin domain that has been implicated in this role (Mayfield and

Preuss 2000). The *A. thaliana grp17* mutant produced pollen grains that were slower in initiation of pollen hydration, although the rate of hydration, once initiated, was similar to the wild type. Compared to wild-type pollen, the *grp17* mutant pollen had a visibly similar pollen coat with a similar lipid composition but was lacking the GRP17 protein (Mayfield and Preuss 2000). Lipids in the pollen–stigma interface may also be enzymatically modified by lipases such as extracellular lipase 4 (EXL4) (Updegraff et al. 2009). A mutation in the EXL4 gene also resulted in changes to pollen hydration. The *exl4* mutant pollen had a pollen coat that was normal in appearance and had a similar lipid profile to wild-type pollen but had reduced esterase activity. Interestingly, the *exl4* mutant pollen initiated hydration at a similar time to wild-type pollen but then displayed a slower rate of hydration (Updegraff et al. 2009). Both the *grp17* and *exl4* mutants displayed mild hydration phenotypes, and it may be that multiple members of the corresponding gene families (Mayfield et al. 2001) need to be knocked out to see a more pronounced hydration defect. Other unknown factors may are also involved in controlling pollen hydration.

Following pollen hydration and germination, the emerging pollen tube penetrates the cell wall of the stigmatic papilla. Changes at the *A. thaliana* stigmatic surface were observed after compatible pollen attachment, including the expansion of the outer layer of the cell wall beneath the grain (Elleman et al. 1992; Kandasamy et al. 1994). In *B. oleracea*, the expansion of the outer layer of the stigmatic cell wall appears to be initiated by factors in the pollen coating as application of isolated pollen coat extracts to the stigma was found to cause cell wall expansion (Elleman and Dickinson 1996). To facilitate the penetration of the developing pollen tube into the stigmatic papillar surface, hydrolytic enzymes from both the pollen and the stigma are thought to cause the breakdown of the waxy cuticle and the underlying cell wall of the stigmatic papilla (Dickinson 1995); these may include enzymes such as serine esterases, cutinases, polygalacturonases, pectin esterases, and expansins (Hiscock and Allen 2008). *A. thaliana* microarray experiments have identified predicted genes for these various enzymes to be enriched in their expression in the pollen (Honys and Twell 2003) or the stigma (Swanson et al. 2005; Tung et al. 2005). A specific example is the *A. thaliana* VANGUARD1 (VGD1) gene, which encodes a pectin methylesterase and is expressed in the pollen grain and pollen tube. VGD1 is required for pollen tube growth through the pistil as the *vgd1* mutant displayed reduced levels of pectin methylesterase activity in the pollen grain and the pollen tube growth proceeded at a much slower rate compared to wild-type pollen (Jiang et al. 2005). After a pollen tube grows through the stigmatic papillar cell wall to the base of the stigma, it enters into the transmitting tract, growing down to an ovule, where fertilization takes place (Lennon et al. 1998; Cheung et al. 2010). A number of factors have also been identified for these later stages (Kessler and Grossniklaus 2011; Takeuchi and Higashiyama 2011).

22.2.2   Signaling Events in the Stigmatic Papilla Regulating Pollen Hydration and Germination

With the stigmatic papilla controlling the very early postpollination stages, starting with pollen adhesion, a specific signaling event is proposed to occur upon contact with compatible pollen. A number of small pollen coat proteins could potentially act as signaling molecules for putative receptors in the stigma papilla (Mayfield et al. 2001; Vanoosthuyse et al. 2001), and several *Brassica* candidates for promoting pollen adhesion have been proposed (Doughty et al. 1998; Luu et al. 1997, 1999; Takayama et al. 2000a). However, there are likely other unknown signaling proteins responsible for activating a cellular response in the stigmatic papilla to promote acceptance of the compatible pollen grain. These secretory vesicles to the specified sites responses include Ca^{2+} spikes, actin polymerization, and microtubule depolymerization, and these events may be linked to polarized exocytosis toward the pollen attachment site, as described next. In *Brassica rapa*, both an actin network and vacuolar network were observed to be established in a direction toward the compatible pollen grain during pollen hydration (Iwano et al. 2007). In *B. napus* and *A. thaliana*, disruption of the microtubule network resulted in increased acceptance of compatible pollen, indicating that microtubule depolymerization is important to this process (Samuel et al. 2011). Ca^{2+} dynamics in the stigmatic papillae were monitored in vivo using transgenic *A. thaliana* plants expressing the yellow cameleon 3.1, a Ca^{2+} indicator that can be used to monitor rapid changes in Ca^{2+} cytoplasmic concentrations (Iwano et al. 2004). Although no increase of $[Ca^{2+}]$ was observed in unpollinated stigmatic papillae, several increases were observed in the stigmatic papillae following pollination. The first increase in $[Ca^{2+}]$ took place in the stigmatic papilla underneath the pollen contact site during the pollen hydration period. A second local increase in $[Ca^{2+}]$ occurred in the stigmatic papilla, again under the pollen attachment site, with pollen germination. Finally, the third and strongest increase in local $[Ca^{2+}]$ occurred with pollen tube penetration of the stigmatic papilla (Iwano et al. 2004). Thus, these Ca^{2+} spikes in the stigmatic papilla underneath the pollen–pistil interface support the premise that the stigmatic papilla responds to the compatible pollen and controls these early postpollination stages.

More recently, Exo70A1 has been identified as a factor required in the stigma for the early responses of the stigmatic papilla to the compatible pollen (Samuel et al. 2009). Exo70A1 is a subunit of the exocyst, an evolutionary conserved protein complex in eukaryotes consisting of eight subunits: Sec3, Sec5, Sec6, Sec8, Sec10, Sec15, Exo70, and Exo84 (Hsu et al. 1996; TerBush et al. 1996; Hala et al. 2008). Some of the exocyst subunit genes are present in multiple copies in plant genomes, and Exo70A1 is one member of the Exo70 gene family in *A. thaliana* (Synek et al. 2006; Chong et al. 2010). In yeast and animal systems, the exocyst was determined to act as a tethering complex to dock secretory vesicles to the plasma membrane for polar secretion, and various small GTPases have been found to interact with exocyst subunits to regulate the assembly, localization, and function of this complex (reviewed in He and Guo 2009; Heider and Munson 2012). Once the vesicles have been tethered at the plasma membrane by the exocyst complex, the SNARE complex catalyzes the fusion of the secretory vesicle to the plasma membrane (Whyte and Munro 2002).

Exo70A1 was identified through work on the self-incompatibility pathway (described in the next section), and this led to the establishment of its role in compatible pollen response in both *B. napus* and *A. thaliana* (Samuel et al. 2009). The *A. thaliana exo70A1* mutant displayed a loss of pollen hydration and pollen tube growth when wild-type pollen was placed on the *exo70A1* mutant stigma (Samuel et al. 2009). Furthermore, this stigmatic defect was rescued by the stigma-specific expression of an RFP:Exo70A1 fusion protein in the *A. thaliana exo70A1* mutant. Confocal microscopy revealed that RFP:Exo70A1 protein was localized to the apical plasma membrane of mature stigmatic papillae (Samuel et al. 2009).

In yeast, Sec3 and Exo70 were found to be located at the plasma membrane before exocyst assembly (Finger et al. 1998; Boyd et al. 2004). Thus, Exo70A1 may play a similar role of being present at the stigmatic papillar plasma membrane before pollination and exocyst assembly. In yeast and mammalian cells, Sec3 and Exo70 were also found to be recruited to the plasma membrane via binding to phosphatidylinositol-4,5-bisphosphate, located at the inner leaflet of the plasma membrane (He et al. 2007; Liu et al. 2007; Zhang et al. 2008). Interestingly, when *A. thaliana* mutants with altered phosphoinositide pools were tested in pollen hydrations assays, wild-type pollen grains were found to have reduced hydration rates on the mutant stigmas, suggesting that specific membrane lipids are also important for these pollen–stigma interactions (Chapman and Goring 2011). Finally, in yeast, the polarized localization of the exocyst has been found to be controlled by the Rho GTPases, Cdc42, Rho1, and Rho3, through interactions with the Sec3 and Exo70 subunits (Robinson et al. 1999; Guo et al. 2001; Zhang et al. 2001; Wu et al. 2010). Exocyst assembly then occurs through an actin-dependent recruitment of the remaining exocyst subunits with the secretory vesicles to the specified sites on the plasma membrane for exocytosis (Boyd et al. 2004; Zhang et al. 2005).

Although RFP:Exo70A1 was localized to the entire stigmatic papillar apical plasma membrane, one would predict that exocyst assembly and vesicle docking would occur in a more localized area, just under the pollen contact site (Samuel et al. 2009). Consistent with this, we have observed, with compatible pollinations in *A. thaliana* and *A. lyrata*, that vesicle-like structures were observed fusing to the papillar plasma membrane under the pollen contact site (Safavian and Goring 2013). This step would require some type of unknown signal, possibly through Exo70A1, to regulate exocyst assembly and secretory vesicle docking under the pollen contact site (Fig. 22.1). The cargo of these secretory vesicles is also unknown, but they presumably contain stigmatic resources for pollen hydration and pollen tube penetration that are released upon fusion with the papillar plasma membrane and with the subsequent discharge of the contents into the apoplastic space. One such candidate cargo could be plasma membrane aquaporins, which could facilitate water transfer (Verdoucq et al. 2008; Postaire et al. 2010) from the stigmatic papilla to the

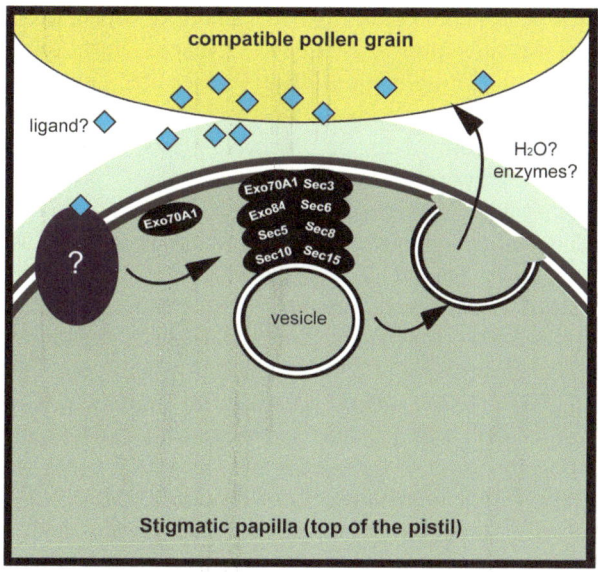

Fig. 22.1 Model of compatible pollen–pistil interactions in *Arabidopsis*. When a compatible pollen grain lands on the stigmatic papilla, an unknown factor or ligand is proposed to be detected by an unknown receptor in the papilla. Polarized vesicle secretion is then targeted to the pollen contact site by the exocyst complex. These vesicles are hypothesized to deliver factors to allow for the transfer of water from the papilla to the pollen grain and to deliver cell wall modification enzymes. The compatible pollen grain is then able to hydrate, and the pollen tube penetrates the stigma to continue onto an ovule or fertilization

pollen grain for hydration. As well, the vesicles may deliver cell wall-modifying enzymes for stigmatic papillar cell wall loosening and pollen tube penetration (Elleman and Dickinson 1996; Samuel et al. 2009).

22.3 Self-Incompatibility in the Genus *Arabidopsis*

Because plants are sessile in nature, flowering plants have evolved various mechanisms to aid in mate selection, and one of those methods is through the development of a self-incompatibility system. In the Brassicaceae, the self-incompatibility system allows the stigma to reject self-pollen and allows for the acceptance of genetically diverse non-self pollen (reviewed in Charlesworth and Vekemans 2005; Iwano and Takayama 2012). Two components control this system: the female pistil determinant, *S* receptor kinase (SRK; Takasaki et al. 2000; Silva et al. 2001), and the male pollen ligand, *S*-locus cysteine-rich/*S*-locus protein 11 (SCR/SP11; Schopfer et al. 1999; Takayama et al. 2000b). The two genes encoding these proteins are highly polymorphic (referred to as *S*-haplotypes), first identified in

Brassica species (Haasen and Goring 2010), and have been characterized in other Brassicaceae species, including *A. lyrata* (Kusaba et al. 2001; Schierup et al. 2001; Mable et al. 2005) and *Capsella grandiflora* (Paetsch et al. 2006; Boggs et al. 2009; Guo et al. 2009). Although much of the research into the self-incompatibility pathway has been performed in the *Brassica* genus, recent discoveries in the *Arabidopsis* genus are described here.

A. *thaliana* is a self-compatible species that has lost its self-incompatibility system by the pseudogenization of the *SCR/SP11* and *SRK* genes (Kusaba et al. 2001; Bechsgaard et al. 2006; Tang et al. 2007; Shimizu et al. 2008; Guo et al. 2011). As *A. thaliana* is easily transformed, a number of studies have reintroduced the *SCR/ SP11* or *SRK* genes from other Brassicaceae species in an attempt to reintroduce the self-incompatibility trait (Bi et al. 2000; Nasrallah et al. 2002). One approach was to transform *SCR/SP11-SRK* S-haplotypes from *A. lyrata* and *C. grandiflora* into different *A. thaliana* ecotypes. In some ecotypes such as C24, Cvi-0, Hodja, Kas-2, and Shadara, the expression of these *SCR/SP11* and *SRK* genes was able to cause self-pollen rejection while other ecotypes such as Col-0, Mt-0, Nd-0, No, RLD, and Ws-0 remained self-compatible (Nasrallah et al. 2004; Boggs et al. 2009). Although both the *SCR/SP11* and *SRK* genes are disrupted in many of the *A. thaliana* ecotypes, Tsuchimatsu et al. (2010) identified some *A. thaliana* ecotypes carrying an intact *SRK* gene. Wei-1 was one of these ecotypes with a functional copy of *SRKa*, but *SCRa* was nonfunctional because of an inversion in the gene coding region (Tsuchimatsu et al. 2010). Transformation of a restored *SCRa* gene into Wei-1 plants resulted in these plants displaying a self-incompatibility phenotype. Interestingly, these plants displayed a change over the course of development, going from being self-incompatible to becoming pseudo-self-compatible; that is, as the flowers became older they were able to accept self-pollen grains that were previously rejected (Tsuchimatsu et al. 2010). As a result of these studies, it was concluded that *SCR/SP11* and *SRK* were the only components required to restore the self-incompatibility trait in *A. thaliana* ecotypes. However, the stability of the self-incompatibility trait was variable, depending on the ecotype being used for the transformation studies. This lack of a completely stable self-incompatibility phenotype in *A. thaliana* may be the result of variation in the *A. thaliana* ecotypes studied, or additional factors such as the *ARC1* gene may be required, as discussed in more detail next.

In *Brassica*, the pollen SCR/SP11 ligand, present in the pollen coat, crosses the pollen–papillar interface to bind to the papillar membrane-localized SRK, and SRK becomes phosphorylated (Kachroo et al. 2001; Takayama et al. 2001; Shimosato et al. 2007). SRK exists as a homodimer and is proposed to interact as a transient complex with the *M* locus protein kinase (MLPK) (Murase et al. 2004; Kakita et al. 2007a, b). *B. rapa* MLPK is a receptor-like cytoplasmic kinase that is also localized to the plasma membrane in the stigma (Murase et al. 2004). MLPK exists as two splice variants with different N-terminal ends. MLPKf1 is generally expressed in a broad range of tissues and encodes a protein with an N-terminal myristoylation site, whereas MLPKf2 was found to be stigma specific in expression and has an

N-terminal hydrophobic domain (Kakita et al. 2007a). *A. thaliana APK1b* is the orthologue to *B. rapa* MLPK, and it has a similar pattern of expression with the two different isoforms that encode the same protein variants, one with an N-terminal myristoylation site (*APK1bf1*) and the other with the N-terminal hydrophobic region (*APK1bf2*). Despite the different N-terminal motifs between MLPKf1 and MLPKf2, both the MLPK variants localized to the plasma membrane through their respective N-terminal domains. Both isoforms also interacted with SRK at the plasma membrane, but the interaction was lost if the N-terminal domain was removed (Kakita et al. 2007a, b). Finally, both isoforms could rescue the *mlpk* mutation in *B. rapa*, restoring the self-incompatibility response (Kakita et al. 2007a). A similar role for *A. thaliana* APK1b in the self-incompatibility response has yet to be established. An *apk1b* knockout mutant in the *A. thaliana* Col-0 ecotype did not show any differences from the wild type when the *SCR/SP11* and *SRK* transgenes were expressed (Rea et al. 2010; Kitashiba et al. 2011). However, because the transgenic *SCR/ SP11-SRK A. thaliana* Col-0 plants only display a very weak self-incompatibility response and remain self-compatible (Nasrallah et al. 2002, 2004), it would be difficult to make a definitive conclusion on the role of *APK1b* in *A. thaliana* self-incompatibility.

In the Brassicaceae, the self-incompatibility signaling pathway is rapidly activated in the stigmatic papilla following contact with a self-pollen grain. The end result of this pathway is pollen rejection by preventing compatible pollen responses such as pollen grain hydration and pollen tube penetration into the stigma (described under compatible pollinations; Dickinson 1995). Following SCR/SP11 binding to SRK, the SRK-MLPK complex is proposed to recruit the ARM repeat-containing 1 (ARC1) E3-ubiquitin ligase (Stone et al. 2003; Samuel et al. 2008). *B. napus* ARC1 was the first downstream signaling component identified for SRK and was shown to be required for the rejection of self-pollen (Gu et al. 1998; Stone et al. 1999). Our most recent research into the *ARC1* orthologue in *A. lyrata* has revealed that it is necessary for the self-incompatibility response in *A. lyrata*, similar to *B. napus* (Indriolo et al. 2012). Previously, ARC1 was proposed to be not required in transgenic SCR/SRK *A. thaliana* plants for restoring the self-incompatibility trait because the ARC1 gene was deleted from the *A. thaliana* genome (Kitashiba et al. 2011). Given that the strength of the self-incompatibility trait varied depending which *A. thaliana* ecotype was used to transform with the SCR and SRK genes, we further explored the extent of the *ARC1* gene deletion. An additional 355 ecotypes were surveyed by polymerase chain reaction (PCR), and in all cases, *ARC1* was determined to be deleted (Indriolo et al. 2012), including ecotypes such as Wei-1 that contained a functional SRK gene (Tsuchimatsu et al. 2010). These data demonstrated that the functional *ARC1* gene was likely lost in *A. thaliana* before the loss of a functional *SRK* and *SCR/SP11* and perhaps aided in the breakdown of self-incompatibility to self-compatibility in *A. thaliana*. A broader bioinformatics survey of the presence or absence of *ARC1* in several Brassicaceae sequenced genomes gave further support to this idea. This survey included the self-incompatible species of *B. rapa*, *A. lyrata*, and *Capsella grandiflora* and the self-compatible species of

A. thaliana, Aethionema arabicum, Capsella rubella, Leavenworthia alabamica, Thellungiella halophila, Thellungiella parvula, and *Sysimbrium irio.* It was determined that the *ARC1* gene was frequently deleted in self-compatible species including *A. thaliana, A. arabicum, L. alabamica, T. halophila,* and *T. parvula* whereas it was always found in the genomes of self-incompatible species (Indriolo et al. 2012). This observation was specific to the *ARC1* gene as the most closely related *ARC1* paralogue, *PUB17,* was completely conserved in all the genomes surveyed (Indriolo et al. 2012). Therefore, the presence or absence of *ARC1* appeared to correlate with a switch from self-incompatibility to self-compatibility in the Brassicaceae.

To directly address the requirement of ARC1 for the self-incompatibility trait in *Arabidopsis,* ARC1 was characterized in a natural self-incompatible species, *A. lyrata.* Basic characterization determined that *ARC1* was found to be more highly expressed in the pistil, similar to *B. napus ARC1,* which was shown to be most highly expressed in the stigma (Gu et al. 1998; Indriolo et al. 2012). The expression of *ARC1* was then knocked down in *A. lyrata* by the use of *ARC1* RNAi construct. Similar to previous observations in *B. napus,* transgenic *ARC1* RNAi *A. lyrata* plants exhibited a partial breakdown in the self-incompatibility response, leading to the acceptance of self-pollen and partial seed set (Indriolo et al. 2012). As a result of the aforementioned data regarding self-incompatibility in *Brassica* spp. and *A. lyrata,* one can conclude that the role of *ARC1* is conserved in the self-incompatibility signaling pathway in the core Brassicaceae (Indriolo et al. 2012).

Because the self-incompatibility pathway caused pollen rejection by inhibiting the early postpollination stages (pollen hydration, pollen tube penetration), compatibility factors in the stigmatic papilla would be predicted to be inhibited as part of this response. Following from this, the role of ARC1 in the self-incompatibility pathway was proposed to promote the ubiquitination and degradation of the proposed compatibility factors by the 26S proteasome (Stone et al. 2003). Through an ARC1-interaction screen, Exo70A1 was identified as a candidate compatibility factor targeted by ARC1 (Samuel et al. 2009). As described earlier in the compatible pollen–stigma interactions section, Exo70A1 is a component of the exocyst complex for tethering secretory vesicles at the plasma membrane and is required in the stigmatic papilla for accepting compatible pollen (Samuel et al. 2009). Therefore, the inhibition or removal of Exo70A1 through ubiquitination via ARC1 would block secretory vesicles from fusing to the papillar plasma membrane underneath the pollen contact site, resulting in pollen rejection (Fig. 22.2). This model was tested in *A. lyrata* where both self-incompatible and cross-compatible pollinations can be followed. At 10 min after a cross-compatible pollination, vesicle-like structures were observed fusing to the papillar plasma membrane under the pollen contact site (Safavian and Goring 2013). However, at 10 min after a self-incompatible pollination, there was a complete absence of these vesicle-like structures at the papillar plasma membrane, and autophagic bodies were detected in the vacuole; thus supporting the inhibition of vesicle secretion as part of the self-incompatibility response (Safavian and Goring 2013).

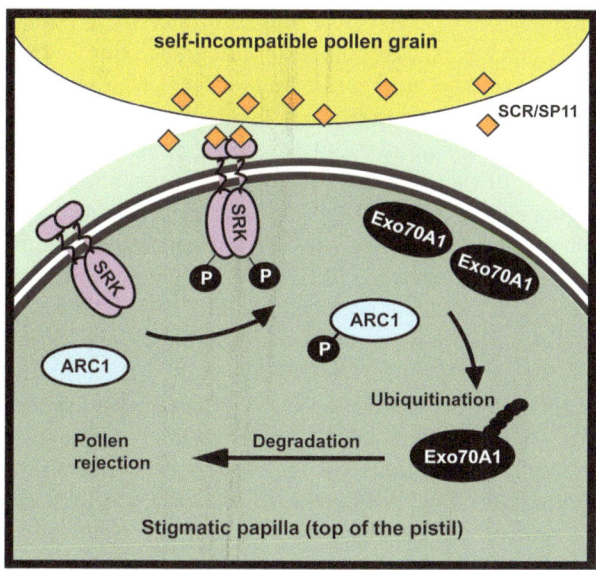

Fig. 22.2 Model of self-incompatible pollen–pistil interactions in *Arabidopsis*. When a self-incompatible pollen grain lands on the stigmatic papilla, the SCR/SP11 pollen ligand binds to the homodimeric receptor, *SRK*, in the stigmatic papilla. SRK then becomes phosphorylated and activated, and is proposed to phosphorylate ARC1. Phosphorylated ARC1 is then proposed to find its target, Exo70A1, at the plasma membrane. ARC1 ubiquinates Exo70A1, and Exo70A1 is targeted to the 26S proteosome where it is then degraded. The removal of Exo70A1 prevents exocyst assembly and the delivery of vesicles to the pollen contact site. As a result, the self-pollen grain is rejected as it is unable to hydrate and the pollen tube cannot penetrate the stigmatic surface

22.4 Conclusions and Future Directions

Much progress has been made recently into the mechanisms underlying pollen–pistil interactions in the *Arabidopsis* genus, but it is quite clear that much still remains unknown. For example, it is surprising that so little is known about what underlies a compatible interaction, such as a putative ligand and receptor to signal a compatible pollen recognition pathway in the stigma. What is clear is that downstream of this pathway, the exocyst complex is required for hydration and pollen tube penetration of compatible pollen in *A. thaliana* and *B. napus* (Samuel et al. 2009). Interestingly, following self-incompatible pollination in *B. napus*, moderate changes in the microtubule network were seen at the apical region of stigmatic papillae where longer microtubule bundles were observed instead of the dense network in unpollinated papillae, although no microtubule shortening or depolymerization was documented. However, a more dramatic localized depolymerization of the microtubule network was observed during compatible pollinations (Samuel et al. 2011). This distinct localized breakdown of the microtubule network was proposed to be regulated by Exo70A1, triggering successful pollination. Thus, the relationships between the exocyst complex, vesicle secretion, actin polymerization, and microtubule depolymerization is an area for further investigation.

The role of Exo70A1 in compatible pollinations came from its identification in the self-incompatibility signaling pathway in *B. napus* as a target for degradation by ARC1, downstream of the SRK-MLPK complex. Interestingly, the SRK-MLPK-ARC1 module appears to be used in other members in these gene families. For example, in plant innate immunity, the BIK1 receptor-like cytoplasmic kinase is part of a complex that includes the FLS2 receptor kinase, the BAK1 receptor-like kinase, and two ARC1-related proteins, PUB12 and PUB13 (Lu et al. 2011). Following bacterial flagellin recognition by FLS2, BIK1 was shown to enhance the ability of BAK1 to phosphorylate PUB13 (Lu et al. 2011). Similarly, using an in vitro assay, MPLK was found to be much more efficient at phosphorylating *B. napus* ARC1 compared to SRK (Samuel et al. 2008).

Other future research directions that may follow from recent discoveries include further investigating the role of ARC1, the *MLPK* orthologue, APK1b, and Exo70A1 in *Arabidopsis* species. For example, *A. lyrata ARC1* could be transformed into *A. thaliana* expressing the *A. lyrata* SRK and SCR/SP11 genes to determine if a more robust self-incompatibility phenotype is generated. *A. thaliana* Col-0 was previously found to lack a strong self-incompatibility response with only *A. lyrata* SRK and SCR/SP11 being expressed (Nasrallah et al. 2002, 2004), and we have some preliminary data that show a much stronger self-incompatibility response when all three *A. lyrata* genes are expressed in the *A. thaliana* Col-0 ecotype. Another direction is to further examine the role of the Exo70A1 orthologue in *A. lyrata* to investigate if Exo70A1 is targeted for degradation by ARC1 following a self-incompatible pollination. Finally, the ability of some *A. thaliana* ecotypes to mount varying degrees of self-incompatibility responses in the absence of ARC1 support that there are still other signaling components to be discovered in this pathway. With the greater access to bioinformatics and genomics tools in the Brassicaceae, it will be much easier to find candidate genes and perform research in a similar fashion to the *Arabidopsis* species in regard to both compatible and self-incompatible pollinations.

References

Bechsgaard JS, Castric V, Charlesworth D, Vekemans X, Schierup MH (2006) The transition to self-compatibility in *Arabidopsis thaliana* and evolution within S-haplotypes over 10 Myr. Mol Biol Evol 23(9):1741–1750

Bi YM, Brugiere N, Cui Y, Goring DR, Rothstein SJ (2000) Transformation of *Arabidopsis* with a *Brassica* SLG/SRK region and ARC1 gene is not sufficient to transfer the self-incompatibility phenotype. Mol Gen Genet 263(4):648–654

Boggs NA, Dwyer KG, Shah P, McCulloch AA, Bechsgaard J, Schierup MH, Nasrallah ME, Nasrallah JB (2009) Expression of distinct self-incompatibility specificities in *Arabidopsis thaliana*. Genetics 182(4):1313–1321

Boyd C, Hughes T, Pypaert M, Novick P (2004) Vesicles carry most exocyst subunits to exocytic sites marked by the remaining two subunits, Sec3p and Exo70p. J Cell Biol 167(5):889–901

Chapman LA, Goring DR (2010) Pollen–pistil interactions regulating successful fertilization in the Brassicaceae. J Exp Bot 61(7):1987–1999

Chapman LA, Goring DR (2011) Misregulation of phosphoinositides in *Arabidopsis thaliana* decreases pollen hydration and maternal fertility. Sex Plant Reprod 24(4):319–326

Charlesworth D, Vekemans X (2005) How and when did *Arabidopsis thaliana* become highly self-fertilising. Bioessays 27(5):472–476

Cheung AY, Boavida LC, Aggarwal M, Wu HM, Feijo JA (2010) The pollen tube journey in the pistil and imaging the in vivo process by two-photon microscopy. J Exp Bot 61(7):1907–1915

Chong YT, Gidda SK, Sanford C, Parkinson J, Mullen RT, Goring DR (2010) Characterization of the *Arabidopsis thaliana* exocyst complex gene families by phylogenetic, expression profiling, and subcellular localization studies. New Phytol 185(2):401–419

Dickinson HG (1995) Dry stigmas, water and self-incompatibility in *Brassica*. Sex Plant Reprod 8:1–10

Dobritsa AA, Geanconteri A, Shrestha J, Carlson A, Kooyers N, Coerper D, Urbanczyk-Wochniak E, Bench BJ, Sumner LW, Swanson R, Preuss D (2011) A large-scale genetic screen in *Arabidopsis* to identify genes involved in pollen exine production. Plant Physiol 157(2): 947–970

Doughty J, Dixon S, Hiscock SJ, Willis AC, Parkin IA, Dickinson HG (1998) PCP-A1, a defensin-like *Brassica* pollen coat protein that binds the S locus glycoprotein, is the product of gametophytic gene expression. Plant Cell 10(8):1333–1347

Elleman CJ, Dickinson HG (1986) Pollen–stigma interactions in *Brassica*. IV. Structural reorganization in the pollen grains during hydration. J Cell Sci 80:141–157

Elleman CJ, Dickinson HG (1990) The role of the exine coating in pollen–stigma interactions in *Brassica oleracea*. New Phytol 114:511–518

Elleman CJ, Dickinson HG (1996) Identification of pollen components regulating pollination-specific responses in the stigmatic papillae of *Brassica oleracea*. New Phytol 133(2):197–205

Elleman CJ, Franklin-Tong V, Dickinson HG (1992) Pollination in species with dry stigmas: the nature of the early stigmatic response and the pathway taken by pollen tubes. New Phytol 121:413–424

Finger FP, Hughes TE, Novick P (1998) Sec3p is a spatial landmark for polarized secretion in budding yeast. Cell 92(4):559–571

Gu T, Mazzurco M, Sulaman W, Matias DD, Goring DR (1998) Binding of an arm repeat protein to the kinase domain of the S-locus receptor kinase. Proc Natl Acad Sci USA 95(1):382–387

Guo W, Tamanoi F, Novick P (2001) Spatial regulation of the exocyst complex by Rho1 GTPase. Nat Cell Biol 3(4):353–360

Guo YL, Bechsgaard JS, Slotte T, Neuffer B, Lascoux M, Weigel D, Schierup MH (2009) Recent speciation of *Capsella rubella* from *Capsella grandiflora*, associated with loss of self-incompatibility and an extreme bottleneck. Proc Natl Acad Sci USA 106(13):5246–5251

Guo YL, Zhao X, Lanz C, Weigel D (2011) Evolution of the S-locus region in *Arabidopsis* relatives. Plant Physiol 157(2):937–946

Haasen KE, Goring DR (2010) The recognition and rejection of self-incompatible pollen in the Brassicaceae. Bot Stud 51(1):1–6

Hala M, Cole R, Synek L, Drdova E, Pecenkova T, Nordheim A, Lamkemeyer T, Madlung J, Hochholdinger F, Fowler JE, Zarsky V (2008) An exocyst complex functions in plant cell growth in *Arabidopsis* and tobacco. Plant Cell 20(5):1330–1345

He B, Guo W (2009) The exocyst complex in polarized exocytosis. Curr Opin Cell Biol 21(4):537–542

He B, Xi F, Zhang X, Zhang J, Guo W (2007) Exo70 interacts with phospholipids and mediates the targeting of the exocyst to the plasma membrane. EMBO J 26(18):4053–4065

Heider MR, Munson M (2012) Exorcising the exocyst complex. Traffic 13(7):898–907

Hiscock SJ, Allen AM (2008) Diverse cell signalling pathways regulate pollen-stigma interactions: the search for consensus. New Phytol 179(2):286–317

Honys D, Twell D (2003) Comparative analysis of the *Arabidopsis* pollen transcriptome. Plant Physiol 132(2):640–652

Hsu SC, Ting AE, Hazuka CD, Davanger S, Kenny JW, Kee Y, Scheller RH (1996) The mammalian brain rsec6/8 complex. Neuron 17(6):1209–1219

Hulskamp M, Kopczak SD, Horejsi TF, Kihl BK, Pruitt RE (1995) Identification of genes required for pollen-stigma recognition in *Arabidopsis thaliana*. Plant J 8(5):703–714

Indriolo E, Tharmapalan P, Wright SI, Goring DR (2012) The ARC1 E3 ligase gene is frequently deleted in self-compatible Brassicaceae species and has a conserved role in *Arabidopsis lyrata* self-pollen rejection. Plant Cell 24(11):4607–4620

Iwano M, Takayama S (2012) Self/non-self discrimination in angiosperm self-incompatibility. Curr Opin Plant Biol 15(1):78–83

Iwano M, Shiba H, Miwa T, Che FS, Takayama S, Nagai T, Miyawaki A, Isogai A (2004) Ca^{2+} dynamics in a pollen grain and papilla cell during pollination of *Arabidopsis*. Plant Physiol 136(3):3562–3571

Iwano M, Shiba H, Matoba K, Miwa T, Funato M, Entani T, Nakayama P, Shimosato H, Takaoka A, Isogai A, Takayama S (2007) Actin dynamics in papilla cells of *Brassica rapa* during self- and cross-pollination. Plant Physiol 144(1):72–81

Jiang L, Yang SL, Xie LF, Puah CS, Zhang XQ, Yang WC, Sundaresan V, Ye D (2005) VANGUARD1 encodes a pectin methylesterase that enhances pollen tube growth in the *Arabidopsis* style and transmitting tract. Plant Cell 17(2):584–596

Kachroo A, Schopfer CR, Nasrallah ME, Nasrallah JB (2001) Allele-specific receptor–ligand interactions in *Brassica* self-incompatibility. Science 293:1824–1826

Kakita M, Murase K, Iwano M, Matsumoto T, Watanabe M, Shiba H, Isogai A, Takayama S (2007a) Two distinct forms of M-locus protein kinase localize to the plasma membrane and interact directly with S-locus receptor kinase to transduce self-incompatibility signaling in *Brassica rapa*. Plant Cell 19(12):3961–3973

Kakita M, Shimosato H, Murase K, Isogai A, Takayama S (2007b) Direct interaction between the S-locus receptor kinase and M-locus protein kinase involved in *Brassica* self-incompatibility signaling. Plant Biotechnol 24:185–190

Kandasamy M, Nasrallah J, Nasrallah M (1994) Pollen–pistil interactions and developmental regulation of pollen tube growth in *Arabidopsis*. Development (Camb) 120:3405–3418

Kessler SA, Grossniklaus U (2011) She's the boss: signaling in pollen tube reception. Curr Opin Plant Biol 14(5):622–627

Kitashiba H, Liu P, Nishio T, Nasrallah JB, Nasrallah ME (2011) Functional test of *Brassica* self-incompatibility modifiers in *Arabidopsis thaliana*. Proc Natl Acad Sci USA 108(44): 18173–18178

Kusaba M, Dwyer K, Hendershot J, Vrebalov J, Nasrallah JB, Nasrallah ME (2001) Self-incompatibility in the genus *Arabidopsis*: characterization of the S locus in the outcrossing *A. lyrata* and its autogamous relative *A. thaliana*. Plant Cell 13(3):627–643

Lennon KA, Roy S, Hepler PK, Lord EM (1998) The structure of the transmitting tissue of *Arabidopsis thaliana* (L.) and the path of pollen tube growth. Sex Plant Reprod 11(1):49–59

Liu J, Zuo X, Yue P, Guo W (2007) Phosphatidylinositol 4,5-bisphosphate mediates the targeting of the exocyst to the plasma membrane for exocytosis in mammalian cells. Mol Biol Cell 18(11):4483–4492

Lolle SJ, Cheung AY (1993) Promiscuous germination and growth of wild-type pollen from *Arabidopsis* and related species on the shoot of the *Arabidopsis* mutant, *fiddlehead*. Dev Biol 155(1):250–258

Lolle SJ, Berlyn GP, Engstrom EM, Krolikowski KA, Reiter WD, Pruitt RE (1997) Developmental regulation of cell interactions in the *Arabidopsis fiddlehead-1* mutant: a role for the epidermal cell wall and cuticle. Dev Biol 189(2):311–321

Lu D, Lin W, Gao X, Wu S, Cheng C, Avila J, Heese A, Devarenne TP, He P, Shan L (2011) Direct ubiquitination of pattern recognition receptor FLS2 attenuates plant innate immunity. Science 332(6036):1439–1442

Luu DT, Heizmann P, Dumas C, Trick M, Cappadocia M (1997) Involvement of SLR1 genes in pollen adhesion to the stigmatic surface in Brassicaceae. Sex Plant Reprod 10(4):227–235

Luu DT, Marty-Mazars D, Trick M, Dumas C, Heizmann P (1999) Pollen-stigma adhesion in *Brassica* spp. involves SLG and SLR1 glycoproteins. Plant Cell 11(2):251–262

Ma JF, Liu ZH, Chu CP, Hu ZY, Wang XL, Zhang XS (2012) Different regulatory processes control pollen hydration and germination in *Arabidopsis*. Sex Plant Reprod 25(1):77–82

Mable BK, Robertson AV, Dart S, Di Berardo C, Witham L (2005) Breakdown of self-incompatibility in the perennial *Arabidopsis lyrata* (Brassicaceae) and its genetic consequences. Evolution 59(7):1437–1448

Mayfield JA, Preuss D (2000) Rapid initiation of *Arabidopsis* pollination requires the oleosin-domain protein GRP17. Nat Cell Biol 2(2):128–130

Mayfield JA, Fiebig A, Johnstone SE, Preuss D (2001) Gene families from the *Arabidopsis thaliana* pollen coat proteome. Science 292(5526):2482–2485

Murase K, Shiba H, Iwano M, Che FS, Watanabe M, Isogai A, Takayama S (2004) A membrane-anchored protein kinase involved in *Brassica* self-incompatibility signaling. Science 303(5663):1516–1519

Murphy DJ (2006) The extracellular pollen coat in members of the Brassicaceae: composition, biosynthesis, and functions in pollination. Protoplasma 228(1-3):31–39

Nasrallah ME, Liu P, Nasrallah JB (2002) Generation of self-incompatible *Arabidopsis thaliana* by transfer of two S locus genes from *A. lyrata*. Science 297(5579):247–249

Nasrallah ME, Liu P, Sherman-Broyles S, Boggs NA, Nasrallah JB (2004) Natural variation in expression of self-incompatibility in *Arabidopsis thaliana*: implications for the evolution of selfing. Proc Natl Acad Sci USA 101(45):16070–16074

Nishikawa S, Zinkl GM, Swanson RJ, Maruyama D, Preuss D (2005) Callose (beta-1,3 glucan) is essential for *Arabidopsis* pollen wall patterning, but not tube growth. BMC Plant Biol 5:22

Paetsch M, Mayland-Quellhorst S, Neuffer B (2006) Evolution of the self-incompatibility system in the Brassicaceae: identification of S-locus receptor kinase (SRK) in self-incompatible *Capsella grandiflora*. Heredity (Edinb) 97(4):283–290

Postaire O, Tournaire-Roux C, Grondin A, Boursiac Y, Morillon R, Schaffner AR, Maurel C (2010) A PIP1 aquaporin contributes to hydrostatic pressure-induced water transport in both the root and rosette of *Arabidopsis*. Plant Physiol 152(3):1418–1430

Preuss D, Lemieux B, Yen G, Davis RW (1993) A conditional sterile mutation eliminates surface components from *Arabidopsis* pollen and disrupts cell signaling during fertilization. Genes Dev 7(6):974–985

Pruitt RE, Vielle-Calzada JP, Ploense SE, Grossniklaus U, Lolle SJ (2000) FIDDLEHEAD, a gene required to suppress epidermal cell interactions in *Arabidopsis*, encodes a putative lipid biosynthetic enzyme. Proc Natl Acad Sci USA 97(3):1311–1316

Rea AC, Liu P, Nasrallah JB (2010) A transgenic self-incompatible *Arabidopsis thaliana* model for evolutionary and mechanistic studies of crucifer self-incompatibility. J Exp Bot 61(7): 1897–1906

Robinson NG, Guo L, Imai J, Toh EA, Matsui Y, Tamanoi F (1999) Rho3 of *Saccharomyces cerevisiae*, which regulates the actin cytoskeleton and exocytosis, is a GTPase which interacts with Myo2 and Exo70. Mol Cell Biol 19(5):3580–3587

Safavian D, Goring DR (2013) Secretory activity is rapidly induced in stigmatic papillae by compatible pollen, but inhibited for self-incompatible pollen in the Brassicaceae. PLoS One, in press.

Samuel MA, Mudgil Y, Salt JN, Delmas F, Ramachandran S, Chilelli A, Goring DR (2008) Interactions between the S-domain receptor kinases and AtPUB-ARM E3 ubiquitin ligases suggest a conserved signaling pathway in *Arabidopsis*. Plant Physiol 147(4):2084–2095

Samuel MA, Chong YT, Haasen KE, Aldea-Brydges MG, Stone SL, Goring DR (2009) Cellular pathways regulating responses to compatible and self-incompatible pollen in *Brassica* and

Arabidopsis stigmas intersect at Exo70A1, a putative component of the exocyst complex. Plant Cell 21(9):2655–2671

Samuel MA, Tang W, Jamshed M, Northey J, Patel D, Smith D, Siu KW, Muench DG, Wang ZY, Goring DR (2011) Proteomic analysis of *Brassica* stigmatic proteins following the self-incompatibility reaction reveals a role for microtubule dynamics during pollen responses. Mol Cell Proteomics 10(12), M111.011338

Schierup MH, Mable BK, Awadalla P, Charlesworth D (2001) Identification and characterization of a polymorphic receptor kinase gene linked to the self-incompatibility locus of *Arabidopsis lyrata*. Genetics 158(1):387–399

Schopfer CR, Nasrallah ME, Nasrallah JB (1999) The male determinant of self-incompatibility in *Brassica*. Science 286(5445):1697–1700

Shimizu KK, Shimizu-Inatsugi R, Tsuchimatsu T, Purugganan MD (2008) Independent origins of self-compatibility in *Arabidopsis thaliana*. Mol Ecol 17(2):704–714

Shimosato H, Yokota N, Shiba H, Iwano M, Entani T, Che FS, Watanabe M, Isogai A, Takayama S (2007) Characterization of the SP11/SCR high-affinity binding site involved in self/nonself recognition in *Brassica* self-incompatibility. Plant Cell 19(1):107–117

Silva NF, Stone SL, Christie LN, Sulaman W, Nazarian KAP, Burnett LA, Arnoldo MA, Rothstein SJ, Goring DR (2001) Expression of the S receptor kinase in self-compatible *Brassica napus* cv. Westar leads to the allele-specific rejection of self-incompatible *Brassica napus* pollen. Mol Genet Genomics 265(3):552–559

Stone SL, Arnoldo M, Goring DR (1999) A breakdown of *Brassica* self-incompatibility in ARC1 antisense transgenic plants. Science 286(5445):1729–1731

Stone S, Anderson E, Mullen R, Goring D (2003) ARC1 is an E3 ubiquitin ligase and promotes the ubiquitination of proteins during the rejection of self-incompatible *Brassica* pollen. Plant Cell 15(4):885–898

Swanson R, Clark T, Preuss D (2005) Expression profiling of *Arabidopsis* stigma tissue identifies stigma-specific genes. Sex Plant Reprod 18(4):163–171

Synek L, Schlager N, Elias M, Quentin M, Hauser MT, Zarsky V (2006) AtEXO70A1, a member of a family of putative exocyst subunits specifically expanded in land plants, is important for polar growth and plant development. Plant J 48(1):54–72

Takasaki T, Hatakeyama K, Suzuki G, Watanabe M, Isogai A, Hinata K (2000) The S receptor kinase determines self-incompatibility in *Brassica* stigma. Nature (Lond) 403(6772):913–916

Takayama S, Shiba H, Iwano M, Asano K, Hara M, Che FS, Watanabe M, Hinata K, Isogai A (2000a) Isolation and characterization of pollen coat proteins of *Brassica campestris* that interact with S locus-related glycoprotein 1 involved in pollen-stigma adhesion. Proc Natl Acad Sci USA 97(7):3765–3770

Takayama S, Shiba H, Iwano M, Shimosato H, Che FS, Kai N, Watanabe M, Suzuki G, Hinata K, Isogai A (2000b) The pollen determinant of self-incompatibility in *Brassica campestris*. Proc Natl Acad Sci USA 97(4):1920–1925

Takayama S, Shimosato H, Shiba H, Funato M, Che FS, Watanabe M, Iwano M, Isogai A (2001) Direct ligand–receptor complex interaction controls *Brassica* self-incompatibility. Nature (Lond) 413(6855):534–538

Takeuchi H, Higashiyama T (2011) Attraction of tip-growing pollen tubes by the female gametophyte. Curr Opin Plant Biol 14(5):614–621

Tang C, Toomajian C, Sherman-Broyles S, Plagnol V, Guo YL, Hu TT, Clark RM, Nasrallah JB, Weigel D, Nordborg M (2007) The evolution of selfing in *Arabidopsis thaliana*. Science 317(5841):1070–1072

TerBush DR, Maurice T, Roth D, Novick P (1996) The exocyst is a multiprotein complex required for exocytosis in *Saccharomyces cerevisiae*. EMBO J 15(23):6483–6494

Tsuchimatsu T, Suwabe K, Shimizu-Inatsugi R, Isokawa S, Pavlidis P, Stadler T, Suzuki G, Takayama S, Watanabe M, Shimizu KK (2010) Evolution of self-compatibility in *Arabidopsis* by a mutation in the male specificity gene. Nature (Lond) 464(7293):1342–1346

Tung CW, Dwyer KG, Nasrallah ME, Nasrallah JB (2005) Genome-wide identification of genes expressed in *Arabidopsis* pistils specifically along the path of pollen tube growth. Plant Physiol 138(2):977–989

Updegraff EP, Zhao F, Preuss D (2009) The extracellular lipase EXL4 is required for efficient hydration of *Arabidopsis* pollen. Sex Plant Reprod 22(3):197–204

Vanoosthuyse V, Miege C, Dumas C, Cock JM (2001) Two large *Arabidopsis thaliana* gene families are homologous to the *Brassica* gene superfamily that encodes pollen coat proteins and the male component of the self-incompatibility response. Plant Mol Biol 46(1):17–34

Verdoucq L, Grondin A, Maurel C (2008) Structure-function analysis of plant aquaporin AtPIP2;1 gating by divalent cations and protons. Biochem J 415(3):409–416

Whyte JR, Munro S (2002) Vesicle tethering complexes in membrane traffic. J Cell Sci 115(Pt 13):2627–2637

Wu H, Turner C, Gardner J, Temple B, Brennwald P (2010) The Exo70 subunit of the exocyst is an effector for both Cdc42 and Rho3 function in polarized exocytosis. Mol Biol Cell 21(3):430–442

Yephremov A, Wisman E, Huijser P, Huijser C, Wellesen K, Saedler H (1999) Characterization of the *FIDDLEHEAD* gene of *Arabidopsis* reveals a link between adhesion response and cell differentiation in the epidermis. Plant Cell 11(11):2187–2201

Zhang X, Bi E, Novick P, Du L, Kozminski KG, Lipschutz JH, Guo W (2001) Cdc42 interacts with the exocyst and regulates polarized secretion. J Biol Chem 276(50):46745–46750

Zhang X, Zajac A, Zhang J, Wang P, Li M, Murray J, TerBush D, Guo W (2005) The critical role of Exo84p in the organization and polarized localization of the exocyst complex. J Biol Chem 280(21):20356–20364

Zhang X, Orlando K, He B, Xi F, Zhang J, Zajac A, Guo W (2008) Membrane association and functional regulation of Sec3 by phospholipids and Cdc42. J Cell Biol 180(1):145–158

Zinkl GM, Zwiebel BI, Grier DG, Preuss D (1999) Pollen-stigma adhesion in *Arabidopsis*: a species–specific interaction mediated by lipophilic molecules in the pollen exine. Development (Camb) 126(23):5431–5440

Chapter 23
Papaver rhoeas S-Determinants and the Signaling Networks They Trigger

Vernonica E. Franklin-Tong

Abstract Higher plants use specific interactions between pollen and pistil to achieve pollination. Self-incompatibility (SI) is an important mechanism used by many species to prevent inbreeding. It is controlled by a multi-allelic *S* locus. "Self" (incompatible) pollen is discriminated from "non-self" (compatible) pollen by interaction of pollen and pistil *S* locus components and is subsequently inhibited. Our studies of the SI system in *Papaver rhoeas* have revealed that the pistil *S* locus protein, PrsS, is a small novel secreted protein that interacts with the pollen *S* locus protein, PrpS, which is a small novel transmembrane protein. This interaction of PrsS with incompatible pollen induces a SI response, involving a Ca^{2+}-dependent signaling network, resulting in pollen inhibition and programmed cell death; this provides a neat way to destroy "self"-pollen. Several SI-induced events have been identified, including Ca^{2+} and K^+ influx, increases in cytosolic free Ca^{2+}, activation of a MAP kinase, alterations to the cytoskeleton, and phosphorylation of a soluble inorganic pyrophosphatase. Here we present an overview of our knowledge of the novel cell–cell recognition *S*-determinants and the signals, targets, and mechanisms triggered by an incompatible interaction. We hope this review is of interest to those involved in the origins and evolution of cell–cell recognition systems involved in discrimination between "self" and "non-self," which include histocompatibility systems in primitive chordates and vertebrates as well as plant self-incompatibility.

Keywords *Papaver rhoeas* • Pollen • Programmed cell death (PCD) • *S*-determinants • Self-incompatibility

V.E. Franklin-Tong (✉)
School of Biosciences, University of Birmingham, Edgbaston, Birmingham B15 2TT, UK
e-mail: v.e.franklin-tong@bham.ac.uk

H. Sawada et al. (eds.), *Sexual Reproduction in Animals and Plants*,
DOI 10.1007/978-4-431-54589-7_23, © The Author(s) 2014

23.1   Introduction

The ability to discriminate between self and non-self (allorecognition) is important to
most multicellular organisms. Self–non-self discrimination and other recognition
systems, controlled by a highly polymorphic locus, are found in a wide range of
organisms. Allorecognition is widespread and is integral to the animal immune
response (Hughes 2002), fusion histocompatibility in lower animals (Scofield et al.
1982; De Tomaso et al. 2005; Nyholm et al. 2006), vegetative incompatibility in fungi
(Glass et al. 2000), disease resistance in plants (Dangl and Jones 2001), and self-
incompatibility (SI) systems found in many flowering plants (Takayama and Isogai
2005; Franklin-Tong 2008). Parallels between nonanalogous recognition systems
were recognized, and their importance appreciated, many years ago (Burnet 1971),
long before the molecular basis of these systems were elucidated. Allorecognition
systems rely upon loci with multiple alleles with high levels of polymorphism. The
nature of their polymorphism has intrigued population and evolutionary biologists for
decades. Molecular and cellular studies are beginning to provide insights into these
systems and provide interesting parallels and similarities. Here we provide an overview of our knowledge of the SI system in *Papaver rhoeas*, the field poppy.

Many plants are hermaphrodites, which provides a strategy that increases the
chances of having progeny, but also the possibility of self-fertilization and concomitant problems with inbreeding depression. To avoid this problem, many plants utilize SI, controlled by the multi-allelic *S*- (Self-sterility or Self-incompatibility)
locus, to prevent inbreeding (Takayama and Isogai 2005; Franklin-Tong 2008). The
pollen and pistil *S* determinants must be physically linked to the *S* locus to maintain
a functional SI system and are expected to have co-evolved. Other characteristics
are high levels of allelic polymorphism and tissue-specific expression. Most importantly, they mediate pollen rejection by inhibition of some stage of the pollination
process. Key aspects of understanding how SI systems operate are the identification
and characterization of the pistil and pollen *S* determinants, coupled with establishing mechanisms involved in pollen inhibition.

Analysis of various pollen and pistil *S*-determinants gene sequences has shown
that SI has evolved independently several times (Allen and Hiscock 2008). Well-
characterized SI systems include the *Brassica* system, where the pistil *S*-determinant
is a *S*-receptor kinase (SRK) and the pollen *S* determinant is SCR/SP11; and the
S-RNase based SI systems with an S-RNase as the pistil *S*-determinant and F-box
proteins SLF/SFB as the pollen *S*-determinant (Stein et al. 1991; Qiao et al. 2004;
Sijacic et al. 2004; Takayama and Isogai 2005; McClure and Franklin-Tong 2006).
The SI system in *Papaver* is distinct from these systems.

23.2   Self-Incompatibility Pistil and Pollen *S*-Locus
Determinants in *Papaver*

Cell–cell communication is often controlled by the interactions of secreted protein
ligands with cell-surface receptors. The *Papaver* pollen and pistil *S*-determinants
are proposed to act in this manner as extracts from stigmas inhibited incompatible,
but not compatible, pollen tubes grown in vitro (Franklin-Tong et al. 1988).

23.2.1   The Pistil S-Locus Determinants

The *Papaver* pistil *S*-locus determinants, PrsS (*Papaver rhoeas stigmatic S*), are expressed in low abundance, secreted specifically by the stigma. The *PrsS* genes are single copy and encode small, secreted, hydrophilic proteins of ~15 kDa. They have no obvious close homologues (Foote et al. 1994), although subsequent analysis of the *Arabidopsis* genome has revealed a large gene family (*S Protein Homologues, SPHs*) with similar predicted secondary structures, despite very low homology. The primary amino-acid sequence of the PrsS proteins is highly polymorphic, with as much as 40–46 % divergence between alleles (Walker et al. 1996). However, their predicted secondary structures are very similar; they all have a highly conserved predicted secondary structure. Site-directed mutagenesis has identified variable and conserved amino acids in a hydrophilic predicted surface loop that play a role in recognition and inhibition of incompatible pollen (Kakeda et al. 1998; Jordan et al. 1999).

   Demonstration that PrsS was the pistil *S*-locus determinant was achieved using a pollen SI bioassay. Addition of recombinant PrsS protein to pollen growing in vitro exhibited the expected *S*-haplotype-specific pollen inhibitory activity, (Foote et al. 1994). Because PrsS encodes a novel protein, their function was unknown. However, studies investigated whether they might be possible candidates for signaling ligands, as they were novel, small secreted proteins, and this was demonstrated using Ca^{2+} imaging studies (see later).

23.2.2   The Pollen S-Locus Determinants

Three alleles of the *Papaver* pollen *S*-determinant, *PrpS*, have been identified and cloned (Wheeler et al. 2009). *PrpS* is a single-copy gene linked to the pistil *S*-determinant, *PrsS*, and displays extensive polymorphism. *PrpS* encodes a novel, small highly hydrophobic protein with a predicted M_r ~20 kDa. Examination of the *PrpS* sequences for nonsynonymous to synonymous (*Ka/Ks*) substitutions revealed no significant difference between substitution rates in *PrpS* and *PrsS* genes (Wheeler et al. 2009). This observation suggests that the pollen and pistil *S* alleles coevolved, and are likely to be equally ancient, which is an expected characteristic of *S*-determinants. Analysis of the three *PrpS* allelic sequences indicates that they share a similar topology, with three or four predicted transmembrane domains and a small predicted extracellular loop. Extensive database searches failed to identify orthologues of *PrpS* genes, so *PrpS* is a novel gene. PrpS has been shown to be expressed at the plasma membrane, and the predicted extracellular loop of PrpS interacts with PrsS. Use of *PrpS* antisense oligonucleotides in the pollen in vitro SI bioassay allowed demonstration that PrpS is involved in *S*-specific inhibition of incompatible pollen, providing evidence that it has the biological function expected (Wheeler et al. 2009).

   So, one big question is "What is PrpS?" As the PrpS amino-acid sequence contains no known catalytic domains, can PrpS be defined as a "receptor?" It clearly is not a "classic" defined/identified receptor. However, PrpS appears to act as a novel class of "receptor" that interacts with PrsS in a very specific manner, triggering an intracellular signaling network and resulting in a highly specific biological response.

As PrpS is novel, we have no real idea of how it functions, which presents a problem in moving forward in our understanding of how it interacts with PrsS to mediate SI in incompatible pollen. An intriguing possibility has emerged. A novel *Drosophila* protein, Flower, which functions in presynaptic vesicle endocytosis, has recently been characterized (Yao et al. 2009). Both Flower and PrpS are novel proteins with no obvious homologues and have very little primary sequence homology. However, they have similar topological predictions, so are "topological homologues." The Flower protein functions as a Ca^{2+}-permeable channel (Yao et al. 2009). Voltage-gated calcium channels have acidic amino acids (either glutamic acid or aspartic acid) in the transmembrane domains that form a pore. Flower has a single glutamic acid residue in a proposed transmembrane domain, and it has been shown to make a homo-multimeric complex that could form this pore (Yao et al. 2009). Excitingly, examination of PrpS sequences reveal that it has three aspartic acids and three glutamic acids conserved across all three PrpS proteins; several are close to the edges of putative predicted transmembrane domains. These are good candidates for a pore/channel selectivity generating amino-acid residue, and this provides a basis for a testable model for PrpS function.

23.3 Mechanisms Involved in SI in the *Papaver* System

An area that has received considerable attention is the downstream signals and targets when incompatible PrsS and PrpS interact. The remainder of this review focuses on this important aspect.

23.3.1 Calcium Signaling Mediates **Papaver** SI

Ca^{2+}-dependent signaling networks are used to generate responses to a huge variety of stimuli in both animal and plant cells. Ca^{2+} imaging established that $[Ca^{2+}]_i$ acted as a second messenger in *Papaver* pollen, triggered specifically by interaction with incompatible PrsS proteins (Franklin-Tong et al. 1993, 1995). These studies formed the basis of a hypothesis proposing that only in an incompatible situation would PrsS-PrpS interaction occur, and this initiates a Ca^{2+}-dependent signaling cascade in incompatible pollen. Unchallenged pollen tubes have low basal $[Ca^{2+}]_i$ with high apical $[Ca^{2+}]_i$ (Fig. 23.1a). Incompatible pollen tubes rapidly exhibit increases in $[Ca^{2+}]_i$ with

---

Fig. 23.1 (continued) rapidly depolymerized (indicated by *dotted lines*). F-actin subsequently starts to form small foci. Phosphorylation of sPPases occurs within minutes, resulting in inhibition of activity and biosynthesis. At 5–10 min, a MAPK is activated. Increases in ROS (~1–5 min) are observed. (**c**) An incompatible pollen tube at ~5 h. Normal pollen tube cytosolic pH is ~7.0 (*green*). A drop in cytosolic pH (*yellow*) occurs several hours after initiation of SI. This acidification allows activation of several caspase-like activities, resulting in programmed cell death (PCD). Actin foci have become larger. Ultimately, DNA fragmentation and cellular dismantling occurs. (Adapted from Fig. 11.2, Chap. 11. Copyright: Franklin-Tong 2008)

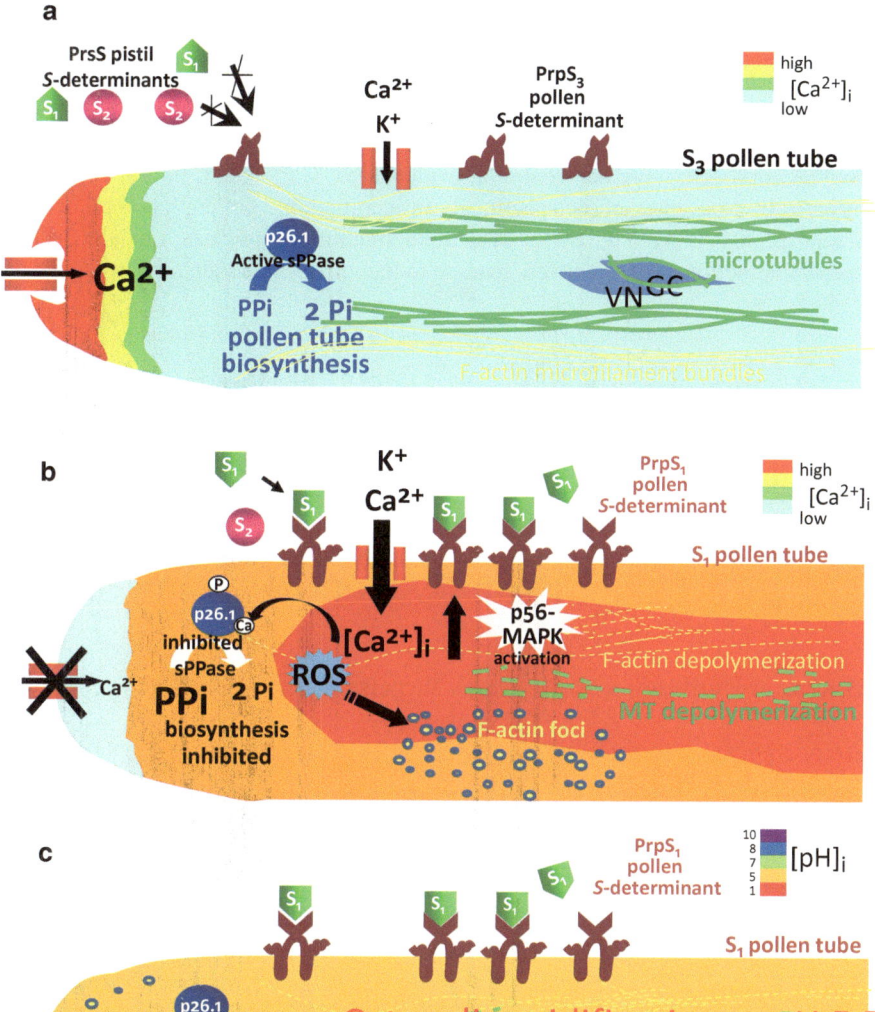

Fig. 23.1 Major changes in a pollen tube during different phases of self-incompatibility (SI). The pollen tubes are shown in pseudo-color to show $[Ca^{2+}]_i$ (a, b), with "hot" colors indicating high $[Ca^{2+}]_i$ and "cool" colors (*green*) indicating low, basal levels of $[Ca^{2+}]_i$. c Pseudo-color shows pH, with "cool" colors (*green*) indicating normal neutral levels of $[pH]_i$ and "hot" colors indicating low, acidic $[pH]_i$. (**a**) An untreated pollen tube, or a compatible situation, where pollen and pistil *S*-determinants do not match. Pistil *S*-determinants PrsS$_1$ and PrsS$_2$ do not interact with the pollen *S*-determinant PrsS, which has an S$_3$ specificity. Pollen germinates and grows, appearing indistinguishable from unchallenged pollen tubes. The pollen tube has a high apical $[Ca^{2+}]_i$ gradient and low, basal levels of $[Ca^{2+}]_i$ in the "shank" of the pollen tube; normal levels of sPPase activity enables biosynthesis; actin microfilaments enable delivery of vesicles to the tip allowing growth. (**b**) An incompatible pollen tube at ~1–10 min. The pistil *S*-determinant PrsS$_1$ matches and interacts with the pollen *S*-determinant, PrpS$_1$; this results in rapid Ca^{2+} influx and high $[Ca^{2+}]_i$ in the "shank" (indicated by *orange/red*), whereas $[Ca^{2+}]_i$ at the tip decreases. F-actin and microtubules are

loss of the apical gradient (Fig. 23.1b). Some of these increases in $[Ca^{2+}]_i$ were from extracellular sources (Franklin-Tong et al. 2002), implicating Ca^{2+} influx as being triggered by an incompatible SI response. More recent studies, using an electrophysiological approach, have demonstrated evidence for Ca^{2+} influx, and also K^+ influx triggered by PrsS interaction, specifically in incompatible pollen, has been obtained (Wu et al. 2011); see Fig. 23.1b. This rapid signaling implicates a receptor–ligand type of interaction being involved, leading to incompatible pollen tube inhibition.

23.3.2   Phosphorylation Events Identified in Papaver SI

Ca^{2+} signaling often causes altered protein kinase activity that results in posttranslational modification such as phosphorylation. Investigations revealed that alterations in protein phosphorylation were stimulated by SI in incompatible *Papaver* pollen with two major targets for SI-specific phosphorylation identified: a soluble inorganic pyrophosphatase (sPPase) and a mitogen-activated protein kinase (MAPK).

23.3.2.1   Identification of sPPases as Targets for Phosphorylation

Use of a phospho-proteomic-type approach revealed rapid *S*-specific phosphorylation of a 26-kDa cytosolic protein named Pr-p26.1, which was phosphorylated in a Ca^{2+}-dependent manner in incompatible pollen (Rudd et al. 1996). Analysis and cloning identified *Pr-p26.1a/b* as soluble inorganic pyrophosphatases, sPPases (de Graaf et al. 2006). sPPases are important enzymes involved in hydrolysis of inorganic pyrophosphate (PPi). During biopolymer synthesis PPi is generated, and sPPase hydrolysis of PPi generates inorganic phosphate (Pi), providing a thermodynamic pull, driving biosynthesis (Kornberg 1962).

The *Papaver* pollen sPPases have classic sPPase activities, which are Mg^{2+} dependent. Although it is well known that sPPases are inhibited by Ca^{2+}, it had not been previously established that phosphorylation could modify sPPase activity (de Graaf et al. 2006). Thus, identification of these sPPases as an early target of SI-induced Ca^{2+}-dependent signaling and kinase-dependent phosphorylation revealed mechanisms for regulating sPPase activity. It also suggested a possible mechanism to inhibit incompatible pollen tube growth. As sPPases are key enzymes that regulate biosynthesis, in principle, high sPPase activity results in faster biosynthesis (Fig. 23.1a), and low sPPase activity will result in a decrease in biosynthesis. SI would be expected to inhibit sPPase activity, which would therefore be predicted to inhibit pollen tube growth (Fig. 23.1b), which requires extensive biosynthesis of membrane and cell wall components. Use of antisense oligonucleotides based on *Pr-p26.1a/b* sequences to downregulate Pr-p26.1 resulted in significantly inhibited pollen tubes, suggesting that sPPases are necessary for pollen tube growth. SI-induced incompatible pollen tubes also had increased [PPi], which is predicted if sPPase activity is inhibited (de Graaf et al. 2006).

Together, these studies identified sPPases as novel, early targets for the Ca^{2+}-dependent SI signals, provided evidence for SI-stimulated modification of the Pr-p26.1a/b sPPase activity by Ca^{2+} and phosphorylation, and suggested that these sPPases play a role in regulating *Papaver* pollen tube growth. These findings provide an important part of the jigsaw puzzle of how SI operates.

23.3.2.2   A MAPK Is Activated by SI

MAPK cascades are involved in triggering numerous signaling networks. They are activated by dual phosphorylation of threonine and tyrosine residues in a TXY motif via a MAPKKK cascade. Investigations of SI-induced phosphorylation revealed the activation of a MAPK, p56, specifically in incompatible, but not compatible, pollen (Rudd et al. 2003; Li et al. 2007) (see Fig. 23.1b). Data suggested that the p56-MAPK plays a role in integrating SI signals. However, as its activation peaked 10 min after SI induction, it could not be involved in the rapid arrest of pollen tube inhibition. Studies have provided evidence that this MAPK is involved in signaling to PCD (see later).

23.3.3   Papaver *SI Triggers Alterations to the Cytoskeleton*

The actin cytoskeleton is both a major target and effector of signaling networks, and $[Ca^{2+}]_i$ is known to have a key role in mediating these responses (Staiger 2000). It has been found to be a target for SI signals, which involves both Ca^{2+} signaling and inhibition of pollen tube growth. The major findings are summarized here.

23.3.3.1   SI-Induced Alterations to F-Actin

Dramatic alterations in F-actin organization, specifically in incompatible pollen tubes, have been observed (Geitmann et al. 2000) (see Fig. 23.1a,b. Depolymerization of the actin filament bundles occurs within a few minutes of SI induction (Snowman et al. 2002), which provides a highly effective mechanism to rapidly inhibit incompatible pollen tube growth. Two Ca^{2+}-dependent actin-binding proteins (ABPs) that could potentially act synergistically in SI-induced depolymerization are profilin and PrABP80 (a putative gelsolin), which has potent Ca^{2+}-dependent severing activity (Huang et al. 2004). Subsequently, and within 30 min after SI induction, F-actin begins to aggregate to form highly stable punctate actin foci. Two ABPS, actin depolymerizing factor (ADF/cofilin) and cyclase-associated protein (CAP), have been identified as being involved in punctate actin foci formation, as they rapidly colocalize to the actin foci (Poulter et al. 2010). It is thought that these unusually stable actin structures play a role in signaling to programmed cell death (PCD) (see later).

23.3.3.2 SI-Induced Alterations to Microtubules

The pollen tube microtubule cytoskeleton is also a target for the SI signals. Rapid apparent depolymerization of cortical microtubules, within 1 min of SI induction, is observed in incompatible pollen tubes (Poulter et al. 2008) (see Fig. 23.1a,b). Interestingly, the distinctive spindle-shaped microtubules around the generative cell remained relatively intact until much later. Artificially depolymerizing actin (using the drug LatB) resulted in microtubule depolymerization, but stabilizing actin before SI prevented total microtubule depolymerization. This result suggested that the SI signals stimulating actin depolymerization trigger microtubule depolymerization (Poulter et al. 2008). Similar to actin, the cortical microtubules have also been shown to have another role, relating to signaling to PCD (see later).

23.3.4 SI Triggers Programmed Cell Death

Unwanted eukaryotic cells are often removed by apoptosis or PCD. Perhaps one of the most important findings relating to events triggered by SI in *Papaver* pollen was the demonstration of the involvement of PCD, which provides a highly effective way to prevent fertilization by incompatible pollen.

23.3.4.1 PCD Involving Caspase-Like Activities Is Triggered by SI

Animal cells utilize caspases to initiate and mediate apoptosis or PCD. Their activation is a classic feature often used to diagnose apoptosis/PCD, using tetrapeptide caspase-specific substrates and inhibitors. Caspase-3, a key executioner caspase in animal cells, has a tetrapeptide recognition motif DEVD. It is often called a DEVDase; its activity is inhibited by DEVD inhibitors, and substrates that act as fluorogenic indicators for caspase activities allow a direct analysis of caspase-like activities. Using these approaches, it has been shown that caspase-like activities are activated in plants. However, the nature of most plant caspases is a mystery, as there are no known caspase homologues (Lam and del Pozo 2000; Woltering et al. 2002; Woltering 2004).

  PCD has been shown to be a key event in the *Papaver* SI response. Evidence includes leakage of cytochrome *c* into the cytosol, and SI-induced DNA fragmentation, which is inhibited by the caspase-3 inhibitor Ac-DEVD-CHO but not the caspase-1 inhibitor Ac-YVAD-CHO (Thomas and Franklin-Tong 2004); this implicated a caspase-3-like/DEVDase activity involved in mediating the SI-induced DNA fragmentation. Moreover, extracts from SI-induced incompatible pollen can cleave the classic caspase substrate, poly(ADP-ribose) polymerase (PARP) (Thomas and Franklin-Tong 2004). Use of a fluorescent caspase-3 substrate, Ac-DEVD-AMC, that acts as an indicator for DEVDase activity has provided more direct evidence for SI-induced DEVDase activity (Bosch and Franklin-Tong 2007; Li et al. 2007) and allowed detailed characterization. Live-cell imaging of the DEVDase activity has revealed that this caspase-like activity first appears in the cytosol around 1–2 h after SI induction and increases over time (Bosch and Franklin-Tong 2007). Other probes have revealed the presence of SI-stimulated VEIDase (caspase-6) and LEVDase (caspase-4) activities in incompatible pollen tubes (Bosch and Franklin-Tong 2007) (see Fig. 23.1c). DEVDase and VEIDase had similar temporal activation profiles, with activity peaking at 5 h post-SI, suggesting that they are involved in early PCD. LEVDase activity, in contrast, was activated rather more slowly and later, with activity still increasing at 8 h, when the DEVDase and VEIDase activities had significantly decreased (Bosch and Franklin-Tong 2007).

Surprisingly, all three of the SI-activated caspase-like activities exhibit peak activity at pH 5, which hinted that a possible early SI-induced event was acidification of the cytosol. Investigation demonstrated that there was, indeed, a dramatic and rapid acidification of the pollen cytosol triggered in the first 1–2 h of SI (Bosch and Franklin-Tong 2007). What is involved in this phenomenon is not yet clear, but it points to major alterations to the nature of the cytosol being elicited by SI. Current studies are investigating this aspect further (Wilkins and Franklin-Tong, unpublished data).

23.3.5 Attempting to Integrate SI-Triggered Events That Signal to PCD

The interaction of the secreted pistil PrsS with the "self"-pollen PrpS causes Ca^{2+} influx and rapid increases in $[Ca^{2+}]_i$ and inactivation of the p26 sPPase. A major focus for the SI-induced signaling network is the activation of PCD. Alterations to the cytoskeleton are implicated in mediating PCD. The SI-induced acidification of the pollen cytosol (Fig. 23.1c) is also likely to be an important decision-making step, as this will allow PCD to proceed by allowing caspase-like proteins to be activated. Next, we summarize data relating to integration of the signals and targets triggered by this interaction. Figure 23.1 shows a cartoon of key events in pollen in three phases: (a) unchallenged or compatible, (b) early and rapid signaling events triggered by SI, and (c) late SI, constituting commitment to PCD.

23.3.5.1 Involvement of Actin Signaling in SI-Mediated PCD

Both actin and microtubule alterations are implicated in mediating PCD (Thomas et al. 2006; Poulter et al. 2008). A number of studies have shown that either actin depolymerization or stabilization can influence whether a eukaryotic cell enters into an apoptotic pathway; see (Franklin-Tong and Gourlay 2008; Smertenko and Franklin-Tong 2011). Investigations examining the effect of the actin-depolymerizing drug, latrunculin B (LatB), and the actin-stabilizing/polymerizing drug, jasplakinolide (Jasp), in *Papaver* pollen revealed that both treatments stimulated high levels of DNA fragmentation, which was mediated by a caspase-3 like/DEVDase activity (Thomas et al. 2006). This finding suggested that disturbance of actin polymer dynamics could trigger PCD in pollen. Data implicated actin depolymerization being functionally involved in the initiation of SI-induced PCD and established a causal link between actin polymerization status and initiation of PCD in plant cells (Thomas et al. 2006). Thus, the rapid and substantial actin depolymerization triggered by SI signaling not only results in the rapid inhibition of incompatible pollen tip growth, but it also activates a caspase-3-like/ DEVDase activity, triggering PCD (see Fig. 23.1b,c).

It seems likely that SI-induced cytosolic acidification plays an important role in the formation of the highly stable actin foci (Fig. 23.1c), as the actin-binding protein ADF/cofilin was found to be associated with the F-actin foci (Poulter et al. 2010). It has been shown that when F-actin is decorated with ADF, it does not exhibit its usual filament severing activity; this may explain why the foci are unusually stable. It is proposed that the formation of the actin foci and their association with ADF is an active process that is also involved in signaling to PCD (Fig. 23.1c), especially as they appear to be a marker for SI-induced PCD that can be alleviated concomitantly with alleviation of PCD.

23.3.5.2 Involvement of MAPK Signaling in SI-Mediated PCD

MAPKs are known to play a key role in mediating signaling to PCD in the plant–pathogen hypersensitive response. Studies based primarily on use of the MAPK cascade inhibitor U0126 in combination with PCD markers implicated a key role for MAPKs in signaling to PCD in incompatible pollen (Li et al. 2007) (see Fig. 23.1b). U0126 prevented the SI-induced activation of p56 in incompatible pollen and "rescued" incompatible pollen and also significantly reduced caspase-3-like (DEVDase) activity and later DNA fragmentation (Li et al. 2007). As p56 appears to be the only MAPK activated by SI, this suggests that p56 could involved in mediating SI-induced PCD and implicated involvement of a MAPK in signaling to PCD.

23.3.5.3 A Role for ROS and NO Signaling in SI-Mediated PCD

It has recently been shown, using live-cell imaging to visualize reactive oxygen species (ROS) and nitric oxide (NO) in growing *Papaver* pollen tubes, that SI induces relatively rapid and transient increases in ROS and NO, which have distinct temporal "signatures" in incompatible pollen tubes (Wilkins et al. 2011). Investigating how these signals integrate with the SI responses, using ROS/NO scavengers, revealed the alleviation of both the formation of SI-induced actin punctate foci and the activation of a DEVDase/caspase-3-like activity, providing evidence that ROS and NO act upstream of these key SI markers and suggesting they signal to these SI events (Wilkins et al. 2011) (see Fig. 23.1b). As the actin foci appear to be an integral part of the PCD response, this suggests that actin stabilization also plays a role in mediating PCD (Fig. 23.1c). These data represent the first steps in understanding ROS/NO signaling triggered by SI in pollen tubes.

23.4   Recruitment of Signaling for SI Events in Other Species

Recent studies, exploring the possibility for transfer of the SI determinants into other species, have demonstrated that the pollen male determinant, PrpS, can be expressed in *Arabidopsis thaliana* pollen. The PrpS-GFP-expressing transgenic pollen exhibits key features of the *Papaver* pollen SI response, including inhibition, formation of actin foci, and increases in caspase-3-like activity when cognate recombinant *Papaver* PrsS is added to it (de Graaf Barend et al. 2012). This observation represents the first demonstration that a SI system can be transferred into a distantly related species and function to trigger appropriate responses. Until now, it was thought that transfer of SI would be limited to close relatives, as so far the only movement of SI systems has been within species or closely related species, *Arabidopsis lyrata* and *Capsella grandiflora* into *A. thaliana* (Nasrallah et al. 2002; Boggs et al. 2009). Although these are important demonstrations, they diverged only ~5 million (Koch et al. 2000), and ~6.2–9.8 million years ago (Acarkan et al. 2000), respectively, and so do not provide major insights into the evolution of SI signaling across angiosperm families because of their close relationship and their possession of a mechanistically common SI system. Our findings provide a breakthrough in this area, as transferral of the *Papaver* pollen *S*-determinant into *A. thaliana* is between highly diverged species with ~144 million years separating them (Bell et al. 2010), and they do not share a common SI system.

Our data provide good evidence that *A. thaliana* possesses proteins that can be recruited to form new signaling networks and targets that are used for a function that does not normally operate in this species. These data suggest that the *Papaver* SI system uses common cellular targets, and that endogenous signaling components can be recruited to elicit a response that most likely never operated in this species, which has potentially important implications. Studies of the evolution of self/non-self recognition systems have generally focused on the receptors and ligands involved in recognition, rather than the downstream signaling networks triggered by their interaction. Our findings suggest either conservation of an ancient signaling system or recruitment of signaling components to mediate the downstream responses for SI recognition. Also, our data suggest that the postulated parallels between SI and plant–pathogen resistance (Hodgkin et al. 1988; Sanabria et al. 2008), with the idea that SI may utilize some of these signaling networks, may not be as unlikely as it initially seems. It appears that the *Papaver* SI system works in *A. thaliana* as a result of "multi-tasking" of endogenous components that can "plug and play" to act in signaling networks in which they do not normally operate. The signaling networks (e.g., Ca^{2+}) and targets for *Papaver* SI (e.g., the actin cytoskeleton) appear to be "universal," unspecialized, and ancient and may be present in all angiosperm cells. If these "common" cellular elements can be recruited to operate under the control of a newly introduced system (as we have shown with PrpS), it appears that a novel and functional signaling network can be set up that results in a specific, predictable physiological outcome; this could be similar to situations where gene redundancy and plasticity operate. MAPK cascade components are classic examples of signaling components that can participate in more than one signaling network in certain situations. For example, in *Saccharomyces cerevisiae* (yeast), components can play a role in more than one pathway; see Widmann et al. (1999), Asai et al. (2002), and Eckardt (2002) for further discussion of this phenomenon. Examples of dual functioning or "multi-tasking" have been cited in the context of innate immune signaling pathways (Ausubel 2005). We think this is a likely explanation of why PrpS functions in *A. thaliana* pollen.

23.5   Summary

Research into the *Papaver* SI system, as outlined in this review, has shown that several major cellular components are targets for SI signals in pollen, which are triggered when the male and female *S*-determinants interact (see Fig. 23.1a for a cartoon of the situation in an unstimulated pollen tube). These events contribute, first, to initiation of signaling events (Fig. 23.1b) and the inhibition of incompatible pollen tube growth, and second, to key mechanisms involved in steps to commitment to PCD (Fig. 23.1c). Together, these mechanisms ensure that self-fertilization does not occur. *Papaver* clearly has a complex network of signaling events that are integrated to contribute to SI-mediated inhibition and death of incompatible pollen tubes. However, there still remain many unanswered questions about mechanisms to be investigated. For the future, the demonstration of wide transgenera functionality of the *Papaver* SI system opens up the possibility that this may, in the longer term, provide a tractable SI system to transfer to crop plants to make F_1 hybrids more efficiently. Demonstration that interaction of PrsS with *A. thaliana* PrpS-GFP pollen to elicit a "SI" response suggests there is scope for transfer of the *Papaver* SI system to completely unrelated crop species. This possibility will be a major challenge for the future.

Acknowledgments Work in the laboratory of V.E. F-T. has been funded long-term by the Biotechnology and Biological Sciences Research Council (B.B.S.R.C.). Other sources of funding include China Scholarship Council, Commonwealth, Conacyt, Conacit, PBL, Royal Society China Fellowship, and Wellcome Trust. This work would not have been possible without the help of my long-term collaborators Chris Staiger and Chris Franklin, and the many members of my lab who were involved in these studies.

References

Acarkan A, Roßberg M, Koch M, Schmidt R (2000) Comparative genome analysis reveals extensive conservation of genome organisation for *Arabidopsis thaliana* and *Capsella rubella*. Plant J 23:55–62

Allen AM, Hiscock SJ (2008) Evolution and phylogeny of self-incompatibility systems in angiosperms. In: Franklin-Tong VE (ed) Self-incompatibility in flowering plants. Springer, Berlin, pp 73–102

Asai T, Tena G, Plotnikova J, Willmann MR, Chiu W-L, Gomez-Gomez L, Boller T, Ausubel FM, Sheen J (2002) MAP kinase signalling cascade in *Arabidopsis* innate immunity. Nature (Lond) 415:977–983

Ausubel FM (2005) Are innate immune signaling pathways in plants and animals conserved? Nat Immunol 6:973–979

Bell CD, Soltis DE, Soltis PS (2010) The age and diversification of the angiosperms re-revisited. Am J Bot 97:1296–1303

Boggs NA, Dwyer KG, Shah P, McCulloch AA, Bechsgaard J, Schierup MH, Nasrallah ME, Nasrallah JB (2009) Expression of distinct self-incompatibility specificities in *Arabidopsis thaliana*. Genetics 182:1313–1321

Bosch M, Franklin-Tong VE (2007) Temporal and spatial activation of caspase-like enzymes induced by self-incompatibility in *Papaver* pollen. Proc Natl Acad Sci USA 104: 18327–18332

Burnet FM (1971) "Self-recognition" in colonial marine forms and flowering plants in relation to the evolution of immunity. Nature (Lond) 232:230–235

Dangl JL, Jones JDG (2001) Plant pathogens and integrated defence responses to infection. Nature (Lond) 411:826–833

de Graaf BH, Vatovec S, Juárez-Díaz JA, Chai L, Kooblall K, Wilkins KA, Zou H, Forbes T, Franklin FC, Franklin-Tong VE (2012) The *Papaver* self-incompatibility pollen S-determinant, PrpS, functions in *Arabidopsis thaliana*. Curr Biol 22:154–159

de Graaf BHJ, Rudd JJ, Wheeler MJ, Perry RM, Bell EM, Osman K, Franklin FCH, Franklin-Tong VE (2006) Self-incompatibility in *Papaver* targets soluble inorganic pyrophosphatases in pollen. Nature (Lond) 444:490–493

De Tomaso AW, Nyholm SV, Palmeri KJ, Ishizuka KJ, Ludington WB, Mitchel K, Weissman IL (2005) Isolation and characterization of a protochordate histocompatibility locus. Nature (Lond) 438:454–459

Eckardt NA (2002) Good things come in threes: a trio of triple kinases essential for cell division in *Arabidopsis*. Plant Cell Online 14:965–967

Foote HCC, Ride JP, Franklin-Tong VE, Walker EA, Lawrence MJ, Franklin FCH (1994) Cloning and expression of a distinctive class of self-incompatibility (*S*) gene from *Papaver rhoeas* L. Proc Natl Acad Sci USA 91:2265–2269

Franklin-Tong VE (ed) (2008) Self-incompatibility in flowering plants: evolution, diversity, and mechanisms. Springer, Berlin

Franklin-Tong VE, Gourlay CW (2008) A role for actin in regulating apoptosis/programmed cell death: evidence spanning yeast, plants and animals. Biochem J 413:389–404

Franklin-Tong VE, Lawrence MJ, Franklin FCH (1988) An *in vitro* bioassay for the stigmatic product of the self-incompatibility gene in *Papaver rhoeas* L. New Phytol 110:109–118

Franklin-Tong VE, Ride JP, Read ND, Trewavas AJ, Franklin FCH (1993) The self-incompatibility response in *Papaver rhoeas* is mediated by cytosolic free calcium. Plant J 4:163–177

Franklin-Tong VE, Ride JP, Franklin FCH (1995) Recombinant stigmatic self-incompatibility-(S-) protein elicits a Ca^{2+} transient in pollen of *Papaver rhoeas*. Plant J 8:299–307

Franklin-Tong VE, Holdaway-Clarke TL, Straatman KR, Kunkel JG, Hepler PK (2002) Involvement of extracellular calcium influx in the self-incompatibility response of *Papaver rhoeas*. Plant J 29:333–345

Geitmann A, Snowman BN, Emons AMC, Franklin-Tong VE (2000) Alterations in the actin cytoskeleton of pollen tubes are induced by the self-incompatibility reaction in *Papaver rhoeas*. Plant Cell 12:1239–1251

Glass NL, Jacobson DJ, Shiu PK (2000) The genetics of hyphal fusion and vegetative incompatibility in filamentous ascomycete fungi. Annu Rev Genet 34:165–186

Hodgkin T, Lyon GD, Dickinson HG (1988) Recognition in flowering plants: a comparison of the *Brassica* self-incompatibility system and plant–pathogen interactions. New Phytol 110:557–569

Huang S, Blanchoin L, Chaudhry F, Franklin-Tong VE, Staiger CJ (2004) A gelsolin-like protein from *Papaver rhoeas* pollen (PrABP80) stimulates calcium-regulated severing and depolymerization of actin filaments. J Biol Chem 279:23364–23375

Hughes AL (2002) Natural selection and the diversification of vertebrate immune effectors. Immunol Rev 190:161–168

Jordan ND, Kakeda K, Conner A, Ride JP, Franklin-Tong VE, Franklin FCH (1999) S-protein mutants indicate a functional role for SBP in the self-incompatibility reaction of *Papaver rhoeas*. Plant J 20:119–125

Kakeda K, Jordan ND, Conner A, Ride JP, Franklin-Tong VE, Franklin FCH (1998) Identification of residues in a hydrophilic loop of the *Papaver rhoeas* S protein that play a crucial role in recognition of incompatible pollen. Plant Cell 10:1723–1731

Koch M, Haubold B, Mitchell-Olds T (2000) Comparative evolutionary analysis of chalcone synthase and alcohol dehydrogenase loci in *Arabidopsis*, *Arabis*, and related genera (Brassicaceae). Mol Biol Evol 17:1483–1498

Kornberg A (1962) On the metabolic significance of phosphorolytic and pyrophosphorolytic reactions. Academic, New York

Lam E, del Pozo O (2000) Caspase-like protease involvement in the control of plant cell death. Plant Mol Biol 44:417–428

Li S, Samaj J, Franklin-Tong VE (2007) A mitogen-activated protein kinase signals to programmed cell death induced by self-incompatibility in *Papaver* pollen. Plant Physiol 145:236–245

McClure B, Franklin-Tong V (2006) Gametophytic self-incompatibility: understanding the cellular mechanisms involved in "self" pollen tube inhibition. Planta (Berl) 224:233–245

Nasrallah ME, Liu P, Nasrallah JB (2002) Generation of self-incompatible *Arabidopsis thaliana* by transfer of two *S* locus genes from *A. lyrata*. Science 297:247–249

Nyholm SV, Passegue E, Ludington WB, Voskoboynik A, Mitchel K, Weissman IL, De Tomaso AW (2006) *fester*, a candidate allorecognition receptor from a primitive chordate. Immunity 25:163–173

Poulter NS, Vatovec S, Franklin-Tong VE (2008) Microtubules are a target for self-incompatibility signaling in *Papaver* pollen. Plant Physiol 146:1358–1367

Poulter NS, Staiger CJ, Rappoport JZ, Franklin-Tong VE (2010) Actin-binding proteins implicated in formation of the punctate actin foci stimulated by the self-incompatibility response in Papaver. Plant Physiol 10(1104):109.152066

Qiao H, Wang F, Zhao L, Zhou J, Lai Z, Zhang Y, Robbins TP, Xue Y (2004) The F-box protein AhSLF-S$_2$ controls the pollen function of S-RNase-based self-incompatibility. Plant Cell 16:2307–2322

Rudd JJ, Franklin FCH, Lord JM, FranklinTong VE (1996) Increased phosphorylation of a 26-kD pollen protein is induced by the self-incompatibility response in *Papaver rhoeas*. Plant Cell 8:713–724

Rudd JJ, Osman K, Franklin FCH, Franklin-Tong VE (2003) Activation of a putative MAP kinase in pollen is stimulated by the self-incompatibility (SI) response. FEBS Lett 547:223–227

Sanabria N, Goring D, Nürnberger T, Dubery I (2008) Self/nonself perception and recognition mechanisms in plants: a comparison of self-incompatibility and innate immunity. New Phytol 178:503–514

Scofield VL, Schlumpberger JM, West LA, Weissman IL (1982) Protochordate allorecognition is controlled by a MHC-like gene system. Nature (Lond) 295:499–502

Sijacic P, Wang X, Skirpan A, Wang Y, Dowd P, McCubbin A, Huang S, Kao T-H (2004) Identification of the pollen determinant of S-RNase-mediated self-incompatibility. Nature (Lond) 429:302–305

Smertenko A, Franklin-Tong VE (2011) Organisation and regulation of the cytoskeleton in plant programmed cell death. Cell Death Differ 18:1263–1270

Snowman BN, Kovar DR, Shevchenko G, Franklin-Tong VE, Staiger CJ (2002) Signal-mediated depolymerization of actin in pollen during the self-incompatibility response. Plant Cell 14:2613–2626

Staiger CJ (2000) Signalling to the actin cytoskeleton in plants. Annu Rev Plant Physiol Plant Mol Biol 51:257–288

Stein JC, Howlett B, Boyes DC, Nasrallah ME, Nasrallah JB (1991) Molecular cloning of a putative receptor protein kinase gene encoded at the self-incompatibility locus of *Brassica oleracea*. Proc Natl Acad Sci USA 88:8816–8820

Takayama S, Isogai A (2005) Self-incompatibility in plants. Annu Rev Plant Biol 56:467–489

Thomas SG, Franklin-Tong VE (2004) Self-incompatibility triggers programmed cell death in *Papaver* pollen. Nature (Lond) 429:305–309

Thomas SG, Huang S, Li S, Staiger CJ, Franklin-Tong VE (2006) Actin depolymerization is sufficient to induce programmed cell death in self-incompatible pollen. J Cell Biol 174:221–229

Walker EA, Ride JP, Kurup S, Franklin-Tong VE, Lawrence MJ, Franklin FC (1996) Molecular analysis of two functional homologues of the S3 allele of the *Papaver rhoeas* self-incompatibility gene isolated from different populations. Plant Mol Biol 30:983–994

Wheeler MJ, de Graaf BHJ, Hadjiosif N, Perry RM, Poulter NS, Osman K, Vatovec S, Harper A, Franklin FCH, Franklin-Tong VE (2009) Identification of the pollen self-incompatibility determinant in *Papaver rhoeas*. Nature (Lond) 459:992–995

Widmann C, Gibson S, Jarpe MB, Johnson GL (1999) Mitogen-activated protein kinase: conservation of a three-kinase module from yeast to human. Physiol Rev 79:143–180

Wilkins KA, Bancroft J, Bosch M, Ings J, Smirnoff N, Franklin-Tong VE (2011) ROS and NO mediate actin reorganization and programmed cell death in the self-incompatibility response of *Papaver*. Plant Physiol 156(1):404–416

Woltering EJ (2004) Death proteases come alive. Trends Plant Sci 9:469–472

Woltering EJ, van der Bent A, Hoeberichts FA (2002) Do plant caspases exist? Plant Physiol 130:1764–1769

Wu J, Wang S, Gu Y, Zhang S, Publicover SJ, Franklin-Tong VE (2011) Self-incompatibility in *Papaver rhoeas* activates nonspecific cation conductance permeable to Ca^{2+} and K^+. Plant Physiol 155:963–973

Yao C-K, Lin YQ, Ly CV, Ohyama T, Haueter CM, Moiseenkova-Bell VY, Wensel TG, Bellen HJ (2009) A synaptic vesicle-associated Ca^{2+} channel promotes endocytosis and couples exocytosis to endocytosis. Cell 138:947–960

Chapter 24
S-RNase-Based Self-Incompatibility in *Petunia*: A Complex Non-Self Recognition System Between Pollen and Pistil

Penglin Sun, Justin Stephen Williams, Shu Li, and Teh-hui Kao

Abstract Self-incompatibility (SI) is an intraspecific reproductive barrier that allows many families of flowering plants to prevent inbreeding and promote out-crosses. Extensive studies of SI in five families during the past more than two decades have revealed three distinct SI mechanisms. This chapter focuses on the mechanism employed by the Solanaceae, using mostly results obtained from *Petunia*. We first discuss the identification of two polymorphic genes at the *S*-locus, the *S-RNase* gene, which controls pistil specificity, and the *S-locus F-box* (*SLF*) gene, now named *SLF1*. For several years after its identification, *SLF1* was thought to be solely responsible for pollen specificity, and biochemical models were developed based on this assumption. However, results inconsistent with this assumption were subsequently obtained, which led to the recent finding that pollen specificity is controlled by multiple, but an as yet unknown number of, polymorphic *SLF* genes located at the *S*-locus. A new model, named collaborative non-self recognition, has been proposed to explain the biochemical basis of specific inhibition of self-pollen tube growth. Based on this model, compatible pollination results from ubiquitination and subsequent degradation of non-self S-RNases collectively mediated by all SLF proteins, with each SLF responsible for detoxifying a subset of non-self S-RNases.

P. Sun • S. Li
Intercollege Graduate Degree Program in Plant Biology, The Pennsylvania State University, University Park, PA 16802, USA

J.S. Williams
Department of Biochemistry and Molecular Biology, The Pennsylvania State University, University Park, PA 16802, USA

T.H. Kao (✉)
Intercollege Graduate Degree Program in Plant Biology, The Pennsylvania State University, University Park, PA 16802, USA

Department of Biochemistry and Molecular Biology, The Pennsylvania State University, University Park, PA 16802, USA
e-mail: txk3@psu.edu

H. Sawada et al. (eds.), *Sexual Reproduction in Animals and Plants*,
DOI 10.1007/978-4-431-54589-7_24, © The Author(s) 2014

We conclude this chapter with a discussion of some of the new questions raised by the finding that pollen specificity is controlled by multiple *SLF* genes.

Keywords Collaborative non-self recognition • *Petunia* • Self-incompatibility • *S*-locus F-box proteins • S-RNase

24.1   Introduction

In contrast to animals, plants cannot freely move about to select appropriate mates. To preserve species identity and to generate genetic diversity within species, it is imperative that plants possess mechanisms by which their female reproductive tissues, the pistils, can prevent unwanted or unsuitable pollen from delivering sperm cells to the ovary to effect fertilization. For example, many flowering plants that produce bisexual flowers have adopted a genetically controlled pre-zygotic barrier, called self-incompatibility (SI). SI involves a self/non-self recognition process between pollen and pistil. As a result of the recognition, self-pollen is rejected by the pistil to prevent inbreeding, and only non-self pollen is accepted to promote outcrosses. The specific recognition process involves the interaction between pistil specificity determinant and pollen specificity determinant, both of which are encoded by genes tightly linked at a highly polymorphic locus, named the *S*-locus. Variants of the *S*-locus are referred to as *S*-haplotypes and designated S_1, S_2, S_3, etc.

SI is quite common in flowering plants, but to date our understanding of the molecular and biochemical basis is limited to 5 of the estimated more than 60 families that possess SI. Extensive studies on these 5 families since the mid-1980s have revealed three distinct mechanisms (Takayama and Isogai 2005). In this chapter, we focus on the mechanism that is employed by the Solanaceae family (and likely by two other families, the Rosaceae and Plantaginaceae) using mostly the results obtained from the study of *Petunia*. We first describe the identification and characterization of the gene that controls pistil specificity, the *S-RNase*, and then describe the identification of the first *S-locus F-box* (*SLF*) gene, now named type-1 *SLF* (or *SLF1*), that is involved in controlling pollen specificity. After the identification of *SLF1*, it was thought to be the only gene that controls pollen specificity, as the prevailing thought then was that pollen specificity must be controlled by a single polymorphic gene, just like the prior finding that pistil specificity is controlled by a single polymorphic gene, *S-RNase*. Moreover, it is already difficult to envision how a single pollen specificity gene has coevolved with a single pistil specificity gene to maintain SI over the course of millions of years since this self/non-self recognition system was established. We thus next turn to the discussion of several lines of evidence, from the evolutionary perspective and experimental data, that are inconsistent with the assumption that a single *SLF* gene is solely responsible for pollen specificity, and describe the identification of additional *SLF* genes that are involved in pollen specificity. We present the latest model, named "collaborative non-self recognition," that explains how the polymorphic *S-RNase* gene and multiple polymorphic *SLF* genes function in SI, allowing cross-compatible pollen tubes to grow through the pistil, but inhibiting the growth of self-pollen tubes in the pistil. Finally, we discuss some of the questions regarding the evolution, maintenance, and operation of S-RNase-based SI that were raised by the unexpected finding that pollen specificity is controlled by multiple *SLF* genes.

24.2 Identification and Characterization of the *S-RNase* Gene

To identify the gene that controls pistil specificity, pistil proteins that showed *S*-haplotype-specific differences in molecular mass and/or isoelectric point were first identified (Bredemeijer and Blaas 1981; Anderson et al. 1986). These proteins were deemed likely candidates, as allelic products of the pistil gene involved in SI were expected to be different in their amino-acid sequences such that they might be different in molecular mass or isoelectric point. This approach led to the identification of the *S-RNase* gene in *Nicotiana alata* (Anderson et al. 1986). However, the biochemical nature of S-RNase was not known until several years later when the amino-acid sequence of the fungal RNase T2 was determined (Kawata et al. 1988) and found to share a significant degree of sequence homology between its catalytic domain and a conserved domain of several allelic variants of S-RNase (McClure et al. 1989). Subsequently, S-RNase was shown to indeed possess RNase activity in vitro (McClure et al. 1989; Broothaerts et al. 1991; Singh et al. 1991). The *S-RNase* gene has been shown to be solely responsible for pistil specificity in SI by gain- and loss-of-function experiments (Lee et al. 1994; Murfett et al. 1994). For example, in *Petunia inflata*, expression of the S_3-*RNase* gene in pistils of wild-type plants of S_1S_2 genotype conferred on the pistils of the transgenic plants the ability to reject S_3 pollen, whereas expression of an antisense S_3-*RNase* gene in wild-type plants of S_2S_3 genotype led to the inability of the transgenic plants to reject S_3 pollen, but did not affect their ability to reject S_2 pollen (Lee et al. 1994). These experiments suggest that S-RNase is necessary and sufficient for the pistil to recognize and reject self-pollen.

The characteristics of S-RNase are also consistent with its involvement in pistil specificity. For example, S-RNase is specific to the pistil. It is first synthesized in the transmitting cell of the style and then secreted into the extracellular space of the transmitting tract. It is most abundant in the upper third segment of the style, where inhibition of incompatible pollen tubes occurs (Ai et al. 1990). Moreover, S-RNase has a high degree of allelic sequence diversity, with the highest divergent pair sharing only 38 % sequence identity (Tsai et al. 1992; McCubbin and Kao 2000), which is expected for a protein that is involved in self/non-self recognition. Sequence comparison revealed the presence of five conserved regions, C1–C5 (with C2 and C3 similar to the corresponding domains of RNase T2), and two hypervariable regions, HVa and HVb (Ioerger et al. 1991; Tsai et al. 1992; Takayama and Isogai 2005). Crystallographic analysis of a Solanaceae S-RNase showed that the protein-folding topology is typical for the RNase T2 family (Ida et al. 2001). Moreover, the results also showed that both HVa and HVb are exposed on the surface of the S-RNase protein and accessible to solvent, raising the possibility that these two regions have the allelic-specific function (Ida et al. 2001; Matsuura et al. 2001).

To further understand the function of S-RNase in SI, site-directed mutagenesis was used to replace the codon for one of the two catalytic His residues (located in C3) of S_3-RNase of *P. inflata* with an Asn codon. The mutant S_3-*RNase* gene was introduced into *P. inflata* plants of S_1S_2 genotype, and the transgenic plants producing this mutated S_3-RNase failed to reject S_3 pollen, suggesting that the RNase activity of S-RNase is an integral part of its function and that the biochemical mechanism of SI involves degradation of RNAs in incompatible pollen tubes (Huang et al. 1994). Moreover, as all S-RNases are glycoproteins with various numbers of N-linked glycan chains, a question was raised as to whether the recognition function of S-RNase resides in the glycan moiety, the protein backbone, or both. To address this question, the codon for the only potential N-glycosylation site of S_3-RNase was replaced with a codon for Asp, and the mutant S_3-*RNase* gene was introduced into wild-type plants of S_1S_2 genotype to examine whether expression of the non-glycosylated S_3-RNase could still confer on the transgenic plants the ability to reject S_3 pollen. The results showed that the non-glycosylated S_3-RNase retained the full ability to reject self-pollen, suggesting that the recognition function of S-RNase resides in its amino-acid sequence (Karunanandaa et al. 1994). Domain-swapping experiments were performed to examine the role of the two hypervariable domains in allelic specificity (Kao and McCubbin 1996; Matton et al. 1997; Zurek et al. 1997; Matton et al. 1999). Swapping HVa and HVb between S_{11}-RNase and S_{13}-RNase of *Solanum chacoense* (which differ by four amino acids in the hypervariable regions) was sufficient to switch the allelic specificity (Matton et al. 1997, 1999). However, in the case of *P. inflata* and *N. alata* S-RNases examined, the two hypervariable regions are necessary but not sufficient for allele specificity. This conclusion is consistent with the finding that two *P. inflata* S-RNases, S_6-RNase and S_9-RNase, have identical sequences in HVa and differ by only two amino acids in HVb (Wang et al. 2001; Kao and Tsukamoto 2004).

For S-RNase to inhibit growth of self-pollen tubes, it must be able to enter incompatible pollen tubes to exert its cytotoxicity. Using immunocytochemistry, Luu et al. (2000) showed that S-RNase was localized in the cytoplasm of both self- and non-self pollen tubes. In contrast, Goldraij et al. (2006), using triple antibody-labeling immunolocalization, showed that S-RNase was initially sequestered in a vacuole-like compartment of both self- and non-self pollen tubes, but the compartment was later disrupted only in incompatible pollen tubes, releasing S-RNase into the cytoplasm. Both groups are in agreement that S-RNase is taken up by self- and non-self pollen tubes; however, the mechanism of S-RNase uptake is as yet unknown.

24.3   Identification of the First *S-Locus F-box* Gene, *SLF1*

One would expect that the approach used to successfully identify S-RNase and clone the *S-RNase* gene could be used to identify the pollen specificity determinant. However, no pollen protein that show *S*-haplotype specific differences in molecular mass or isoelectric point was ever identified. Another approach, RNA differential display, was used to identify pollen-specific genes that are tightly linked to the *S*-locus (Wang et al. 2003). This approach was based on the prediction that the gene controlling pollen specificity must be tightly linked to the *S-RNase* gene and shows allelic sequence diversity. Thirteen such genes of *P. inflata* were identified (Dowd et al. 2000; McCubbin et al. 2000). However, none of them was considered a good candidate for the pollen specificity gene because of their low allelic sequence diversity (Wang et al. 2003). Ultimately, it was through sequencing the *S*-locus region containing the *S-RNase* gene that a good candidate was first identified in *Antirrhinum hispanicum* (Plantaginaceae), which also possesses S-RNase-based SI (Lai et al. 2002). This gene is located ~9 kb downstream from the S_2-*RNase* gene, and its deduced amino-acid sequence contains an F-box motif at the N-terminus. Thus, the gene was named *AhSLF* (*A. hispanicum S-locus F-box* gene). Subsequently, *PmSLF*, encoding an F-box protein, was identified in *Prunus mume* of Rosaceae through genomic sequence analysis of an ~60-kb region containing *S-RNase* (Entani et al. 2003), and *SFB* (*S haplotype-specific F-box* gene), also encoding an S-locus F-box protein, was identified in another rosaceous species, *Prunus dulcis* (almond) (Ushijima et al. 2003).

In *P. inflata*, *PiSLF* (*P. inflata SLF*) was identified from sequencing a 328-kb contig of an *S*-locus region containing the S_2-*RNase* gene (Wang et al. 2004). *PiSLF* is located ~161 kb downstream from the S_2-*RNase* gene. RNA blotting results showed that *PiSLF* is specifically expressed in developing pollen, mature pollen and pollen tubes. It also shows *S*-haplotype-specific restriction-fragment length polymorphism, as expected of a gene located at the *S*-locus. The deduced amino-acid sequences of three alleles (S_1, S_2, and S_3) of *PiSLF* show 10.3–11.6 % allelic sequence diversity. These properties of *PiSLF* suggest that it is a good candidate for the pollen specificity gene. *PiSLF* control of pollen specificity was established by a transgenic functional assay of its S_2-allele, $PiSLF_2$ (Fig. 24.1a), designed based on the phenomenon named competitive interaction. This phenomenon was first observed in solanaceous species, including *P. inflata* and *P. hybrida* (Stout and Chandler 1942; Brewbaker and Natarajan 1960; Entani et al. 1999), where it was found that SI breaks down in heteroallelic pollen carrying two different *S*-alleles. For example, S_1S_2 pollen produced by a tetraploid $S_1S_1S_2S_2$ plant derived from self-incompatible plant S_1S_2 loses its SI function and cannot be rejected by the $S_1S_1S_2S_2$ pistil. Heteroallelic pollen can also result from duplication of one of the *S*-loci of self-incompatible plants carrying two different *S*-haplotypes (Golz et al. 1999). It was reasoned that if *PiSLF* controls pollen specificity, introducing $PiSLF_2$ into S_1S_1, S_1S_2, and S_2S_3 plants should cause breakdown of SI in S_1 and S_3 pollen carrying the transgene, as a result of competitive interaction between the endogenous S_1-allele or S_3-allele of *PiSLF* and the introduced S_2-allele. The results of these experiments were precisely as expected (Sijacic et al. 2004). For example (as shown in Fig. 24.1a), expression of a $PiSLF_2$ transgene in an S_2S_3 transgenic plant caused breakdown of SI in S_3 pollen (heteroallelic), but not in S_2 pollen (homoallelic).

As already stated, both self- and non-self S-RNases are taken up by a pollen tube, but only self-S-RNase can exert its cytotoxicity to inhibit the growth of the pollen tube. Before the discovery of the involvement of multiple *SLF* genes in pollen

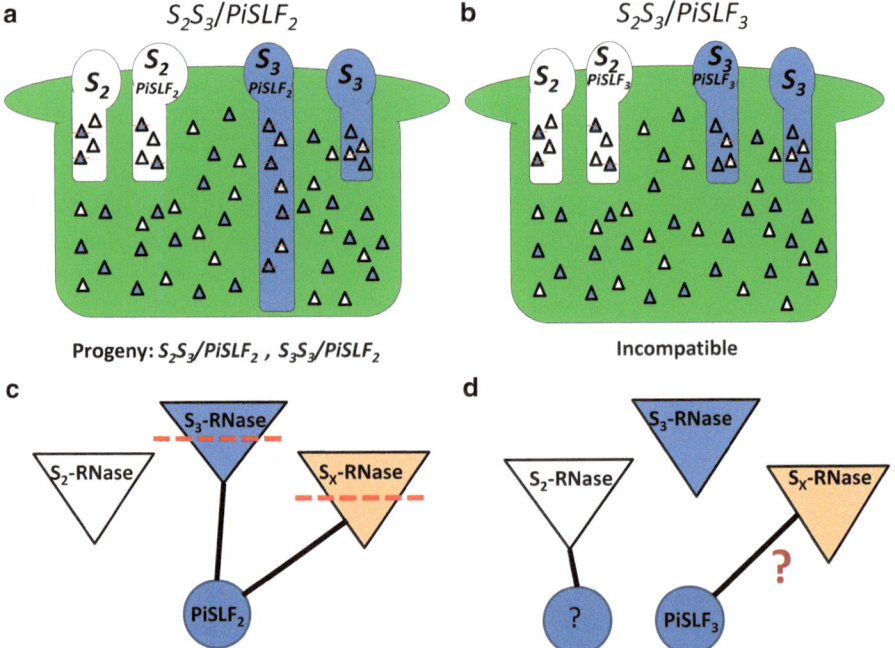

Fig. 24.1 Transgenic functional assay showing that *PiSLF* is not the sole pollen specificity gene. (**a**) Self-pollination of an S_2S_3 transgenic plant carrying a *PiSLF$_2$* transgene. The transgenic plant produces four different genotypes of pollen as indicated. *White triangles* and *blue triangles* denote S_2-RNase and S_3-RNase molecules, respectively. Based on the findings that S_2S_3 and S_3S_3, but no S_2S_2, genotypes were present in the self-progeny and that all progeny inherited the transgene (either one copy or two copies), only S_3 pollen carrying the *PiSLF$_2$* transgene was accepted by the pistil. *White* and *blue triangles* marked with a *broken line* across denote those S_2-RNases and S_3-RNases, respectively, that are inside pollen tubes but fail to exert their cytotoxicity. That *PiSLF$_2$* causes breakdown of SI in S_3, but not in S_2, pollen is consistent with the prediction by competitive interaction that SI breaks down in pollen carrying two different S-alleles. (**b**) Self-pollination of an S_2S_3 transgenic plant carrying a *PiSLF$_3$* transgene. The transgenic plant produces four different genotypes of pollen as indicated. *White triangles* and *blue triangles* denote S_2-RNase and S_3-RNase molecules, respectively. The transgenic plant remained self-incompatible, suggesting that all four genotypes of pollen were rejected by the pistil. *White* and *blue triangles* marked with a *broken line* across denote those S_2-RNases and S_3-RNases, respectively, that are inside pollen tubes but fail to exert their cytotoxicity. The finding that *PiSLF$_3$* does not cause breakdown of SI in S_2 pollen is not consistent with the prediction by competitive interaction. (**c**) A protein degradation model based on the assumption that *PiSLF* is solely responsible for pollen specificity predicts that a PiSLF interacts with all S-RNases, except its self-S-RNase, to mediate their ubiquitination and ultimate degradation by the 26S proteasome. In the example shown, PiSLF$_2$ interacts with S$_3$-RNase and S$_x$-RNase (with S$_x$ denoting S-RNases produced by any additional haplotype except S_2) to mediate their ubiquitination and degradation, as indicated by a *broken line* across S$_3$-RNase and S$_x$-RNase; however, PiSLF$_2$ does not interact with S$_2$-RNase, its self-S-RNase. This model can explain the results of the transgenic functional assay shown in **a**. In S_3 pollen tubes expressing the *PiSLF$_2$* transgene, PiSLF$_3$ produced by the endogenous gene interacts with S$_2$-RNase, and PiSLF$_2$ produced from the transgene interacts with S$_3$-RNase. As a result, both S$_2$-RNase and S$_3$-RNase are ubiquitinated and degraded, thus allowing the pollen tube to circumvent the toxic effect of these S-RNases. (**d**) Interpretation of the results of the transgenic functional assay shown in **b** based on the protein-degradation model shown in **c**. The failure of the *PiSLF$_3$* transgene to cause breakdown of SI in S_2 pollen suggests that PiSLF$_3$ cannot interact with S$_2$-RNase, a non-self S-RNase, raising a question of whether PiSLF$_3$ can interact with any other non-self S-RNases. Most importantly, this finding suggests that there must be additional protein(s) involved in pollen specificity that allow(s) S_3 pollen to interact with and detoxify S$_2$-RNase to result in compatible pollination between S_3 pollen and S_2-carrying pistils

specificity (see Sect. 24.5 below), our lab proposed a protein degradation model, based on the assumption that PiSLF (SLF1) is the sole pollen determinant, to explain why only self-S-RNase can function inside a pollen tube (Hua and Kao 2006; Hua et al. 2008). The model predicts that for pollen of a given S-haplotype, (1) the allelic variant of PiSLF produced interacts with all its non-self S-RNases in the cytoplasm of the pollen tubes to mediate their ubiquitination and ultimate degradation by the 26S proteasome, thus allowing non-self pollen to effect fertilization, and (2) the allelic variant of PiSLF does not interact with its self-S-RNase, allowing the self-S-RNase to degrade pollen tube RNAs to cause growth inhibition of self-pollen tubes. These predictions were largely based on the findings by Hua and Kao (2006) that bacterially expressed S-RNases were ubiquitinated and degraded via the 26S pro-teasomal pathway in an in vitro cell-free system, and that an allelic variant of PiSLF interacted with its non-self S-RNases more strongly than with its self-S-RNase in an in vitro protein binding assay. As most F-box proteins are components of SCF com-plexes, a class of E3 ubiquitin ligase involved in ubiquitin-mediated protein degra-dation, the protein degradation model provided a reasonable explanation for the biochemical basis of the SI process. Moreover, the following findings are consistent with this model. (1) The subcellular localization of AhSLF-S_2 (the S_2-allelic variant of an SLF in *Antirrhinum*) detected by immunocytochemistry suggests that it is localized in the cytoplasm of pollen tubes (Wang and Xue 2005). (2) In *S. chacoense*, the level of S-RNase in compatible pollen tubes was ~30 % lower than that in incompatible pollen tubes (Liu et al. 2009). (3) Six lysine residues in S_3-RNase of *P. inflata* were found to be necessary for targeting ubiquitination and degradation (Hua and Kao 2008).

This protein degradation model can explain why expression of a *PiSLF$_2$* trans-gene in heteroallelic pollen causes breakdown of SI. For example, when a *PiSLF$_2$* transgene is introduced into an S_2S_3 plant, it causes specific breakdown of SI in S_3 transgenic pollen (Fig. 24.1a). As shown in Fig. 24.1c, in the S_3 transgenic pollen, PiSLF$_2$ produced from the transgene interacts with S_3-RNase, a non-self S-RNase, to mediate its ubiquitination and degradation. As PiSLF$_3$ produced by the endoge-nous gene interacts with S_2-RNase, a non-self S-RNase, to mediate its ubiquitina-tion and degradation, both S_2-RNase and S_3-RNase taken up by the transgenic S_3 pollen tube are detoxified, allowing the S_3 transgenic pollen tube to be compatible with the S_3-carrying pistil.

24.4   Evidence Against *SLF1* Being Solely Responsible for Controlling Pollen Specificity

After the transgenic functional assay showing that one allele of *PiSLF*, *PiSLF$_2$*, behaved as precisely as expected for the gene controlling pollen specificity, it was thought that the long search for the pollen specificity gene was over. However, fur-ther analyses of this gene, now named *SLF1*, revealed puzzling results that called into question this notion.

24.4.1 Evolutionary Consideration

First, the degree of sequence diversity of *SLF1* is much lower than that of the *S-RNase* gene. For example, the allelic sequence diversity is only ~10 % for three alleles (S_1, S_2 and S_3) of *PiSLF*, but ranges from 19 % to 26 % for the same three alleles of *S-RNase*. Most puzzlingly, the allelic sequence diversity of four alleles (S_1, S_2, S_4, S_5) of *AhSLF* (*A. hispanicum*) ranges from only 1 % to 3 %, whereas that of the same four alleles of *S-RNase* ranges from 36 % to 51 % (Zhou et al. 2003; Newbigin et al. 2008). The high degree of allelic sequence diversity observed for *S-RNase* is what would be expected for a gene that controls pistil specificity in SI interactions with pollen. During the evolution of the SI system, any new *S*-haplotype that arises in a population has a reproductive advantage over all existing haplotypes, because its frequency is lower to begin with and thus has a higher probability of successful fertilization. As such, the new *S*-haplotype will increase in its frequency in the population. Under this kind of frequency-dependent selection, there is pressure for sequence divergence in the genes that constitute *S*-haplotype specificity to generate new *S*-haplotypes. Thus, the low degree of allelic sequence diversity of the *SLF1* gene is not consistent with its presumed pollen specificity role in SI.

Second, the evolutionary history of *S-RNase* is consistent with the ancient origin of SI, but *SLF1* seemed to have evolved much more recently. Phylogenetic studies of *S-RNase* alleles from several solanaceous species revealed that polymorphism of *S-RNase* predated divergence of these species (e.g., some *P. inflata* alleles are more similar to some *N. alata* alleles than to other *P. inflata* alleles) (Ioerger et al. 1990). Allelic polymorphism of *S-RNase* was shown to exist in the common ancestor of the solanaceous species 30–40 million years ago (Paape et al. 2008). However, phylogenetic studies of *SLF1* alleles from *P. inflata* and *A. hispanicum* showed that *SLF1* has a much shorter evolutionary history (Newbigin et al. 2008).

Third, the evolution of *S-RNase* and *SLF1* seems not to be concordant. Coevolution of genes controlling pistil and pollen specificity has long been a central theory in SI. In Brassicaceae-type SI, the genes encoding pollen and style specificity determinants have been shown to be coevolved (Sato et al. 2002; Takebayashi et al. 2003). However, as stated here, *S-RNase* and *SLF1* seem to have different evolutionary histories, and moreover, genealogies of *S-RNase* and *SLF1* do not show a pattern of coevolution (Newbigin et al. 2008).

24.4.2 Experimental Data

When the same transgenic functional assay used to establish the function of $PiSLF_2$ in pollen specificity was used to examine two other alleles of *PiSLF* and several alleles of *SLF1* of *P. hybrida*, the results showed that none of them behaved as expected for the pollen specificity gene. For example, as shown in Fig. 24.1b, introduction of the S_3-allele of *PiSLF*, *PiSLF3*, into *P. inflata* plants of S_2S_3 genotype did not cause breakdown of SI in S_2 transgenic pollen, even though the transgenic S_2 pollen is heteroallelic (Kubo et al. 2010). Moreover, S_7-allele of *P. hybrida SLF1* caused breakdown of SI in S_9 and S_{17} transgenic pollen, but not in S_5, S_{11}, or S_{19} transgenic pollen (Kubo et al. 2010). These results, on the one hand, further confirm the involvement of *SLF1* in pollen specificity, but on the other hand, suggest that *SLF1* is not the only gene involved in pollen specificity. For example, based on the protein degradation model discussed in Sect. 24.3 earlier, the failure of a *PiSLF3* transgene to cause breakdown of SI in S_2 pollen suggests that PiSLF3 does not interact with S_2-RNase, a non-self S-RNase, to mediate its ubiquitination and degradation (Fig. 24.1d). This notion raises a question as to whether PiSLF3 interacts with any other non-self S-RNases. Most importantly, this finding suggests that there must be additional protein(s) produced in S_3 pollen that can interact with and detoxify S_2-RNase, as S_3 pollen is compatible with S_2-carrying pistils.

The most definitive evidence for the involvement of additional gene(s) in pollen specificity came from the finding that the deduced amino-acid sequences of the *S7-allele* of *SLF1* of *P. hybrida* and S_9-allele of *SLF1* of *P. axillaris* are completely identical, even though S_7 and S_9 are genetically distinct S-haplotypes and the deduced amino-acid sequences of their *S-RNase* alleles are only 45 % identical (Kubo et al. 2010).

24.5   Involvement of Additional *SLF* Genes in Controlling Pollen Specificity

After the identification of the first *SLF* gene, *SLF1*, it was discovered that there exist additional *F-box* genes that are tightly linked to the S-locus. For example, our lab identified four such *PiSLF*-like genes, named *PiSLFLa, PiSLFLb, PiSLFLc*, and *PiSLFLd* (Hua et al. 2007). Interestingly, these genes are also specifically expressed in pollen, and at least three of them show allelic-specific sequence differences. However, these *PiSLF*-like genes were initially thought not to be involved in pollen specificity, as when *PiSLFLc-S1*, *PiSLFLb-S2*, and *PiSLFLd-S2* were introduced into S_1S_2, S_2S_3, and S_2S_3, respectively, none of them caused breakdown of SI in S_2 pollen (in the case of *PiSLFLc-S1*) or S_3 pollen (in the case of *PiSLFLb-S2* and *PiSLFLd-S2*) (Hua et al. 2007). The likely involvement of additional gene(s) in pollen specificity prompted the reexamination of these *PiSLF*-like genes, as well as additional *SLF1*-like genes of *P. hybrida* identified by the laboratory of Seiji Takayama.

Takayama's lab used PCR primers designed based on the sequences of *PiSLFLb-S2*, *PiSLFLc-S1*, and *PiSLFLd-S2* to identify 30 *SLF-like* genes from different S-haplotypes of *P. hybrida*, and classified them into five types, type 2 *SLF* (or *SLF2*) to type 6 *SLF* (*SLF6*) (Kubo et al. 2010). Within each type, the sequence identity between different alleles is high, ranging from 70.3 % to 99 %. However, the sequence identity between different types is only ~50 %. To date, *SLF2* and

SLF3 have been shown to be involved in pollen specificity as well. For example, S_7-*SLF2* (*SLF2* of S_7 haplotype) caused breakdown of SI in S_9, S_{11}, and S_{19} pollen, and S_{11}-*SLF3* (*SLF3* of S_{11} haplotype) caused breakdown of SI in S_7 pollen. The results also suggest that additional *SLF* genes are required for pollen specificity. For example, *SLF1*, *SLF2*, and *SLF3* of S_7-haplotype did not cause breakdown of SI in S_5 pollen (Kubo et al. 2010).

24.6   Collaborative Non-Self Recognition Model

After the discovery that multiple SLF proteins are involved in pollen specificity, a modified protein degradation model, named collaborative non-self recognition, was proposed. This model proposes that, for a given *S*-haplotype, each type of SLF protein interacts with a subset of non-self S-RNases and that multiple types of SLF proteins are required to recognize the entire suite of non-self S-RNases to mediate their degradation. Similar to the original protein degradation model, the new model also predicts that none of the SLF proteins produced in pollen of a given *S*-haplotype interacts with their self-S-RNase. The prediction that an SLF interacts with a subset of its non-self S-RNases was confirmed by a co-immunoprecipitation (co-IP) experiment. This experiment showed that S_7-SLF2:FLAG produced in transgenic pollen coprecipitated, by an anti-FLAG antibody, with S_9-RNase and S_{11}-RNase, but not with S_5-RNase or S_7-RNase, in style extracts. Moreover, the transgenic functional assay showed that expression of S_7-*SLF2* caused breakdown of SI in S_9 and S_{11} pollen, but did not cause breakdown of SI in S_5 or S_7 pollen (Kubo et al. 2010). Thus, the co-IP results not only provide strong support for the protein degradation aspect of collaborative non-self recognition, but also are entirely consistent with the results of the transgenic functional assay.

The collaborative non-self recognition model is graphically illustrated in Fig. 24.2. This figure depicts the outcomes of pollination of S_5-, S_{17}-, S_{11}-, and S_9-carrying pistils by S_5 pollen, as explained by the relationships between two types of SLF proteins produced in S_5 pollen, S_5-SLF1 and S_5-SLF2, and four S-RNases, S_5-, S_{17}-, S_{11}-, and S_9-RNases, based on the results obtained from the transgenic functional assay (Kubo et al. 2010). When S_5-RNase is taken up by the S_5 pollen tube, none of the SLF proteins produced in S_5 pollen can interact with their self-S-RNase, S_5-RNase; thus, pollen RNAs are degraded by S_5-RNase and pollen tube growth is inhibited. When S_{17}-RNase is taken up by the S_5 pollen tube, S_5-SLF1 interacts with S_{17}-RNase to mediate its ubiquitination/degradation. When S_{11}-RNase is taken up by the S_5 pollen tube, S_5-SLF2 interacts with S_{11}-RNase to mediate its ubiquitination/degradation. When S_9-RNase is taken up by the S_5 pollen tube, both S_5-SLF1 and S_5-SLF2 will be able to interact with S_9-RNase to mediate its ubiquitination/degradation. Thus, in cross-pollinations, at least one SLF protein can interact with and detoxify a non-self S-RNase, allowing the pollen tube to grow down through the style to effect fertilization.

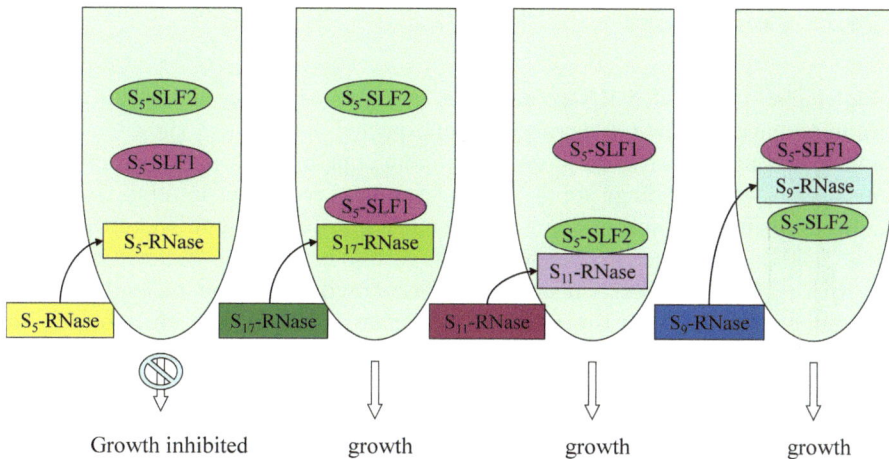

Fig. 24.2 Collaborative non-self recognition model. For pollen of a given *S*-haplotype, multiple types of SLF proteins collaboratively interact with all their non-self S-RNases, with each type of SLF only interacting with a subset of the non-self S-RNases. For the sake of simplicity, only two different types of SLF proteins, SLF1 and SLF2, are shown for S_5 pollen. When S_5-RNase is taken up by the S_5 pollen tube (*leftmost panel*), no SLF proteins can interact with their self-S-RNase, S_5-RNase. Pollen RNA is degraded by S_5-RNase and pollen tube growth is inhibited. When S_{17}-RNase is taken up by the S_5 pollen tube (*second panel from left*), S_5-SLF1 interacts with S_{17}-RNase to mediate its degradation. When S_{11}-RNase is taken up by the S_5 pollen tube (*second panel from right*), S_5-SLF2 interacts with S_{11}-RNase to mediate its degradation. When S_9-RNase is taken up by the S_5 pollen tube (*rightmost panel*), both S_5-SLF1 and S_5-SLF2 interact with S_9-RNase to mediate its degradation. Degradation of S_{17}-RNase, S_{11}-RNase, and S_9-RNase inside S_5 pollen tubes mediated by one or more types of SLF proteins explains compatible pollinations between S_5 pollen and S_{17}-, S_{11}-, and S_9-carrying pistils

The collaborative non-self recognition model can explain the puzzles about the properties of *SLF1* mentioned in Sect. 24.4.1, for example, the low allelic sequence diversity of SLF1. As multiple types of SLF proteins act collaboratively to interact with all non-self S-RNases, each SLF only has to interact with a small number of non-self S-RNases, and thus it is unnecessary to have a high degree of allelic sequence diversity. In extreme cases, some allelic variants of an SLF can have identical amino-acid sequences, as is the case for S_7-SLF1 and S_{19}-SLF1 of *P. axillaris* (Kubo et al. 2010). In this case, S_7 and S_{19} pollen most likely use their SLF1 proteins (with identical sequences) to interact with the same subset of their common non-self S-RNases (i.e., not including S_7-RNase or S_{19}-RNase). S_7 and S_{19} pollen could then use some other SLF protein(s) with allelic sequence diversity, or use different types of SLF proteins, to recognize and detoxify S_{19}-RNase and S_7-RNase, respectively. Indeed, the degree of amino-acid sequence diversity between different types of SLF proteins is comparable to that of S-RNase, allowing different types of SLF proteins to interact with different S-RNases.

24.7 Conclusions

Molecular studies of S-RNase-based SI over the preceding more than a quarter century have revealed a number of unexpected findings, suggesting that this inbreeding-prevention genetic system is more complex than initially thought. The most unexpected findings are (1) pollen specificity comprises multiple *SLF* genes, and (2) rejection of self-pollen is the result of non-self recognition between SLFs and S-RNases, and not by specific self-interaction between an SLF and its cognate S-RNase. Thus, the mechanism of S-RNase-based SI is fundamentally different from the mechanisms of Brassicaceae and Papaveraceae SI, as each employs a single male *S*-gene and a single female *S*-gene to control SI, and both involve the interaction between products of matching alleles of the pollen and pistil *S*-genes (Iwano and Takayama 2012).

The finding that pollen specificity is controlled by multiple SLF proteins has also raised a number of interesting and challenging questions about S-RNase-based SI. We list some of the questions here. For a given *S*-haplotype, how many *SLF* genes constitute pollen specificity? Do all *S*-haplotypes employ the same number of *SLF* genes? What is the mechanism for the generation of *SLF* genes? What is the biochemical basis for the differential interactions between SLF proteins and S-RNases? Why does this SI system employ a single gene with a very high degree of allelic sequence polymorphism for pistil specificity, and multiple genes, each with a much lower degree of allelic sequence polymorphism for pollen specificity, but not vice versa? How have a single *S-RNase* gene and multiple polymorphic *SLF* genes coevolved to maintain SI?

Efforts are under way to address some of these questions. For example, our laboratory is using RNA-seq to characterize the complete pollen transcriptome of S_2 and S_3 haplotypes of *P. inflata* to identify all the F-box genes that are linked to the *S*-locus. The genes identified could then be examined by the well-established transgenic functional assay (Sijacic et al. 2004; Hua et al. 2007; Kubo et al. 2010) to see whether they can cause breakdown of SI in pollen of any non-self *S*-haplotype. As more *SLF* genes are identified, comparison of deduced amino-acid sequences of different alleles of the same type of *SLF*, and deduced amino-acid sequences of different types of *SLF* genes will likely reveal amino-acid sequences/domains of SLFs that are involved in allele specificity and in type specificity. The approach of chimeric genes could then be used to identify these sequence features. S-RNase-based SI also provides a model for F-box protein-mediated protein degradation, a key regulatory mechanism in many cellular and developmental processes in eukaryotes. Identification of most, if not all, *SLF* genes of S_2- and S_3-haplotype involved in pollen specificity and understanding of how these SLF proteins interact with their specific S-RNases will lay the foundation for studying the structural basis of interactions between F-box proteins and their substrates.

Acknowledgments The work from the authors' laboratory was supported by grants to T.-H. K. from the U.S. National Science Foundation.

References

Ai YJ, Singh A, Coleman CE, Ioerger TR, Kheyrpour A, Kao T-H (1990) Self-incompatibility in *Petunia inflata*—isolation and characterization of cDNAs encoding 3 *S*-allele-associated proteins. Sex Plant Reprod 3(2):130–138

Anderson MA, Cornish EC, Mau SL, Williams EG, Hoggart R, Atkinson A, Bonig I, Grego B, Simpson R, Roche PJ, Haley JD, Penschow JD, Niall HD, Tregear GW, Coghlan JP, Crawford RJ, Clarke AE (1986) Cloning of cDNA for a stylar glycoprotein associated with expression of self-incompatibility in *Nicotiana alata*. Nature (Lond) 321(6065):38–44

Bredemeijer GMM, Blaas J (1981) S-specific proteins in styles of self-incompatible *Nicotiana alata*. Theor Appl Genet 59(3):185–190. doi:10.1007/bf00264974

Brewbaker JL, Natarajan AT (1960) Centric fragments and pollen-part mutation of incompatibility alleles in *Petunia*. Genetics 45(6):699–704

Broothaerts W, Vanvinckenroye P, Decock B, Vandamme J, Vendrig JC (1991) *Petunia hybrida* S-proteins: ribonuclease activity and the role of their glycan side chains in self-incompatibility. Sex Plant Reprod 4(4):258–266

Dowd PE, McCubbin AG, Wang X, Verica JA, Tsukamoto T, Ando T, Kao T-H (2000) Use of *Petunia inflata* as a model for the study of solanaceous type self-incompatibility. Ann Bot 85:87–93. doi:10.1006/anbo.1999.1032

Entani T, Takayama S, Iwano M, Shiba H, Che FS, Isogai A (1999) Relationship between polyploidy and pollen self-incompatibility phenotype in *Petunia hybrida* Vilm. Biosci Biotechnol Biochem 63(11):1882–1888

Entani T, Iwano M, Shiba H, Che FS, Isogai A, Takayama S (2003) Comparative analysis of the self-incompatibility (*S*-) locus region of *Prunus mume*: identification of a pollen-expressed F-box gene with allelic diversity. Genes Cells 8(3):203–213

Goldraij A, Kondo K, Lee CB, Hancock CN, Sivaguru M, Vazquez-Santana S, Kim S, Phillips TE, Cruz-Garcia F, McClure B (2006) Compartmentalization of S-RNase and HT-B degradation in self-incompatible *Nicotiana*. Nature (Lond) 439(7078):805–810. doi:10.1038/Nature04491

Golz JF, Su V, Clarke AE, Newbigin E (1999) A molecular description of mutations affecting the pollen component of the *Nicotiana alata S* locus. Genetics 152(3):1123–1135

Hua ZH, Kao T-H (2006) Identification and characterization of components of a putative *Petunia* *S*-locus F-box-containing E3 ligase complex involved in S-RNase-based self-incompatibility. Plant Cell 18(10):2531–2553. doi:10.1105/tpc.106.041061

Hua Z, Kao T-H (2008) Identification of major lysine residues of S_3-RNase of *Petunia inflata* involved in ubiquitin-26S proteasome-mediated degradation in vitro. Plant J 54(6):1094–1104. doi:10.1111/j.1365-313X.2008.03487.x

Hua ZH, Meng XY, Kao T-H (2007) Comparison of *Petunia inflata* S-locus F-box protein (Pi SLF) with Pi SLF-like proteins reveals its unique function in S-RNase-based self-incompatibility. Plant Cell 19(11):3593–3609. doi:10.1105/tpc.107.055426

Hua ZH, Fields A, Kao T-H (2008) Biochemical models for S-RNase-based self-incompatibility. Mol Plant 1(4):575–585. doi:10.1093/Mp/Ssn032

Huang S, Lee HS, Karunanandaa B, Kao T-H (1994) Ribonuclease activity of *Petunia inflata* S-proteins is essential for rejection of self-pollen. Plant Cell 6(7):1021–1028

Ida K, Norioka S, Yamamoto M, Kumasaka T, Yamashita E, Newbigin E, Clarke AE, Sakiyama F, Sato M (2001) The 1.55 Å resolution structure of *Nicotiana alata* S_{F11}-RNase associated with gametophytic self-incompatibility. J Mol Biol 314(1):103–112. doi:10.1006/jmbi.2001.5127

Ioerger TR, Clark AG, Kao T-H (1990) Polymorphism at the self-incompatibility locus in Solanaceae predates speciation. Proc Natl Acad Sci USA 87(24):9732–9735. doi:10.1073/pnas.87.24.9732

Ioerger TR, Gohlke JR, Xu B, Kao T-H (1991) Primary structural features of the self-incompatibility protein in Solanaceae. Sex Plant Reprod 4(2):81–87

Iwano M, Takayama S (2012) Self/non-self discrimination in angiosperm self-incompatibility. Curr Opin Plant Biol 15(1):78–83. doi:10.1016/j.pbi.2011.09.003

Kao T-H, McCubbin AG (1996) How flowering plants discriminate between self and non-self pollen to prevent inbreeding. Proc Natl Acad Sci USA 93(22):12059–12065. doi:10.1073/pnas.93.22.12059

Kao T-H, Tsukamoto T (2004) The molecular and genetic bases of S-RNase-based self-incompatibility. Plant Cell 16:S72–S83. doi:10.1105/Tpc.016154

Karunanandaa B, Huang S, Kao T-H (1994) Carbohydrate moiety of the *Petunia inflata* S₃ protein is not required for self-incompatibility interactions between pollen and pistil. Plant Cell 6(12):1933–1940. doi:10.1105/tpc.6.12.1933

Kawata Y, Sakiyama F, Tamaoki H (1988) Amino-acid sequence of ribonuclease T2 from *Aspergillus oryzae*. Eur J Biochem 176(3):683–697

Kubo K, Entani T, Takara A, Wang N, Fields AM, Hua ZH, Toyoda M, Kawashima S, Ando T, Isogai A, Kao T-H, Takayama S (2010) Collaborative non-self recognition system in S-RNase-based self-incompatibility. Science 330(6005):796–799. doi:10.1126/science.1195243

Lai Z, Ma W, Han B, Liang L, Zhang Y, Hong G, Xue Y (2002) An F-box gene linked to the self-incompatibility (*S*) locus of *Antirrhinum* is expressed specifically in pollen and tapetum. Plant Mol Biol 50(1):29–42

Lee HS, Huang S, Kao T-H (1994) S proteins control rejection of incompatible pollen in *Petunia inflata*. Nature (Lond) 367(6463):560–563. doi:10.1038/367560a0

Liu B, Morse D, Cappadocia M (2009) Compatible pollinations in *Solanum chacoense* decrease both S-RNase and S-RNase mRNA. PLoS One 4(6):e5774. doi:10.1371/journal.pone.0005774

Luu DT, Qin XK, Morse D, Cappadocia M (2000) S-RNase uptake by compatible pollen tubes in gametophytic self-incompatibility. Nature (Lond) 407(6804):649–651

Matsuura T, Sakai H, Unno M, Ida K, Sato M, Sakiyama F, Norioka S (2001) Crystal structure at 1.5-Å resolution of *Pyrus pyrifolia* pistil ribonuclease responsible for gametophytic self-incompatibility. J Biol Chem 276(48):45261–45269. doi:10.1074/jbc.M107617200

Matton DP, Maes O, Laublin G, Xike Q, Bertrand C, Morse D, Cappadocia M (1997) Hypervariable domains of self-incompatibility RNases mediate allele-specific pollen recognition. Plant Cell 9(10):1757–1766. doi:10.1105/tpc.9.10.1757

Matton DP, Luu DT, Xike Q, Laublin G, O'Brien M, Maes O, Morse D, Cappadocia M (1999) Production of an S-RNase with dual specificity suggests a novel hypothesis for the generation of new *S* alleles. Plant Cell 11(11):2087–2097

McClure BA, Haring V, Ebert PR, Anderson MA, Simpson RJ, Sakiyama F, Clarke AE (1989) Style self-incompatibility gene products of *Nicotiana alata* are ribonucleases. Nature (Lond) 342(6252):955–957

McCubbin AG, Kao T-H (2000) Molecular recognition and response in pollen and pistil interactions. Annu Rev Cell Dev Biol 16:333–364. doi:10.1146/annurev.cellbio.16.1.333

McCubbin AG, Wang X, Kao T-H (2000) Identification of self-incompatibility (*S*-) locus linked pollen cDNA markers in *Petunia inflata*. Genome 43(4):619–627. doi:10.1139/Gen-43-4-619

Murfett J, Atherton TL, Mou B, Gasser CS, McClure BA (1994) S-RNase expressed in transgenic *Nicotiana* causes *S*-allele-specific pollen rejection. Nature (Lond) 367(6463):563–566. doi:10.1038/367563a0

Newbigin E, Paape T, Kohn JR (2008) RNase-based self-incompatibility: puzzled by *Pollen S*. Plant Cell 20(9):2286–2292. doi:10.1105/tpc.108.060327

Paape T, Igic B, Smith SD, Olmstead R, Bohs L, Kohn JR (2008) A 15-Myr-old genetic bottleneck. Mol Biol Evol 25(4):655–663. doi:10.1093/molbev/msn016

Sato K, Nishio T, Kimura R, Kusaba M, Suzuki T, Hatakeyama K, Ockendon DJ, Satta Y (2002) Coevolution of the *S*-locus genes *SRK, SLG* and *SP11/SCR* in *Brassica oleracea* and *B. rapa*. Genetics 162(2):931–940

Sijacic P, Wang X, Skirpan AL, Wang Y, Dowd PE, McCubbin AG, Huang S, Kao T-H (2004) Identification of the pollen determinant of S-RNase-mediated self-incompatibility. Nature (Lond) 429(6989):302–305. doi:10.1038/Nature02523

Singh A, Ai Y, Kao T-H (1991) Characterization of ribonuclease activity of three *S*-allele-associated proteins of *Petunia inflata*. Plant Physiol 96(1):61–68

Stout AB, Chandler C (1942) Hereditary transmission of induced tetraploidy and compatibility in fertilization. Science 96:257–258. doi:10.1126/science.96.2489.257-a

Takayama S, Isogai A (2005) Self-incompatibility in plants. Annu Rev Plant Biol 56:467–489. doi:10.1146/annurev.arplant.56.032604.144249

Takebayashi N, Brewer PB, Newbigin E, Uyenoyama MK (2003) Patterns of variation within self-incompatibility loci. Mol Biol Evol 20(11):1778–1794. doi:10.1093/molbev/msg209

Tsai DS, Lee HS, Post LC, Kreiling KM, Kao T-H (1992) Sequence of an S-protein of *Lycopersicon peruvianum* and comparison with other solanaceous S-proteins. Sex Plant Reprod 5(4): 256–263

Ushijima K, Sassa H, Dandekar AM, Gradziel TM, Tao R, Hirano H (2003) Structural and transcriptional analysis of the self-incompatibility locus of almond: identification of a pollen-expressed F-box gene with haplotype-specific polymorphism. Plant Cell 15(3):771–781

Wang H-Y, Xue Y-B (2005) Subcellular localization of the *S* locus F-box protein AhSLF-S₂ in pollen and pollen tubes of self-incompatible *Antirrhinum*. J Integr Plant Biol 47(1):76–83. doi:10.1111/j.1744-7909.2005.00014.x

Wang X, Hughes AL, Tsukamoto T, Ando T, Kao T-H (2001) Evidence that intragenic recombination contributes to allelic diversity of the *S-RNase* gene at the self-incompatibility (*S*) locus in *Petunia inflata*. Plant Physiol 125(2):1012–1022. doi:10.1104/pp. 125.2.1012

Wang Y, Wang X, McCubbin AG, Kao T-H (2003) Genetic mapping and molecular characterization of the self-incompatibility (*S*) locus in *Petunia inflata*. Plant Mol Biol 53(4):565–580. doi:10.1023/B:Plan.0000019068.00034.09

Wang Y, Tsukamoto T, Yi KW, Wang X, Huang SS, McCubbin AG, T-H K (2004) Chromosome walking in the *Petunia inflata* self-incompatibility (*S*-) locus and gene identification in an 881-kb contig containing S₂-*RNase*. Plant Mol Biol 54(5):727–742. doi:10.1023/B:Plan. 0000040901.98982.82

Zhou JL, Wang F, Ma WS, Zhang YS, Han B, Xue YB (2003) Structural and transcriptional analysis of *S*-locus F-box genes in *Antirrhinum*. Sex Plant Reprod 16(4):165–177. doi:10.1007/s00497-003-0185-5

Zurek DM, Mou BQ, Beecher B, McClure B (1997) Exchanging sequence domains between S-RNases from *Nicotiana alata* disrupts pollen recognition. Plant J 11(4):797–808. doi:10.1046/j.1365-313X.1997.11040797.x

Chapter 25
Self-Incompatibility System of *Ipomoea trifida*, a Wild-Type Sweet Potato

Tohru Tsuchiya

Abstract Diploid *Ipomoea trifida* (Convolvulaceae) is a close relative of the cultivated hexaploid *Ipomoea batatas*, the cultivated sweet potato. These plants have sporophytic self-incompatibility that is regulated by a single multiallelic locus, designated as the *S*-locus. Genetic analyses of *I. trifida* plants collected from Central America identified about 50 different *S*-haplotypes with a linear dominance hierarchy having some codominance relationships. A linkage map of DNA markers around the *S*-locus indicated that the *S*-locus is delimited to 0.23 cM. Within the *S*-locus genomic region, a hypervariable genomic region of 35–95 kbp was identified, and we designated this region SDR (*S*-locus-specific divergent region). Of the several genes located within the SDR, one anther-specific gene, *AB2*, and three stigma-specific genes, *SE1*, *SE2*, and *SEA*, are candidate *S*-genes that may encode male and female *S*-determinants of self-incompatibility.

Keywords *Ipomoea trifida* • Self-incompatibility • Sweet potato

25.1 Introduction

Self-incompatibility (SI) is a genetic mechanism to prevent self-fertilization (and thus encourage outcrossing) in angiosperms. In self-incompatible plants, when a pollen grain is recognized as the same type as self, some stage of pollen germination, pollen tube elongation, ovule fertilization, or embryo development is halted, and no seeds are produced. SI is classified into several groups: homomorphic SI, heteromorphic SI, cryptic SI (CSI), and late-acting SI. Heteromorphic SI is classified into two groups: distyly is determined by a single locus, which has two alleles,

T. Tsuchiya (✉)
Division of Plant Functional Genomics, Life Science Research Center, Mie University,
1577, Kurima-Machiya, Tsu, Mie 514-8507, Japan
e-mail: tsuchiya@gene.mie-u.ac.jp

H. Sawada et al. (eds.), *Sexual Reproduction in Animals and Plants*,
DOI 10.1007/978-4-431-54589-7_25, © The Author(s) 2014

and tristyly is determined by two loci, each with two alleles. Heteromorphic SI is sporophytic, in that both alleles in the male plant determine the SI response in the pollen. CSI exists in a limited number of taxa (for example, there is evidence for CSI in *Silene vulgaris* in the Caryophyllaceae; Glaettli 2004). In this mechanism, the simultaneous presence of cross- and self-pollen on the same stigma results in higher seed set from cross-pollen (Bateman 1956). Late-acting SI is also termed ovarian SI. In this mechanism, self-pollen germinates and reaches the ovules, but no fruit is set (Seavey and Bawa 1986; Sage et al. 1994).

Homomorphic SI is classified into sporophytic SI (SSI) and gametophytic SI (GSI) . The SSI system is found in species of several plant families, such as the Brassicaceae, Asteraceae, Malvaceae, Betulaceae, Sterculiaceae, Polemoniaceae, and Convolvulaceae (de Nettancourt 2001; Allen and Hiscock 2008). In the plants in these families, SI is genetically regulated by a single multi-allelic locus, the *S*-locus. Within the *S*-locus, a pair of genes (named *S*-genes), one encoding the male-determinant molecules and the other the female-determinant molecules, is localized. These genes are essential for the SI reaction, because these gene products contribute to self/non-self recognition. The sets of *S*-genes are tightly linked at the *S*-locus, and the *S*-locus (also called *S*-haplotype: Nasrallah and Nasrallah 1993), is inherited as a single unit to maintain the SI system. In the SSI system, self-pollen rejection is observed as the arrest of pollen germination, or pollen tube penetration into the stigma cell, and therefore self/non-self recognition occurs on the surface of the stigma. The male *S*-gene is sporophytically expressed in the tapetum of the anther, and the product of the male *S*-gene (*S*-protein) is deposited onto the pollen surface. The male phenotype of the SSI plant is determined by the diploid *S*-haplotypes of the pollen-producing plant; therefore, determination of the male *S*-phenotype is under the control of dominant–recessive relationships. On the other hand, the female *S*-gene is expressed in the papilla cells of the stigma. SSI of the Brassicaceae is well characterized at the molecular level; *SP11/SCR* and *SRK* are the male and female *S*-genes, respectively. SP11/SCR acts as a ligand of the membrane-anchored protein kinase SRK; these two determinants interact when self-pollination occurs, which induces a phosphorylation pathway and inhibits self-pollen germination.

The GSI system is found in the Solanaceae, Rosaceae, Fabaceae, Papavaraceae, and Poaceae, and in most cases it is regulated by a single *S*-locus, except in the Poaceae, which has both *S*- and *Z*-loci. In *S*-RNase-mediated GSI, the female *S*-determinant, *S*-RNase, is taken up into the elongating pollen tube and degrades RNA molecules that are recognized as self. In the Papavaraceae, the female *S*-determinant, PrsS, acts as a ligand of the pollen tube membrane-anchored receptor/channel male *S*-determinant, PrpS. After recognition of the pollen grain as self, the Ca^{2+} concentration in the pollen tube is increased, and actin is depolymerized, resulting in programmed cell death.

Ipomoea trifida is a close relative of the cultivated sweet potato *Ipomoea batatas*. Self-incompatible plants of the genus *Ipomoea* show SSI; however, this SI is strong, and is active even in hexaploid species, such as *I. batatas* (Fig. 25.1). Moreover, the SSI of *Ipomoea* may be regulated by a different mechanism than SSI in *Brassica*. In this chapter, this unique SSI system of *I. trifida* is described.

Fig. 25.1 Flowers of diploid *Ipomoea trifida*. Floral structures of plants in the Convolvulaceae are similar to each other, with funnel-shaped, radially symmetrical corolla, five sepals, five fused petals, and five epipetalous stamens. The flowering of this plant is enhanced under short-day conditions with relatively high temperature

25.2 Origin, Domestication, and Compatibility of *Ipomoea* Plants

Cultivated sweet potato, *I. batatas*, and its wild relatives belong to the section Batatas. The center of the origin and domestication of the sweet potato is thought to be either in Central America or South America. In Central America, domestication of the sweet potato might have started at least 5,000 years ago, and in South America, Peruvian sweet potato remnants dating as far back as 8000 BC have been found. Austin (1988) postulated that the center of origin of *I. batatas* was between the Yucatán Peninsula of Mexico and the mouth of the Orinoco River in Venezuela. Zhang et al. (1998) provided strong supporting evidence that the geographic zone postulated by Austin is the primary center of diversity. The much lower molecular diversity found in Peru and Ecuador suggests this region should be considered a secondary center of sweet potato diversity. The sweet potato was also grown in Polynesia. Sweet potato was cultivated in the Cook Islands in 1000 AD, and it may have been brought to central Polynesia around 700 AD, and spread across Polynesia to Hawaii and New Zealand from there (van Tilburg 1994; Bassett et al. 2004). A theory that the plant could have spread by seeds floating across the ocean is not supported by evidence. Another point is that the sweet potato in Polynesia is the cultivated *I. batatas*, which is generally spread by vine cuttings and not by seeds (Fig. 25.2).

Charles Darwin described outcrossing in higher plants and the diversity of reproduction modes (Darwin 1876). He described inbreeding depression based on his experiments with morning glory, *Ipomoea purpurea*. The family Convolvulaceae contains 55 genera and more than 1,500 species (Austin 1997), and the genus *Ipomoea* is the largest member of the Convolvulaceae, with more than 500 species. The genus occurs throughout the tropical and subtropical regions of the world, including both annual and perennial herbaceous plants, and most species are twining climbing plants. The floral structure of the genus *Ipomoea* is almost the same, a

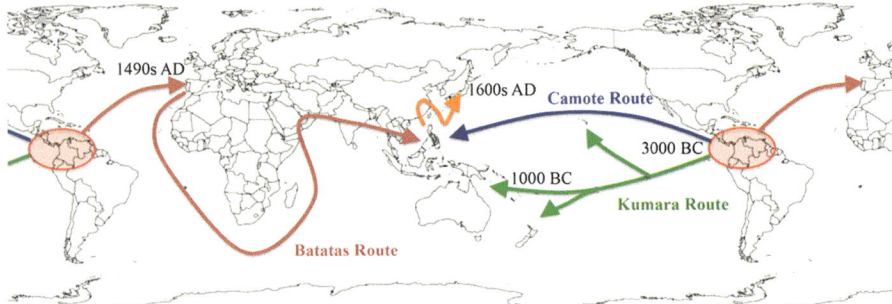

Fig. 25.2 Domestication and distribution of sweet potato. The center of origin and domestication of the sweet potato is thought to be in either Central America or South America. In Central America, sweet potatoes were domesticated at least 5,000 years ago. During the ancient era, sweet potato was thought to be distributed from Central America to the Pacific Islands by two routes, the Camote and the Kumara routes. After Columbus introduced sweet potato to Europe, the sweet potato was distributed to South and Southeast Asian countries, including Japan

gamopetalous flower, resembling that of the morning glory. This genus *Ipomoea* is classified into several sections based on reproductive and morphological traits (Austin and Huáman 1996). For instance, the section Pharbitis includes both the self-incompatible *Ipomoea serifera* and the self-compatible species *I. purpurea*; both horticultural varieties are known as morning glory. This classification is also supported by genetic analysis based on chloroplast DNA restriction site variations (McDonald and Mabry 1992) in section Batatas, which includes sweet potato and its wild relatives, series Pharbitis, which includes Japanese morning glory, and section Tricolor, which includes western morning glory, and others.

About 50 species belong to section Batatas. The site of origin of these plants is considered to be the same region as sweet potato. Plants in the section Batatas can be classified into two groups, one cross-compatible with sweet potato, *I. batatas*, and the other not. For example, South American species *Ipomoea triloba* ($2n = 2x = 30$), *Ipomoea tiliacea* ($4x$), and *I. trifida* ($2x$, $4x$, $6x$) are cross-compatible with *I. batatas* ($6x$); however, *Ipomoea umbraticola* ($2x$) is not (Table 25.1). Plants belonging to section Batatas can be alternatively classified as belonging to a self-incompatible (group A) and a self-incompatible (group B) (Table 25.2) (Nishiyama et al. 1975). In group B, the diploid species *I. trifida* and the tetraploid species *Ipomoea tabascana* are most closely related to the hexaploid sweet potato *I. batatas* (Rajapakse et al. 2004). However, according to its chromosome number and SI, *I. trifida* seems to be the wild species most closely related to *I. batatas*, and Huang and Sun (2000) revealed that *I. trifida* is the ancestral species of *I. batatas*, according to inter-simple sequence repeat and restriction analyses of chloroplast DNA. Artificial hexaploid *I. trifida* has also been produced (Shiotani and Kawase 1987); currently, it is understood that *I. batatas* is an autohexaploid of diploid *I. trifida*, produced by several chromosomal duplication and crossing events (Fig. 25.3) (Shiotani and Kawase 1989).

Table 25.1 Cross-incompatibility of sweet potato (*Ipomoea batatas*) with its wild relatives

	Origin		
	North America	South America	Caribbean region
Cross-compatible with *I. batatas* (6*x*)	*I. lacunosa* (2*x*) *I. cordatotriloba* (2*x*)		*I. triloba* (2*x*) *I. tiliacea* (4*x*) *I. trifida* (2*x*, 4*x*, 6*x*)
Cross-incompatible with *I. batatas* (6*x*)	*I. tenuissima* (2*x*)	*I. ramosissima* (2*x*) *I. grandifolia* (2*x*)	*I. umbraticola* (2*x*)

The wild population of plants that belong to section Batatas are classified into two major groups, one cross-compatible with cultivated sweet potato, and the other cross-incompatible. *I. trifida*, the close relative to cultivated sweet potato, is cross-compatible with sweet potato, despite the varied ploidy level

Table 25.2 Classification of several species of *Ipomoea* in section Batatas according to their self-incompatibility

	Scientific name	Chromosome numbers
Group A: self-compatible	*I. lacunosa*	2*n*=30
	I. cordatotriloba	2*n*=30
	I. tenuissima	2*n*=30
	I. triloba	2*n*=30
	I. ramosissima	2*n*=30
	I. cynanchifolia	2*n*=30
Group B: self-incompatible	*I. perviana*	2*n*=30
	I. gracilis	2*n*=30
	I. tiliacea	2*n*=60
	I. trifida	2*n*=30, 60, 90
	I. batatas	2*n*=90

Plants belonging to section Batatas are also classified under their mode of reproduction: self-compatibility (group A) and self-incompatibility (group B)

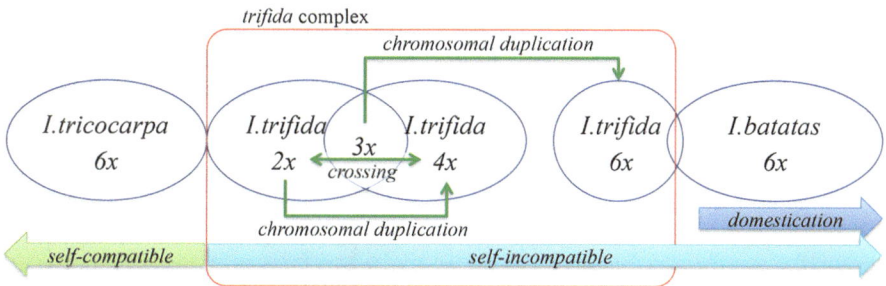

Fig. 25.3 Establishment of *Ipomoea batatas* from *Ipomoea trifida*. *I. batatas* (2*n*=6*x*=90) is thought to be an autohexaploid of diploid *I. trifida*. Hexaploid *I. trifida* was produced through two chromosomal duplications and crossing in the wild population, and the resulting plant became the ancestral species of *I. batatas*, the cultivated sweet potato. During these stages, self-incompatibility did not break down. However, once self-incompatibility was broken down, the reproduction mode was not resumed, and a self-compatible species was established

   In the practical breeding of sweet potato, cross-incompatibility between parental plants is a genetic barrier to producing hybrids because the choice of parents for cross-pollination is sometimes limited to a small number of lines. *I. trifida*, a diploid species of the section Batatas, is a useful genetic resource in sweet potato breeding (Shiotani and Kawase 1987). *I. trifida* is an herbaceous insect-pollinated weed native to Central America and has a relatively small genome (532 Mbp, 831 Mbp per haploid) (Arumuganathan and Earle 1991; Ozias-Akins and Jarret 1994).

25.3   SI of *Ipomoea* Plants

Compatible cross-pollinating pollen grains germinate about 10–20 min after pollination (Kowyama et al. 2008), with pollen grains attaching to stigma surface cells called papilla cells (Fig. 25.4). However, in the case of incompatible self-pollination, pollen germination is arrested, not resulting in seed formation. In the SI system of *I. trifida*, the self-recognition reaction between male and female *S*-determinants (*S*-gene products) occurs rapidly after pollination. Kowyama et al. (1980, 1994) reported that the SI phenotype of *I. trifida* segregates as a single multi-allelic *S*-locus, and the pollen phenotype of SI is regulated sporophytically with clear dominant–recessive relationships between *S*-haplotypes. These features are consistent with the general features of SSI. In *Brassica*, CO_2 treatment and bud pollination can permit self-pollination (Hinata et al. 1994); however, these artificial techniques are not applicable in *Ipomoea*. From this point of view, it is conceivable that the SSI mechanism of *Ipomoea* is stronger and clearer than in *Brassica* plants.

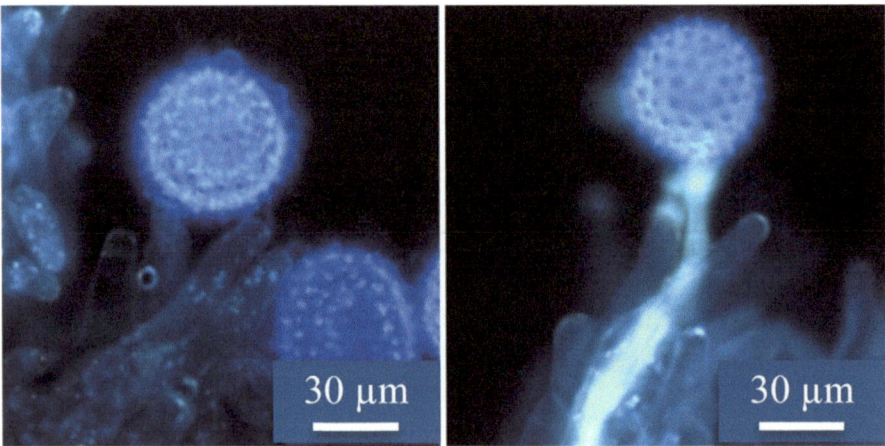

Fig. 25.4 Pollen behavior during self- and cross-pollination of *I. trifida*. Pollen tube germination is completely arrested during self-pollination (*left*), including crossing between plants with the same *S*-phenotype; however, pollen tube germination is not inhibited during cross-pollination (*right*)

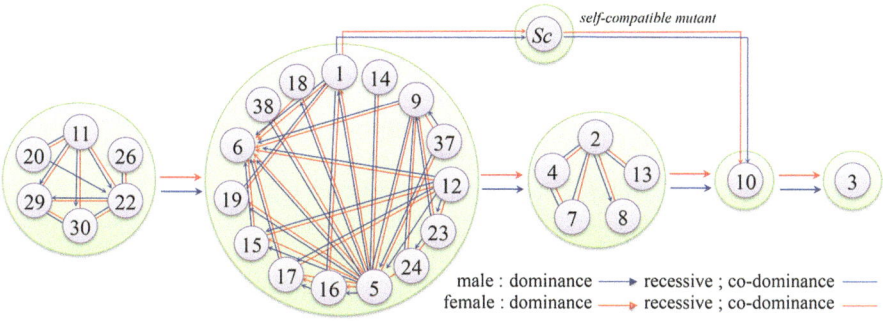

Fig. 25.5 Dominant–recessive hierarchy among 28 self-incompatible and 1 self-compatible mutant of *I. trifida*. Twenty-eight *S*-haplotypes are classified into five linear dominant–recessive groups. In each group, dominant–recessive relationships are slightly different between *S*-haplotypes in male and female reproductive organs. *Arrows* are drawn from dominant to recessive; *lines* indicate codominant relationships. The original *S*-haplotype of the self-compatible (Sc) allele is unknown; however, *Sc* is recessive to the S_1 haplotype and dominant over the S_{10} haplotype. (Figure is redrawn from Kowyama et al. 2008)

Kowyama et al. (1994) identified 49 different *S*-haplotypes from 224 individuals collected from six natural populations in Central America. *S*-haplotypes of *I. trifida* showed a linear dominant–recessive hierarchy between *S*-alleles and could be placed into five classes (Fig. 25.5). The *S*-alleles belonging to each class are the same for the male and female sides of the interaction (Fig. 25.5); however, several pairs of *S*-alleles show different interactions on the male and female sides. For example, S_{22} and S_{29} are codominant on the female side (stigma), and S_{22} is dominant over S_{29} on the male side (pollen). The linear hierarchy of dominant–recessive relationships suggests that new dominant *S*-haplotypes were created over recessive *S*-haplotypes, and the variation in allelic interaction suggests that the genetic mechanism for determination of the dominant–recessive hierarchy is regulated in different ways on the male and female sides.

In surveys for self-compatible (Sc) mutants to understand the genetic mechanism underlying SI in *I. trifida*, only one self-compatible plant (MX1) has been found as a spontaneous mutant from a natural population in Central America; its *S*-haplotype is designated *Sc* (Kakeda et al. 2000). Genetic analysis of the F_1 progeny derived from crosses between MX1 and several *S*-homozygous plants indicated that the *Sc-allele* is also within the dominant–recessive hierarchy; *Sc* is recessive to S_1 and dominant over S_{10} (Kakeda et al. 2000). The original *S*-allele is unknown; however, the fact that the *Sc* haplotype is within the dominant–recessive hierarchy suggests that *Sc* lacks one or more genes that contribute to SI but maintains genes that contribute to determination of dominance among *S*-haplotypes. Analyses of *Sc* mutants have provided important information for understanding the genetic features of SI in *Brassica* (Watanabe et al. 1997), *Pyrus* (Sassa et al. 1997), *Solanum* (Royo et al. 1994), and other plant species. In the case of *Ipomoea*, analyzing the *Sc* mutant may provide information to allow elucidation of the genetic features of SI.

25.4   Several Approaches to Identifying *S*-Genes in *I. trifida*

Because *I. trifida* has SSI, the *S*-genes have to meet several conditions. Male and female *S*-genes have to be expressed in the tapetum of the anther and papilla cells of the stigma, respectively, and these *S*-genes have to be linked tightly to maintain the *S*-haplotype at the *S*-locus. Moreover, these *S*-genes have to be present in a single copy per genome, and *S*-genes have to be polymorphic between *S*-haplotypes. In initial studies of gene products correlating with SI of *I. trifida*, reproductive organ-specific genes and genes similar to *S*-genes of other plant species were surveyed. *Ipomoea* stigma protein 11 (ISP11) has been isolated from a mature stigma cDNA library; however, this gene cannot be an *S*-gene, because it is expressed not only in stigma but also in the anther, and it is not linked to the *S*-locus, although it is a single-copy gene (Kowyama et al. 1995b). Because *I. trifida* has the same genetic SI system as *Brassica* species, it might be expected that products of genes homologous to *SLG* or *SRK* would be present in the reproductive tissues of *I. trifida*. *SRK* is known as a female *S*-gene (Stein et al. 1991; Takasaki et al. 2000), and *SLG* is an *S*-gene-related gene in *Brassica* (Kandasamy et al. 1989). *Ipomoea secreted glycoprotein* genes (*ISG1, -2*, and *-3*) were isolated from a mature stigma cDNA library, and the sequences of these genes showed structural similarities to *Brassica* SLGs. However, these genes were not linked to the *S*-locus of *I. trifida*, and these cDNAs have been considered truncated derivatives or modified genes of membrane-anchored protein kinase genes that are expressed predominantly in various vegetative tissues of *I. trifida* (Kowyama et al. 1995a; Kakeda and Kowyama 1996), similar to the *SLR3* gene in *Brassica oleracea* (Cock et al. 1995). *Ipomoea receptor kinase 1* (*IRK1*) was isolated as a gene homologous to *SRK* in *Brassica*, and its predicted amino-acid sequence was more similar to that of *SRK6* (Stein et al. 1996) in *Brassica*, *ARK1* (Tobias et al. 1992) in *Arabidopsis* than to *ZmPK1* (Walker and Zhang 1990) in *Zea mays*. However, both the pattern of gene expression and the results of RFLP analysis indicate that the *IRK1* gene does not have any major involvement in the SI system of *Ipomoea* (Kowyama et al. 1996).

In initial studies of the gene products associated with SI, an *S*-locus glycoprotein (SLG) in *Brassica* (Nishio and Hinata 1977; Nasrallah et al. 1985) and an *S*-locus ribonuclease (*S*-RNase) in *Nicotiana* (Anderson et al. 1986) were identified as pistil proteins that co-segregate with *S*-haplotypes. SLG and *S*-RNase are major proteins that are expressed abundantly in the pistil and are detectable by protein electrophoresis of tissue extracts. Currently, the functional role of SLG is somewhat controversial, because some functional *S*-haplotypes of *Brassica rapa* lack the *SLG* gene (Suzuki et al. 2003). Because these pioneering studies stimulated molecular studies on SI in many plant species, a similar approach was attempted with *Ipomoea*. After two-dimensional polyacrylamide gel electrophoresis (2D-PAGE) of proteins extracted from mature stigmas of *I. trifida*, one highly polymorphic protein spot was detected (Kowyama et al. 2000). This protein is about 70 kDa, with a pI of 4–6. The protein spots were associated with *S*-haplotypes, so we designated them *S-locus-linked* stigma proteins (SSPs). The *SSP* gene encodes a short-chain alcohol dehydrogenase family protein, and this gene is expressed abundantly in mature papilla cells of the stigma. However, the amino-acid sequence of the SSPs from several *S*-haplotypes showed more than 95 % identity, and the *SSP* gene is located about 1.1 cM from the *S*-locus of *I. trifida* (Tomita et al. 2004a). Our surveys of 2D-PAGE profiles from stigma and pollen extracts have so far identified no *S-haplotype-specific* proteins other than SSP, which suggests that the *S*-locus gene products of *I. trifida* are minor proteins that might be present in amounts too small to be detectable by standard 2D-PAGE analysis.

25.5 Isolating and Analyzing the *S*-Locus in *I. trifida*

Fine-scale mapping of a gene locus is necessary to start positional cloning to identify a gene of interest. To obtain DNA markers, AFLP (amplified fragment length polymorphism: Vos et al. 1995) and AMF (AFLP-based mRNA fingerprinting: Money et al. 1996) methods are useful to identify molecular markers that are tightly linked to or co-segregate with a genetic trait (Agrama et al. 2002; Simoes-Araujo et al. 2002). These methods were attempted in *I. trifida* for identification of DNA markers around the *S*-locus with DNA and cDNA from 10 to 15 plants for each of the four *S* genotypes in the F_1 progeny from a single cross of $S_1S_{22} \times S_{10}S_{29}$. Based on the AFLP and AMF analyses, eight DNA markers were linked to the *S*-locus, and three were mapped closely to the *S*-locus (SAM-23, AAM-68, and AF-41; Fig. 25.6: Tomita et al. 2004a). The SAM-23 marker, derived from the stigma AMF analysis, contains a partial sequence of the *SSP* gene. The AAM-68 marker was obtained from the anther AMF analysis, and is tightly linked to the *S*-locus, with no recombinants among 873 F_1 plants. The AAM-68 marker is a partial sequence of a glycosyltransferase family member and is expressed in the anther and pollen; however, the predicted amino-acid sequences of this gene from different *S*-haplotypes exhibit high similarity. This finding suggests that the *AAM-68* gene is unlikely to be the *S*-gene. One AFLP marker, AF-41, was located on the opposite side of the *S*-locus to SAM-23 at an interval of 0.11 cM, and the sequence of AF-41 was similar to that of the histone deacetylase gene of *Arabidopsis*. Using these DNA markers as probes, BAC, cosmid and lambda phage clones covering the *S*-locus were screened from genomic libraries of the S_1 haplotype. The terminal sequences of the BAC clones 682-T7 and 681-SP6 map to 0.74 and 0.46 cM from the *S*-locus, respectively.

Up to now, no recombination between male and female determinant genes of SI has been observed in any plant species. If recombination between these two genes occurs, it leads to breakdown of SI. In the case of *I. trifida*, no recombination between the two components was observed among 873 F_1 plants, a result similar to that observed in *Brassica* (Casselman et al. 2000) and sweet cherry (Ikeda et al. 2005), so the *S*-locus may be located within a region where recombination is suppressed. Analysis of the DNA sequences amplified near the DNA markers among 873 F_1 siblings allowed identification and mapping of four recombination breakpoints (Fig. 25.6; Rahman et al. 2007a). According to DNA marker analysis, the *S*-locus of *I. trifida* is delimited to between 0.23 and 0.57 cM. The physical size of the *S*-locus may be estimated at about 212 kb in the S_1 haplotype (Kowyama et al. 2008).

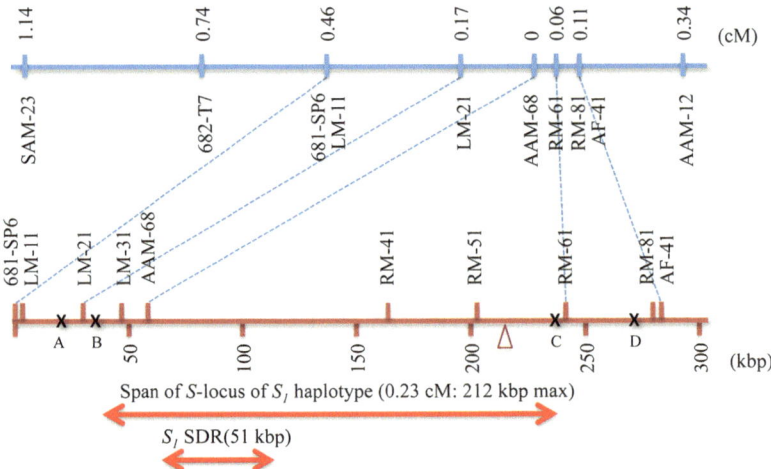

Fig. 25.6 Linkage map around the S-locus and location of recombination breakpoints in the S_1 haplotype. *Upper*: Linkage map of DNA markers derived from AFLP and AMF analyses and from PCR-amplified fragments. *Numbers* above the line indicate genetic distances (in cM) from the S-locus. *Lower*: Physical map of the DNA markers and positions of the recombination breakpoints. The *numbers* below the line indicate physical distance (in kbp). Breakpoints A–D are marked with Xs on the map. The physical span of the S-locus (about 212 kbp) was estimated from the distance between breakpoints B and C. The S-haplotype-specific divergent region (SDR) was estimated to be about 51 kbp on the basis of sequence comparison between the S_1 and S_{10} haplotypes. (Figure redrawn from Rahman et al. 2007a)

In the case of *Brassica campestris*, S-locus of the S_8 haplotype is 70 kb long (Casselman et al. 2000). Within this region, two recombination breakpoints were identified that are 0.3 cM apart. The relative amount of recombination on the chromosome is calculated by the ratio of DNA length per recombination unit (kb/cM). A ratio of about 920 kb/cM was calculated for the S-locus of the S_1 haplotype (Rahman et al. 2007a). This value is higher than in *B. campestris*, in which the ratio appears to be 233 kb/cM (Casselman et al. 2000), so recombination within the S-locus is highly suppressed in *I. trifida*. In the S-locus region of *Petunia inflata*, the ratio was calculated as 17.6 Mb/cM because of its centromeric localization (Wang et al. 2003). Fluorescence in situ hybridization analysis of metaphase chromosomes in *I. trifida* indicated that the S-locus region is localized at the distal end of a chromosome (Suzuki et al. 2004). This result supports the idea that suppression of recombination at the S-locus of *I. trifida* is not caused by the location of the locus. Prevention of recombination within the S-locus may be regulated because of the necessity of preventing outbreeding through SI. Therefore, the two genes that encode recognition molecules for SI have to be located within a delimited S-locus, SP11/SCR [S-locus protein 11/S-locus cysteine (Cys)-rich protein] and SRK (S-receptor kinase) for *Brassica* plants, and two unidentified genes for *I. trifida*. The suppression of recombination around the S-locus might contribute to the maintenance of the gene complex as a single genetic unit.

25.6   *S*-Locus of *I. trifida*

The structure of the *S*-locus has been analyzed in several plant species by genomic sequencing. These analyses provide evidence to determine male- and female-determinant genes at the *S*-locus that contribute to the self/non-self recognition in SI, and to polymorphism between *S*-haplotypes among these genes. *S*-determinant genes (*S*-genes) of the SSI system in *Brassica* species, *SP11/SCR* and *SRK*, show polymorphism between *S*-haplotypes (Sherman-Broyles and Nasrallah 2008). To determine the genomic sequence of the S-locus region of *I. trifida*, sequence contigs of about 300 kbp were constructed by map-based cloning using BAC, cosmid, and lambda libraries from S_1 homozygotes (Tomita et al. 2004b), and 68 kbp from S_{10} homozygotes (Rahman et al. 2007a). Comparison of the S_1 and S_{10} *S*-locus regions revealed high variability in the *S*-locus region, which was designated the *S-haplotype-specific* divergent region (SDR; Fig. 25.7). The length of the SDR is about 50 kbp in the S_{10} haplotype, 35 kbp in the S_1 haplotype, and about 95 kbp in the S_{29} haplotype. The flanking region of the SDR is highly conserved between *S*-haplotypes. In other self-incompatible plant species, a highly divergent region has been identified. In *Brassica* species, highly divergent regions from 30 to 56 kbp are present in the *S*-locus region, and this region is flanked by sequences with high similarity (Fukai et al. 2003; Shiba et al. 2003). This polymorphic region is also observed in *Prunus*

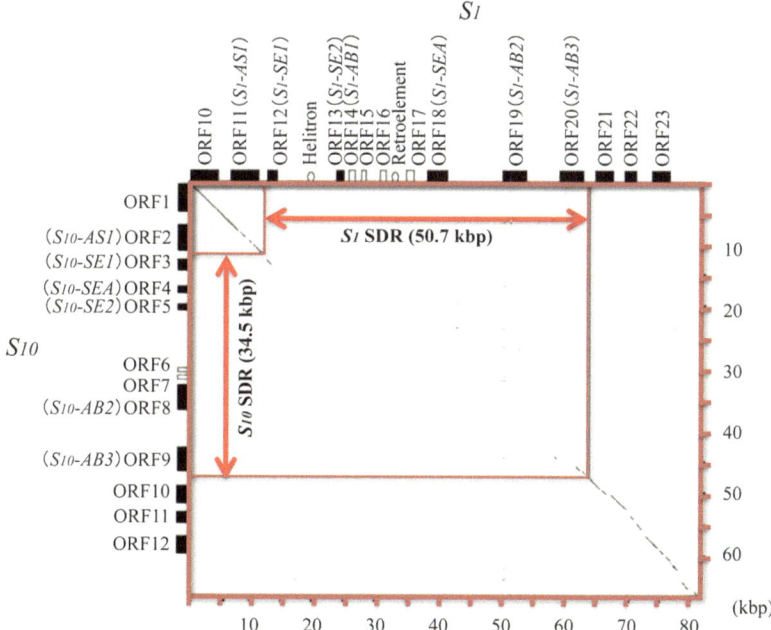

Fig. 25.7 Comparison of the nucleotide sequences spanning the *S* loci of the S_1 and S_{10} haplotypes by ARR plot analysis. Highly polymorphic regions are shown as SDRs. Estimated sizes of the SDRs are 50.7 and 34.5 kbp for the S_1 and S_{10} haplotypes, respectively. (Figure from Rahman et al. 2007a)

and *Malus* (Ushijima et al. 2001; Entani et al. 2003; Sassa et al. 2007). In these plants, male- and female-determinant genes involved in SI are located within these polymorphic regions in the *S*-locus region. Therefore, the *S*-determinant genes of *I. trifida* may be located in the SDR of the *S*-locus region.

The size differences in the SDR of *I. trifida* among *S*-haplotypes may be caused by the insertion of transposon-like sequences, retroelement-like sequences, and simple sequence repeats that accumulate in the larger SDRs (Rahman et al. 2007a). An interesting correlation between allelic dominance and SDR size has emerged from sequence comparisons of *S*-locus regions: the more dominant the *S*-haplotype, the larger the SDR. This finding suggests that acquisition of sequence complexity by recessive *S*-haplotypes is responsible for the differentiation of more dominant *S*-haplotypes.

25.7 Genes Located at or Near the *S*-Locus of *I. trifida*

Reproductive organ-specific genes at the *S*-locus may be directly correlated with self/non-self recognition in SI, one acting as male and the other as female determinant genes. In *S*-RNase-based GSI, *S*-RNases are abundantly expressed in the transmitting tract of the style, and they directly inhibit the growth of self-pollen tubes (Lee et al. 1994). The pollen determinant genes, *SFB/SLF* and related genes, are expressed in developing pollen grains and pollen tubes (Ushijima et al. 2003; Entani et al. 2003; Sijacic et al. 2004; Sassa et al. 2007, 2010; Meng et al. 2010; Kakui et al. 2011), and they recognize self or non-self *S*-RNases (Meng et al. 2010; Sassa et al. 2010). In *Papaver* plants, the female *S*-gene product, PrsS, is an extracellular signaling molecule that acts as a ligand to the self male *S*-gene product, PrpS, a transmembrane ion-channel/receptor. When self PrsS is recognized by self PrpS, Ca^{2+} flux is triggered and results in the depolymerization of actin fibers (Geitmann et al. 2000), activating several signaling pathways (Li et al. 2007) and inducing programmed cell death through activation of caspase-like activity in the growing pollen tube (Bosch and Franklin-Tong 2007). In the SSI system of *Brassica*, *SLG* and *SRK* genes are expressed in the mature papilla cells of the stigma (Nasrallah and Nasrallah 1993). The expression of *SRK* is significantly lower than *SLG*; however, *SRK* plays a key role in the self-pollen recognition of *Brassica* (Takasaki et al. 2000). The male *S*-gene in *Brassica*, *SP11/SCR*, is tightly linked to the *SRK* gene at the *S*-locus and is expressed in the anther tapetum (the sporophytic tissue) and microspores at a late developmental stage of pollen grains (Schopfer et al. 1999; Takayama et al. 2000; Shiba et al. 2002). According to previous experiments, the *S*-genes are specifically expressed in reproductive organs, and the expression pattern of these two components coincides with the characters and types of SI.

To identify the genes located in the *S*-locus region of *I. trifida*, Northern blot analyses were carried out with shotgun clones as probes that were used to determine the whole sequence of the S_1 *S*-locus (Fig. 25.8). More than ten genes were identified in the S_1 *S*-locus region; however, three stigma-specific (*SE1*, *SE2*, and *SEA*)

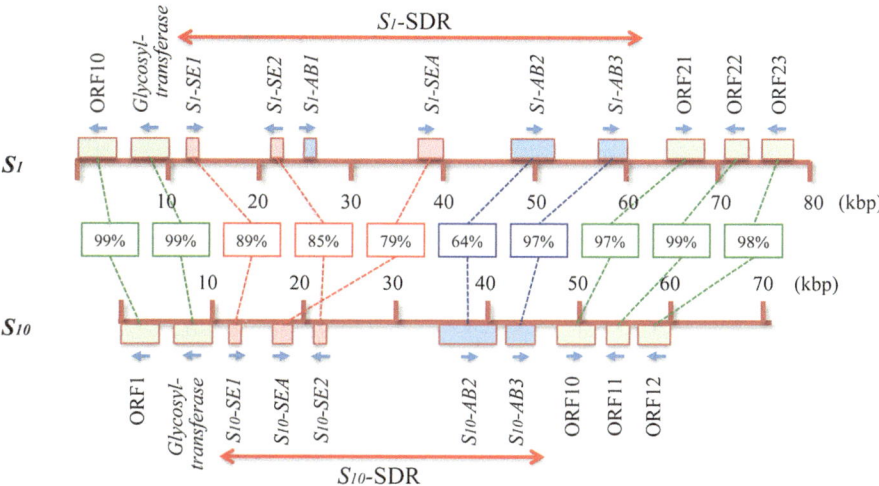

Fig. 25.8 Alignment of *S*-locus genes in the S_1 and S_{10} haplotypes showing locations of the SDRs in the S_1 and the S_{10} haplotype. *Red boxes* indicate stigma-specific genes (*SE1*, *SE2*, *SEA*) and *blue boxes* anther-specific genes (*AB1*, *AB2*, *AB3*) identified in the SDR. *Green boxes* indicate open reading frames (*ORF*s) outside the SDR. *Arrows* beside the boxes show orientation of transcription and *numbers* below each line show physical distances (in kbp). Genes or ORFs in common between the two *S*-haplotypes are connected by *dotted lines* with percentage nucleotide identity shown in *boxes*. (Figure redrawn from Rahman et al. 2007a)

and four anther-specific genes (*AB1*, *AB2*, *AB3*, and *AB4*) were located in the region (Rahman et al. 2007b). All three *SE* genes and three *AB* genes (*AB1–AB3*) were located in the SDR of the S_1 haplotype. Northern blot analyses using total RNA prepared from reproductive organs at several developmental stages and from vegetative organs indicated that all *AB* genes were expressed in the anthers 1–2 weeks before flowering. During this developmental period, pollen grains are at the microspore stage, and thus the tapetum of the anther is viable, not degraded. All *SE* genes were expressed in the stigma beginning 1 week before anthesis through the day before flowering. The transcripts of these six genes were not detected in other reproductive organs (*AB* genes were not expressed in the stigma, and *SE* genes were not expressed in the anther) or vegetative organs. According to Southern blot analysis, the *AB1* gene is present in at least two copies per genome; however, the other five genes are present as single copies. Therefore, these five genes (*AB2*, *AB3*, and three *SE* genes) are currently candidate *S*-genes based on their localization in the SDR, expression patterns, and copy number.

Among the anther-specific genes, S_1-*AB1* showed 95 % similarity to S_1-*AB3*; however, *AB1* is not located in the SDR of the S_{10}-haplotype. Furthermore, the sequence similarity of *AB1* and *AB3* genes was more than 95 % between *S*-haplotypes. From these results, *AB1* and *AB3* are not likely to be involved in the determination of *S*-haplotype specificity on the male side of *I. trifida*. On the other hand, *AB2* is located in the *S*-locus genomic region of all tested *S*-haplotypes, S_1, S_{10}, and S_{29}.

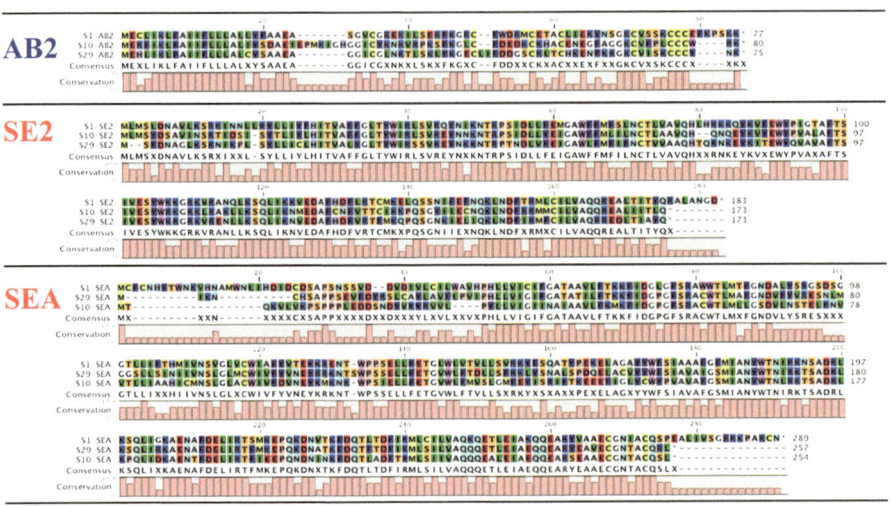

Fig. 25.9 Alignment of amino-acid sequences of *S*-candidates from S_1, S_{10}, and S_{29} alleles. Similar residues are indicated by the *same colors*

Sequence comparison of the predicted amino acid sequences of these genes showed 46–58 % identity between tested *S*-haplotypes (Fig. 25.9). According to temporal and spatial expression analyses, the *AB2* gene is only expressed in the tapetum of the anther at 14 to 7 days before anthesis, not in other reproductive organs or vegetative organs. A prediction program for protein structural features suggested that AB2 proteins have extracellular signaling domains at their N-termini that are highly conserved among *S*-haplotypes.

In addition, the predicted AB2 protein sequence shows homology to plant defensins, a class of Cys-rich proteins that are members of the gamma-thionin protein family. The defensins are small peptides (about 100 amino acids or less) with antimicrobial activities and are widely distributed in plants and animals (Boman 2003). Genes for defensin-like proteins such as PCP-A1 (Doughty et al. 1998) and SP11/SCR (Suzuki et al. 1999; Schopfer et al. 1999) are also expressed in the tapetum of developing *Brassica* anthers. Amino-acid sequence comparison of *Ipomoea* AB2 proteins with *Brassica* SP11/SCR proteins showed that only the eight Cys residues are conserved; however, sequences interlaid between the conserved Cys residues are not conserved in length or amino-acid residues (Fig. 25.10). Because the number of Cys residues is only conserved between AB2 and SP11/SCR, these two proteins may be structurally different. However, the possibility remains that this small AB2 protein acts as male *S*-determinant in the SI of *I. trifida*. If true, the AB2 protein may act as a ligand for the female *S*-determinant, which may be a receptor-like protein, as in *Brassica*.

cDNA clones of the stigma-specific genes (the *SE* genes), also showed a high level of allelic polymorphism among *S*-haplotypes and were located on the SDR of the *S*-locus. The predicted amino-acid sequences of the SE1, SE2, and SEA proteins

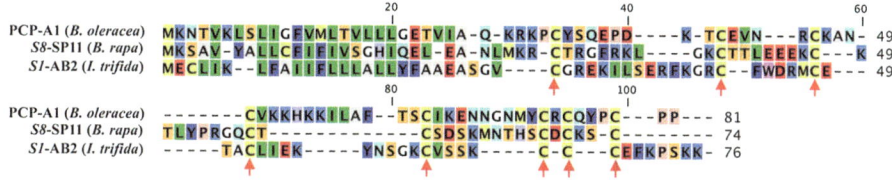

Fig. 25.10 Similarity between AB2 protein sequence and that of other PCP-A1 family proteins. Protein sequences of PCP-A1 from *Brassica oleracea* and S_8-SP11/SCR from *B. rapa*, and the predicted protein sequence of S_1-AB2 from *I. trifida*, were aligned. Conserved Cys residues in mature peptides of all three proteins are indicated with *arrows*

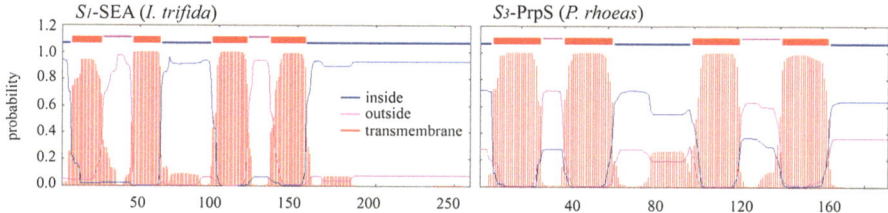

Fig. 25.11 Predicted structures of S_1-SEA protein of *I. trifida* and S_3-PrpS protein of *Papaver rhoeas*. Female *S*-candidate proteins, SE2 and SEA, of *I. trifida*, and male *S*-protein, PrpS, of *P. rhoeas* are predicted as membrane-anchored proteins with three or four membrane-spanning domains by the TMHMM program (http://www.cbs.dtu.dk/services/TMHMM/). However, these protein sequences showed no significant similarity to other reported proteins

are, respectively, 53–76 %, 67–69 %, and 52–62 % identical among the tested *S*-haplotypes (Fig. 25.9).

Expression analyses indicated that *SE1*, *SE2*, and *SEA* are expressed in the papilla cells of the stigma at 7 to 1 days before anthesis. The predicted amino-acid sequences of all *SE* genes showed no similarity to known proteins in the database. However, hydropathy plot analysis indicated that these proteins may share three to four membrane-spanning domains, and may localize to the plasma membrane of papilla cells, where the interaction between pollen and stigma cells occurs. The structural features of SE proteins are similar to the predicted structure of PrpS proteins, the female determinant of SI of *Papaver* (Wheeler et al. 2009), and the flower protein of *Drosophila* (Yao et al. 2009; Brose and Neher 2009), which may act as a Ca^{2+} channel in synaptic endocytosis (Fig. 25.11). In the GSI of *Papaver* plants, Ca^{2+} influx into self-pollen tubes is observed, and may regulated by PrpS. If the SE proteins play the same role in the SI of *I. trifida*, some ion, such as Ca^{2+}, may induce signal transduction to inhibit germination of self-pollen on the stigma surface, as in *Papaver* plants.

Taken together, the available data suggest that the *SE1*, *SE2*, and *SEA* genes that are expressed in the papilla cells of the stigma are candidates for the female *S*-determinant genes, and the *AB2* gene that is expressed in the tapetum of the anther is a strong candidate for the male *S*-determinant gene in the SI of *I. trifida*. To determine the true *S*-genes of *I. trifida*, both functional analyses of these genes by creating transgenic plants expressing these genes and molecular–molecular interaction analyses of the gene products may be necessary to obtain definitive evidence. Further study is also necessary to determine an outline of the self-incompatible reaction in *I. trifida*.

25.8 Conclusion

SI of angiosperms is not regulated by a single and simple genetic mechanism. In the GSI system of the Plantaginaceae, Solanaceae, and Rosaceae, *S-RNase* has been identified as a female determinant and *SFB/SLF* and related genes as male determinants of recognition specificity. In this system, RNA and protein degradation are involved during pollen tube growth in the style (McClure and Franklin-Tong 2006). In another gametophytic system in the Papavaraceae, PrsS acts as a ligand from the stigma and PrpS as a receptor or channel molecule on the plasma membrane of the pollen tube, and their interaction induces a Ca^{2+}-mediated signaling cascade to induce programmed cell death and inhibit pollen tube growth (Franklin-Tong and Franklin 2003; Thomas and Franklin-Tong 2004; Wheeler et al. 2010). On the other hand, in SSI of the Brassicaceae, SRK acts as receptor kinase on the papilla cells of stigma and SP11/SCR as its cognate ligand on the pollen grain, resulting in self-pollen tube germination through activation of a phosphorylation pathway of downstream signaling molecules (Kachroo et al. 2002; Takayama and Isogai 2005). In the other plant species with SSI, such as the Convolvulaceae and Asteraceae, SRK-mediated self-recognition has been postulated (Hiscock and McInnis 2003). However, at least in *I. trifida* of the Convolvulaceae, the SRK-mediated signaling pathway is not recruited.

Interestingly, in the SI system in *I. trifida*, a candidate molecule for the male determinant, AB2, is similar to SP11/SCR protein, the male *S*-determinant molecule of *Brassica* plants with SSI; however, candidate molecules for the female determinant, SE1, SE2, and SEA, resemble PrpS protein, the male *S*-determinant of *Papaver* plants with GSI. This finding supports the hypothesis that SI systems may have arisen independently in the evolution of angiosperms (Allen and Hiscock 2008). Our research with *I. trifida* may provide important information about the evolution of the system of SI and may contribute to the breeding of crop plants in the Convolvulaceae, including cultivated sweet potato. The classification of SI into two types, namely GSI and SSI, is based simply on differences in gene expression of pollen determinants, not on the molecular mechanisms that underlie the SI systems in these plant families. Investigation of a wide range of plant taxa is required to understand the evolutionary lineage of the SI systems in flowering plants.

Acknowledgments Our research was supported by a Grant-in-Aid for Scientific Research on Innovative Areas (No. 22112513) and Grants-in-Aid for Scientific Research C (Nos. 18580004, 20580004 and 23580005) from the Ministry of Education, Culture, Sports, Science and Technology, Japan.

References

Agrama HA, Houssin SF, Tarek MA (2002) Cloning of AFLP markers linked to resistance to *Peronosclerospora sorghi* in maize. Mol Genet Genomics 267:814–819

Allen AM, Hiscock SJ (2008) Evolution and phylogeny of self-incompatibility systems in angio-sperms. In: Franklin-Tong VE (ed) Self-incompatibility in flowering plants: evolution, diversity and mechanisms. Springer, Berlin/Heidelberg, pp 73–101

Anderson MA, Cornish EC, Mau S-L, Williams EG, Hoggart R, Atkinson A, Bonig I, Grego B, Simpson R, Roche PJ, Haley JD, Penschow JD, Niall HD, Tregear GW, Coughlan JP, Crawford RJ, Clarke AE (1986) Cloning of cDNA for a stylar glycoprotein associated with expression of self-incompatibility in *Nicotiana alata*. Nature (Lond) 321:38–44

Arumuganathan K, Earle ED (1991) Nuclear DNA content of some important plant species. Plant Mol Biol Rep 9:208–218

Austin DF (1988) The taxonomy, evolution and genetic diversity of sweetpotatoes and related wild species. In: Gregory P (ed) Exploration, maintenance, and utilization of sweetpotato genetic resources. CIP, Lima, pp 27–60

Austin DF (1997) Convolvulaceae (morning glory family). http://www.fau/divdept/biology/people/convolv.htm

Austin DF, Huáman Z (1996) A synopsis of *Ipomoea* (Convolvulaceae) in the Americas. Taxon 45:3–38

Bassett KN, Gordon HW, Nobes DC, Jacomb C (2004) Gardening at the edge: documenting the limits of tropical Polynesian kumara horticulture in Southern New Zealand. Geoarchaeology 19:185–218

Bateman AJ (1956) Cryptic self-incompatibility in the wallflower: *Cheiranthus cheiri* L. Heredity 10(2):257–261

Boman HG (2003) Antibacterial peptides: basic facts and emerging concepts. J Intern Med 254:197–215

Bosch M, Franklin-Tong VE (2007) Temporal and spatial activation of caspase-like enzymes induced by self-incompatibility in *Papaver* pollen. Proc Natl Acad Sci USA 104: 18327–18332

Brose N, Neher E (2009) Flowers for synaptic endocytosis. Cell 138:836–837

Casselman AL, Vrebalov J, Conner JA, Singhal A, Giovannoni J, Nasrallah ME, Nasrallah JB (2000) Determining the physical limits of *Brassica S*-locus by recombinational analysis. Plant Cell 12:23–33

Cock JM, Stanchev B, Delorme V, Croy RRD, Dumas C (1995) *SLR3*: a modified receptor kinase gene that has been adapted to encode a putative secreted glycoprotein similar to the S locus glycoprotein. Mol Gen Genet 248:151–161

Darwin CR (1876) The effects of cross- and self-fertilisation in the vegetable kingdom. Murray, London

de Nettancourt D (2001) Incompatibility and incongruity in wild and cultivated plants. Springer-Verlag Berlin Heidelberg New York

Doughty J, Dixon S, Hiscock SJ, Willis AC, Parkin IAP, Dickinson HG (1998) PCP-A1, a defensin-like *Brassica* pollen coat protein that binds the *S*-locus glycoprotein, is the product of gameto-phytic gene expression. Plant Cell 10:1333–1347

Entani T, Iwano M, Shiba H, Che FS, Isogai A, Takayama S (2003) Comparative analysis of the self-incompatibility *S*-locus region of *Prunus mume*: identification of a pollen-expressed F-box gene with allelic diversity. Genes Cells 8:203–213

Franklin-Tong VE, Franklin FCH (2003) Gametophytic self-incompatibility inhibits pollen tube growth using different mechanisms. Trends Plant Sci 8:598–605

Fukai E, Fujimoto R, Nishio T (2003) Genomic organization of the *S* core region and the *S* flanking regions of a class-II *S*-haplotype in *Brassica rapa*. Mol Genet Genomics 269:361–369

Geitmann A, Snowman BN, Emons AMC, Franklin-Tong VE (2000) Alterations in the actin cytoskeleton of pollen tubes are induced by the self-incompatibility reaction in *Papaver rhoeas*. Plant Cell 12:1239–1251

Glaettli M (2004) Mechanisms involved in the maintenance of inbreeding depression in gynodioecious *Silene vulgaris* (Caryophyllaceae): an experimental investigation. PhD dissertation, University of Lausanne

Hinata K, Isogai A, Isuzugawa K (1994) Manipulation of sporophytic self-incompatibility in plant breeding. In: Williams EG, Clarke AE, Knox RB (eds) Genetic control of self-incompatibility and reproductive development in flowering plants. Kluwer, Dordrecht, pp 102–115

Hiscock SJ, McInnis S (2003) Pollen recognition and rejection during the sporophytic self-incompatibility response: *Brassica* and beyond. Trends Plant Sci 8:606–613

Huang JC, Sun M (2000) Genetic diversity and relationships of sweet potato and its wild relatives in *Ipomoea* series Batatas (Convolvulaceae) as revealed by inter-simple sequence repeat (ISSR) and restriction analysis of chloroplast DNA. Theor Appl Genet 100:1050–1060

Ikeda K, Ushijima K, Yamane H, Tao R, Hauck NR, Sebolt AM, Iezzoni AF (2005) Linkage and physical distances between the *S*-haplotype S-RNase and SFB genes in sweet cherry. Sex Plant Reprod 17:289–296

Kachroo A, Nasrallah ME, Nasrallah JB (2002) Self-incompatibility in the Brassicaceae: receptor–ligand signaling and cell-to-cell communication. Plant Cell 14(suppl):s227–s238

Kakeda K, Kowyama Y (1996) Sequences of *Ipomoea trifida* cDNAs related to the *Brassica* *S*-locus genes. Sex Plant Reprod 9:309–310

Kakeda K, Tsukada H, Kowyama Y (2000) A self-compatible mutant *S* allele conferring a dominant negative effect on the functional *S* allele in *Ipomoea trifida*. Sex Plant Reprod 13:119–125

Kakui H, Kato M, Ushijima K, Kitaguchi M, Kato S, Sassa H (2011) Sequence divergence and loss-of-function phenotypes of *S* locus F-box brothers genes are consistent with non-self recognition by multiple pollen determinants in self-incompatibility of Japanese pear (*Pyrus pyrifolia*). Plant J 68:1028–1038

Kandasamy MK, Paolillo DJ, Faraday CD, Nasrallah JB, Nasrallah ME (1989) The *S*-locus specific glycoproteins of *Brassica* accumulate in the cell wall of developing stigma papillae. Dev Biol 134:462–472

Kowyama Y, Shimano N, Kawase T (1980) Genetic analysis of incompatibility in the diploid *Ipomoea* species closely related to the sweet potato. Theor Appl Genet 58:149–155

Kowyama Y, Takahashi H, Muraoka K, Tani T, Hara K, Shiotani I (1994) Number, frequency and dominance relationships of *S*-alleles in diploid *Ipomoea trifida*. Heredity 73:275–283

Kowyama Y, Kakeda K, Nakano R, Hattori T (1995a) *SLG/SRK*-like genes are expressed in the reproductive tissues of *Ipomoea trifida*. Sex Plant Reprod 8:333–338

Kowyama Y, Morikami A, Furusawa T, Hattori T, Nakamura K (1995b) Molecular characterization of a reproductive organ-specific cDNA clone, *ISP11* from *Ipomoea trifida*. Breed Sci 45:497–501

Kowyama Y, Kakeda K, Kondo K, Imada T, Hattori T (1996) A putative receptor protein kinase gene in *Ipomoea trifida*. Plant Cell Physiol 37(5):681–685

Kowyama Y, Tsuchiya T, Kakeda K (2000) Sporophytic self-incompatibility in *Ipomoea trifida*, a close relative of sweet potato. Ann Bot 85(Suppl A):191–196

Kowyama Y, Tsuchiya T, Kakeda K (2008) Molecular genetics of sporophytic self-incompatibility in *Ipomoea*, a member of the Convolvulaceae. In: Franklin-Tong VE (ed) Self-incompatibility in flowering plants: evolution, diversity and mechanisms. Springer, Berlin/Heidelberg, pp 259–274

Lee H-S, Huang S, Kao T-H (1994) *S* proteins control rejection of incompatible pollen in *Petunia inflata*. Nature (Lond) 367:560–563

Li S, Samaj J, Franklin-Tong VE (2007) A mitogen-activated protein kinase signals to programmed cell death induced by self-incompatibility in *Papaver* pollen. Plant Physiol 145:236–245

McClure BA, Franklin-Tong V (2006) Gametophytic self-incompatibility: understanding the cellular mechanisms involved in "self" pollen tube inhibition. Planta (Berl) 224:233–245

McDonald JA, Mabry TJ (1992) Phylogenetic systematics of New World *Ipomoea* (Convolvulaceae) based on chloroplast DNA restriction site variation. Plant Syst Evol 180:243–259

Meng X, Sun P, Kao TH (2010) *S*-RNase-based self-incompatibility in *Petunia inflata*. Ann Bot 108:637–646

Money T, Reader R, Qu LJ, Dunford RP, Moore G (1996) AFLP-based mRNA fingerprinting. Nucleic Acids Res 24(13):2616–2617

Nasrallah JB, Nasrallah ME (1993) Pollen-stigma signaling in the sporophytic self-incompatibility response. Plant Cell 5:1325–1335

Nasrallah JB, Kao T-H, Goldberg ML, Nasrallah ME (1985) A cDNA clone encoding an *S*-locus specific glycoprotein from *Brassica oleracea*. Nature (Lond) 318:263–267

Nishio T, Hinata K (1977) Analysis of *S*-specific proteins in stigma of *Brassica oleracea* L. by isoelectric focusing. Heredity 38:391–396

Nishiyama I, Miyazaki T, Sakamoto S (1975) Evolutionary autoploidy in the sweet potato (*Ipomoea batatas* (L.) Lam.) and its progenitors. Euphytica 24:197–208

Ozias-Akins P, Jarret RL (1994) Nuclear DNA content and ploidy levels in the genus *Ipomoea*. J Am Soc Hortic Sci 119:110–115

Rahman MH, Tsuchiya T, Suwabe K, Kohori J, Tomita RN, Kagaya Y, Kobayashi I, Kakeda K, Kowyama Y (2007a) Physical size of the *S*-locus region defined by genetic recombination and genome sequencing in *Ipomoea trifida*, Convolvulaceae. Sex Plant Reprod 20:63–72

Rahman MH, Uchiyama M, Kuno M, Hirashima N, Suwabe K, Tsuchiya T, Kagaya Y, Kobayashi I, Kakeda K, Kowyama Y (2007b) Expression of stigma- and anther-specific genes located in the *S*-locus region of *Ipomoea trifida*. Sex Plant Reprod 20:73–85

Rajapakse S, Nilmalgoda SD, Molnar M, Ballard RE, Austin DF, Bohac JR (2004) Phylogenetic relationships of the sweet potato in *Ipomoea* series Batatas (Convolvulaceae) based on nuclear β-amylase gene sequences. Mol Phylogenet Evol 30:623–632

Royo J, Kunz C, Kowyama Y, Anderson MA, Clarke AE, Newbigin E (1994) Loss of histidine residue at the active site of *S*-ribonuclease leads to self-compatibility in *Lycopersicon peruvianum*. Proc Natl Acad Sci USA 91:6511–6514

Sage TL, Bertin RI, Williams EG (1994) Ovarian and other late-acting self-incompatibility systems. In: Williams EG, Knox RB, Clarke AE (eds) Genetic control of self-incompatibility and reproductive development in flowering plants. Kluwer, Amsterdam, pp 116–140

Sassa H, Hirano H, Nishio T, Koba T (1997) Style-specific self-compatible mutation caused by deletion of the *S-RNase* gene in Japanese pear (*Pyrus serotina*). Plant J 12:223–227

Sassa H, Kakui H, Miyamoto M, Suzuki Y, Hanada T, Ushijima K, Kusaba M, Hirano H, Koba T (2007) *S*-locus F-box brothers; multiple and pollen-specific F box genes with *S-haplotype-specific* polymorphisms in apple and Japanese pear. Genetics 175:1869–1881

Sassa H, Kakui H, Minamikawa M (2010) Pollen-expressed F-box gene family and mechanism of *S*-RNase-based gametophytic self-incompatibility (GSI) in Rosaceae. Sex Plant Reprod 23:39–43

Schopfer CR, Nasrallah ME, Nasrallah JB (1999) The male determinant of self-incompatibility in *Brassica*. Science 286:1697–1700

Seavey SF, Bawa KS (1986) Late-acting self-incompatibility in angiosperms. Bot Rev 52(2): 195–218

Sherman-Broyles S, Nasrallah JB (2008) Self-incompatibility and evolution of mating systems in the Brassicaceae. In: Franklin-Tong VE (ed) Self-incompatibility in flowering plants: evolution, diversity and mechanisms. Springer, Berlin/Heidelberg, pp 123–147

Shiba H, Iwano M, Entani T, Ishimoto K, Shimosato H, Che F-S, Satta Y, Ito A, Takada Y, Watanabe M, Isogai A, Takayama S (2002) The dominance of alleles controlling self-incompatibility in *Brassica* pollen is regulated at the RNA level. Plant Cell 14:491–504

Shiba H, Kenmochi M, Sugihara M, Iwano M, Kawasaki S, Suzuki G, Watanabe M, Isogai A, Takayama S (2003) Genomic organization of the S-locus region of *Brassica*. Biosci Biotechnol Biochem 67:622–626

Shiotani I, Kawase T (1987) Synthetic hexaploids derived from wild species related to sweet potato. Jpn J Breed 37:367–376

Shiotani I, Kawase T (1989) Genetic structure of the sweet potato and hexaploids in *Ipomoea trifida* (H.B.K.) Don. Jpn J Breed 39:57–66

Sijacic P, Wang X, Skirpan AL, Wang Y, Dowd PE, McCubbin AG, Huang S, Kao T-H (2004) Identification of the pollen determinant of S-RNase-mediated self-incompatibility. Nature (Lond) 429:302–305

Simoes-Araujo JL, Rodrigues RL, de A Gerhardt LB, Mondego JM, Alves-Ferreira M, Rumjanek NG, Margis-Pinheiro (2002) Identification of differentially expressed genes by cDNA-AFLP technique during heat stress in cowpea nodules. FEBS Lett 515:44–50

Stein JC, Dixit R, Nasrallah ME, Nasrallah JB (1996) SRK, the stigma-specific S locus receptor kinase of *Brassica*, is targeted to the plasma membrane in transgenic tobacco. Plant Cell 8:429–445

Stein JC, Howlett B, Boyes DC, Nasrallah ME, Nasrallah JB (1991) Molecular cloning of a putative receptor protein kinase gene encoded at the self-incompatibility locus of *Brassica oleracea*. Proc Natl Acad Sci USA 88:8816–8820

Suzuki G, Kai K, Hirose T, Fukui K, Nishio T, Takayama S, Isogai A, Watanabe M, Hinata K (1999) Genomic organization of the S-locus: identification and characterization of genes in *SLG/SRK* region of S_9 haplotype of *Brassica campestris* (syn. *rapa*). Genetics 153:391–400

Suzuki G, Kakizaki T, Takada Y, Shiba H, Takayama S, Isogai A, Watanabe M (2003) The S-haplotypes lacking *SLG* in the genome of *Brassica rapa*. Plant Cell Rep 21:911–915

Suzuki G, Tanaka S, Yamamoto M, Tomita RN, Kowyama Y, Mukai Y (2004) Visualization of the S-locus region in *Ipomoea trifida*: toward positional cloning of self-incompatibility genes. Chromosome Res 12:475–481

Takasaki T, Hatakeyama K, Suzuki G, Watanabe M, Isogai A, Hinata K (2000) The S receptor kinase determines self-incompatibility in *Brassica* stigma. Nature (Lond) 403:913–916

Takayama S, Isogai A (2005) Self-incompatibility in plants. Annu Rev Plant Biol 56:467–489

Takayama S, Shiba H, Iwano M, Shimosato H, Che FS, Kai N, Watanabe M, Suzuki G, Hinata K, Isogai A (2000) The pollen determinant of self-incompatibility in *Brassica campestris*. Proc Natl Acad Sci USA 97:1920–1925

Thomas SG, Franklin-Tong VE (2004) Self-incompatibility triggers programmed cell death in *Papaver* pollen. Nature (Lond) 429:305–309

Tobias CM, Howlett B, Nasrallah JB (1992) An *Arabidopsis thaliana* gene with sequence similarity to the S-locus receptor kinase of *Brassica oleracea*. Plant Physiol 99:284–290

Tomita RN, Fukami K, Takayama S, Kowyama Y (2004a) Genetic mapping of AFLP/AMF-derived DNA markers in the vicinity of the self incompatibility locus in *Ipomoea trifida*. Sex Plant Reprod 16:265–272

Tomita RN, Suzuki G, Yoshida K, Yano Y, Tsuchiya T, Kakeda K, Mukai Y, Kowyama Y (2004b) Molecular characterization of a 313-kb genomic region containing the self-incompatibility locus of *Ipomoea trifida*, a diploid relatives of sweet potato. Breed Sci 54:165–175

Ushijima K, Sassa H, Tamura M, Kusaba M, Tao R, Gradziel TM, Dandekar AM, Hirano H (2001) Characterization of the S-locus region of almond (*Prunus dulcis*): analysis of a somaclonal mutant and a cosmid contig for an S-haplotype. Genetics 158:379–386

Ushijima K, Sassa H, Dandekar AM, Gradziel TM, Tao R, Hirano H (2003) Structural and transcriptional analysis of the self-incompatibility locus of almond: identification of a pollen-expressed F-box gene with haplotype-specific polymorphism. Plant Cell 15:771–781

van Tilburg JA (1994) Easter Island: archaeology, ecology and culture. Smithsonian Institution Press, Washington, DC

Vos P, Hogers R, Bleeker M, Reijans M, van de Lee T, Hornes M, Friters A, Pot J, Paleman J, Kuiper M, Zabeau M (1995) AFLP: a new technique for DNA fingerprinting. Nucleic Acids Res 23(21):4407–4414

Walker JC, Zhang R (1990) Relationship of a putative receptor protein kinase from maize to the *S*-locus glycoproteins of *Brassica*. Nature (Lond) 345:743–746

Wang Y, Wang X, McCubbin AG, Kao T-H (2003) Genetic mapping and molecular characterization of the self-incompatibility *S*-locus in *Petunia inflata*. Plant Mol Biol 53:565–580

Watanabe M, Ono T, Hatakeyama K, Takayama S, Isogai A, Hinata K (1997) Molecular characterization of *SLG* and *S*-related genes in a self-compatible *Brassica campestris* L. var. yellow sarson. Sex Plant Reprod 10:332–340

Wheeler MJ, de Graaf BH, Hadjiosif N, Perry RM, Poulter NS, Osman K, Vatovec S, Harper A, Franklin FC, Franklin-Tong VE (2009) Identification of the pollen self-incompatibility determinant in *Papaver rhoeas*. Nature (Lond) 459:992–995

Wheeler MJ, Vatovec S, Franklin-Tong VE (2010) The pollen *S*-determinant in *Papaver*: comparisons with known plant receptors and protein ligand partners. J Exp Bot 6:2015–2025

Yao CK, Lin YQ, Ly CV, Ohyama T, Haueter CM, Moiseenkova-Bell VY, Wensel TG, Bellen HJ (2009) A synaptic vesicle-associated Ca^{2+} channel promotes endocytosis and couples exocytosis to endocytosis. Cell 138:947–960

Zhang DP, Ghislain M, Huamán Z, Golmirzaie A, Hijmans RJ (1998) RAPD variation in sweet potato [*Ipomoea batatas* (L.) Lam] cultivars from South America and Papua New Guinea. Genet Res Crop Evol 45:271–277

Part IV
Male–Female Interaction and Gamete Fusion

Chapter 26
Profiling the GCS1-Based Gamete Fusion Mechanism

Toshiyuki Mori

Abstract Angiosperm double fertilization is composed of two sets of gamete fusion, in which two sperm cells released from a pollen tube fuse with egg and central cells to produce an embryo and an endosperm, respectively. GCS1 was, for the first time, identified as a sperm membrane protein essential for plant gamete fusion, and this indicates that plant fertilization is regulated by molecular systems on the gamete surface. Interestingly, GCS1 is conserved in various phylogenetically distant organisms, such as algae, unicellular parasites, slime molds, and some invertebrate species, and *Plasmodium* and *Chlamydomonas* GCS1 orthologues have proved to similarly function in their gamete fusion as a male factor. This finding indicates that GCS1 is an ancient fertilization regulator; thus, elucidating GCS1-based gamete fusion mechanism should lead to understanding of conserved eukaryotic fertilization system. Because GCS1 is a novel transmembrane protein in which no well-known functional domains are present, the molecular mechanics of GCS1-based fertilization remain elusive.

Keywords Gamete fusion • GCS1 • Male gamete • Transmembrane protein

26.1 Introduction

In a previous study using male gametic cells (generative cells) isolated from the trumpet lily (*Lilium longiflorum*), GENERATIVE CELL SPECIFIC 1 (GCS1), also called HAP2, was identified as a critical factor for flowering plant gamete fusion (Mori et al. 2006). GCS1 is a novel transmembrane protein composed of

T. Mori (✉)
Waseda Institute for Advanced Study, Waseda University,
1-6-1, Nishiwaseda, Shinjuku, Tokyo 169-8050, Japan
e-mail: moritoshi@aoni.waseda.jp

H. Sawada et al. (eds.), *Sexual Reproduction in Animals and Plants*,
DOI 10.1007/978-4-431-54589-7_26, © The Author(s) 2014

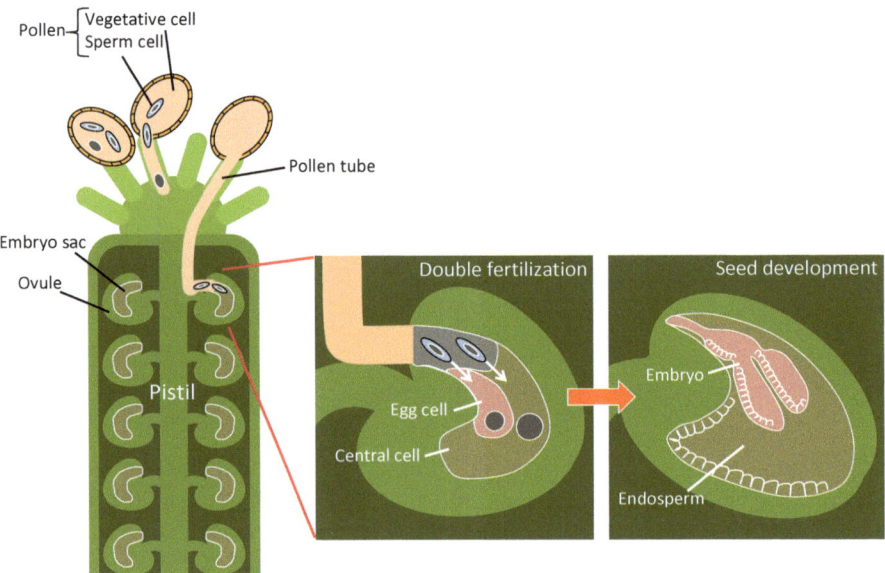

Fig. 26.1 Double fertilization of flowering plants. The pollen grain contains a pair of sperm cells. After the journey through the pollen tube, those sperm cells fuse with the egg and central cells in the ovule. The fertilized egg and central cells develop into an embryo and an endosperm, respectively

approximately 700–1,000 amino-acid residues, and it possesses a single transmembrane domain (Mori et al. 2006). In flowering plants (*L. longiflorum* and *Arabidopsis thaliana*), functional *GCS1* expression is specific to male gametes and GCS1 protein is detected exclusively in male gametes (generative cells and sperm cells) suspended in a pollen grain (Mori et al. 2006; von Besser et al. 2006). The typical fertilization of flowering plants involves two sets of gamete fusion and is performed by a pair of sperm cells delivered by pollen tube elongation. When the sperm cells are released into an embryo sac enclosed in an ovule, one of them fuses with an egg cell, and the other fuses with a central cell to produce an embryo and an endosperm, respectively (Fig. 26.1). This fertilization type, which is specific to angiosperm species, is known as "double fertilization." Reverse-genetic approaches using *Arabidopsis GCS1* mutants clearly demonstrated that no *GCS1*-knockout sperm cells are able to fuse with female gametes (egg and central cells) (Mori et al. 2006). This finding became the first with respect to the plant gamete fusion factor. Interestingly, GCS1 orthologues have been detected in various organisms from unicellular protists (e.g., green and red algae, slime molds, unicellular parasites) to some animal species (e.g., arthropods, cnidarians) in addition to plants (Mori et al. 2006; von Besser et al. 2006; Liu et al. 2008; Hirai et al. 2008; Steele and Dana 2009). Of the GCS1-possessing protists, the malaria parasite (*Plasmodium berghei*) and *Chlamydomonas reinhardtii* show functional GCS1 expression specific to a particular sex gamete corresponding to the male, similar to flowering plants (Liu et al. 2008; Hirai et al. 2008). Furthermore, *GCS1*-knockout lines of those organisms show serious sterility, resulting in unsuccessful zygote formation (Liu et al. 2008; Hirai et al. 2008). Because the *GCS1-knockout* male gametes in both *Plasmodium* and *Chlamydomonas* succeed in gamete attachment, the GCS1 molecule may function in gamete membrane fusion itself or in processes close to the membrane fusion step following gamete attachment. In addition, a recent study has begun to analyze animal GCS1 function and showed that *Hydra magnipapillata GCS1* is specifically expressed in its testis (Steele and Dana 2009). These findings strongly suggest that GCS1 is a highly conserved gamete fusion factor and characterizes male fertility in plants, animals, and protists.

26.2 GCS1 Functions at the Gamete Attachment Site

The expression pattern of GCS1 appears to vary by species. In *Chlamydomonas*, GCS1 expression is exclusive to the mt^- (male)-side mating structure (mating protrusion), in which gamete membrane fusion occurs, and this is consistent with the fact that mt^+ (female)-specific gamete fusion factor FUS1 is expressed specifically in mt^+-side mating structures (Liu et al. 2008) (Fig. 26.2). In contrast to *Chlamydomonas*, GCS1 molecules are uniformly distributed in male gametes of plants and the malaria parasite (Mori et al. 2006; von Besser et al. 2006; Liu et al. 2008; Hirai et al. 2008). The *Plasmodium* male gamete enters the female gamete until initiating fusion (Sinden and Croll 1975), and therefore gamete fusion via GCS1 function may occur on the entire male gamete surface after the male gamete is engulfed by the female gamete (Fig. 26.2). Interestingly, a recent study showed detailed changes in the distribution of *Arabidopsis* GCS1 in sperm cells (Sprunck et al. 2012). Some studies showed that *Arabidopsis* GCS1 is mainly detected in the internal membrane components surrounding the sperm nucleus, although slight GCS1 localization is also observed on the sperm surface (von Besser et al. 2006; Igawa et al. 2013). Sprunck et al. (2012) identified a novel egg cell-specific secretory protein, named EGG CELL 1 (EC1), and showed that EC1 family peptides secreted into the gamete fusion space of the embryo sac induce a shift of GCS1 distribution toward the sperm surface membrane, likely to ensure that the GCS1 molecules directly contact the female gametes (Fig. 26.2). As EC1 is specific to plant species, this EC1-based sperm–egg crosstalk system may have been developed for correct double-fertilization progression. At the same time, this finding strongly supports GCS1 as a vital player for gamete fusion at the membrane attachment site in plant species.

26.3 Characterization of the GCS1 Molecule

Since the identification of GCS1, researchers have tried to elucidate the functional mechanism of GCS1 because it is a novel protein with no characterized domains (Mori et al. 2006). Recently, a pair of reports on *Arabidopsis* GCS1 structures were

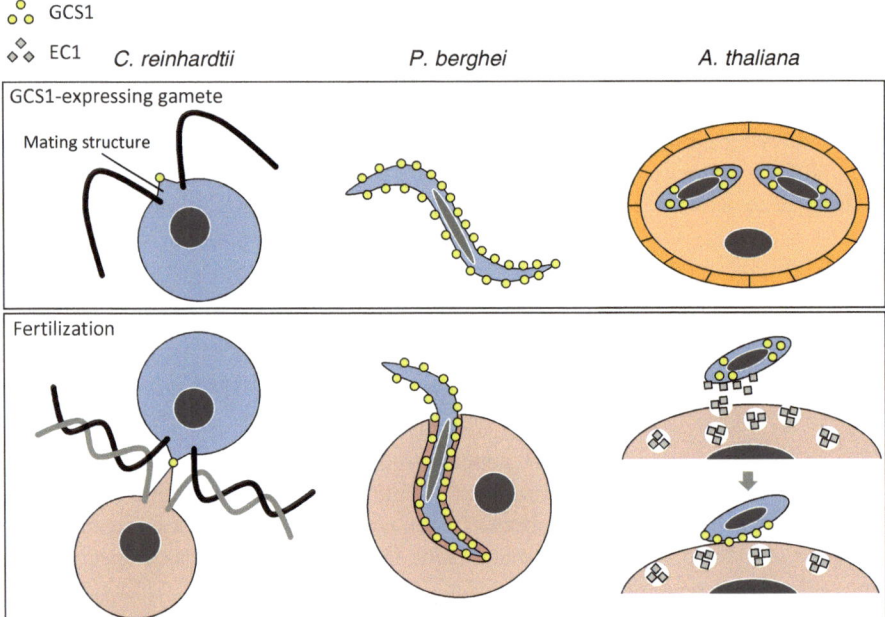

Fig. 26.2 GENERATIVE CELL SPECIFIC 1 (GCS1)-possessing organisms and their gamete fusion method based on GCS1. In *C. reinhardtii*, GCS1 is localized specifically on the male-side mating structure, and the GCS1 molecules face the female-side mating structure during fertilization. In *Plasmodium (P.) berghei*, GCS1 is uniformly distributed on the male gamete surface, and the male gamete enters the female gamete during fertilization. In *A. thaliana*, the GCS1 molecules are mainly kept in the sperm cytoplasm. When the sperm cells arrive at an egg cell, they are activated by EC1 molecules secreted from the egg cell. The activated sperm cells transfer GCS1 toward the cell surface

released (Wong et al. 2010; Mori et al. 2010). The general GCS1 structure is composed of an N-terminal signal sequence, a highly conserved domain (HAP2-GCS1 domain), a transmembrane (TM) domain, and a positively charged C-terminal domain (Fig. 26.3a) (Wong et al. 2010; Mori et al. 2010). Wong et al. (2010) produced various *Arabidopsis GCS1* constructs, in which the entire N-terminal (AtN) sequence before the TM domain or the entire C-terminal (AtC) sequence after the TM domain was deleted, and introduced these constructs into an *Arabidopsis GCS1* mutant to determine whether such truncated GCS1s are functional. As a result, both the N- and C-terminus-deleted GCS1s were shown to be nonfunctional. In addition, they showed that an N-terminal GCS1 sequence derived from *Sisymbrium irio*, which is phylogenetically near *A. thaliana*, complemented the *Arabidopsis GCS1* mutation but that of *Oryza sativa* GCS1 did not. However, the *Oryza* GCS1 C-terminus did not affect the *Arabidopsis* GCS1 function. Furthermore, they produced *Arabidopsis GCS1* constructs in which the positively charged C-terminal domain was modified and showed that these positively charged amino-acid residues are also important for GCS1 function. However, Mori et al. (2010) reported that the

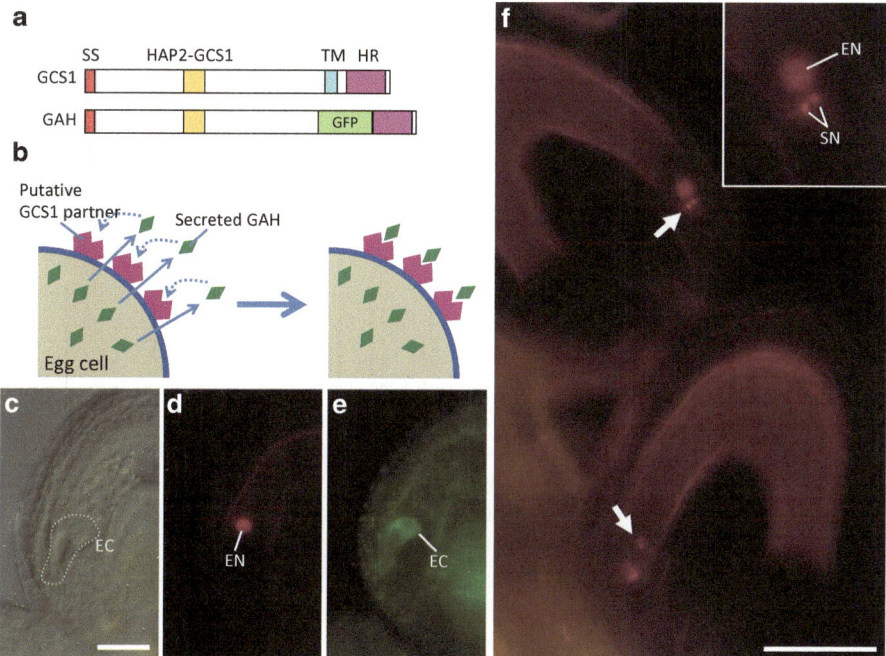

Fig. 26.3 Expression of secretory-type GCS1 (GAH) and gamete fusion inhibition. (**a**) *Arabidopsis* GCS1 is composed of an N-terminal signal sequence (*SS*), a highly conserved domain (*HAP2-GCS1*), a transmembrane domain (*TM*), and a positively charged histidine-rich domain (*HR*). In the *GAH* protein, the TM domain has been exchanged for a *GFP* sequence. (**b**) GAH was expressed specifically in an egg cell by the *DD45* promoter. The GAH molecules were expected to be secreted from the egg cell and then to block the putative GCS1 receptor on the egg surface. (**c–e**) Observation of GAH expression. The egg cell (*EC*) detected by differential interference contrast microscopy (**c**) showed both egg nucleus (*EN*) RFP (**d**) and GFP signals (**e**) from the GAH molecules. (**f**) In the GAH-expressing lines, unfused sperm cells were frequently detected (*arrows*). *Insert* is a magnification of the region indicated by the *top arrow* and shows a pair of RFP signals from unfused sperm cell nuclei (*SN*). *Bars* **c** 20 μm; **f** 50 μm

majority of the GCS1 C-terminus is dispensable for GCS1 function. In their paper, various modified *Arabidopsis GCS1* constructs, in which each of the characteristic domains is disrupted by a *GFP* cDNA insertion, were produced and similarly introduced into *Arabidopsis GCS1* mutants. As a result, although most of the C-terminal sequence was exchanged for the GFP (green fluorescent protein) sequence, the modified GCS1 retained its gamete fusion function at a level comparable with that of normal GCS1. In the same paper, a similar result was also obtained in the case of modified *P. berghei* GCS1, in which the entire C-terminal sequence was deleted. Although the detailed cause of the conflict between the findings of these two groups remains unclear, the contribution of the C-terminus to gamete fusion function may not be generalized among all GCS1-possessing organisms.

26.4   A New Attempt to Elucidate the Mechanism of GCS1-Based Gamete Fusion

The N-terminal GCS1 sequence is expected to be exposed on the male gamete surface as an extracellular domain because of the presence of the N-terminal signal sequence and C-terminal single-pass transmembrane domain. These features also suggest that the C-terminus following the TM domain should be a cytosolic domain. Based on this expectation and the aforementioned findings of the Mori group, GCS1 may function as a type of ligand attached on the male gamete surface, because if it is a receptor-like molecule, the cytosolic C-terminus should function in internal signal transduction similar to a receptor kinase. However, as shown by the Mori group, the C-terminus is dispensable for GCS1 function, suggesting that only the extracellular N-terminus is functional in gamete fusion. In addition, if GCS1 functions as a ligand, the female gamete may express a partner molecule functioning as a GCS1 receptor to produce a gamete fusion complex on the surface. Recently, a modified *Arabidopsis GCS1* gene construct, in which the TM domain was exchanged for *GFP* cDNA under an egg cell-specific promoter (*pDD45*), was newly produced and named *pDD45::GAH* (unpublished data) (Fig. 26.3a). Because the GAH peptide does not possess the TM domain, the GAH peptide produced in the egg cell was expected to be secreted into the gamete fusion space of the embryo sac and to prevent sperm–egg fusion by blocking the GCS1 partner (Fig. 26.3b). As a result, when the *pDD45::GAH* construct was introduced into an *Arabidopsis* egg nucleus marker line expressing *pEC1::H2B-RFP*, the GFP signal derived from the GAH peptide was detected specifically in the egg cell (unpublished data) (Fig. 26.3c–e). Next, the GAH-expressing line was pollinated with pollen from a sperm nucleus marker line expressing *pHTR10::HTR10-RFP* to determine whether gamete fusion was prevented. Unfused sperm cells were occasionally detected in the ovules (unpublished data) (Fig. 26.3f). Notably, both sperm cells were prevented from fusing, suggesting that the GAH peptide blocked not only the sperm–egg fusion but also the sperm–central cell fusion. These observations also support GAH secretion. Based on these results, GCS1 molecules are expected to bind female counterpart(s) most likely expressed on both the egg and central cells. The GAH-expressing line may be useful to isolate such GCS1-partner molecules in the future.

26.5   Perspectives

Even before the identification of GCS1, some transmembrane proteins critical to gamete fusion had been identified and characterized in some species, such as *Chlamydomonas* FUS1, *Caenorhabditis elegans* Spe-9, and mammalian IZUMO1 (Twell 2006). Because of species specificity, those proteins may have been developed to facilitate species-specific gamete fusion processes. However, given that the final process of gamete fusion is based upon lipid bilayer membrane contact in any species, there must be gamete fusion processes shared by different sexual reproductive organisms. GCS1 is a promising protein to help elucidate such conserved gamete fusion systems, and any approaches used to study GCS1 should lead to a fundamental understanding of these systems.

Acknowledgments This study is funded by a Grant-in-Aid for Scientific Research on Innovative Areas (21112008) and a Grant-in-Aid for Young Scientists (B) (24770062), to T. M., from the Ministry of Education, Culture, Sports, Science and Technology of Japan.

References

Hirai M, Arai M, Mori T et al (2008) Male fertility of malaria parasites is determined by GCS1, a plant-type reproduction factor. Curr Biol 18:607–613

Igawa T, Yanagawa Y, Miyagishima SY et al (2013) Analysis of gamete membrane dynamics during double fertilization of *Arabidopsis*. J Plant Res 126:387–394

Liu Y, Tewari R, Ning J et al (2008) The conserved plant sterility gene HAP2 functions after attachment of fusogenic membranes in *Chlamydomonas* and *Plasmodium* gametes. Genes Dev 22:1051–1068

Mori T, Kuroiwa H, Higashiyama T et al (2006) GENERATIVE CELL SPECIFIC 1 is essential for angiosperm fertilization. Nat Cell Biol 8:64–71

Mori T, Hirai M, Kuroiwa T et al (2010) The functional domain of GCS1-based gamete fusion resides in the amino terminus in plant and parasite species. PLoS One 5:e15957

Sinden RE, Croll NA (1975) Cytology and kinetics of microgametogenesis and fertilization in *Plasmodium yoelii nigeriensis*. Parasitology 70:53–65

Sprunck S, Rademacher S, Vogler F et al (2012) Egg cell-secreted EC1 triggers sperm cell activation during double fertilization. Science 338:1093–1097

Steele RE, Dana CE (2009) Evolutionary history of the HAP2/GCS1 gene and sexual reproduction in metazoans. PLoS One 4:e7680

Twell D (2006) A blossoming romance: gamete interactions in flowering plants. Nat Cell Biol 8:14–16

von Besser K, Frank AC, Johnson MA et al (2006) *Arabidopsis* HAP2 (GCS1) is a sperm-specific gene required for pollen tube guidance and fertilization. Development (Camb) 133:4761–4769

Wong JL, Leydon AR, Johnson MA (2010) HAP2(GCS1)-dependent gamete fusion requires a positively charged carboxy-terminal domain. PLoS Genet 6:e1000882

Chapter 27
Fertilization Mechanisms of the Rodent Malarial Parasite *Plasmodium berghei*

Makoto Hirai

Abstract Malaria, caused by *Plasmodium* spp., is transmitted by anopheline mosquitoes. When male and female gametes are introduced into the mosquito midgut, they reproduce sexually and proliferate, at which time the mosquito becomes infectious to vertebrates. It has been proposed that fertilization is a critical target in the parasite life cycle for the reduction of malarial prevalence. Although understanding parasite fertilization is crucial for the control of malaria, the precise molecular mechanisms involved have long remained unknown. Generative cell-specific 1 (GCS1) has been reported to be a critical fertilization factor in angiosperms. It was subsequently shown that the function of GCS1 is conserved in both the rodent malaria parasite *Plasmodium berghei* and the green alga *Chlamydomonas reinhardtii*. Moreover, a *GCS1*-like gene has been detected in the genomes of various organisms, suggesting that it plays a conserved role in gamete interaction. As GCS1 is thought to act as a membrane-anchoring protein in male gametes, a female counterpart is assumed to exist. To reveal the mechanisms involved in parasite fertilization, it is important to clarify the function of GCS1 and to identify GCS1 partners and other fertilization factors. In this review, I first describe the life cycle of malaria parasites, focusing on gametogenesis and fertilization and the underlying mechanisms. I then discuss the functions of GCS1 at the time of gamete interaction. Finally, I consider whether parasite fertilization factors, including GCS1, might be utilized in the development of antimalarial vaccines.

Keywords Fertilization • Gametogenesis • GCS1 • Malarial parasite

M. Hirai (✉)
Department of Parasitology, Gunma University,
3-39-22 Showa, Maebashi City, Gunma 371-8511, Japan
e-mail: makotohirai@gunma-u.ac.jp

H. Sawada et al. (eds.), *Sexual Reproduction in Animals and Plants*,
DOI 10.1007/978-4-431-54589-7_27, © The Author(s) 2014

27.1 Sexual Differentiation and Gametogenesis of Malaria Parasites

In vertebrates, mature malaria parasites infect erythrocytes by egressing from older cells and invading newer ones. In asexually proliferating parasites, a subset of the parasites escapes from the asexual cycle and differentiate to produce male and female gametocytes (gametocytogenesis). When these gametocytes are introduced into the midgut of mosquitoes through feeding on malaria-infected vertebrates, they differentiate into gametes (gametogenesis). During gametogenesis, gametocytes shed the erythrocytic membrane within a few minutes of mosquito feeding. Subsequently, male gametes undergo three rounds of DNA replication within 2 to 10 min. Between 10 and 12 min, eight motile flagella (the motile form of male gametes) are released (Sinden et al. 1996) in a phenomenon termed exflagellation (Ross and Smyth 1997). For *Plasmodium*, an in vitro fertilization (IVF) assay has been established involving three factors required to induce gametogenesis: (1) a drop in temperature from 37 °C to 21 °C; (2) the presence of xanthurenic acid (XA), a previously determined mosquito-derived factor (Billker et al. 1998); and (3) alkaline conditions (pH 8.0). Through IVF, it has been demonstrated that XA activates membrane-bound guanylate cyclase (GC) (Muhia et al. 2001) and cGMP-dependent protein kinase (PKG) (McRobert et al. 2008), which are essential for male exflagellation. Moreover, N-methyl-hydroxylamine, an inhibitor of GC, inhibits exflagellation (Kawamoto et al. 1990). These findings demonstrate that cGMP is involved in XA-mediated signaling. In addition to cGMP signaling, it has been reported that intracellular Ca^{2+} concentrations in gametocytes are increased within 10 s after XA stimulation. Parasites lacking the *calcium-dependent protein kinase 4* gene are defective in genomic replication and mitosis, resulting in an absence of exflagellation (Billker et al. 2004). This finding suggests that XA may also activate calcium signaling. Furthermore, it has been reported that *P. berghei* homologues of atypical MAP kinase 2 (Map-2) (Rangarajan et al. 2005; Tewari et al. 2005) and cdc20 (Guttery et al. 2012) are required for male gametogenesis. Although the aforementioned molecules have been demonstrated to be involved in XA-mediated signal transduction, the most important question remains unanswered: How does XA induce gametogenesis? Given that membrane-bound GC in gametocytes is activated by XA in vitro (Muhia et al. 2001), a membrane-coupled XA receptor may exist to activate PKG and calcium signaling, thereby triggering gametogenesis.

27.2 Recognition, Attachment, and Membrane Fusion of *Plasmodium* Gametes

In general, the process of fertilization can be divided into three steps: recognition, attachment, and membrane fusion of gametes. In malaria parasites, it has been shown that male flagella swim freely, exhibiting no directed movement toward female gametes in vitro (Sinden 1983). Therefore, although many plants and animals utilize chemoattractants for gamete recognition when they are spatially isolated, malaria parasites may not possess such a system. Nevertheless, *Plasmodium* gametes must possess a molecular system for species-specific recognition. A recent study demonstrated that the gamete membrane-surface proteins P48/45, P47, and P230, which belong to the 6-cys family, are critical in gamete recognition. Mutant parasites lacking either P48/45 or P47 and P230 show male or female subfertility, respectively (van Dijk et al. 2010). Further phenotypic observation revealed that these mutants are unable to either recognize or attach to each other, demonstrating the vital role of this family of proteins in gamete recognition/attachment. Importantly, the fertility of these mutants is not completely impaired, suggesting that alternative pathways may compensate for the function of these proteins. Thus, definitive fertilization factors for malaria parasites have not been identified until recently.

In *Plasmodium*, the male gametes swim in a fashion similar to mammalian sperm and attach vigorously to female gametes. Interestingly, after entering female cells, the male gametes continue flagellar movement inside the cell for 1 min after fusion (Sinden and Croll 1975). A similar phenomenon has been observed in the fruit fly *Drosophila melanogaster*, whose spermatic membrane is detected in eggs until immediately after fertilization. Sneaky (Snky), a sperm membrane-specific protein, is responsible for sperm membrane degradation and is essential for successful fertilization (Wilson et al. 2006). Putative Snky homologues are present in vertebrate species, implying their functional conservation (Wilson et al. 2006). However, *Plasmodium* species do not possess an obvious homologue of this protein, suggesting that the molecular mechanism of *Plasmodium* fertilization is, at least in part, distinct from that of animals.

27.3 GCS1: An Ancient Fertilization Factor Conserved in Animals and Plants

Mori et al. reported a male gamete-specific protein, designated generative cell-specific 1 (GCS1), as a novel fertilization factor in angiosperms (Mori et al. 2006). GCS1 is a putative single-pass transmembrane protein with a putative N-terminal signal sequence but no functional domain (Mori et al. 2006; von Besser et al. 2006). *GCS1* knockout (KO) *Arabidopsis thaliana* plants show a severe form of male sterility in which male gametes can access the egg cell normally but cannot fuse with female gametes, suggesting that only the gamete interaction process is impaired in GCS1 KO plants (Mori et al. 2006; von Besser et al. 2006). Interestingly, functional GCS1 homologues are present in green algae and rodent malaria parasites (Mori et al. 2006; von Besser et al. 2006; Liu et al. 2008; Hirai et al. 2008). Similar to angiosperm GCS1s, the expression of *P. berghei* GCS1 (PbGCS1) shows male gamete specificity, and loss of PbGCS1 results in male sterility, whereas female gametes remain fertile (Hirai et al. 2008; Liu et al. 2008). This situation is in sharp contrast to that in mutant parasites lacking any of the 6-cys family proteins, whose fertility is only partially impaired, suggesting that no complementary molecule for GCS1 exists in the parasite. In green algae, Liu et al. demonstrated the exact timing of GCS1 function. Algal GCS1 acts during the few seconds between the pre-fusion adhesion and fusion steps, providing evidence that GCS1 may play a role in gamete membrane fusion following gamete recognition (Liu et al. 2008). It is generally accepted that the molecules that function in species recognition evolve rapidly, leading to reproductive isolation and speciation among eukaryotes (Swanson and Vacquier 2002). In this context, GCS1 has been shown to be conserved in various organisms, and the functional timing observed during gamete recognition suggests that GCS1 may have a role in membrane fusion following species recognition. A previous study by our group demonstrated that male gametes lacking PbGCS1 can attach to their female counterparts, suggesting that PbGCS1 is unnecessary for gamete recognition (Hirai et al. 2008). To further confirm that GCS1 is not involved in gamete recognition, we generated transgenic *P. berghei* in which the endogenous *PbGCS1* gene was replaced with that of *P. yoelii* (PyGCS1), another rodent malaria parasite. Because *P. berghei* cannot fertilize *P. yoelii*, the PyGCS1-expressing *P. berghei* male gametes are expected to fail in the fertilization of wild-type (WT) *P. berghei* female gametes if the GCS1 protein plays any role in species-specific gamete recognition. Our results clearly demonstrated that the transgenic *P. berghei* male gametes could fertilize WT *P. berghei* female gametes, indicating that *Plasmodium* GCS1 is not involved in gamete recognition. Moreover, heterogeneous expression of PyGCS1 in *P. berghei* did not result in any adverse effects on parasite development, even after the zygote stage, suggesting functional complementation of PyGCS1 in *P. berghei* (Hirai et al., unpublished data).

Given that GCS1 contains no predictable functional motif except for a signal sequence and transmembrane domain, there are no available clues to allow speculation regarding its function. As an initial step toward understanding the function of PbGCS1, we generated a series of transgenic parasites in which endogenous PbGCS1 was partially deleted to characterize the critical functional domains of the PbGCS1 protein. Our results revealed that the putative cytosolic region in the C-terminus of the protein is not necessary for successful fertilization but that all other regions (e.g., the single-peptide transmembrane domain and recently defined motifs such as HAP2/GCS1) are essential. Taken together, the results of this study revealed that functional domains of PbGCS1 involved in gamete fusion reside in putative extracellular locations and that PbGCS1 must be localized to the cell surface to function properly (Mori et al. 2010). Interestingly, a similar analysis was performed for *Arabidopsis* GCS1, generating results similar to those found for PbGCS1 (Mori et al. 2010) and suggesting a conserved topology of GCS1 among GCS1-possessing organisms. Additionally, the extracellular localization of GCS1 allows us to speculate that GCS1 may function on the surface of male gametes by interacting with a partner molecule on female gametes. It is clear that the identification of GCS1 partners and other fertilization factors will accelerate our understanding of fertilization systems not only for malaria parasites but also for all GCS1-possessing organisms.

27.4 Application of Fertilization Factors to Anti-Malarial Vaccine Development

Malaria is a serious disease that was responsible for 655,000 deaths and 216 million reported cases in 2010 alone (Murray et al. 2012). The emergence of parasites that are resistant to drugs and vaccines necessitates the development of new strategies for malaria control. However, the development of antimalarial vaccines and drugs is difficult because both the antibodies elicited by vaccination and drugs provide selective pressure and induce recombination/mutation in the parasite genome, resulting in antigenic polymorphisms and the emergence of parasites that are tolerant to these treatments. In addition to antimalarial strategies aimed at the human blood-stage parasite, a transmission-blocking vaccine (TBV) that attacks the insect-stage parasite in mosquitoes has been proposed. Mosquitoes do not exhibit adaptive immunity and rely solely on innate immunity (Faye 1990), suggesting that insect-stage parasites do not face antibody-based immune pressure. It has therefore been speculated that an antibody-based immune attack could be effective against mosquito-stage parasites because they may not evolve any strategy for combating antibodies during the mosquito stage. Based on this assumption, TBV candidates have been selected from molecules expressed in gametocytes, gametes, zygotes, and ookinetes. Several cell-surface molecules (Pfs230, P48/45, P25/28) have been tested as TBV targets thus far, and the antibodies raised against them have succeeded in reducing the rate of parasite transmission (Barr et al. 1991; Healer et al. 1999; Williamson 2003). However, none of these studies has demonstrated complete blockage of transmission, most likely because the TBV candidate molecules are quite abundant in parasites and because the small quantity of antibodies administered may not be sufficient to completely block the activity of the candidate molecules. Even if the antibodies were able to fully block this activity, it is possible that the parasites would still be fertile, because parasites lacking any of these genes can still reproduce sexually in mosquitoes (van Dijk et al. 2010). Therefore, other molecules that display a definitive and crucial function in sexual reproduction represent ideal TBV candidates. We believe GCS1 best fits this criterion because PbGCS1 protein expression levels are quite low in the parasite, and a lack of PbGCS1 protein completely abrogates parasite fertility (Khan et al. 2005; Hirai et al. 2008). Recently, Blagborough et al. and our group performed experiments demonstrating that an anti-PbGCS1 antibody significantly, but not completely, blocked parasite reproduction based on IVF assays (Blagborough et al. 2013; Hirai et al. unpublished data). This failure to completely inhibit reproduction may arise from PbGCS1 possessing many cysteine residues and displaying a complex conformation that is not recapitulated by the recombinant protein, resulting in less effective antibody production. Such a requirement for correct folding of recombinant TBV proteins for effective antibody production has been observed in cases in which the TBV candidate is produced in yeast and algal systems (Kaslow et al. 1994; Gregory et al. 2013). In addition to this problem, the immunization protocol needs to be improved to generate higher antibody titers (Kubler-Kielb et al. 2007; Outchkourov et al. 2008).

It is reasonable to assume that the use of multiple TBV targets would be effective because the antibodies raised would simultaneously neutralize or block multiple parasite molecules, leading to complete blockage of parasite fertilization. Thus, we have employed a transcriptomic approach to identify new fertilization factors for malaria parasites, and TBV candidates will be screened from this source.

27.5   Concluding Remarks

Since the draft genome sequences of the human malaria parasite, *P. falciparum*, and its vector, *Anopheles gambiae*, have been published and reverse genetics has been applied to the parasite, our knowledge of malaria biology has expanded. However, this infectious disease is still life threatening. As I mentioned earlier, the main reason for failure in eradication of this disease is the appearance of parasites tolerant to drug and vaccine treatments, and we need to find new ways to eradicate this disease. In this context, TBV might be a new weapon to combat the parasites. Because parasite fertilization is the first stage in the mosquito life cycle, it is assumed that the parasite fertilization factor would be the most effective TBV candidate. We understand that there are many hurdles to overcome before a long-lasting and highly effective TBV is obtained. Nevertheless, we believe that our attempt will result in not only a discovery of new TBV but also in clearer understanding of the molecular mechanisms underlying malarial parasite sexual reproduction.

Acknowledgments This research was supported by a Grant-in-Aid for Scientific Research on Innovative Areas (24112705) a grant for research into emerging and re-emerging infectious diseases from the Ministry of Health, Labor and Welfare of Japan (H23-Shinkosaiko-ippan-014; http://www.mhlw.go.jp/english/).

References

Barr PJ, Green KM, Gibson HL, Bathurst IC, Quakyi IA, Kaslow DC (1991) Recombinant Pfs25 protein of *Plasmodium falciparum* elicits malaria transmission-blocking immunity in experimental animals. J Exp Med 174:1203–1208

Billker O, Lindo V, Panico M, Etienne AE, Paxton T, Dell A, Rogers M, Sinden RE, Morris HR (1998) Identification of xanthurenic acid as the putative inducer of malaria development in the mosquito. Nature (Lond) 392:289–292

Billker O, Dechamps S, Tewari R, Wenig G, Franke-Fayard B, Brinkmann V (2004) Calcium and a calcium-dependent protein kinase regulate gamete formation and mosquito transmission in a malaria parasite. Cell 117:503–514

Blagborough AM, Churcher TS, Upton LM, Ghani AC, Gething PW, Sinden RE (2013) Transmission-blocking interventions eliminate malaria from laboratory populations. Nat Commun 4:1812

Faye I (1990) Acquired immunity in insects: the recognition of nonself and the subsequent onset of immune protein genes. Res Immunol 141:927–932

Gregory JA, Topol AB, Doerner DZ, Mayfield S (2013) Algae-produced cholera toxin-Pfs25 fusion proteins as oral vaccines. Appl Environ Microbiol 79(13):3917–3925

Guttery DS, Ferguson DJ, Poulin B, Xu Z, Straschil U, Klop O, Solyakov L, Sandrini SM, Brady D, Nieduszynski CA et al (2012) A putative homologue of CDC20/CDH1 in the malaria parasite is essential for male gamete development. PLoS Pathog 8:e1002554

Healer J, McGuinness D, Carter R, Riley E (1999) Transmission-blocking immunity to *Plasmodium falciparum* in malaria-immune individuals is associated with antibodies to the gamete surface protein Pfs230. Parasitology 119(pt 5):425–433

Hirai M, Arai M, Mori T, Miyagishima SY, Kawai S, Kita K, Kuroiwa T, Terenius O, Matsuoka H (2008) Male fertility of malaria parasites is determined by GCS1, a plant-type reproduction factor. Curr Biol 18:607–613

Kaslow DC, Bathurst IC, Lensen T, Ponnudurai T, Barr PJ, Keister DB (1994) *Saccharomyces cerevisiae* recombinant Pfs25 adsorbed to alum elicits antibodies that block transmission of *Plasmodium falciparum*. Infect Immun 62:5576–5580

Kawamoto F, Alejo-Blanco R, Fleck SL, Kawamoto Y, Sinden RE (1990) Possible roles of Ca^{2+} and cGMP as mediators of the exflagellation of *Plasmodium berghei* and *Plasmodium falciparum*. Mol Biochem Parasitol 42:101–108

Khan SM, Franke-Fayard B, Mair GR, Lasonder E, Janse CJ, Mann M, Waters AP (2005) Proteome analysis of separated male and female gametocytes reveals novel sex-specific *Plasmodium* biology. Cell 121:675–687

Kubler-Kielb J, Majadly F, Wu Y, Narum DL, Guo C, Miller LH, Shiloach J, Robbins JB, Schneerson R (2007) Long-lasting and transmission-blocking activity of antibodies to *Plasmodium falciparum* elicited in mice by protein conjugates of Pfs25. Proc Natl Acad Sci USA 104:293–298

Liu Y, Tewari R, Ning J, Blagborough AM, Garbom S, Pei J, Grishin NV, Steele RE, Sinden RE, Snell WJ et al (2008) The conserved plant sterility gene HAP2 functions after attachment of fusogenic membranes in *Chlamydomonas* and *Plasmodium* gametes. Genes Dev 22:1051–1068

McRobert L, Taylor CJ, Deng W, Fivelman QL, Cummings RM, Polley SD, Billker O, Baker DA (2008) Gametogenesis in malaria parasites is mediated by the cGMP-dependent protein kinase. PLoS Biol 6:e139

Mori T, Kuroiwa H, Higashiyama T, Kuroiwa T (2006) Generative cell specific 1 is essential for angiosperm fertilization. Nat Cell Biol 8:64–71

Mori T, Hirai M, Kuroiwa T, Miyagishima SY (2010) The functional domain of GCS1-based gamete fusion resides in the amino terminus in plant and parasite species. PLoS One 5:e15957

Muhia DK, Swales CA, Deng W, Kelly JM, Baker DA (2001) The gametocyte-activating factor xanthurenic acid stimulates an increase in membrane-associated guanylyl cyclase activity in the human malaria parasite *Plasmodium falciparum*. Mol Microbiol 42:553–560

Murray CJ, Rosenfeld LC, Lim SS, Andrews KG, Foreman KJ, Haring D, Fullman N, Naghavi M, Lozano R, Lopez AD (2012) Global malaria mortality between 1980 and 2010: a systematic analysis. Lancet 379:413–431

Outchkourov NS, Roeffen W, Kaan A, Jansen J, Luty A, Schuiffel D, van Gemert GJ, van de Vegte-Bolmer M, Sauerwein RW, Stunnenberg HG (2008) Correctly folded Pfs48/45 protein of *Plasmodium falciparum* elicits malaria transmission-blocking immunity in mice. Proc Natl Acad Sci USA 105:4301–4305

Rangarajan R, Bei AK, Jethwaney D, Maldonado P, Dorin D, Sultan AA, Doerig C (2005) A mitogen-activated protein kinase regulates male gametogenesis and transmission of the malaria parasite *Plasmodium berghei*. EMBO Rep 6:464–469

Ross R, Smyth J (1997) On some peculiar pigmented cells found in two mosquitoes fed on malarial blood. 1897. Indian J Malariol 34:47–55

Sinden RE (1983) Sexual development of malarial parasites. Adv Parasitol 22:153–216

Sinden RE, Croll NA (1975) Cytology and kinetics of microgametogenesis and fertilization in *Plasmodium yoelii* nigeriensis. Parasitology 70:53–65

Sinden RE, Butcher GA, Billker O, Fleck SL (1996) Regulation of infectivity of *Plasmodium* to the mosquito vector. Adv Parasitol 38:53–117

Swanson WJ, Vacquier VD (2002) The rapid evolution of reproductive proteins. Nat Rev Genet 3:137–144

Tewari R, Dorin D, Moon R, Doerig C, Billker O (2005) An atypical mitogen-activated protein kinase controls cytokinesis and flagellar motility during male gamete formation in a malaria parasite. Mol Microbiol 58:1253–1263

van Dijk MR, van Schaijk BC, Khan SM, van Dooren MW, Ramesar J, Kaczanowski S, van Gemert GJ, Kroeze H, Stunnenberg HG, Eling WM et al (2010) Three members of the 6-cys protein family of *Plasmodium* play a role in gamete fertility. PLoS Pathog 6:e1000853

von Besser K, Frank AC, Johnson MA, Preuss D (2006) *Arabidopsis* HAP2 (GCS1) is a sperm-specific gene required for pollen tube guidance and fertilization. Development (Camb) 133:4761–4769

Williamson KC (2003) Pfs230: from malaria transmission-blocking vaccine candidate toward function. Parasite Immunol 25:351–359

Wilson KL, Fitch KR, Bafus BT, Wakimoto BT (2006) Sperm plasma membrane breakdown during *Drosophila* fertilization requires sneaky, an acrosomal membrane protein. Development (Camb) 133:4871–4879

Chapter 28
Sexual Reproduction of a Unicellular Charophycean Alga, *Closterium peracerosum-strogosum-littorale* Complex

Hiroyuki Sekimoto, Yuki Tsuchikane, and Jun Abe

Abstract The genus *Closterium* is the best characterized charophycean green alga with respect to the process of sexual reproduction. Two sex pheromones, named PR-IP Inducer and PR-IP, that are involved in the progress of these processes were physiologically and biochemically characterized and the corresponding genes were cloned. These pheromones function in most steps of sexual reproduction. The timing after mixing, appropriate concentrations of the pheromones, and conditions of the cells are all essential for pheromones to be functional. To elucidate the molecular mechanisms of sexual reproduction in detail, molecular tools such as expressed sequence tag (EST), microarray analysis, and genetic transformation systems have been established. These methods will enable us to clarify the details of sexual reproduction in the near future.

Keywords *Closterium* • Conjugation • Pheromone

28.1　Introduction

In the process of sexual reproduction, two sexually competent cells recognize each other, followed by conjugation or fertilization. In some algae, dormant zygospores are formed as a result of sexual reproduction and show resistance to severe environmental conditions, such as drought stress. In the case of *Chlamydomonas reinhardtii*, one of the best characterized models in green algae, sexual adhesion between the gametes is mediated by agglutinin molecules on their flagellar membranes. The plus and minus agglutinins are sex specifically displayed by nitrogen-starved

H. Sekimoto (✉) • Y. Tsuchikane • J. Abe
Department of Chemical and Biological Sciences, Faculty of Science, Japan Women's University, 2-8-1 Mejirodai, Bunkyo-ku, Tokyo 112-8681, Japan
e-mail: sekimoto@fc.jwu.ac.jp

H. Sawada et al. (eds.), *Sexual Reproduction in Animals and Plants*,
DOI 10.1007/978-4-431-54589-7_28, © The Author(s) 2014

mating-type *plus* (mt[+]) and mating-type *minus* (mt[−]) gametes, respectively (Adair et al. 1983; Goodenough et al. 1985; Ferris et al. 2005). Once an agglutinin molecule directly binds to the agglutinin molecule on the flagellum of an opposite mating type as a consequence of agglutination, a gamete-specific flagellar adenylyl cyclase is activated and the intracellular cAMP level is elevated nearly tenfold, triggering dramatic alterations in the cell (Pasquale and Goodenough 1987; Saito et al. 1993; Zhang and Snell 1994). First, flagellar motility is altered, and the adhesiveness of the flagellar surface is increased (Saito et al. 1985; Goodenough 1989; Hunnicutt et al. 1990). Second, a matrix-degrading enzyme is activated (Buchanan et al. 1989; Snell et al. 1989; Kinoshita et al. 1992) and the cell wall is degraded so that the gametes are able to fuse. Third, mt[+] gametes erect an actin-filled microvillus ("fertilization tube") as a mating structure and the mt[−] gametes also erect a small, dome-like, actin-free mating structure. Cell fusion initiates with an adhesive interaction between mt[+] and mt[−] mating structures, followed by localized membrane fusion. Two proteins, FUS1 and GCS1/HAP2, are known to be essential for the membrane fusion reaction (Ferris et al. 1996; Misamore et al. 2003; Liu et al. 2008; Mori et al. 2006). Both proteins are degraded rapidly upon fusion, as would be expected for a block to polygamy (Liu et al. 2010).

The desmid *Closterium*, which belongs to Zygnematophyceae, is the most successfully characterized unicellular charophycean in terms of the maintenance of strains and sexual reproduction (Ichimura 1971). Charophyceans, which are most closely related to land plants, form a relevant monophyly with land plants. Recently, it was suggested that either the Zygnematophyceae or a clade consisting of Zygnematophyceae and Coleochaetophyceae might be the most likely sister group of land plants (Turmel et al. 2006; Wodniok et al. 2011).

In this review, the sexual reproductive processes of *Closterium peracerosum-strigosum-littorale* complex (*C. psl.* complex) are described in detail. Molecular tools for analyses of the processes are also presented.

28.2 Sexual Reproduction Controlled by Specific Sex Pheromones in the *C. psl.* Complex

28.2.1 Overview of Sexual Reproduction in Closterium

The sexual reproduction of species in the genus *Closterium* has been of interest to many investigators for more than 100 years, and the morphological details and modes of sexual reproduction are well documented (Cook 1963; Ichimura 1973; Lippert 1967; Noguchi 1988; Noguchi and Ueda 1985; Pickett-Heaps and Fowke 1971). *Closterium* has no flagellum-like machinery for active movement. Therefore,

Fig. 28.1 Schematic illustrations of the sexual reproduction of *Closterium peracerosum-strigosum-littorale* complex: (*1*) mucilage secretion, (*2*) sexual cell division, (*3*) sexual pair formation induced by unknown chemoattractant pheromone(s), (*4*) protoplast release, (*5*) zygospore. Most processes are induced by the protoplast-release-inducing protein (PR-IP) and the PR-IP Inducer. *Gray arrows* indicate pheromonal communication

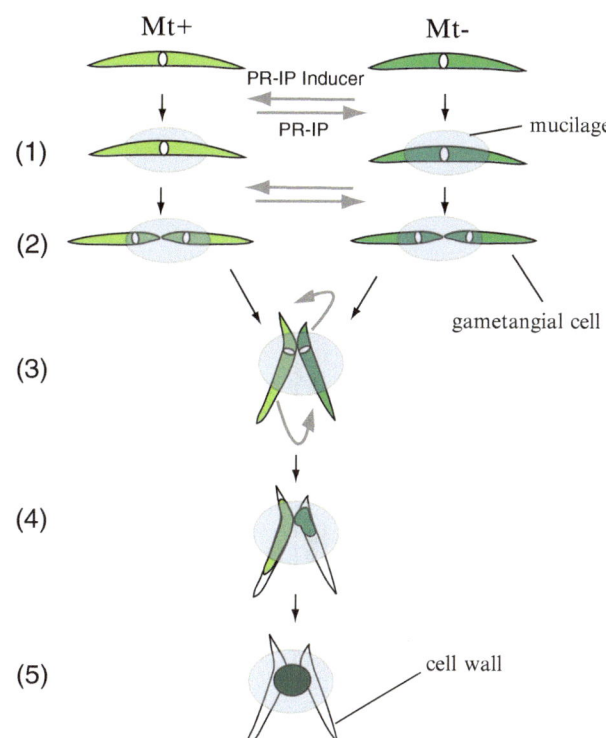

it is thought that the cells of this alga exploit some diffusible substances for the intercellular communication.

In the case of *C. psl.* complex, there are two types of conjugation to produce zygospores: that between two complementary mating-type cells (mt⁺ and mt⁻) and that between clonal cells. The former is called heterothallism and the latter is called homothallism (Graham and Wilcox 2000). The conjugation process can be divided into several steps: sexual cell division (SCD), which produces sexually competent gametangial cells; pairing, formation of conjugation papillae; condensing of their cytoplasm; release and fusion of gametic protoplasts (gametes); and formation of zygospores (Fig. 28.1). Zygospores become dormant and acquire resistance to dry conditions. Exposure to dry conditions and subsequent water supply lead to the start of meiosis. Two non-sister nuclei of the second meiotic division survive and the other two degenerate. As a result, the two surviving nuclei carry opposite mating-type genes in the absence of crossing over, and a pair of mt⁺ and mt⁻ cells arise from one zygospore, in the case of heterothallic strains (Brandham and Godward 1965; Hamada et al. 1982; Lippert 1967; Watanabe and Ichimura 1982).

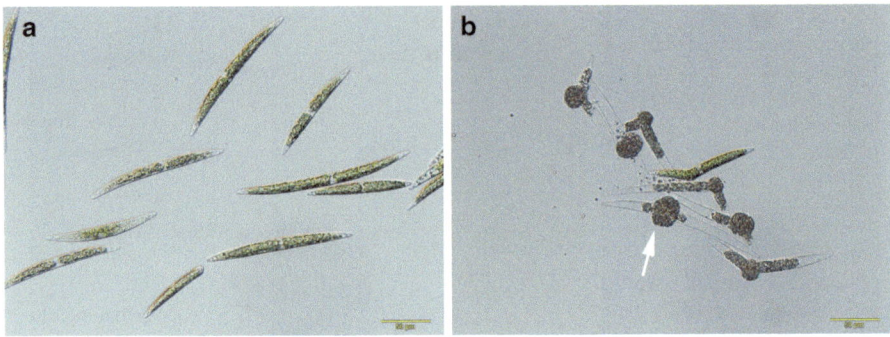

Fig. 28.2 Effect of PR-IP on the release of protoplasts from mt⁻ cells. In contrast to be untreated mt⁻ cells (**a**), mt⁻ cells incubated with PR-IP (**b**) formed conjugation papillae, condensed their cytoplasm, and finally released their protoplasts without pairing (*arrow*), through the effect of PR-IP. *Bars* 50 μm

28.2.2 Characters of Sex Pheromones in the Heterothallic C. psl. *Complex*

28.2.2.1 Sex Pheromones Responsible for Protoplast Release During Sexual Reproduction

When mt⁺ and mt⁻ cells are mixed together in a nitrogen-depleted mating medium in the light, cells of both types differentiate into gametangial cells as a result of SCD and become paired. Then, paired cells release their protoplasts to form zygospores (Fig. 28.1). Sekimoto et al. (1990) successfully isolated the first *Closterium* pheromone from the *C. psl.* complex and designated it as protoplast-release-inducing protein (PR-IP) (Sekimoto et al. 1990). The PR-IP is a glycoprotein that consists of subunits of 42 and 19 kDa: it is released by mt⁺ cells (NIES-67) and is responsible for inducing the release of protoplasts from mt⁻ cells (NIES-68) (Fig. 28.2). PR-IP receptors have not yet been isolated; however, specific binding of the biotinylated 19-kDa subunit of PR-IP to sexually competent mt⁻ cells has been clearly demonstrated (Sekimoto and Fujii 1992; Sekimoto et al. 1993b).

Secretion of PR-IP by mt⁺ cells is induced in medium in which only mt⁻ cells have been cultured (Sekimoto et al. 1993a). Another pheromone that induces the synthesis and release of PR-IP was detected and named PR-IP Inducer (Sekimoto et al. 1993a). The pheromone is also a glycoprotein with a molecular mass of 18.7 kDa (Nojiri et al. 1995). PR-IP Inducer is released constitutively from mt⁻ cells in the presence of light and directly induces the production and release of PR-IP from mt⁺ cells.

cDNAs encoding the subunits of PR-IP (Sekimoto et al. 1994a, b) and PR-IP Inducer (Sekimoto et al. 1998) have been isolated. A computer search using the nucleotide sequences and deduced amino-acid sequences failed to reveal any homologies to known proteins. Genes for these pheromones can be detected in cells of both mating types using genomic Southern hybridization analysis, but they are only expressed in cells of the respective mating type, suggesting the presence of sex-specific regulation of gene expression by sex-limited *trans*-acting factors (Sekimoto et al. 1998; Sekimoto et al. 1994c; Endo et al. 1997).

28.2.2.2 Sex Pheromones Involved in SCD and Mucilage Secretion

In the sexual reproductive processes of *Closterium* species, gametangial cells are produced from haploid vegetative cells. Ichimura (1971) reported that vegetative cells of the *C. psl.* complex divided at once before formation of sexual pairs when the two mating-type cells were mixed (Ichimura 1971). This type of cell division is SCD.

The SCD-inducing activities specific to the two mating-type cells have been detected and characterized physiologically (Tsuchikane et al. 2003). Mt⁻ cells release an SCD-inducing pheromone (SCD-IP) specific for mt⁺ cells, and are designated SCD-IP-minus, whereas an mt⁻-specific pheromone released from mt⁺ cells is designated SCD-IP-plus. Recent time-lapse video analyses revealed that SCD was not always required for successful pairing (Y. Tsuchikane, M. Sato, H. Sekimoto, personal communication).

Closterium exhibits gliding locomotory behavior, mediated by the forceful extrusion of mucilage from one pole of the cell that causes the cell to glide in the opposite direction (Domozych et al. 1993). Substances with the ability to stimulate secretion of uronic acid-containing mucilage from mt⁺ and mt⁻ cells were detected in media in which mt⁻ and mt⁺ cells had been cultured separately and were designated mucilage secretion-stimulating pheromone (MS-SP)-minus and MS-SP-plus, respectively (Akatsuka et al. 2003).

28.2.2.3 Multifunction of Sex Pheromones

Both MS-SP-minus and SCD-IP-minus show quite similar characteristics to PR-IP Inducer, whereas both MS-SP-plus and SCD-IP-plus show quite similar characteristics to PR-IP, with respect to molecular weight, heat stability, and dependency on light for their secretion and function, indicating close relationships among these pheromones. Recombinant PR-IP Inducer produced in yeast cells was assayed for both production of PR-IP and induction of SCD (Sekimoto 2002; Tsuchikane et al. 2005). Although both biological activities were observed by treating recombinant pheromone with mt⁺ cells, SCD could be induced by exposure to a relatively lower concentration of recombinant PR-IP Inducer. Moreover, SCD was induced by a shorter period of treatment with the pheromone than the production of PR-IP (Tsuchikane et al. 2005). In addition, purified native PR-IP Inducer showed mucilage secretion-stimulating activity against mt$^+$ cells (Akatsuka et al. 2003). These results strongly indicate that previously characterized PR-IP Inducer has mucilage secretion-stimulating, SCD-inducing, and PR-IP-inducing activities for mt$^+$ cells, although the induction mechanisms seem to differ.

Purified PR-IP showed not only protoplast-releasing activity, but also mucilage secretion-stimulating and SCD-inducing activities against mt$^-$ cells (Akatsuka et al. 2006). Minimum concentrations required for the respective activities were quite different: 5×10^{-16} M PR-IP stimulated mucilage secretion, 5×10^{-10} M PR-IP was required for protoplast release, and 5×10^{-11} M PR-IP resulted in the induction of SCD as well as mucilage secretion. These results strongly suggest that PR-IP is also a multifunctional pheromone that independently promotes multiple steps in conjugation at the appropriate times through different induction mechanisms.

28.2.3 Summary of Sexual Reproduction in the Heterothallic C. psl. Complex

Based on the results described here, sexual reproductive events, postulated at this time, are summarized (Fig. 28.1). The PR-IP Inducer is released from mt$^-$ cells when cells are exposed to nitrogen-depleted conditions in the light. Then, mt$^+$ cells receive the signal and begin to release the PR-IP into the medium. During this communication, mucilage is secreted into the surrounding medium. Concentrations of these pheromones are gradually elevated, leading to the induction of SCD and the respective formation of gametangial cells. Then, mt$^+$ and mt$^-$ gametangial cells move together and become paired through the effect of unknown chemotactic pheromones. After the final communication by PR-IP and PR-IP Inducer, mt$^-$ cells begin to release their protoplasts. Then the release of protoplasts from mt$^+$ cells is eventually induced by direct adhesion of cells, and these protoplasts fuse to form a zygospore.

Information concerning physical cell–cell recognition and fusion of cells involved in conjugation processes has not yet been clarified; however, fluorescein isothiocyanate (FITC)-labeled lectins, *Lycopersicon esculentum* lectin (LEL) and Concanavalin A (ConA), accumulated on the conjugation papillae and inhibited the progress of zygote formation (Hori et al. 2012). These results suggest that different carbohydrates specifically recognized by these lectins are involved in cell recognition or fusion during conjugation processes in the *C. psl.* complex.

28.3   Molecular Biological Approaches to Sexual Reproduction

28.3.1   Expressed Sequence Tag (EST) and Microarray Analyses

To elucidate the molecular mechanism of intercellular communication during sexual reproduction, a normalized cDNA library was established from a mixture of cDNA libraries prepared from cells at various stages of sexual reproduction and from a mixture of vegetative mt$^+$ and mt$^-$ cells. The aim was to reduce redundancy, and 3,236 ESTs were generated, which were classified into 1,615 nonredundant groups (Sekimoto et al. 2003; Sekimoto et al. 2006). The EST sequences were compared with nonredundant protein sequence databases in the public domain using the BLASTX program, and 1,045 nonredundant sequences displaying similarity to previously registered genes in the public databases were confirmed. The source group with the highest similarity was land plants, including *Arabidopsis thaliana*.

A cDNA microarray was then constructed and expression profiles were obtained using mRNA isolated from cells in various stages of the life cycle. Finally, 88 pheromone-inducible, conjugation-related, or sex-specific genes were identified (Sekimoto et al. 2006), although their functions during sexual reproduction have not been characterized.

Of the 88 genes identified, a gene encoding receptor-like protein kinase (RLK) was the most notable and was named *CpRLK1*. The gene is expressed during sexual reproduction, and treatment of mt$^+$ cells with the PR-IP Inducer also induces its expression, indicating that the CpRLK1 protein probably functions during sexual reproduction (Sekimoto et al. 2006). The full-length cDNA has been isolated, and an amino-acid sequence containing an extracellular domain (ECD) was obtained (unpublished data). In *A. thaliana*, the RLK family is the largest gene family with more than 600 family members (Shiu and Bleecker 2001, 2003; Shiu et al. 2004), although the functions of most of these genes are still unknown. Only two RLK genes have been found in the genome of *Chlamydomonas reinhardtii*; however, the predicted proteins do not have recognizable ECDs. No RLK gene was found in the genome of *Ostreococcus tauri* (Lehti-Shiu et al. 2009). In contrast, RLKs having transmembrane domains or ECDs have been isolated from two charophyceans (*Nitella axillaris* and *Closterium ehrenbergii*) (Sasaki et al. 2007), indicating that the receptor configuration was likely established before the divergence of land plants from charophyceans but after the divergence of charophyceans from chlorophytes (Graham and Wilcox 2000; Karol et al. 2001). The receptor configuration is likely to function for intercellular communication, especially during sexual reproduction; however, the confirmation of genomic information from early diversified nonsexual charophyceans such as Klebsormidiophyceae and Chlorokybophyceae is necessary to confirm this assumption.

A gene named *CpRLP1* (receptor-like protein-1) was also fascinating. Several leucine-rich repeats and a transmembrane domain were found in the deduced protein, but a kinase domain was not involved. As in the case of the CLV2 protein of *A.*

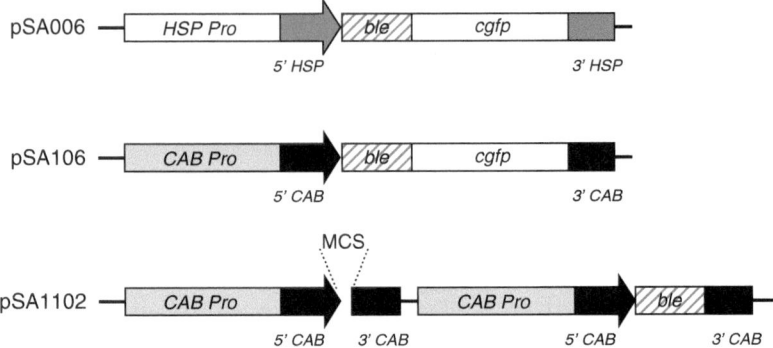

Fig. 28.3 Constructs for *Closterium* nuclear transformation: *HSP/CAB Pro*, *5'HSP/CAB*, and *3'HSP/CAB* indicate promoters and 5'/3'-untranslated regions of *CpHSP70* and *CpCAB1* genes, respectively. *ble*, gene for phleomycin resistance; *cgfp*, *Chlamydomonas*-adapted *GFP* gene; MCS, multi-cloning sites

thaliana, the CpRLP1 protein may form a heterodimer with another protein, such as a receptor-like protein kinase (Zhu et al. 2010a; Zhu et al. 2010b), to transduce the unknown extracellular signal into the intracellular compartment.

28.3.2 Genetic Transformation

Establishment of a nuclear transformation system for genes of interest obtained from transcriptome analyses greatly enhances the understanding of molecular mechanisms for sexual reproduction in *C. psl.* complex. Particle bombardment was used for gene delivery into *C. psl.* complex cells. In general, it is most important to choose efficient promoters to drive the introduced genes. However, expression using the CaMV 35S promoter in the *C. psl.* complex was quite low (Abe et al. 2008a). Two endogenous promoters derived from the highly and constitutively expressed genes *CpHSP70* and *CpCAB1*, encoding a heat shock protein 70 (HSP70) and a chlorophyll *a/b*-binding protein (CAB) in the *C. psl.* complex, respectively, were selected and isolated to drive the transgenes. In the *C. psl.* complex, codons are highly biased in G and C, resulting in synonymous codons favoring G and C at the third position (Abe et al. 2008a). Because this feature is very similar to that of *Chlamydomonas reinhardtii*, the marker and reporter genes used in *Chlamydomonas reinhardtii* are applicable to transformation in the *C. psl.* complex.

Two constructs, pSA006 and pSA106, were successfully transformed in the *C. psl.* complex (Fig. 28.3). These constructs consisted of the *Chlamydomonas* selectable marker gene *ble* encoding a phleomycin-resistant protein (Stevens et al. 1996) and the *cgfp* gene encoding a *Chlamydomonas*-adapted green fluorescent protein (GFP) (Fuhrmann et al. 1999). These genes were mutually fused in-frame and linked either to the *CpHSP70* (pSA006) or the *CpCAB1* (pSA106) promoters. Finally, approximately 250 and 100 of the transiently GFP-expressed cells were obtained in a plate (in one trial of particle bombardment) using plasmid pSA006 and pSA106, respectively (Abe et al. 2008a).

Phleomycin is a useful antibiotic for the selection of stable transformants in the *C. psl.* complex because the drug inhibits cell proliferation at low concentrations both in liquid media and on solid media (Abe et al. 2008b). The overexpression vector pSA1102, which allowed direct selection by phleomycin and the overexpression of the arbitrary genes, was constructed (Fig. 28.3, Abe et al. 2011). In the case of *CpPI* (encoding PR-IP Inducer), the expression level in transformed mt+ cells displayed about a 16-fold increase compared with wild-type cells. In addition, both transcripts encoding the respective PR-IP subunits (*Cp19ksu* and *Cp42ksu*) also displayed an approximately 67-fold increase in the same transformants, indicating that the ectopically expressed PR-IP Inducer would be functional in vivo in the *C. psl.* complex. Further improvements such as the selection of more powerful promoters and the application of gene silencing will provide useful information to enhance our understanding of sexual reproduction in *Closterium*.

28.4 Perspective

In this chapter, regulation of sexual reproduction in the unicellular charophycean alga *C. psl.* complex was described in detail. In the sexual reproduction processes, two sex pheromones (PR-IP Inducer and PR-IP), released from mt− and mt+ cells, respectively, were indispensable. These exerted multiple functions, such as stimulation of mucilage secretion, induction of SCD, and release of PR-IP from mt+ cells or release of protoplasts from mt− cells. Moreover, timing after mixing, appropriate concentrations of the pheromones, and conditions of the cells are all essential for pheromones to be functional.

Using microarray analyses, cDNAs encoding a receptor-like kinase (CpRLK1) and a leucine-rich repeat containing receptor-like protein (CpRLP1) were identified, which may function as sex-specific receptors for recognition of unknown signals from opposite mating-type cells. To characterize sex-specific and sexual reproduction-related genes, including *CpRLK1* and *CpRLP1*, genetic transformation systems have recently been established. Further improvements, such as selection of more powerful promoters, will enable us to analyze the function of unknown genes in the near future. In addition, large-scale EST analysis and draft genome sequencing of the *C. psl.* complex are now in progress.

Using the *C. psl.* complex as a model, the problem of speciation of organisms can be approached. One species of the *C. psl.* complex can be subclassified into several reproductively isolated groups (biological species). The reasons for the isolation could be partly explained as the loss of pheromonal communication (Tsuchikane et al. 2008; Sekimoto et al. 2012). Pheromones are also involved in the sexual reproduction of the homothallic strain (Tsuchikane et al. 2010a). In this strain, conjugation of two sister gametangial cells derived from one vegetative cell was predominant (Tsuchikane et al. 2010b). SCD of one vegetative cell into two sister gametangial cells seemed to be a segregative process that was required for the production of complementary mating types observed in the heterothallic cells (Tsuchikane et al. 2012).

As mentioned previously, the algal genus *Closterium* is one of the closest living organisms to land plants. The present studies concerning sexual reproduction of *C. psl.* complex are useful when considering the mechanisms and evolution of sexual reproduction in terrestrial plants.

Acknowledgments The research projects were partly supported by Grants-in-Aid for Scientific Research (nos. 22405014, 23657161, 24370038, and 24247042 to H.S., no. 23770277 to J.A., no. 23770093 to Y.T.) from the Japan Society for the Promotion of Science, Japan, Grants-in-Aid for Scientific Research on Innovative Areas "Elucidating common mechanisms of allogenic authentication" (nos. 22112521 and 24112713 to H.S.) from the Ministry of Education, Culture, Sports, Science and Technology, Japan, a grant from the New Technology Development Foundation to Y.T.

References

Abe J, Hiwatashi Y, Ito M, Hasebe M, Sekimoto H (2008a) Expression of exogenous genes under the control of endogenous *HSP70* and *CAB* promoters in the *Closterium peracerosum-strigosum-littorale* complex. Plant Cell Physiol 49:625–632

Abe J, Sakayori K, Sekimoto H (2008b) Effect of antibiotics on cell proliferation in the *Closterium peracerosum-strigosum-littorale* complex (Charophyceae, Chlorophyta). Biologia 63:932–936

Abe J, Hori S, Tsuchikane Y, Kitao N, Kato M, Sekimoto H (2011) Stable nuclear transformation of the *Closterium peracerosum-strigosum-littorale* complex. Plant Cell Physiol 52:1676–1685. doi:10.1093/pcp/pcr103

Adair WS, Hwang C, Goodenough UW (1983) Identification and visualization of the sexual agglutinin from the mating-type plus flagellar membrane of *Chlamydomonas*. Cell 33:183–193

Akatsuka S, Sekimoto H, Iwai H, Fukumoto R, Fujii T (2003) Mucilage secretion regulated by sex pheromones in *Closterium peracerosum-strigosum-littorale* complex. Plant Cell Physiol 44:1081–1087

Akatsuka S, Tsuchikane Y, Fukumoto R, Fujii T, Sekimoto H (2006) Physiological characterization of the sex pheromone protoplast-release-inducing protein from the *Closterium peracerosum-strigosum-littorale* complex (Charophyta). Phycol Res 54:116–121

Brandham PE, Godward MBE (1965) The inheritance of mating type in desmids. New Phytol 64:428–435

Buchanan MJ, Imam SH, Eskue WA, Snell WJ (1989) Activation of the cell wall degrading protease, lysin, during sexual signalling in *Chlamydomonas*: the enzyme is stored as an inactive, higher relative molecular mass precursor in the periplasm. J Cell Biol 108:199–207

Cook PA (1963) Variation in vegetative and sexual morphology among the small curved species of *Closterium*. Phycologia 3:1–18

Domozych CR, Plante K, Blais P (1993) Mucilage processing and secretion in the green alga *Closterium*. 1. Cytology and biochemistry. J Phycol 29:650–659

Endo B, Fujii T, Kamiya Y, Sekimoto H (1997) Analysis of genomic sequences encoding a sex pheromone from the *Closterium peracerosum-strigosum-littorale* complex. J Plant Res 110:463–467

Ferris PJ, Woessner JP, Goodenough UW (1996) A sex recognition glycoprotein is encoded by the *plus* mating-type gene *fus1* of *Chlamydomonas reinhardtii*. Mol Biol Cell 7:1235–1248

Ferris P, Waffenschmidt S, Umen JG, Lin H, Lee J-H, Ishida K, Kubo T, Lau J, Goodenough UW (2005) Plus and minus sexual agglutinins from *Chlamydomonas reinhardtii*. Plant Cell 17:597–615

Fuhrmann M, Oertel W, Hegemann P (1999) A synthetic gene coding for the green fluorescent protein (GFP) is a versatile reporter in *Chlamydomonas reinhardtii*. Plant J 19:353–361

Goodenough UW (1989) Cyclic AMP enhances the sexual agglutinability of *Chlamydomonas* flagella. J Cell Biol 109:247–252

Goodenough UW, Adair WS, Collin-Osdoby P, Heuser JE (1985) Structure of the *Chlamydomonas* agglutinin and related flagellar surface proteins in vitro and in situ. J Cell Biol 101:924–941

Graham LE, Wilcox LW (2000) Algae. Prentice-Hall, Upper Saddle River

Hamada J, Yoshizawa-Katoh T, Tsunewaki K (1982) Genetic study on mating type genes by a new type of tetrad analysis in *Closterium ehrenbergii*. Bot Mag Tokyo 95:101–108

Hori S, Sekimoto H, Abe J (2012) Properties of cell surface carbohydrates in sexual reproduction of the *Closterium peracerosum–strigosum–littorale* complex (Zygnematophyceae, Charophyta). Phycol Res 60:254–260. doi:10.1111/j.1440-1835.2012.00656.x

Hunnicutt GR, Kosfiszer MG, Snell WJ (1990) Cell body and flagellar agglutinins in *Chlamydomonas reinhardtii*: the cell body plasma membrane is a reservoir for agglutinins whose migration to the flagella is regulated by a functional barrier. J Cell Biol 111:1605–1616

Ichimura T (1971) Sexual cell division and conjugation-papilla formation in sexual reproduction of *Closterium strigosum*. In: Nishizawa K (ed) Proceedings of the 7th international seaweed symposium. University of Tokyo Press, Tokyo, pp 208–214

Ichimura T (1973) The life cycle and its control in some species of *Closterium*, with special reference to the biological species problem. D. Science thesis, University of Tokyo

Karol KG, McCourt RM, Cimino MT, Delwiche CF (2001) The closest living relatives of land plants. Science 294:2351–2353. doi:10.1126/science.1065156

Kinoshita T, Fukuzawa H, Shimada T, Saito T, Matsuda Y (1992) Primary structure and expression of a gamete lytic enzyme in *Chlamydomonas reinhardtii*: similarity of functional domains to matrix metalloproteases. Proc Natl Acad Sci USA 89:4693–4697

Lehti-Shiu MD, Zou C, Hanada K, Shiu SH (2009) Evolutionary history and stress regulation of plant receptor-like kinase/pelle genes. Plant Physiol 150:12–26. doi:10.1104/pp. 108.134353

Lippert BE (1967) Sexual reproduction in *Closterium moniliferum* and *Closterium ehrenbergii*. J Phycol 3:182–198

Liu Y, Tewari R, Ning J, Blagborough AM, Garbom S, Pei J, Grishin NV, Steele RE, Sinden RE, Snell WJ, Billker O (2008) The conserved plant sterility gene HAP2 functions after attachment of fusogenic membranes in *Chlamydomonas* and *Plasmodium* gametes. Genes Dev 22: 1051–1068. doi:10.1101/gad.1656508

Liu Y, Misamore MJ, Snell WJ (2010) Membrane fusion triggers rapid degradation of two gamete-specific, fusion-essential proteins in a membrane block to polygamy in *Chlamydomonas*. Development (Camb) 137:1473–1481. doi:10.1242/dev.044743

Misamore MJ, Gupta S, Snell WJ (2003) The *Chlamydomonas* Fus1 protein is present on the mating type plus fusion organelle and required for a critical membrane adhesion event during fusion with minus gametes. Mol Biol Cell 14:2530–2542

Mori T, Kuroiwa H, Higashiyama T, Kuroiwa T (2006) Generative cell specific 1 is essential for angiosperm fertilization. Nat Cell Biol 8:64–71. doi:10.1038/ncb1345

Noguchi T (1988) Numerical and structural changes in dictyosomes during zygospore germination of *Closterium ehrenbergii*. Protoplasma 147:135–142

Noguchi T, Ueda K (1985) Cell walls, plasma membranes, and dictyosomes during zygote maturation of *Closterium ehrenbergii*. Protoplasma 128:64–71

Nojiri T, Fujii T, Sekimoto H (1995) Purification and characterization of a novel sex pheromone that induces the release of another sex pheromone during sexual reproduction of the heterothallic *Closterium peracerosum-strigosum-littorale* complex. Plant Cell Physiol 36:79–84

Pasquale SM, Goodenough UW (1987) Cyclic AMP functions as a primary sexual signal in gametes of *Chlamydomonas reinhardtii*. J Cell Biol 105:2279–2292

Pickett-Heaps JD, Fowke LC (1971) Conjugation in the desmid *Closterium littorale*. J Phycol 7:37–50

Saito T, Tsubo Y, Matsuda Y (1985) Synthesis and turnover of cell body agglutinin as a pool of flagellar surface agglutinin in *Chlamydomonas reinhardtii* gamete. Arch Microbiol 142:207–210

Saito T, Small L, Goodenough UW (1993) Activation of adenylyl cyclase in *Chlamydomonas reinhardtii* by adhesion and by heat. J Cell Biol 122:137–147

Sasaki G, Katoh K, Hirose N, Suga H, Kuma K, Miyata T, Su ZH (2007) Multiple receptor-like kinase cDNAs from liverwort *Marchantia polymorpha* and two charophycean green algae, *Closterium ehrenbergii* and *Nitella axillaris*: Extensive gene duplications and gene shufflings in the early evolution of streptophytes. Gene (Amst) 401:135–144. doi:10.1016/j.gene.2007.07.009

Sekimoto H (2002) Production and secretion of a biologically active *Closterium* sex pheromone by *Saccharomyces cerevisiae*. Plant Physiol Biochem 40:789–794

Sekimoto H, Fujii T (1992) Analysis of gametic protoplast release in the *Closterium peracerosum-strigosum-littorale* complex (Chlorophyta). J Phycol 28:615–619

Sekimoto H, Satoh S, Fujii T (1990) Biochemical and physiological properties of a protein inducing protoplast release during conjugation in the *Closterium peracerosum-strigosum-littorale* complex. Planta (Berl) 182:348–354

Sekimoto H, Inoki Y, Fujii T (1993a) Detection and evaluation of an inducer of diffusible mating pheromone of heterothallic *Closterium peracerosum-strigosum-littorale* complex. Plant Cell Physiol 37:991–996

Sekimoto H, Satoh S, Fujii T (1993b) Analysis of binding of biotinylated protoplast-release-inducing protein that induces release of gametic protoplasts in the *Closterium peracerosum-strigosum-littorale* complex. Planta (Berl) 189:468–474

Sekimoto H, Sone Y, Fujii T (1994a) cDNA cloning of a 42-kilodalton subunit of protoplast-release-inducing protein from *Closterium*. Plant Physiol 104:1095–1096

Sekimoto H, Sone Y, Fujii T (1994b) A cDNA encoding a 19-kilodalton subunit of protoplast-release-inducing protein from *Closterium*. Plant Physiol 105:447

Sekimoto H, Sone Y, Fujii T (1994c) Regulation of expression of the genes for a sex pheromone by an inducer of the sex pheromone in the *Closterium peracerosum-strigosum-littorale* complex. Planta (Berl) 193:137–144

Sekimoto H, Fukumoto R, Dohmae N, Takio K, Fujii T, Kamiya Y (1998) Molecular cloning of a novel sex pheromone responsible for the release of a different sex pheromone in *Closterium peracerosum-strigosum-littorale* complex. Plant Cell Physiol 39:1169–1175

Sekimoto H, Tanabe Y, Takizawa M, Ito N, Fukumoto R, Ito M (2003) Expressed sequence tags from the *Closterium peracerosum-strigosum-littorale* complex, a unicellular charophycean alga, in the sexual reproduction process. DNA Res 10:147–153

Sekimoto H, Tanabe Y, Tsuchikane Y, Shirosaki H, Fukuda H, Demura T, Ito M (2006) Gene expression profiling using cDNA microarray analysis of the sexual reproduction stage of the unicellular charophycean alga *Closterium peracerosum-strigosum-littorale* complex. Plant Physiol 141:271–279

Sekimoto H, Abe J, Tsuchikane Y (2012) New insights into the regulation of sexual reproduction in *Closterium*. Int Rev Cell Mol Biol 297:309–338. doi:10.1016/B978-0-12-394308-8.00014-5

Shiu SH, Bleecker AB (2001) Plant receptor-like kinase gene family: diversity, function, and signaling. Sci STKE 2001:re22. doi:10.1126/stke.2001.113.re22

Shiu SH, Bleecker AB (2003) Expansion of the receptor-like kinase/Pelle gene family and receptor-like proteins in *Arabidopsis*. Plant Physiol 132:530–543. doi:10.1104/pp. 103.021964

Shiu SH, Karlowski WM, Pan R, Tzeng YH, Mayer KF, Li WH (2004) Comparative analysis of the receptor-like kinase family in Arabidopsis and rice. Plant Cell 16:1220–1234. doi:10.1105/tpc.020834

Snell WJ, Eskue WA, Buchanan MJ (1989) Regulated secretion of a serine protease that activates an extracellular matrix-degrading metalloprotease during fertilization in *Chlamydomonas*. J Cell Biol 109:1689–1694

Stevens DR, Rochaix JD, Purton S (1996) The bacterial phleomycin resistance gene *ble* as a dominant selectable marker in *Chlamydomonas*. Mol Gen Genet 251:23–30

Tsuchikane Y, Fukumoto R, Akatsuka S, Fujii T, Sekimoto H (2003) Sex pheromones that induce sexual cell division in the *Closterium peracerosum-strigosum-littorale* complex (Charophyta). J Phycol 39:303–309

Tsuchikane Y, Ito M, Fujii T, Sekimoto H (2005) A sex pheromone, protoplast-release-inducing protein (PR-IP) Inducer, induces sexual cell division and production of PR-IP in *Closterium*. Plant Cell Physiol 46:1472–1476

Tsuchikane Y, Ito M, Sekimoto H (2008) Reproductive isolation by sex pheromones in the *Closterium peracerosum-strigosum-littorale* complex (Zynematales, Charophyceae). J Phycol 44:1197–1203

Tsuchikane Y, Kokubun Y, Sekimoto H (2010a) Characterization and molecular cloning of conjugation-regulating sex pheromones in homothallic *Closterium*. Plant Cell Physiol 51:1515–1523

Tsuchikane Y, Sato M, Ootaki T, Kokubun Y, Nozaki H, Ito M, Sekimoto H (2010b) Sexual processes and phylogenetic relationships of a homothallic strain in the *Closterium peracerosum-strigosum-littorale* complex (Zygnematales, Charophyceae). J Phycol 46:278–284

Tsuchikane Y, Tsuchiya M, Hindak F, Nozaki H, Sekimoto H (2012) Zygospore formation between homothallic and heterothallic strains of *Closterium*. Sex Plant Reprod 25:1–9. doi:10.1007/s00497-011-0174-z

Turmel M, Otis C, Lemieux C (2006) The chloroplast genome sequence of *Chara vulgaris* sheds new light into the closest green algal relatives of land plants. Mol Biol Evol 23:1324–1338. doi:10.1093/molbev/msk018

Watanabe MM, Ichimura T (1982) Biosystematic studies of the *Closterium peracerosum-strigosum-littorale* complex. IV. Hybrid breakdown between two closely related groups, group II-A and group II-B. Bot Mag Tokyo 95:241–247

Wodniok S, Brinkmann H, Glockner G, Heidel AJ, Philippe H, Melkonian M, Becker B (2011) Origin of land plants: do conjugating green algae hold the key? BMC Evol Biol 11:104. doi:10.1186/1471-2148-11-104

Zhang YH, Snell WJ (1994) Flagellar adhesion-dependent regulation of *Chlamydomonas* adenylyl cyclase in vitro: a possible role for protein kinases in sexual signaling. J Cell Biol 125:617–624

Zhu Y, Wan Y, Lin J (2010a) Multiple receptor complexes assembled for transmitting CLV3 signaling in *Arabidopsis*. Plant Signal Behav 5:300–302

Zhu Y, Wang Y, Li R, Song X, Wang Q, Huang S, Jin JB, Liu CM, Lin J (2010b) Analysis of interactions among the CLAVATA3 receptors reveals a direct interaction between CLAVATA2 and CORYNE in Arabidopsis. Plant J 61:223–233. doi:10.1111/j.1365-313X.2009.04049.x

Chapter 29
Fertilization of Brown Algae: Flagellar Function in Phototaxis and Chemotaxis

Gang Fu, Nana Kinoshita, Chikako Nagasato, and Taizo Motomura

Abstract Sexual reproduction of brown algae includes isogamy, anisogamy, and oogamy. The mechanisms of fertilization events remain largely unknown despite the diverse reproduction patterns. It is thought that the flagella of brown algal reproduction cells play crucial roles in not only cellular motility but also signal transduction in the aquatic habitat. Flagella of brown algae are composed of $9+2$ axonemes and several appendage structures, such as mastigonemes and a paraflagellar body, which have close associations with flagellar function. We observed flagellar activities during recognition and fusion of male and female gametes. We also investigated flagellar proteins involved in phototaxis of brown algal motile cells.

Keywords Brown algae • Chemotaxis • Fertilization • Flagella • Phototaxis

29.1 Flagellar Structure of Brown Algal Swarmers

Stramenopiles, including heterokonts, comprise an independent group of eukaryotes and are phylogenetically distinct from red and green algae, land plants, and animals (Baldauf 2003). One of the characteristic features of this group is that they have heterogeneous flagella, one of which is decorated with mastigonemes, tripartite fine hairs (van den Hoek et al. 1995). Brown algae, such as *Saccharina*, *Undaria*, and *Sargassum*, are a group of heterokonts and the only group in heterokonts having a complex multicellular organization.

In the life cycle of many brown algae, sporophyte and gametophyte generations exist independently, and sexual reproduction (male and female gametes, or sperm

G. Fu • N. Kinoshita • C. Nagasato • T. Motomura (✉)
Muroran Marine Station, Field Science Center for Northern Biosphere,
Hokkaido University, Muroran 051-0013, Japan
e-mail: motomura@fsc.hokudai.ac.jp

H. Sawada et al. (eds.), *Sexual Reproduction in Animals and Plants*,
DOI 10.1007/978-4-431-54589-7_29, © The Author(s) 2014

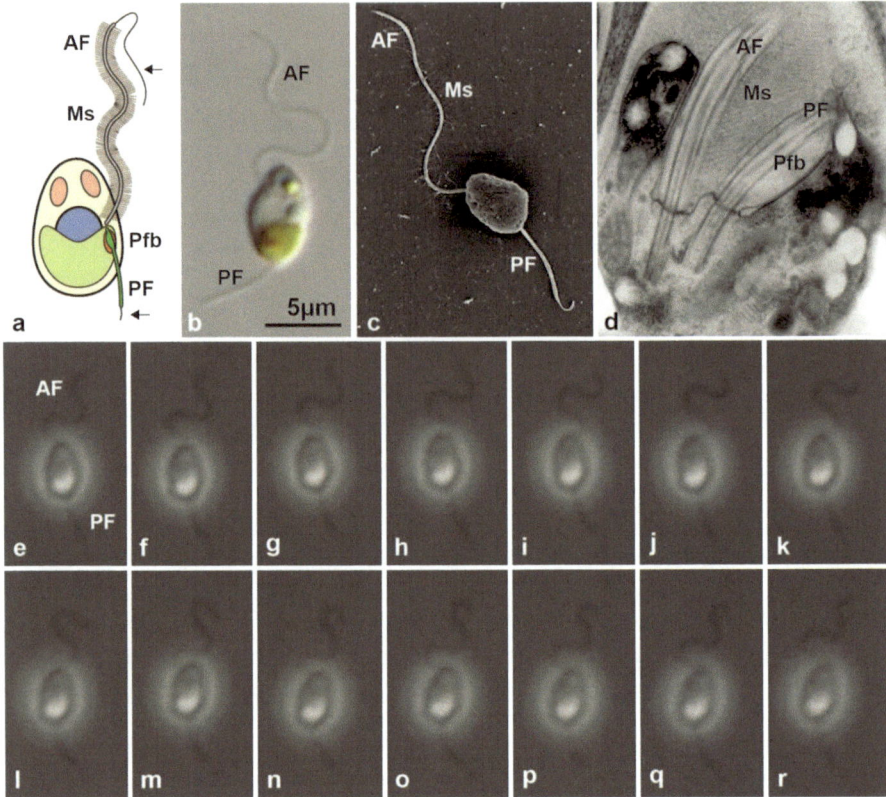

Fig. 29.1 Two heterogeneous flagella of brown algal motile cells. (**a**) Schematic representation of a motile cell. AF anterior flagellum, Ms mastigonemes, PF posterior flagellum, Pfb paraflagellar body. *Arrows* show the acronema, tip of the AF, and PF. (**b**) DIC image of a male gamete of *Scytosiphon lomentaria*. (**c**) SEM image of a male gamete of *S. lomentaria*. Note that mastigonemes are only on the AF. (**d**) TEM image of the AF and PF of *Ectocarpus siliculosus*. Note that Ms is on the AF and Pfb is on the PF. (**e–r**) High-speed video images (600 frame/s) of free swimming of a *S. lomentaria* male gamete

and egg) and asexual reproduction (zoospore or tetraspore after meiosis) connect the two generations. Swarmers, namely flagellated motile cells (gametes, sperm, and zoospores), have two heterogeneous flagella (Fig. 29.1a–d), a long anterior flagellum (AF) decorated with mastigonemes and a short posterior flagellum (PF) with a basal swollen part called the paraflagellar body, which is composed of crystallized materials and electron-dense materials (Green et al. 1989; Andersen 2004; Fu et al. 2012). The acronema, which is composed of two central microtubules of axonemes, makes the tip of each flagellum. The AF and PF are laterally inserted into the cell body in two opposite directions corresponding to the swimming orientation (O'Kelly 1989). Swimming force of motile cells of brown algae is produced by the AF, not the PF (Fig. 29.1e–r), because motile sperm of the brown alga *Dictyota* only have an

AF (Manton 1959), and it is thought that mastigonemes on the AF would be related to tractive force (Jahn et al. 1964). It has remained unclear how mastigonemes regularly attach only to the AF during flagellar elongation (Fu et al. 2012). The PF of brown algal motile cells functions as steering against signals of light (phototaxis) and chemicals (chemotaxis) (Geller and Müller 1981; Matsunaga et al. 2010).

29.2 Phototaxis of Brown Algae and a Putative Photoreceptor Protein

Phototaxis is widespread among motile cells of the major eukaryotic lineages, and a similar mechanism by which phototactic reorientation is complemented is shared despite the great diversity of cell types. When considering fertilization of algae living in an aquatic environment, several processes, including phototaxis, chemotaxis, and recognition, would be critical. Swarmers including gametes and zoospores of algae show a strong response to direction of light from the sun. For example, male and female gametes of the green alga *Monostroma angicava*, which have two equal-length flagella, swim toward the seawater surface after liberation from gametangia, showing positive phototaxis (Togashi et al. 1999). On the other hand, just after fertilization, motile zygotes having four flagella derived from male and female gametes show negative phototaxis and swim toward the bottom of the sea. The change of phototaxis before and after fertilization is necessary for continuous support of successful fertilization, because accumulation of unfertilized male and female gametes at the surface of the sea ensures easy fusion of both gametes. Zygotes settle on boulders on the bottom and develop into the next generation by negative phototaxis (Togashi et al. 1999; Togashi and Cox 2004). In the case of brown algae, for example, isogamous brown algae *Scyotosiphon lomentaria*, *Colpomenia bullosa*, and *Ectocarpus siliculosus*, freshly liberated male and female gametes show strong negative phototaxis, and female gametes settle on the substratum sooner than do male gametes and secrete a sexual pheromone that attracts male gametes. Therefore, in the case of algal fertilization in an aquatic environment, phototaxis reaction is important for dense accumulation of gametes.

In phototaxis reaction of motile cells, a specific photoreceptor senses the light source of restricted wavelength and transduces the signals to downstream molecular modules, which will eventually alter the flagellar beating activity with changing the swimming direction (Jekely 2009). However, different organisms have evolved exclusive strategies in response to light stimuli, for example, employing diverse photoreceptors and the corresponding downstream signaling. In the green alga *Chlamydomonas reinhardtii*, two light-gated cation-channel proteins, channelrhodopsin-1 (ChR1) and ChR2, were identified as photoreceptors regulating phototaxis through depolarizing photoelectric currents (Sineshchekov et al. 2002; Berthold et al. 2008). Similar to other types of rhodopsins, both proteins are 7-TM membrane proteins and bind retinal as a chromophore. An immunofluorescence assay indicated that ChR1 was localized near the eyespot, which is part of the chloroplast

(Suzuki et al. 2003). In the unicellular flagellate *Euglena gracilis*, photoactivated adenylyl cyclase (PAC), a flavoprotein, was shown to be the blue light receptor, which binds flavin adenine dinucleotide (FAD) as a chromophore and localizes to the paraflagellar body (Iseki et al. 2002; Ntefidou et al. 2003).

As another independent eukaryotic group, it has been well known that the PF of swarmers of brown algae has the capability to emit green autofluorescence when excited with blue light. Based on the results of spectral analysis, the substance that caused the green autofluorescence was identified as flavin (Müller et al. 1987; Kawai 1988). Flavin in the PF is widely distributed among chlorophyll *c*-containing algal species having an eyespot or paraflagellar body (Kawai and Inouye 1989; Kawai 1992). Close associations between the autofluorescence substance and presence of an eyespot or paraflagellar body, as well as results of spectral action studies on phototaxis of brown algal swarmers (Kawai et al. 1990, 1991), have suggested that the blue light receptor is a flavoprotein and is likely localized in the PF. However, the photoreceptor protein involved in phototaxis of brown algae has not yet been identified, although a fluorescent flagellar protein homologous to Old Yellow enzyme was found in isolated flagella of *S. lomentaria* (Fujita et al. 2005), which seems to play roles in general redox reactions rather than light-sensing activities.

In our recent flagellar proteomics studies based on the whole-genome sequence of the model brown alga *E. siliculosus* (Cock et al. 2010), a putative blue-light receptor protein was found in the PF of brown algal swarmers and might have a close relationship to phototaxis. Flagella were isolated from swarmers of several brown algal species, including *C. bullosa* and *S. lomentaria*, by vortexing in flagellar isolation buffer (30 mM HEPES, 5 mM $MgSO_4$, 5 mM EGTA, 25 mM KCl, 1 M Sorbitol, pH 7.0). Flagellar proteins were further digested by trypsin and subjected to LC-MS/MS analysis, which yielded about 600 proteins of brown algal flagella. Among PF-specific proteins that were identified by proteomics analysis, an RGS/LOV domain-containing protein was found to be a potential photoreceptor. This protein contained 1,522 amino-acid residues and the predicted molecular weight was 168 kDa. In contrast to known photoreceptors, the protein has a unique domain architecture of two RGS (regulator of G-protein signaling) domains and four LOV (light, oxygen, and voltage sensing) domains. It is well known that the LOV domain is a ubiquitous molecular module capable of binding FMN (flavin mononucleotide) as a chromophore in diverse photoreceptors (Crosson et al. 2003; Losi and Gärtner 2012; Suetsugu and Wada 2013). The RGS domain has a key activity in accelerating GTP hydrolysis by the Gα subunit; therefore, it is likely that heterotrimeric G proteins may be involved in the downstream signaling of blue-light sensing, which eventually modifies the beating pattern of the PF of swarmers.

Although the interactions between inner- and interproteins during phototaxis are far from understood, an antibody against the RGS/LOV domain-containing protein revealed that this protein is widely distributed in brown algal species. An immunofluorescence assay confirmed that this protein is localized throughout the PF with a stronger intensity at the paraflagellar body, corresponding to the distribution of green autofluorescence when observed under blue light. In addition, immunoelectron microscopy analysis revealed that the subcellular localization of this protein is in the compartment between the flagellar membrane and axoneme.

29.3 Chemotaxis of Brown Algae

Regarding sexual pheromones in brown algae, since the first discovery of the pheromone "ectocarpen" in *E. siliculosus* by the German phycologist Dieter G. Müller (Müller 1967, 1968; Müller et al. 1971), eleven sexual pheromones, including lamoxirene, fucoserraten, and hormosirene, have been reported (Maier and Müller 1986; Boland 1995). These brown algal pheromones are volatile, lipophilic, and fragrant and have a low molecular mass with unsaturated C_8 and C_{11} hydrocarbons with biogenetically related structures. Male gametes actively surround the sexual pheromone that female gametes secrete soon after their settlement on the substratum (Fig. 29.2). Motile female gametes never gather around a settled female one. Therefore, it is clear that the pheromone receptor must naturally exist in the male gamete, not in the female one. In the case of isogamous brown algae, such as *E. siliculosus* and *S. lomentaria*, male gametes freely swim in a straight or slightly curved track with maximum velocity in seawater without a pheromone (Fig. 29.1e–r), whereas the PF of male gametes has occasional beats and strong lateral bias with the signal of a pheromone (Fig. 29.2h–k) and, as a result, male gametes swim in a characteristic U-turn (Maier and Müller 1986). Those authors reported chemo-thigmo-klinokinesis, which means that the pheromone has two effects for attracting male gametes around settled female gametes: (1) reducing male gamete velocity by a thigmotactic response and (2) increasing beating frequency of the PF of male gametes in proportion to the pheromone concentration. Unfortunately, the pheromone receptor in male gametes has not yet been identified.

A male gamete attracted to a female gamete by the sexual pheromone shows a characteristic behavior in brown algal fertilization, first making contact with the surface of the female gamete by using the tip of the long AF (Fig. 29.2a–f) (Müller 1966), followed by fusion of both bodies. Therefore, initial recognition and contact between male and female gametes is carried out by using the AF of the male gamete. As already mentioned, the AF of gametes and zoospores of brown algae characteristically bears mastigonemes, which may be involved in the contact between male and female gametes. High-speed video and high-resolution scanning electron microscopy (SEM) reveal the the fusion process of both gametes (Fig. 29.2g–k).

29.4 Cytoplasmic Inheritance of Organelles

Finally, we briefly introduce the cytoplasmic inheritance of mitochondria, chloroplasts, and centrioles during zygote development of brown algae. Three types of sexual reproduction—isogamy, anisogamy, and oogamy—can be observed in brown algae, similar to green algae. Cytoplasmic inheritance of mitochondria, chloroplasts, and centrioles is restrictively regulated in each pattern of sexual reproduction (Motomura et al. 2010). In oogamy, mitochondria and chloroplasts of sperm are selectively digested in the lysosome after fertilization (Motomura 1990). In the case of isogamy, chloroplasts are biparentally inherited, whereas mitochondria

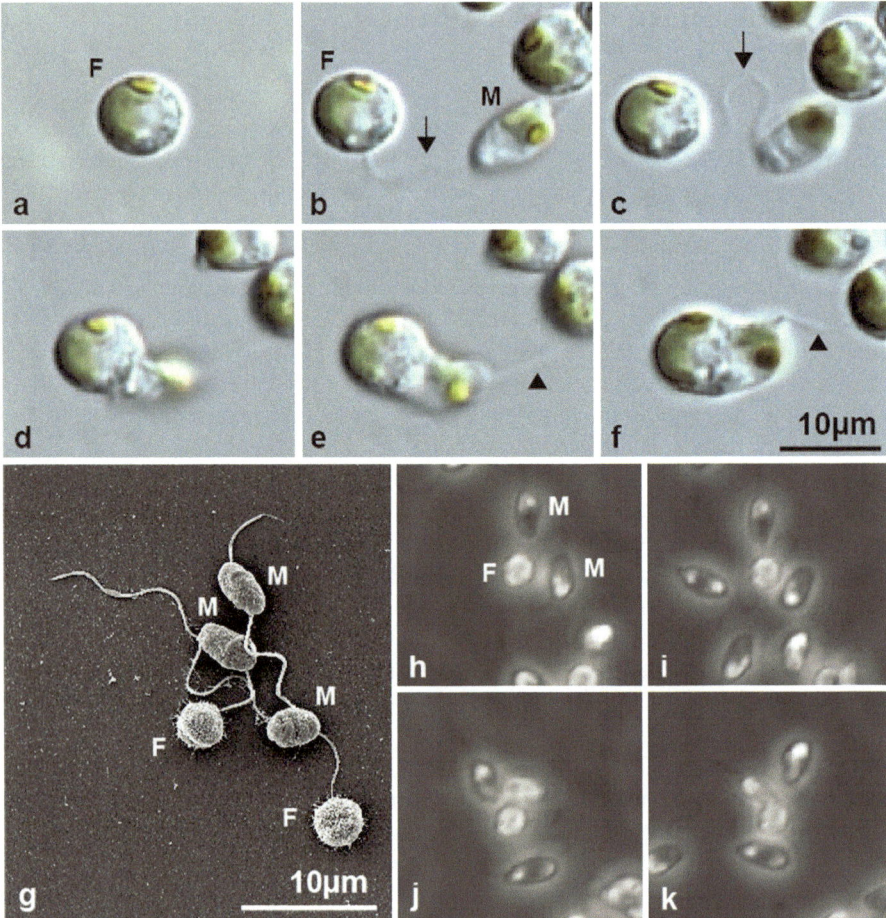

Fig. 29.2 Isogamous fertilization of *Scytosiphon lomentaria*. (**a–f**) Process of fertilization between male (*M*) and female (*F*) gametes. Female gametes settle on the substratum and then release a pheromone, by which male gametes are attracted. The AF of the male gamete (*arrow* in **b, c**) attaches to the cell surface of the female gamete with the flagellar tip. Cell fusion starts, and then the PF (*arrowheads* in **e, f**) of the male gamete is withdrawn. The zygote becomes a spherical shape within a few minutes. (**g**) SEM image. Three male gametes gather around two female gametes. (**h–k**) High-speed video images. Note that the PF of male gamete bends

(or mitochondrial DNA) derived from the female gamete only remain during zygote development (Nagasato and Motomura 2002; Peters et al. 2004; Kato et al. 2006; Kimura et al. 2010). Similar to the paternal inheritance of centrioles in animal fertilization (Schatten 1994), centrioles in zygotes are definitely derived from the male gamete regardless of the sexual reproduction pattern (Nagasato 2005). In isogamous *S. lomentaria*, degeneration of the maternal centrioles was found to start 1 h after fertilization with degradation of triplet MTs from the distal end, and in a 2-h-old zygote, there was no trace of the maternal centrioles ultrastructurally (Nagasato and Motomura 2004).

29.5 Perspectives

The flagellar structure is the most important character for defining the stramenopile (Heterokontae) in the eukaryote groups. Brown algal swarmers have a long AF bearing fine hairs, mastigonemes, and a short PF having the basal swelling, the parabasal body. With these morphological differences, their behaviors in gamete swimming are also characteristic. During the fertilization process, these heterogeneous flagella play crucial roles in phototaxis, chemotaxis, and gamete recognition. Our proteomics analysis on flagella of the brown algae identified first about 600 flagellar proteins, and AF-specific and PF-specific proteins were found. A candidate protein of the new blue-light receptor, RGS/LOV protein, working in phototaxis of gametes, could be also detected in PF-specific proteins. These molecular approaches will expand a new insight for understanding the function of flagella of male and female gametes of the brown algae during fertilization, including the pheromone receptor that may exist in flagella of male gametes and the molecular nature of the flagellar tip of male gametes for the first attachment to the surface of female gametes.

Acknowledgments We thank Drs. Kazuo Inaba and Kogiku Shiba, Shimoda Marine Research Center, University of Tsukuba, for kindly teaching us to operate the high-speed video camera. Dr. Tatyana A. Klochkova, Kamchatka State Technical University, kindly provided photographs for Fig. 29.2a–f. This study was supported by a Grant-in-Aid for Scientific Research on Innovative Areas from the Ministry of Education, Culture, Sports, Science and Technology.

References

Andersen RA (2004) Biology and systematics of heterokont and haptophyte algae. Am J Bot 91:1508–1522
Baldauf SL (2003) The deep roots of eukaryotes. Science 300:1703–1706
Berthold P, Tsunoda SP, Ernst OP, Mages W, Gradmann D, Hegemann P (2008) Channel rhodopsin-1 initiates phototaxis and photophobic responses in *Chlamydomonas* by immediate light-induced depolarization. Plant Cell 20:1665–1677
Boland W (1995) The chemistry of gamete attraction: Chemical structures, biosynthesis, and (a)biotic degradation of algal pheromones. Proc Natl Acad Sci USA 92:37–43
Cock MJ, Sterck L, Rouzé P, Scornet D, Allen AE, Amoutzias G, Anthouard V, Artiguenave F, Aury JM, Badger JH, Beszteri B, Billiau K, Bonnet E, Bothwell JHF, Bowler C, Boyen C, Brownlee C, Carrano CJ, Charrier B, Cho GY, Coelho SM, Collén J, Corre E, Silva CD, Delage L, Delaroque N, Dittami SM, Doulbeau S, Elias M, Farnham G, Gachon CMM, Gschloessl B, Heesch S, Jabbari K, Jubin C, Kawai H, Kimura K, Kloareg B, Küpper FC, Lang D, Le Bail A, Leblanc C, Lerouge P, Lohr M, Lopez PJ, Martens C, Maumus F, Michel G, Miranda-Saavedra D, Morales J, Moreau H, Motomura T, Nagasato C, Napoli CA, Nelson DR, Nyvall-Collén P, Peters AF, Pommier C, Potin P, Poulain J, Quesneville H, Read B, Rensing SA, Ritter A, Rousvoal S, Samanta M, Samson G, Schroeder DC, Ségurens B, Strittmatter M, Tonon T, Tregear J, Valentin L, von Dassow P, Yamagishi T, Van de Peer Y, Wincker P (2010) The

Ectocarpus genome and the independent evolution of multicellularity in the brown algae. Nature (Lond) 465:617–621

Crosson S, Rajagopal S, Moffat K (2003) The LOV domain family: photoresponsive signaling modules coupled to diverse output domains. Biochemistry 42:2–10

Fu G, Nagasato C, Ito T, Müller DG, Motomura T (2012) Ultrastructural analysis of flagellar development in plurilocular sporangia of *Ectocarpus siliculosus* (Phaeophyceae). Protoplasma 250:261–272

Fujita S, Iseki M, Yoshikawa S, Makino Y, Watanabe M, Motomura T, Kawai H, Murakami A (2005) Identification and characterization of a fluorescent flagellar protein from the brown alga *Scytosiphon lomentaria* (Scytosiphonales, Phaeophyceae): A flavoprotein homologous to Old Yellow Enzyme. Eur J Phycol 40:159–167

Geller A, Müller DG (1981) Analysis of the flagellar beat pattern of male *Ectocarpus siliculosus* gametes (Phaeophyta) in relation to chemotactic stimulation by female cells. J Exp Biol 92:53–66

Green JC, Leadbeater BSC, Diver WL (1989) The chromophyte algae: problems and perspectives. Oxford Science, Oxford, p 429

Iseki M, Matsunaga S, Murakami A, Ohno K, Shiga K, Yoshida K, Sugai M, Takahashi T, Hori T, Watanabe M (2002) A blue-light-activated adenylyl cyclase mediates photoavoidance in *Euglena gracilis*. Nature (Lond) 415:1047–1051

Jahn TL, Landman MD, Fonseca JR (1964) The mechanism of locomotion of flagellates. II. Function of the mastigonemes of *Ochromonas*. J Protozool 11:291–296

Jekely G (2009) Evolution of phototaxis. Philos Trans R Soc Lond B Biol Sci 364:2795–2808

Kato Y, Kogame K, Nagasato C, Motomura T (2006) Inheritance of mitochondrial and chloroplast genomes in the isogamous brown alga *Scytosiphon lomentaria* (Phaeophyceae). Phycol Res 54:65–71

Kawai H (1988) A flavin-like autofluorescent substance in the posterior flagellum of golden and brown algae. J Phycol 24:114–117

Kawai H (1992) Green flagellar autofluorescence in brown algal swarmers and their phototactic responses. Bot Mag Tokyo 105:171–184

Kawai H, Inouye I (1989) Flagellar autofluorescence in forty-four chlorophyll *c*-containing algae. Phycologia 28:222–227

Kawai H, Müller DG, Fölster E, Häder DP (1990) Phototactic responses in the gametes of the brown alga, *Ectocarpus siliculosus*. Planta (Berl) 182:292–297

Kawai H, Kubota M, Kondo T, Watanabe M (1991) Action spectra for phototaxis in zoospores of the brown alga *Pseudochorda gracilis*. Mol Cell Biochem 161:17–22

Kimura K, Nagasato C, Kogame K, Motomura T (2010) Disappearance of male mitochondrial DNA after 4-celled stage sporophyte of the isogamous brown alga *Scytosiphon lomentaria* (Scytosiphonales, Phaeophyceae). J Phycol 46:143–152

Losi A, Gärtner W (2012) The evolution of flavin-binding photoreceptors: an ancient chromophore serving trendy blue-light sensors. Annu Rev Plant Biol 63:49–72

Maier I, Müller DG (1986) Sexual pheromones in algae. Biol Bull 170:145–175

Manton I (1959) Observations on the internal structure of the spermatozoid of *Dictyota*. J Exp Bot 10:448–461

Matsunaga S, Uchida H, Iseki M, Watanabe M, Murakami A (2010) Flagellar motions in phototactic steering in a brown algal swarmer. Photochem Photobiol 86:374–381

Motomura T (1990) Ultrastructure of fertilization in *Laminaria angustata* (Phaeophyta, Laminariales) with emphasis on the behavior of centrioles, mitochondria and chloroplasts of the sperm. J Phycol 26:80–89

Motomura T, Nagasato C, Kimura K (2010) Cytoplasmic inheritance of organelles in brown algae. J Plant Res 123:185–192

Müller DG (1966) Untersuchungen zur Entwicklungsgeschichte der Braunalge *Ectocarpus siliculosus* aus Neapel. Planta (Berl) 98:57–68

Müller DG (1967) Ein leicht flüchtiges Gyno-Gamon der Braunalge *Ectocarpus siliculosus*. Naturwissenshaften 54:496–497

Müller DG (1968) Versuche zur Charakterisierung eines Sexuallockstoffers bei Braunalge *Ectocarpus siliculosus*. I. Methoden, Isolierung und gaschromatographischer Nachweis. Planta (Berl) 81:160–168

Müller DG, Jaenicke L, Donike M, Akintobi T (1971) Sex attractant in a brown alga: chemical structure. Science 171:815–817

Müller DG, Maier I, Müller H (1987) Flagellum autofluorescence and photoaccumulation in heterokont algae. Photochem Photobiol 46:1003–1008

Nagasato C (2005) Behavior and function of paternally inherited centrioles in brown algal zygotes. J Plant Res 118:361–369

Nagasato C, Motomura T (2002) Influence of the centrosome in cytokinesis of brown algae: polyspermic zygotes of *Scytosiphon lomentaria* (Scytosiphonales, Phaeophyceae). J Cell Sci 115:2541–2548

Nagasato C, Motomura T (2004) Destruction of maternal centrioles during fertilization of the brown algae, *Scytosiphon lomentaria* (Scytosiphonales, Phaeophyceae). Cell Motil Cytoskeleton 59:109–118

Ntefidou M, Iseki M, Watanabe M, Lebert M, Häder D-P (2003) Photoactivated adenylyl cyclase controls phototaxis in the flagellate *Euglena gracilis*. Plant Physiol 133:1517–1521

O'Kelly CJ (1989) The evolutionary origin of the brown algae: information from studies of motile cell ultrastructure. In: Green JC, Leadbeater BSC, Diver WL (eds) The chromophyte algae: problems and perspectives. Oxford University Press, Oxford, pp 255–278

Peters AF, Scornet D, Müller DG, Kloareg B, Cock JM (2004) Inheritance of organelles in artificial hybrids of the isogamous multicelluar chromist alga *Ectocarpus siliculosus* (Phaeophyceae). Eur J Phycol 39:235–242

Schatten G (1994) The centrosome and its mode of inheritance: the reduction of the centrosome during gametogenesis and its restoration during fertilization. Dev Biol 165:229–335

Sineshchekov OA, Jung K-H, Spudich JL (2002) Two rhodopsins mediate phototaxis to low- and high-intensity light in *Chlamydomonas reinhardtii*. Proc Natl Acad Sci USA 97:8689–8694

Suetsugu N, Wada M (2013) Evolution of three LOV blue light receptor families in green plants and photosynthetic stramenopiles: phototropin, ZTL/FKF1/LKP2 and aureochrome. Plant Cell Physiol 54:8–23

Suzuki T, Yamasaki K, Fujita S, Oda K, Iseki M, Yoshida K, Watanabe M, Daiyasu H, Toh H, Asamizu E (2003) Archaeal-type rhodopsins in *Chlamydomonas*: model structure and intracellular localization. Biochem Biophys Res Commun 301:711–717

Togashi T, Cox PA (2004) Phototaxis and the evolution of isogamy and 'slight anisogamy' in marine green algae: insights from laboratory observations and numerical experiments. Bot J Linn Soc 144:321–327

Togashi T, Motomura T, Ichimura T, Cox PA (1999) Gametic behavior in a marine green alga, *Monostroma angicava*: an effect of phototaxis on mating efficiency. Sex Plant Reprod 12:158–163

van den Hoek C, Mann DG, Jahns HM (1995) Algae: an introduction to phycology. Cambridge University Press, Cambridge, p 623

Chapter 30
Gene and Protein Expression Profiles in Rice Gametes and Zygotes: A Cue for Understanding the Mechanisms of Gametic and Early Zygotic Development in Angiosperms

Takashi Okamoto

Abstract In angiosperms, female gamete differentiation, fertilization, and subsequent zygotic development occur in embryo sacs deeply embedded in the ovaries. Despite their importance in plant reproduction and development, how the egg cell is specialized, fuses with the sperm cell, and converts into an active zygote for early embryogenesis remains unclear. This lack of knowledge is partly attributable to the difficulty of direct analyses of gametes and zygotes in angiosperms. Cell type-specific transcriptomes were obtained by microarray analyses for egg cells, sperm cells and zygotes isolated from rice flowers, and up- or down-regurated genes in zygotes after fertilization were identified as well as genes enriched in male and female gametes. In addition to transcriptome, proteins expressing in egg and sperm cells were globally detected by highly sensitive liquid chromatography coupled with tandem mass spectroscopy technology, and proteins that are specifically or predominantly expressing in gametes were also identified by comparison of protein expression profiles between gametes and somatic cells/pollen grains. Several rice or *Arabidopsis* lines with mutations in genes identified by these proteome/transcriptome analyses showed clear phenotypic defects in seed set or seed development. These findings suggest that the cell type-specific proteome/transcriptome data for gametes/zygotes are foundational information toward understanding the mechanisms of gametic and early zygotic development in angiosperms.

Keywords Egg cell • Fertilization • *Oryza sativa* • Proteome • Sperm cell • Transcriptome • Zygote

T. Okamoto (✉)
Department of Biological Sciences, Tokyo Metropolitan University,
Hachioji, Tokyo 192-0397, Japan
e-mail: okamoto-takashi@tmu.ac.jp

H. Sawada et al. (eds.), *Sexual Reproduction in Animals and Plants*,
DOI 10.1007/978-4-431-54589-7_30, © The Author(s) 2014

30.1 Introduction

In angiosperms, the sporophytic generation is initiated by double fertilization, resulting in the formation of seeds (reviewed in Raghavan 2003). In double fertilization, one sperm cell from the pollen grain fuses with the egg cell, and the resultant zygote develops into an embryo that transmits genetic material from the parents to the next generation. The central cell fuses with the second sperm cell to form a triploid primary endosperm cell, which develops into the endosperm that nourishes the developing embryo/seedling (Nawaschin 1898; Guignard 1899; Russell 1992). The conversion of the egg cell into the zygote is completed by two serial gametic processes: plasmogamy, the fusion of the plasma membrane between male and female gametes, and karyogamy, fusion of the nuclei of the male and female gametes in the fused gamete. Thereafter, the zygotic genome switches on within hours of fertilization for subsequent development of zygotes (Meyer and Scholten 2007; Zhao et al. 2011; Nodine and Bartel 2012).

As for molecular players of plasmogamy, GENERATIVE CELL-SPECIFIC 1/HAPLESS 2 (GCS1/HAP2) and EGG CELL 1 (EC1) have been identified as putative fusiogens for male and female gametes, respectively (Mori et al. 2006; von Besser et al. 2006; Sprunck et al. 2012). GCS1/HAP2 was identified as a key male membrane protein with a single transmembrane domain and a histidine-rich domain in the extracellular region. Recently, Sprunck et al. (2012) indicated that small cysteine-rich EC1 proteins accumulated in storage vesicles in the *Arabidopsis* egg cell are secreted via exocytosis upon sperm cell attachment to the egg cell, and that the secreted EC1 proteins function in redistribution of GCS1/HAP2 proteins to the sperm cell surface, resulting in successful gamete fusion. In addition to these two possible fusiogens, other players should be identified to understand the mechanisms in plasmogamy.

Karyogamy is accompanied by the congression of the male nucleus to the female nucleus and subsequent nuclear fusion. Yeast mating is the most intensively investigated karyogamy event, and cytoskeleton-dependent nuclear congression and chaperone/ER-protein-dependent nuclear fusion have been well studied (Kurihara et al. 1994; Melloy et al. 2009; Tartakoff and Jaiswal 2009). However, in angiosperms, the mechanism in karyogamy is poorly understood, except that Bip, a chaperone in the lumen of endoplasmic reticulum, and NFD1, a component of the mitochondrial ribosome, function in polar nuclei fusion in *Arabidopsis* (Portereiko et al. 2006; Maruyama et al. 2010).

In contrast to animals and lower plants, which have free-living gametes, angiosperm fertilization and subsequent events, such as embryogenesis and endosperm development, occur in the embryo sac, which is deeply embedded in ovular tissue. Difficulties in directly researching the biology of the embedded female gametophyte, zygote, and early embryo have impeded investigations into the molecular mechanisms of fertilization and embryogenesis. Therefore, such studies have been conducted predominantly through analyses of *Arabidopsis* mutants or transformants coupled with live-imaging (Berger 2011; Hamamura et al. 2012). Alternatively, direct analyses using isolated gametes or zygotes are possible because procedures for isolating viable gametes have been established, and an in vitro fertilization (IVF) system using the isolated gametes can be used to observe and analyze fertilization and postfertilization processes directly (Wang et al. 2006).

It has been supposed that genes specifically/predominantly expressing in gamete function in reproductive or developmental processes such as gamete differentiation, gamete fusion, and early zygotic development. Therefore, using isolated gametes or embryos, several studies have successfully identified genes specifically expressed in male gametes, female gametes, or early embryos (Kasahara et al. 2005; Márton et al. 2005; Sprunck et al. 2005; Ning et al. 2006; Yang et al. 2006; Steffen et al. 2007; Borges et al. 2008; Amien et al. 2010; Wang et al. 2010; Wuest et al. 2010; Ohnishi et al. 2011). Moreover, changes in gene expression from prefertilization to postfertilization phases were recently monitored using microarray-based transcriptome analyses of rice sperm cells, egg cells, and zygotes using the same experimental platform (Abiko et al. 2013b). In addition to gene expression profiles, single-cell-type proteomic approaches have been widely employed to dissect the functions of specific cells, because cellular-level information is diluted when organs or tissues, which comprise various differentiated cells, are used as starting materials (Dai and Chen 2012). However, such global proteomic analyses have not been conducted for plant gametes, possibly because of the difficulty in obtaining sufficient highly pure homogeneous cells, especially for egg cells. However, state-of-the-art proteomics technologies enable high-throughput and high-resolution analyses using such limited numbers of cells, and recently proteins expressing in the rice gamete were globally identified (Abiko et al. 2013a).

30.2 Isolation of Plant Gametes and IVF

Procedures for isolating viable gametes have been reported for a wide range of plant species, including maize, wheat, tobacco, rape, rice, barley, *Plumbago zeylanica*, *Alstroemeria*, and *Arabidopsis* (Kranz 1999; Okamoto 2011). IVF systems using isolated male and female gametes have been utilized to dissect fertilization-induced events. The IVF system used for angiosperms includes a combination of three basic microtechniques: (1) the isolation and selection of male and female gametes, (2) the fusion of pairs of gametes, and (3) single-cell culture (Kranz 1999). A complete IVF system was first developed by Kranz and Lörz (1993) using maize gametes and electrical fusion, and a rice IVF system was also established to take advantage of the abundant resources stemming from rice research, including the whole genome sequence and abundant mutant stocks (Fig. 30.1; Uchiumi et al. 2007b). These IVF systems have been successfully used to observe and analyze postfertilization events, such as karyogamy in zygotes (Faure et al. 1993), egg/zygote activation and development (Kranz et al. 1995), decondensation of paternal chromatin in zygotes (Scholten et al. 2002), changes in the microtubular architecture in zygotes (Hoshino et al. 2004), fertilization-induced/suppressed gene expression (Okamoto et al. 2005), epigenetic resetting in early embryos (Jahnke and Scholten 2009), positional relationship between gamete fusion point and zygotic development (Nakajima et al. 2010), and asymmetrical division of zygotes (Sato et al. 2010).

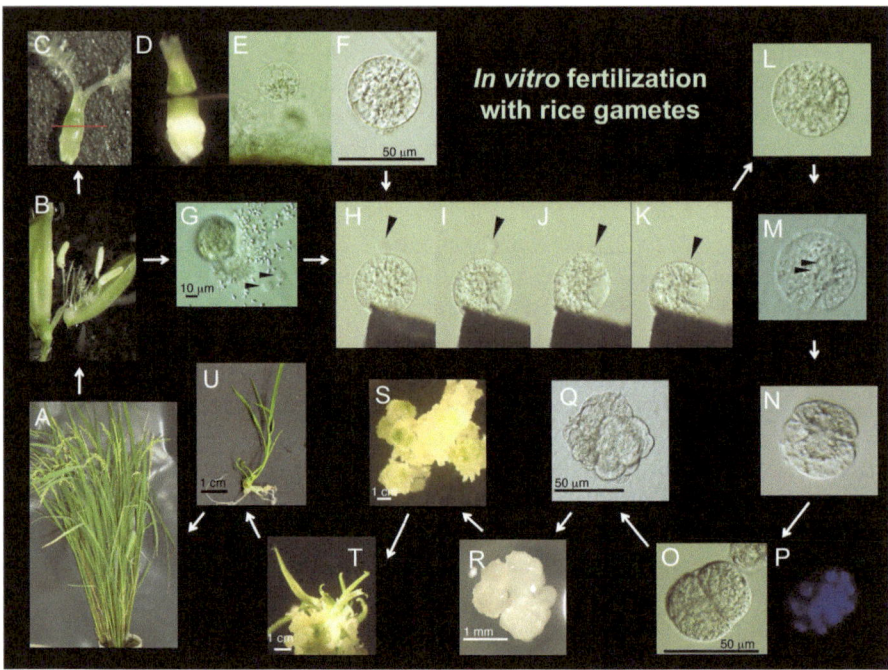

Fig. 30.1 Isolation of gamete from rice flowers (**a–g**), in vitro fusion of gametes (**h–k**), early development of zygote produced by in vitro fertilization into a globular embryo (**l–p**), and development and regeneration of globular embryos (**q–u**). (**a**) Rice plants. (**b**) Dissected rice flower. (**c**) Isolated ovary. *Red line* indicates incision line for egg isolation. (**d**) Cut ovary. (**e**) Rice egg cell being released from basal portion of dissected ovary. (**f**) Isolated egg cell. (**g**) Two sperm cells released from a pollen grain. (**h–k**) Serial images for electro-fusion of a sperm cell with an egg cell. Gametes were aligned on one of the electrodes under an alternating current field, and the aligned egg and sperm cells are fused by a negative direct current pulse. *Arrowheads* indicate sperm cell or fusion point. (**l**) Zygote 1 h after fusion. (**m**) Zygote 4 h after fusion. Two nucleoli are indicated by *arrowheads*. (**n**) Asymmetrical two-celled embryo. (**o, p**) Nuclear staining of globular embryo, visualized by bright-field and fluorescence microscopy, respectively. (**q**) Cell mass developed from globular-like embryo. (**r**) White cell colony from cell mass in **q**. (**s**) Developed cell colony after transferring white cell colony into regeneration medium. *Green spots* are visible on the cell colony. (**t**) Regenerated shoots from *green spots* in **s**. (**u**) A plantlet after subculturing a regenerated shoot

30.3 Gene Expression Profiles in Rice Gametes and Zygotes

30.3.1 Genes Enriched in Rice Gametes

Cell type-specific transcriptomes were obtained by microarray analyses using 33 to 111 egg cells (33, 34, and 111 egg cells; three biological replicates), 30 and 34 zygotes (two biological replicates), approximately 3,000 sperm cells (two biological replicates), and approximately 100 pollen grains (three biological replicates); subsequent

Table 30.1 Genes enriched in rice egg cells

Gene locus	Annotation	Expression in egg cell
Os03g0296600	Similar to ECA1 protein	8.82
Os05g0491400	Similar to LRR protein	6.25
Os07g0574500	Ubiquitin domain-containing protein	5.58
Os04g0289600	Allergen V5/Tpx-1-related family protein	4.65
Os01g0299700	3′-5′-Exonuclease domain-containing protein	4.45
Os06g0602400	Similar to DEAD-box protein 3, X-chromosomal	4.26
Os11g0187600	Similar to heat shock protein 70	3.65
Os07g0108900	Similar to MADS-box transcription factor 15	3.02
Os07g0136300	Conserved hypothetical protein	2.80
Os03g0679800	Similar to TPR domain-containing protein	2.95
Os10g0560200	Protein of unknown function Cys-rich family protein	2.93
Os01g0350500	Conserved hypothetical protein	2.66
Os11g0579900	Armadillo-like helical domain-containing protein	2.38
Os05g0153200	Region of unknown function XH domain-containing protein	2.28

Values are average of binary log values of two biological replicates
Annotations are referred from The Rice Annotation Project Database (RAPDB)
Mean t test P values < 0.05

data processing resulted in identification of 14 and 19 genes with expression profiles specific to egg cells and sperm cells, respectively (Tables 30.1, 30.2).

A gene enriched in egg cells, Os11g0187600, encodes heat shock protein 70 (HSP70). In addition to HSP70, HSP90 was identified as a major protein component of rice egg cells by previous proteomic analysis (Uchiumi et al. 2007a). Interestingly, Calvert et al. (2003) revealed that mouse eggs contain molecular chaperones, including HSP90, HSP70, and protein disulfide isomerase (PDI), as major protein components. An abundance of HSP proteins may be a common characteristic of mammalian and plant eggs. Among pleiotropic functions of HSPs, it will be notable that they play a role in buffering the expression of genetic variation when divergent ecotypes are crossed and profoundly affect developmental plasticity in response to environmental cues (Queitsch et al. 2002; Sangster and Queitsch 2005). HSPs in egg cells may function following fertilization by a sperm cell, because conversion of an egg cell into a zygote represents major genetic and environmental changes. MADS-box proteins, the DNA-binding proteins that regulate their own transcription and that of target genes (West et al. 1998), act early in organ development (Riechmann and Meyerowitz 1997; Theissen et al. 2000). Os07g0108900, encoding MADS-box transcription factor 15 (OsMADS15), was egg enriched, and, *ZmMADS3*, which is orthologous gene to *OsMADS15*, is also strongly expressed in maize egg cells (Heuer et al. 2001). Although their function in egg cells is still unclear, MADS-box proteins that accumulate in female gametes may have roles in egg-cell differentiation during gametophytogenesis or zygotic development after fertilization.

Among 19 genes enriched in sperm cells, 9 were annotated as hypothetical proteins or genes (Table 30.2), being consistent with a previous report indicating

Table 30.2 Genes enriched in rice sperm cells

Gene locus	Annotation	Expression in sperm cell
Os01g0605400	Quinon protein alcohol dehydrogenase-like domain-containing protein	6.21
Os06g0715300	Similar to CEL5	5.72
Os02g0664400	Hypothetical conserved gene	5.04
Os01g0180900	Similar to oxidoreductase	5.08
Os07g0634100	Hypothetical conserved gene	4.60
Os10g0550400	Hypothetical conserved gene	4.13
Os03g0809200	Similar to transcription factor EmBP-1	4.22
Os01g0876100	Similar to chloride channel	3.61
Os01g0853600	Conserved hypothetical protein	3.47
Os03g0661900	Peptidase, trypsin-like serine, and cysteine domain-containing protein	3.18
Os11g0601600	Protein of unknown function DUF248	3.14
Os02g0177400	Conserved hypothetical protein	3.03
Os05g0393800	Protein of unknown function DUF221 domain-containing protein	2.45
Os08g0266700	Rad21/Rec8-like protein, C-terminal domain-containing protein	2.28
Os09g0483200	Similar to UBQ13 (ubiquitin 13)	2.12
Os09g0244200	Conserved hypothetical protein	1.96
Os11g0620800	Hypothetical conserved gene	1.46
Os02g0628100	Hypothetical gene	1.22
Os08g0474400	Hypothetical conserved gene	1.19

Values are average of binary log values of two biological replicates
Annotations are referred from The Rice Annotation Project Database (RAPDB)
Mean t test P values < 0.05

enrichment of genes encoding proteins with unknown function in sperm-specific genes (Russell et al. 2012). Os03g0661900 encodes a trypsin-like serine protease, and, in animals, serine proteases in the trypsin family can be expressed in sperm and involved in fertilization, although their molecular mechanisms during the fertilization process remain unknown (Sawada et al. 1984, 1996; Baba et al. 1994; Adham et al. 1997). Trypsin-like protease may be expressed in male gametes of both plants and animals and perhaps have similar roles in gamete attachment, recognition, or fusion, although the fertilization systems are largely divergent in the kingdoms.

30.3.2 Genes Down- or Up-Regulated in Rice Zygotes After Fertilization

Egg cells are developmentally quiescent, a state that is broken after fertilization and subsequent egg activation. Genes down-regulated after fertilization in zygotes were searched because the expression of genes involved in maintaining egg-cell quiescence should be suppressed in zygotes. Ninety-four genes that had threefold-lower

Table 30.3 Fifteen genes whose expression levels were most putatively up-regurated in zygotes after fertilization

Gene locus	Annotations	Fold change (zygote/egg)
Os01g0840300	Similar to WUSCHEL-related homeobox 5	44.9
Os05g0571200	Similar to WRKY transcription factor 19	41.1
Os01g0841700	Similar to isoform ERG1b of elicitor-responsive protein 1	39.6
Os07g0182900	Similar to cytosine-5-DNA methyltransferase MET1	35.7
Os02g0258200	High-mobility group, HMG1/HMG2 domain-containing protein	35.2
Os02g0462800	WRKY transcription factor 42	29.9
Os01g0895600	Similar to calreticulin 3	29.1
Os03g0279200	Similar to histone H2A	27.5
Os10g0580900	Conserved hypothetical protein	24.8
Os05g0127300	Serine/threonine protein kinase domain-containing protein	23.0
Os08g0562800	Similar to transparent testa 12 protein	19.4
Os03g0214100	Replication protein A1	18.7
Os03g0188500	Glutelin family protein	18.6
Os01g0551000	Conserved hypothetical protein	16.3
Os02g0572600	Protein kinase PKN/PRK1, effector domain-containing protein	16.0

Annotations are referred from The Rice Annotation Project Database (RAPDB)
Mean t test P values < 0.05

expression levels in zygotes than in egg cells were obtained, and most ontologies for these genes were related to metabolic or biosynthetic processes, including terpene, flavonoid, and amino-acid synthetic pathways (Abiko et al. 2013b).

Upon fertilization, the developmentally quiescent egg cell converts to an active zygote, and expression of genes involved in zygotic development should be induced. Comprehensive overviews of metabolism and regulation in zygotes, compared to egg cells, indicated that synthetic pathways for cell wall, auxin and ethylene and signal transduction pathways appeared to be activated via fertilization. A total of 325 genes whose expression levels in zygotes were threefold higher than those in egg cells were identified, and genes related to chromatin and DNA organization and assembly were well represented among these up-regurated genes (Table 30.3). The gene Os07g0182900, encoding DNA methyltransferase 1 (MET1), which functions in maintaining CG DNA methylation (Kankel et al. 2003), was identified among the highly up-regurated genes, and the specific inhibitor for the enzyme partly affected polarity or division asymmetry in rice zygotes (Abiko et al. 2013b). In addition, several genes encoding homeobox protein or transcription factors were strongly induced in zygotes. Os01g0840300 encodes a Wuschel-related homeobox (WOX) protein, the key regulator in determining cell fate in plants (Mayer et al. 1998; Haecker et al. 2004; Zhao et al. 2009), and 15 *WOX* genes, including *WUSCHEL*, have been identified in *Arabidopsis*. Interestingly, Os01g0840300 has been reported as the rice orthologue of *Arabidopsis WOX2* (Deveaux et al. 2008), whose transcripts accumulate in *Arabidopsis* zygotes and are restricted to the apical cell of two-celled proembryos (Haecker et al. 2004). In addition, *WOX2* has been proposed to be the predominant regulator of apical patterning (Jeong et al. 2011), suggesting WOX proteins encoded by *Os01g0840300* may have a role in determining cell fate during early embryogenesis in rice.

30.4   Protein Expression Profiles in Rice Gametes

Lysates from 500 egg cells and 3×10^4 sperm cells were separated by one-dimensional polyacrylamide gel electrophoresis. Proteins in gel were digested with trypsin and identified by a direct nanoflow LC-MS system equipped with an Orbi Trap XL mass spectrometer. The proteins were judged as "identified" if at least two peptides were identified from the protein. Proteome analyses were also conducted for seedlings, callus, and pollen grains to compare their protein expression profiles to those of gametes. By analyzing proteins from egg and sperm cell lysates, 1,276 and 1,076 proteins were identified, respectively. In callus, seedlings, and pollen grains, 1,641, 1,329, and 1,274 proteins were detected, respectively. Putative proteins specifically or predominantly expressing in egg or sperm cells were chosen on a basis of comparison of the number of matched peptides in egg or sperm cells with those in other cell types. In total, 102 and 73 putative proteins were identified as egg- or sperm-specific or predominant proteins, respectively (Abiko et al. 2013a). Table 30.4 presents putative gamete-specific/predominant proteins with more than five matched peptides. Notably, except for HSP 70 (HSP70), none of these proteins has been reported to play a role in reproductive or developmental processes, suggesting that investigating these proteins further may uncover novel molecular mechanisms during gametic development and fusion and early embryogenesis.

30.5   Conclusion

Functional defects in proteins, whose expressions are specific/predominant in gametes or up- or down-regurated after fertilization, are supposed to affect reproductive or developmental processes. In fact, several rice or *Arabidopsis* lines with mutations in genes encoding the putative gamete-specific or -predominant proteins showed clear phenotypic defects in seed set or seed development (Fig. 30.2), suggesting that the cell type-specific proteome and transcriptome data for gametes and zygotes are foundational information toward understanding the mechanisms of gametic and zygotic development in angiosperms.

Table 30.4 Proteins specifically or predominantly expressed in egg or sperm cells with more than five matched peptides

cDNA accession	Gene locus	Number of matched peptide					Annotations
		E	S	C	L	P	
AK106474	Os06g0602400	15	0	0	0	1	Similar to DEAD-box protein 3, X-chromosomal
AK101183	Os05g0168800	11	0	1	0	0	KIP1-like domain-containing protein
AK065887	Os03g0283100	11	0	1	0	0	Similar to In2-1 protein
AK063589	Os05g0115600	9	0	0	0	0	Protein of unknown function DUF674 family protein
AK106371	Os03g0276800	9	0	0	0	0	Heat shock protein Hsp70 family protein
AK063560	Os12g0600100	8	0	0	0	0	Tetratricopeptide-like helical domain-containing protein
AK067215	Os01g0698000	8	0	0	0	0	Conserved hypothetical protein
AK121612	Os02g0717400	8	0	0	0	1	Tetratricopeptide-like helical domain-containing protein
AK058611	Os01g0895100	7	0	0	0	0	Similar to membrane-associated 30-kDa protein, chloroplast
Os06t0706700-01	Os06t0706700-01	7	0	0	0	0	Similar to PsAD1
AK073477	Os01g0369200	7	0	0	0	0	Similar to cullin-1
AK106478	Os01g0771100	7	0	1	0	0	Mitochondrial glycoprotein family protein
AK107844	Os05g0143600	6	0	0	1	0	Similar to jasmonate-induced protein
AK072587	Os05g0164900	6	0	0	0	0	Galactose oxidase/kelch, beta-propeller domain-containing protein
AK072334	Os03g0583900	5	0	0	0	0	DEAD-like helicase, N-terminal domain-containing protein
AK119521	Os06g0175800	5	0	0	0	0	Similar to cystathionine beta-lyase, chloroplast precursor
AK072719	Os10g0574800	5	0	0	0	0	Similar to ARF GAP-like zinc finger-containing protein ZIGA2
AK064995	Os12g0197500	5	0	0	0	0	Putative zinc finger, XS, and XH domain-containing protein
Os01t0267600-01	Os01g0267600	1	10	0	0	0	Sad1/UNC-like, C-terminal domain-containing protein
AK071495	Os11g0255300	0	9	0	0	0	Cysteine endopeptidase
AK071561	Os05g0163700	0	7	0	0	0	Similar to acyl-coenzyme A oxidase 4, peroxisomal
AK065231	Os01g0323100	0	7	0	0	0	Similar to Pto kinase interactor 1
AK107034	Os02g0185200	0	6	0	0	0	Cytochrome P450 family protein
AK065311	Os06g0174400	0	6	0	1	0	Similar to vesicle-associated membrane protein 712
Os04t0611200-00	Os04g0611200	0	6	0	1	0	Similar to OSIGBa0152L12.11 protein
AK069025	Os04g0569000	1	6	0	0	0	Similar to activator 1 40-kDa subunit

(continued)

Table 30.4 (continued)

cDNA accession	Gene locus	Number of matched peptide					Annotations
		E	S	C	L	P	
AK066587	Os03g0220100	0	5	0	0	0	Similar to very long chain fatty acid-condensing enzyme CUT1
AK099178	Os02g0726000	0	5	0	0	0	FAS1 domain-containing protein
AK105867	Os02g0608900	0	5	0	0	0	Epstein–Barr virus U2-IR2 domain-encoding nuclear protein
AB087745	Os05g0595100	0	5	1	0	0	Similar to UDP-glucose-4-epimerase
AK069984	Os02g0775200	1	5	0	0	0	Similar to activator 1 36-kDa subunit

Matched peptides indicate the number of MS-identified tryptic peptides of the protein
Annotations are referred from The Rice Annotation Project Database (RAPDB)
E egg cell, *S* sperm cell, *C* callus, *L* seedling, *P* pollen grain

Fig. 30.2 Rice (**a**) and *Arabidopsis* (**b**) mutants showing defects in seed set or seed development. (**a**) Fertility of a rice TOS17 transposon insertional line (ND8460) for Os11g0143400, a sperm cell-enriched gene. In panicles of wild-type (Nipponbare), more than 95 % of seeds developed fully with light brown color. In the mutants, undeveloped seeds were often observed. Two typical undeveloped seeds are indicated by *arrowheads* in each panel. (**b**) Dissected siliques of wild-type (Colombia 0) and T-DNA insertional lines (SALK_095847) for At4g02060, a gene orthologous to the rice gene (Os12g0560700) encoding a sperm-specific protein. Failed ovules or seeds arrested at early immature stages are visible in the mutant silique

Acknowledgments I thank Drs. M. Abiko (Tokyo Metropolitan University), Y. Nagamura, and M. Motoyama (National Institute of Agrobiological Sciences, Tsukuba, Japan) for assistance with the microarray experiment, Ms. T. Mochizuki and H. Maeda (Tokyo Metropolitan University) for isolating rice egg cells, Drs. M. Abiko, T. Isobe, M. Taoka, Y. Yamauchi, and C. Fujita (Tokyo Metropolitan University) for proteome analyses, RIKEN Bio Resource Center (Tsukuba, Japan) for providing cultured rice cells (Oc line), NIAS Rice Genome Resource Center (Tsukuba, Japan) for providing rice Tos17 mutants, and ABRC for providing *Arabidopsis* mutants (Ohio State University, OH, USA). This work was supported, in part, by a Grant-in-Aid from the Ministry of Education, Culture, Sports, Science and Technology of Japan (No. 21112007 to T.O.) and from the Japan Society for the Promotion of Science (No. 20570206 to T.O.).

References

Abiko M, Furuta K, Yamauchi Y et al (2013a) Identification of proteins enriched in rice egg or sperm cells by single-cell proteomics. PLoS One 8:e69578

Abiko M, Maeda H, Tamura K et al (2013b) Gene expression profiles in rice gametes and zygotes: identification of gamete-enriched genes and up- or down-regulated genes in zygotes after fertilization. J Exp Bot 64:1927–1940

Adham IM, Nayernia K, Engel W (1997) Spermatozoa lacking acrosin protein show delayed fertilization. Mol Reprod Dev 46:370–376

Amien S, Kliwer I, Márton ML et al (2010) Defensin-like ZmES4 mediates pollen tube burst in maize via opening of the potassium channel KZM1. PLoS Biol 8:e1000388

Baba T, Azuma S, Kashiwabara S, Toyoda Y (1994) Sperm from mice carrying a targeted mutation of the acrosin gene can penetrate the oocyte zona pellucida and effect fertilization. J Biol Chem 269:31845–31849

Berger F (2011) Imaging fertilization in flowering plants, not so abominable after all. J Exp Bot 62:1651–1658

Borges F, Gomes G, Gardner R et al (2008) Comparative transcriptomics of *Arabidopsis* sperm cells. Plant Physiol 148:1168–1181

Calvert ME, Digilio LC, Herr JC, Coonrod SA (2003) Oolemmal proteomics-identification of highly abundant heat shock proteins and molecular chaperons in the mature mouse egg and their localization on the plasma membrane. Reprod Biol Endocrinol 14:1–27

Dai S, Chen S (2012) Single-cell-type proteomics: toward a holistic understanding of plant function. Mol Cell Proteomics 11:1122–1130

Deveaux Y, Toffano-Nioche C, Claisse G et al (2008) Genes of the most conserved WOX clade in plants affect root and flower development in *Arabidopsis*. BMC Evol Biol 8:291

Faure JE, Mogensen HL, Dumas C et al (1993) Karyogamy after electrofusion of single egg and sperm cell protoplasts from maize: cytological evidence and time course. Plant Cell 5:747–755

Guignard ML (1899) Sur les antherozoides et la double copulation sexuelle chez les vegetaux angiosperms. Rev Gén Bot 11:129–135

Haecker A, Gross-Hardt R, Geiges B et al (2004) Expression dynamics of WOX genes mark cell fate decisions during early embryonic patterning in *Arabidopsis thaliana*. Development (Camb) 131:657–668

Hamamura Y, Nagahara S, Higashiyama T (2012) Double fertilization on the move. Curr Opin Plant Biol 15:70–77

Heuer S, Hansen S, Bantin J et al (2001) The maize MADS box gene ZmMADS3 affects node number and spikelet development and is co-expressed with ZmMADS1 during flower development, in egg cells, and early embryogenesis. Plant Physiol 127:33–45

Hoshino Y, Scholten S, von Wiegen P et al (2004) Fertilization induced changes in the microtubular architecture of the maize egg cell and zygote—an immunocytochemical approach adapted to single cells. Sex Plant Reprod 17:89–95

Jahnke S, Scholten S (2009) Epigenetic resetting of a gene imprinted in plant embryos. Curr Biol 19:1677–1681

Jeong S, Bayer M, Lukowitz W (2011) Taking the very first steps: from polarity to axial domains in the early *Arabidopsis* embryo. J Exp Bot 62:1687–1697

Kankel MW, Ramsey DE, Stokes TL et al (2003) *Arabidopsis* MET1 cytosine methyltransferase mutants. Genetics 163:1109–1122

Kasahara RD, Portereiko MF, Sandaklie-Nikolova L et al (2005) MYB98 is required for pollen tube guidance and synergid cell differentiation in *Arabidopsis*. Plant Cell 17:2981–2992

Kranz E (1999) In vitro fertilization with isolated single gametes. Methods Mol Biol 111:259–267

Kranz E, Lörz H (1993) In vitro fertilization with isolated, single gametes results in zygotic embryogenesis and fertile maize plants. Plant Cell 5:739–746

Kranz E, von Wiegen P, Lörz H (1995) Early cytological events after induction of cell division in egg cells and zygote development following in vitro fertilization with angiosperm gametes. Plant J 8:9–23

Kurihara LJ, Beh CT, Latterich M, Schekman R, Rose MD (1994) Nuclear congression and membrane fusion: two distinct events in the yeast karyogamy pathway. J Cell Biol 126:911–923

Márton ML, Cordts S, Broadhvest J, Dresselhaus T (2005) Micropylar pollen tube guidance by egg apparatus 1 of maize. Science 307:573–576

Maruyama D, Endo T, Nishikawa S (2010) BiP-mediated polar nuclei fusion is essential for the regulation of endosperm nuclei proliferation in *Arabidopsis thaliana*. Proc Natl Acad Sci USA 107:1684–1689

Mayer KF, Schoof H, Haecker A et al (1998) Role of WUSCHEL in regulating stem cell fate in the Arabidopsis shoot meristem. Cell 95:805–815

Melloy P, Shen S, White E, Rose MD (2009) Distinct roles for key karyogamy proteins during yeast nuclear fusion. Mol Biol Cell 20:3773–3782

Meyer S, Scholten S (2007) Equivalent parental contribution to early plant zygotic development. Curr Biol 17:1686–1691

Mori T, Kuroiwa H, Higashiyama T, Kuroiwa T (2006) GENERATIVE CELL SPECIFIC 1 is essential for angiosperm fertilization. Nat Cell Biol 8:64–71

Nakajima K, Uchiumi T, Okamoto T (2010) Positional relationship between the gamete fusion site and the first division plane in the rice zygote. J Exp Bot 61:3101–3105

Nawaschin S (1898) Revision der Befruchtungsvorgange bei Lilium martagon und Fritillaria tenella. Bull Sci Acad Imp Sci Saint Pétersbourg 9:377–382

Ning J, Peng X-B, Qu L-H et al (2006) Differential gene expression in egg cells and zygotes suggests that the transcriptome is restructed before the first zygotic division in tobacco. FEBS Lett 580:1747–1752

Nodine MD, Bartel DP (2012) Maternal and paternal genomes contribute equally to the transcriptome of early plant embryos. Nature (Lond) 482:94–98

Ohnishi T, Takanashi H, Mogi M et al (2011) Distinct gene expression profiles in egg and synergid cells of rice as revealed by cell type-specific microarrays. Plant Physiol 155:881–891

Okamoto T (2011) In vitro fertilization with isolated rice gametes: production of zygotes and zygote and embryo culture. Methods Mol Biol 710:17–27

Okamoto T, Scholten S, Lörz H, Kranz E (2005) Identification of genes that are up- or down-regulated in the apical or basal cell of maize two-celled embryos and monitoring their expression during zygote development by a cell manipulation- and PCR-based approach. Plant Cell Physiol 46:332–338

Portereiko MF, Sandaklie-Nikolova L, Lloyd A et al (2006) NUCLEAR FUSION DEFECTIVE 1 encodes the *Arabidopsis* RPL21M protein and is required for karyogamy during female gametophyte development and fertilization. Plant Physiol 141:957–965

Queitsch C, Sangster TA, Lindquist S (2002) Hsp90 as a capacitor of phenotypic variation. Nature (Lond) 416:618–624

Raghavan V (2003) Some reflections on double fertilization, from its discovery to the present. New Phytol 159:565–583

Riechmann JL, Meyerowitz EM (1997) MADS domain proteins in plant development. Biol Chem 378:1079–1101

Russell SD (1992) Double fertilization. Int Rev Cytol 40:357–390

Russell SD, Gou X, Wong C et al (2012) Genomic profiling of rice sperm cell transcripts reveals conserved and distinct elements in the flowering plant male germ lineage. New Phytol 195:560–573

Sangster TA, Queitsch C (2005) The HSP90 chaperone complex, an emerging force in plant development and phenotypic plasticity. Curr Opin Plant Biol 8:86–92

Sato A, Toyooka K, Okamoto T (2010) Asymmetric cell division of rice zygotes located in embryo sac and produced by in vitro fertilization. Sex Plant Reprod 23:211–217

Sawada H, Yokosawa H, Ishii S (1984) Purification and characterization of two types of trypsin-like enzymes from sperm of the ascidian (Prochordata) *Halocynthia roretzi*. Evidence for the presence of spermosin, a novel acrosin-like enzyme. J Biol Chem 259:2900–2904

Sawada H, Iwasaki K, Kihara-Negishi F et al (1996) Localization, expression, and the role in fertilization of spermosin, an ascidian sperm trypsin-like protease. Biochem Biophys Res Commun 222:499–504

Scholten S, Lörz H, Kranz E (2002) Paternal mRNA and protein synthesis coincides with male chromatin decondensation in maize zygotes. Plant J 32:221–231

Sprunck S, Baumann U, Edwards K et al (2005) The transcript composition of egg cells changes significantly following fertilization in wheat (*Triticum aestivum* L.). Plant J 41:660–672

Sprunck S, Rademacher S, Vogler F et al (2012) Egg cell-secreted EC1 triggers sperm cell activation during double fertilization. Science 338:1093–1097

Steffen JG, Kang IH, Macfarlane J, Drews GN (2007) Identification of genes expressed in the *Arabidopsis* female gametophyte. Plant J 51:281–292

Tartakoff AM, Jaiswal P (2009) Nuclear fusion and genome encounter during yeast zygote formation. Mol Biol Cell 20:2932–2942

Theissen G, Becker A, DiRosa A et al (2000) A short history of MADS-box genes in plants. Plant Mol Biol 42:115–149

Uchiumi T, Shinkawa T, Isobe T, Okamoto T (2007a) Identification of the major protein components of rice egg cells. J Plant Res 120:575–579

Uchiumi T, Uemura I, Okamoto T (2007b) Establishment of an in vitro fertilization system in rice (*Oryza sativa* L.). Planta (Berl) 226:581–589

von Besser K, Frank AC, Johnson MA, Preuss D (2006) *Arabidopsis* HAP2 (GCS1) is a sperm-specific gene required for pollen tube guidance and fertilization. Development (Camb) 133:4761–4769

Wang YY, Kuang A, Russell SD, Tian HQ (2006) In vitro fertilization as a tool for investigating sexual reproduction of angiosperms. Sex Plant Reprod 19:103–115

Wang D, Zhang CQ, Hearn DJ et al (2010) Identification of transcription factor genes expressed in the *Arabidopsis* female gametophyte. BMC Plant Biol 10:110

West AG, Causier BE, Davies B, Sharrocks AD (1998) DNA binding and dimerization determinants of *Antirrhinum majus* MADS-box transcription factors. Nucleic Acids Res 26:5277–5287

Wuest SE, Vijverberg K, Schmidt A et al (2010) *Arabidopsis* female gametophyte gene expression map reveals similarities between plant and animal gametes. Curr Biol 20:506–512

Yang H, Kaur N, Kiriakopolos S, McCormick S (2006) EST generation and analyses towards identifying female gametophyte-specific genes in *Zea mays* L. Planta (Berl) 224:1004–1014

Zhao Y, Hu Y, Dai M et al (2009) The WUSCHEL-related homeobox gene WOX11 is required to activate shoot-borne crown root development in rice. Plant Cell 21:736–748

Zhao J, Xin H, Qu L et al (2011) Dynamic changes of transcript profiles after fertilization are associated with de novo transcription and maternal elimination in tobacco zygote, and mark the onset of the maternal-to-zygotic transition. Plant J 65:131–145

Chapter 31
Role of CD9 in Sperm–Egg Fusion and Virus-Induced Cell Fusion in Mammals

Keiichi Yoshida, Natsuko Kawano, Yuichiroh Harada, and Kenji Miyado

Abstract In mammals, two integral membrane proteins, sperm IZUMO1 and egg CD9, regulate sperm–egg fusion, and their roles are critical but as yet unknown. In such situation, a recent study has shown that CD9-containing exosome-like vesicles, which are released from wild-type eggs, can induce the fusion between sperm and *Cd9*-deficient egg, even though *Cd9*-deficient eggs are highly refractory to the fusion with sperm. This result provides compelling evidence for the crucial involvement of CD9-containing, fusion-facilitating vesicles in sperm–egg fusion. On the other hand, similarities have been observed between the generation of retroviruses in host cells and the formation of small cellular vesicles, termed exosomes, in mammalian cells. The exosomes are thought to regulate intercellular communication through transfer of proteins and RNAs. These collective studies provide an insight into the molecular mechanisms of gamete fusion and other membrane fusion events.

Keywords CD9 • Exosome • Membrane fusion • Tetraspanin

31.1 Introduction

Fertilization is an event that consists of cell–cell adhesion, cell–cell fusion, and activation of cell signaling to allow the resumption of the egg cell cycle (Fig. 31.1). In mammals, two membrane protein families, a cell adhesion molecule "integrin" (Almeida et al. 1995) and a membrane-anchored protease "ADAM (a disintegrin

K. Yoshida • N. Kawano • Y. Harada • K. Miyado (✉)
Department of Reproductive Biology, National Center for Child Health
and Development, 2-10-1 Okura, Setagaya, Tokyo 157-8535, Japan
e-mail: miyado-k@ncchd.go.jp

H. Sawada et al. (eds.), *Sexual Reproduction in Animals and Plants*,
DOI 10.1007/978-4-431-54589-7_31, © The Author(s) 2014

Fig. 31.1 Series of steps from sperm–egg interaction to fusion during mammalian fertilization: an overview of mammalian fertilization. Fertilization is divided into multiple steps: interaction of sperm-somatic cells (termed cumulus cells), binding of sperm to the extracellular matrix (termed zona pellucida), and penetration of the egg. After the sperm penetrates the zona pellucida, it can bind and fuse to the egg cell membrane. Successful fertilization requires not only that a sperm and egg fuse, but also that polyspermy block occurs

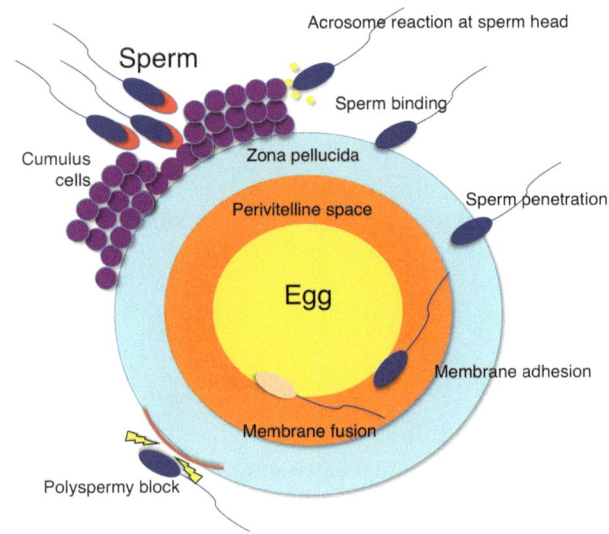

and metalloprotease)" (Blobel et al. 1992), were biochemically identified and immunocytochemically confirmed to localize on the outer cell membranes of egg and sperm, respectively; furthermore, antibodies against these protein families were shown to significantly reduce the rate of sperm–egg binding and fusion in mice (Almeida et al. 1995). Integrins, which are expressed in many types of cells in animals, mediate cell–cell and cell-matrix interaction and intercellular communication, including cell adhesion and cell–cell fusion (Almeida et al. 1995). On the other hand, the ADAM family has a characteristic domain that is homologous to an extracellular region of integrin family (Evans 2001). Thus, the presence of the domain conserved between integrins and ADAMs indicates that these two protein families play a role in sperm–egg adhesion and/or fusion (Evans 2001). Unexpectedly, when their genetically manipulated mice were produced, both male and female mice displayed no overt anomalies in either sperm–egg fusion or adhesion (Miller et al. 2000; He et al. 2003). In the past, many factors predicted to participate in sperm–egg fusion have emerged in mice, but despite expectations, most were found to be unnecessary (Okabe and Cummins 2007). From these studies, to ensure the continuous success of the reproduction cycle in mice, compatible pathways tuned by overlapping functions of multiple proteins seem to regulate the mechanism of fertilization. In other words, there will be more than one way for a sperm and egg to fuse and the network of multiple pathways may minimize the severity of the malfunction that may occur in sperm or eggs lacking a single gene. Concerning sperm–egg fusion, CD9 on the egg membrane (Miyado et al. 2000; Le Naour et al. 2000; Kaji et al. 2000) and IZUMO1 on the sperm membrane (Inoue et al. 2005) are exceptionally essential factors in gene disruption experiments (Fig. 31.2).

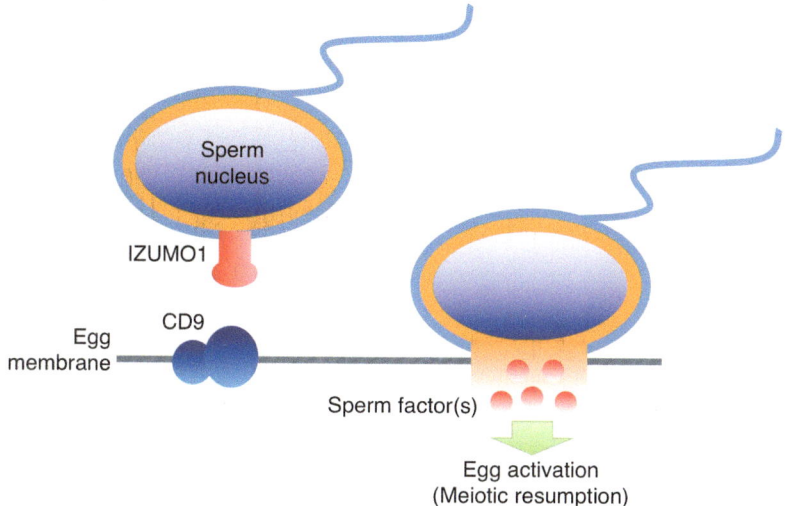

Fig. 31.2 Players identified in sperm–egg fusion. IZUMO1 is expressed on the sperm membrane, and *Izumo1*-deficient sperm show a defect in fusion with the egg cell membrane. CD9 is expressed on the egg cell membrane and functions in fusion with the sperm. Two membrane proteins, IZUMO1 and CD9, are essential for sperm–egg fusion in mice. Direct interaction between CD9 and IZUMO1 has not been identified, and the unidentified sperm and egg factors may be involved in sperm–egg fusion. After sperm–egg membrane fusion occurs, a sperm factor triggers Ca^{2+} oscillations and initiates egg activation in mammals

31.2 CD9 and Its Role in Cellular Function

The CD9 gene encoding a 24-kDa protein is transcribed in all types of mammalian cells (Hemler 2008). This protein is localized on the cell membranes and partially on endosomes, and it is expected to be involved in cell–cell adhesion, because CD9 associates with the integrin family (Hemler 2008). CD9 is known as a motility-related protein 1 (MRP-1), which plays a role in suppressing tumor metastasis (Miyake et al. 1991). CD9 has two extracellular loops, four transmembrane domains, and two short cytoplasmic domains. Its functional domain is expected to be included in a large extracellular loop (LEL) (Fig. 31.3), because CD9 associates with other membrane proteins via LEL in vitro (Hemler 2008). In addition, because the amount of CD9 in mesenchymal and embryonic stem cells is significantly higher than in fibroblastic cells, CD9 is useful as a cell-surface marker for isolating undifferentiated cells from cell pools in mice and humans (Akutsu et al. 2009).

Despite two decades of effort, however, the in vivo role of CD9 was unclear, and therefore three laboratories have independently generated *Cd9*$^{-/-}$ mice (Miyado et al. 2000; Le Naour et al. 2000; Kaji et al. 2000). Consequently, all strains of the *Cd9*$^{-/-}$ mice have shown severe female subfertility whereas the *Cd9*$^{-/-}$ male mice were fertile. Moreover, the *Cd9*$^{-/-}$ female mice exhibited severely reduced sperm

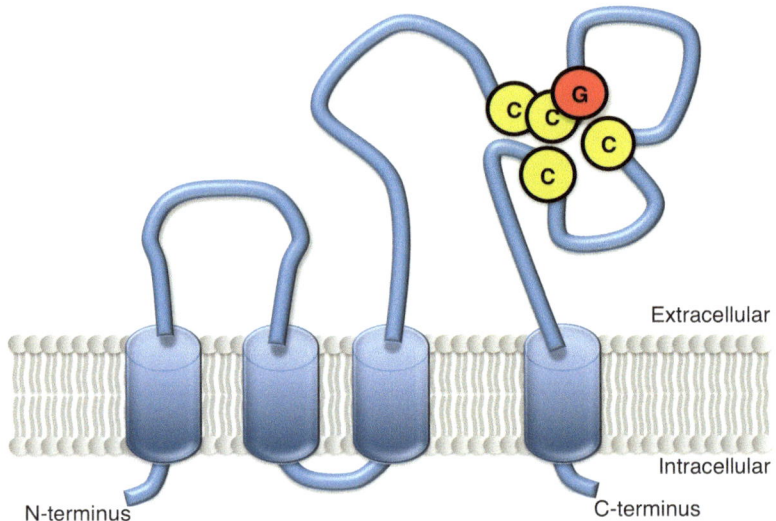

Fig. 31.3 Structural features of tetraspanin CD9, a member of the tetraspan-membrane protein family, termed tetraspanin, and its molecular mass is 24 kDa. The structural features of CD9 include four transmembrane domains, two extracellular loops, short and large extracellular loops (SEL and LEL), and two short cytoplasmic tails. CD9 has cysteine-cysteine-glycine (CCG) residues (amino acids 152–154) as a tetraspanin-specific motif and two other cysteines within LEL

fusion ability. Since these findings, CD9 has been studied as one of the crucial factors in sperm–egg fusion in mammals. A functionally essential domain of CD9 is predicted to be located within the LEL (Zhu et al. 2002; Kaji et al. 2002); however, even though CD9-binding proteins have been identified in non-gamete cells, LEL-binding, potentially fusion-related proteins have not been found yet.

31.3 Tetraspanin

CD9 belongs to a membrane protein superfamily collectively termed "tetraspanin" that encompasses 35 members in mammals (including CD9, CD37, CD53, CD63, CD81, CD82, and CD151) (Hemler 2008), 30 in nematodes (Moribe et al. 2004), and 30 in flies (Todres et al. 2000). In mammals, tetraspanin is related to infectious diseases; in particular, these proteins are involved in cell–cell transmission of human immunodeficiency virus (HIV-1) (Garcia et al. 2005; Wiley and Gummuluru 2006). After its primary infection into host cells, CD9, CD63, CD81, and CD82 are enriched at HIV-1 budding sites of HIV-1 virions. When tetraspanin-containing HIV-1 particles are secondarily formed and released from host cells, they become tenfold more infectious than the cell-free virus particles. In mice, CD81 is linked to infection of hepatocytes with the malaria parasite (Silvie et al. 2003).

Malaria sporozoites, a cell form that infects new hosts, are transmitted into the liver of the mammalian host through bites from infected mosquitoes, but the sporozoites fail to infect $Cd81^{-/-}$ mouse hepatocytes, suggesting that CD81 is involved in sporozoite entry into hepatocytes as a host factor. Otherwise, $Cd81^{-/-}$ female mice are subfertile, because $Cd81$-deficient eggs exhibit impaired sperm fusion ability (Rubinstein et al. 2006; Tanigawa et al. 2008). CD81 is expressed on $Cd9$-deficient eggs, and CD9 is also expressed on $Cd81$-deficient eggs at an expression rate comparable with that of wild-type eggs, indicating that CD9 and CD81 work independently in sperm–egg fusion (Ohnami et al. 2012).

More than 60 members of tetraspanin are known in plants (Huang et al. 2005; Chiu et al. 2007). Tetraspanin-like proteins have also been identified in fungi, and their molecular mass (more than 200 kDa) is greater than those of tetraspanin identified in animals and plants (20–30 kDa) (Lambou et al. 2008). An appressorium is a specialized cell typical of many fungal plant pathogens that is used to infect host plants. By analyzing a nonpathogenic mutant, *punchless*, isolated from the rice blast fungus *Magnaporthe grisea*, tetraspanin-like PLS1 (MgPLS1) has been shown to control the appressorial function, which is essential for fungal penetration into host leaves (Clergeot et al. 2001). Similarly, *Colletotrichum lindemuthianum* PLS1 (ClPLS1) is a functional homologue of MgPLS1. The nonpathogenic *ClPLS1-deficient* mutant to bean leaves exhibits a defect in the formation and positioning of the penetration pore (Veneault-Fourrey et al. 2005). As the invasion of pathogenic fungi into leaves is an event closely related to membrane fusion events, these studies indicate that tetraspanin-like PLS1s are involved in the membrane fusion-related event between fungus and plant.

Taken together, these results suggest that tetraspanin is related to membrane fusion-related events in multicellular organisms. On the other hand, the physiological activities of tetraspanin are still unknown, and its fusogenic activity corresponding to fusogenic transmembrane proteins, such as syncytin identified in human placenta (Mi et al. 2000), and virus envelope proteins (Hernandez et al. 1996), has not been identified.

31.4 Tetraspanin as an Exosome Component

In mammals, cell-cultured media contain nanosized membrane vesicles, but they were not attractive to researchers because they could not be structurally distinguished from the debris of dead cells (Couzin 2005). Recent studies have shown that the vesicles, termed exosomes, are derived from living cells, but not dead cells; furthermore, they have been proven to play a significant role in the mediation of adaptive immune reactions to pathogens and tumors through the enhancement of antigen-specific T-cell responses (Couzin 2005; Simons and Raposo 2009). Besides immune cells, the exosomes are released from a wide range of normal and malignant mammalian cell types, and their diameter is estimated to range from 50 to

90 nm (Simons and Raposo 2009). The protein composition of exosomes varies with the origin of cells, yet the exosomes commonly contain a ganglioside GM3, two kinds of heat shock proteins (HSP70 and HSP90), and tetraspanin (Simons and Raposo 2009). The exosomes also contain transcripts, mRNA and microRNA, which are thought to be shuttled from one cell to another, thereby influencing protein synthesis in recipient cells (Valadi et al. 2007).

31.5   Exosome-Like Vesicles Released from Eggs

In eggs, two reports have demonstrated the contribution of CD9 in the organization of the egg cell membrane. First, CD9 is transferred from the egg to the fertilizing sperm present in the perivitelline space, implying the involvement of a process similar to trogocytosis, which is a mechanism for the cell-to-cell contact-dependent transfer of membrane fragments from antigen-presenting cells to lymphocytes in immune responses for pathogens (Barraud-Lange et al. 2007). Second, CD9 deficiency alters the length and density of microvilli on the egg cell membrane (Runge et al. 2007).

On the other hand, a recent study has reported the potential of enhanced green fluorescent protein-tagged CD9 (CD9-EGFP) as a reporter protein for studying sperm-egg fusion in living mouse eggs (Miyado et al. 2008). Interestingly, in eggs just before fertilization, CD9-EGFP is significantly accumulated within the perivitelline space that completely surrounds the eggs and lies between the egg cell membrane and the zona pellucida. Consistent with the images from CD9-EGFP, immunoelectron microscopic analysis of wild-type eggs has revealed that CD9 is not only present in the perivitelline space but is also incorporated into vesicles of varying size (50–200 nm in diameter) without a sectional profile of a typical lipid bilayer. Furthermore, in opossums (Talbot and DiCarlantonio 1984) and humans (Dandekar et al. 1992), as well as in mice, membrane vesicles have been detected by electron microscopy within the perivitelline space of their eggs. Moreover, recent study has demonstrated that the vesicles identified in mouse eggs share CD9, GM3, and HSP90 with exosomes, and these components are absent in eggs lacking CD9 and are reproduced by CD9-EGFP expression restricted to the eggs (Miyado et al. 2008). These results provide two types of evidence for the nature of CD9 in mouse eggs. First, CD9-incorporated exosome-like vesicles are produced in mouse eggs and are released outside the egg cell membrane just before fertilization. Second, CD9 is essential for the formation and release of the exosome-like vesicles (hereafter referred to as egg exosomes) in mouse eggs.

CD9-containing egg exosomes render sperms capable of fusing with *Cd9-deficient* eggs (Miyado et al. 2008) (Fig. 31.4). *Cd9*-deficient eggs cannot fuse with eggs, but the coexistence of wild-type eggs results in 60–70 % of the *Cd9*-deficient eggs fusing with at least one sperm. Thus, sperm can fuse with *Cd9*-deficient eggs with impaired microvilli via the egg exosomes of wild-type eggs, which means that the egg exosomes, but not the microvilli, are essential for sperm-egg fusion.

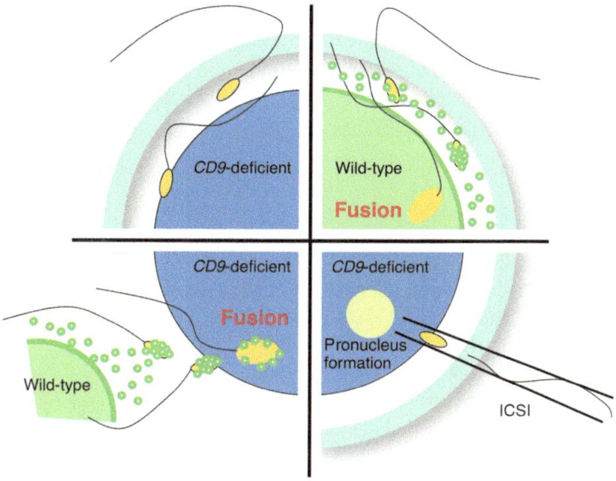

Fig. 31.4 Overview of studies of *Cd9*-deficient eggs. In wild-type eggs, CD9-containing egg exosomes are released from wild-type eggs before any interaction with the sperm (*upper right*). Shortly after the sperm penetrates the perivitelline space, the egg exosomes are transferred on the acrosome-reacted sperm head. Then, a sperm fuses with the egg cell membrane. Interaction between sperm and exosomes is an essential step for sperm-fusing ability. In contrast, *Cd9-deficient* eggs cannot release egg exosomes, which are correlated with the formation of microvilli on the egg cell membrane (*upper left*). Sperm cannot fuse to the cell membrane of the *Cd9*-deficient egg. However, when the zona pellucida is removed from the eggs, the sperm is able to interact with the egg exosomes released from wild-type eggs and can fuse with the *Cd9*-deficient egg (*lower left*). By coincubation with wild-type eggs, the sperm can fuse with a similar number of *Cd9-deficient* and wild-type eggs. Intracytoplasmic sperm injection (ICSI) is an in vitro fertilization procedure in which a single sperm head is injected directly into an egg (*lower right diagram*). This procedure is most commonly used to overcome male infertility and fusion defects in *Cd9*-deficient eggs

31.6 Membrane Fusion and Exosomes

The close relationship between egg exosomes and sperm–egg fusion raises the question of how egg exosomes facilitate fusion. According to a previous report, exosomes contain both functional mRNA and microRNA, which are shuttled from one cell to another, affecting recipient cell ability to produce protein (Valadi et al. 2007). Moreover, HIV-1 utilizes the exosome biogenesis pathway for the formation of infectious particles, and in macrophages, HIV-1 assembles into an intracellular plasma membrane domain-containing tetraspanin (i.e., CD9, CD81, CD53, or CD63) (Wiley and Gummuluru 2006). Thus, the exosomes may play at least two roles in regulating cell function: first, shuttling proteins and RNAs (mRNAs and microRNAs) from one cell to another, and second, forming infectious particles. In fertilization, these two actions may be required for sperm–egg fusion in mammals. The studies of exosomes underscore the relevance of CD9 for healthy and pathogenic cell–cell fusion processes and may present a useful strategy for regulating the cell-to-cell spread of specific viruses and fertilization ability.

References

Akutsu H, Miura T, Machida M et al (2009) Maintenance of pluripotency and self-renewal ability of mouse embryonic stem cells in the absence of tetraspanin CD9. Differentiation 78:137–142

Almeida EA, Huovila AP, Sutherland AE et al (1995) Mouse egg integrin alpha-6-beta-1 functions as a sperm receptor. Cell 81:1095–1104

Barraud-Lange V, Naud-Barriant N, Bomsel M et al (2007) Transfer of oocyte membrane fragments to fertilizing spermatozoa. FASEB J 21:3446–3449

Blobel CP, Wolfsberg TG, Turck CW et al (1992) A potential fusion peptide and an integrin ligand domain in a protein active in sperm-egg fusion. Nature (Lond) 356:248–252

Chiu WH, Chandler J, Cnops G et al (2007) Mutations in the TORNADO2 gene affect cellular decisions in the peripheral zone of the shoot apical meristem of *Arabidopsis thaliana*. Plant Mol Biol 63:731–744

Clergeot PH, Gourgues M, Cots J et al (2001) PLS1, a gene encoding a tetraspanin-like protein, is required for penetration of rice leaf by the fungal pathogen *Magnaporthe grisea*. Proc Natl Acad Sci USA 98:6963–6968

Couzin J (2005) Cell biology: the ins and outs of exosomes. Science 308:1862–1863

Dandekar P, Aggeler J, Talbot P (1992) Structure, distribution and composition of the extracellular matrix of human oocytes and cumulus masses. Hum Reprod 7:391–398

Evans JP (2001) Fertilin beta and other ADAMs as integrin ligands: insights into cell adhesion and fertilization. Bioessays 23:628–639

Garcia E, Pion M, Pelchen-Matthews A et al (2005) HIV-1 trafficking to the dendritic cell–T-cell infectious synapse uses a pathway of tetraspanin sorting to the immunological synapse. Traffic 6:488–501

He ZY, Brakebusch C, Fassler R et al (2003) None of the integrins known to be present on the mouse egg or to be ADAM receptors are essential for sperm–egg binding and fusion. Dev Biol 254:226–237

Hemler ME (2008) Targeting of tetraspanin proteins: potential benefits and strategies. Nat Rev Drug Discov 7:747–758

Hernandez LD, Hoffman LR, Wolfsberg TG et al (1996) Virus–cell and cell–cell fusion. Annu Rev Cell Dev Biol 12:627–661

Huang S, Yuan S, Dong M et al (2005) The phylogenetic analysis of tetraspanins projects the evolution of cell–cell interactions from unicellular to multicellular organisms. Genomics 86:674–684

Inoue N, Ikawa M, Isotani A et al (2005) The immunoglobulin superfamily protein Izumo is required for sperm to fuse with eggs. Nature (Lond) 434:234–238

Kaji K, Oda S, Shikano T et al (2000) The gamete fusion process is defective in eggs of Cd9-deficient mice. Nat Genet 24:279–282

Kaji K, Oda S, Miyazaki S et al (2002) Infertility of CD9-deficient mouse eggs is reversed by mouse CD9, human CD9, or mouse CD81; polyadenylated mRNA injection developed for molecular analysis of sperm-egg fusion. Dev Biol 247:327–334

Lambou K, Tharreau D, Kohler A et al (2008) Fungi have three tetraspanin families with distinct functions. BMC Genomics 9:63

Le Naour F, Rubinstein E, Jasmin C et al (2000) Severely reduced female fertility in CD9-deficient mice. Science 287:319–321

Mi S, Lee X, Li X et al (2000) Syncytin is a captive retroviral envelope protein involved in human placental morphogenesis. Nature (Lond) 403:785–789

Miller BJ, Georges-Labouesse E, Primakoff P et al (2000) Normal fertilization occurs with eggs lacking the integrin alpha6beta1 and is CD9-dependent. J Cell Biol 149:1289–1296

Miyado K, Yamada G, Yamada S et al (2000) Requirement of CD9 on the egg plasma membrane for fertilization. Science 287:321–324

Miyado K, Yoshida K, Yamagata K et al (2008) The fusing ability of sperm is bestowed by CD9-containing vesicles released from eggs in mice. Proc Natl Acad Sci USA 105:12921–12926

Miyake M, Koyama M, Seno M et al (1991) Identification of the motility-related protein (MRP-1), recognized by monoclonal antibody M31-15, which inhibits cell motility. J Exp Med 174:1347–1354

Moribe H, Yochem J, Yamada H et al (2004) Tetraspanin protein (TSP-15) is required for epidermal integrity in *Caenorhabditis elegans*. J Cell Sci 117:5209–5220

Ohnami N, Nakamura A, Miyado M et al (2012) CD81 and CD9 work independently as extracellular components upon fusion of sperm and oocyte. Biology Open 1:640–647

Okabe M, Cummins JM (2007) Mechanisms of sperm–egg interactions emerging from gene-manipulated animals. Cell Mol Life Sci 64:1945–1958

Rubinstein E, Ziyyat A, Prenant M et al (2006) Reduced fertility of female mice lacking CD81. Dev Biol 290:351–358

Runge KE, Evans JE, He ZY et al (2007) Oocyte CD9 is enriched on the microvillar membrane and required for normal microvillar shape and distribution. Dev Biol 304:317–325

Silvie O, Rubinstein E, Franetich JF et al (2003) Hepatocyte CD81 is required for *Plasmodium falciparum* and *Plasmodium yoelii* sporozoite infectivity. Nat Med 9:93–96

Simons M, Raposo G (2009) Exosomes: vesicular carriers for intercellular communication. Curr Opin Cell Biol 21:575–581

Talbot P, DiCarlantonio G (1984) Ultrastructure of opossum oocyte investing coats and their sensitivity to trypsin and hyaluronidase. Dev Biol 103:159–167

Tanigawa M, Miyamoto K, Kobayashi S et al (2008) Possible involvement of CD81 in acrosome reaction of sperm in mice. Mol Reprod Dev 75:150–155

Todres E, Nardi JB, Robertson HM (2000) The tetraspanin superfamily in insects. Insect Mol Biol 9:581–590

Valadi H, Ekstrom K, Bossios A et al (2007) Exosome-mediated transfer of mRNAs and micro-RNAs is a novel mechanism of genetic exchange between cells. Nat Cell Biol 9:654–659

Veneault-Fourrey C, Parisot D, Gourgues M et al (2005) The tetraspanin gene ClPLS1 is essential for appressorium-mediated penetration of the fungal pathogen *Colletotrichum lindemuthianum*. Fungal Genet Biol 42:306–318

Wiley RD, Gummuluru S (2006) Immature dendritic cell-derived exosomes can mediate HIV-1 transinfection. Proc Natl Acad Sci USA 103:738–743

Zhu GZ, Miller BJ, Boucheix C et al (2002) Residues SFQ (173-175) in the large extracellular loop of CD9 are required for gamete fusion. Development (Camb) 129:1995–2002

Chapter 32
The Mechanism of Sperm–Egg Fusion in Mouse and the Involvement of IZUMO1

Naokazu Inoue

Abstract A typical ejaculate contains more than 100 million spermatozoa; however, only 1 spermatozoon participates in fertilization. There might be an ingenious molecular mechanism to ensure that a spermatozoon fertilizes an egg. Recent gene disruption experiments in mice have revealed that there are two key factors in the sperm–egg fusion process. CD9 on eggs and IZUMO1 on spermatozoa have emerged as indispensable factors. However, the molecular mechanism of sperm–egg fusion, that is, how and when sperm–egg fusion occurs, remains virtually unknown. We have recently reported the dynamics of redistribution of IZUMO1 during the acrosome reaction using red fluorescent protein-tagged IZUMO1 for live imaging of the moment of sperm–egg fusion. Consequentially, the result suggests that IZUMO1 diffused from the acrosomal membrane to the sperm surface in the equatorial segment, which is considered to initiate fusion with the oocyte; fusion takes place after the acrosome reaction. This was the first evidence for live imaging to monitor the fusion-related protein at sperm–egg fusion.

Keywords Egg • Fusion • Gene manipulation • Interaction • IZUMO1 • Spermatozoon

32.1   Introduction

Mammalian fertilization is the phenomenon in which a spermatozoon and egg find each other, interact, and fuse. There are many obstacles, including migration into the oviduct and penetration into the cumulus layer and zona pellucida, for the

N. Inoue (✉)
Department of Cell Science, Institutes for Biomedical Sciences, School of Medicine,
Fukushima Medical University, 1 Hikarigaoka, Fukusima-City, Fukushima 960-1295, Japan
e-mail: n-inoue@fmu.ac.jp

H. Sawada et al. (eds.), *Sexual Reproduction in Animals and Plants*,
DOI 10.1007/978-4-431-54589-7_32, © The Author(s) 2014

spermatozoa before reaching the egg. This phenomenon is essential for sexual reproduction. Although many experiments have been performed and papers published about the biological importance of fertilization so far, the molecular mechanism of fertilization remains substantially unknown. However, gene-knockout experiments made us aware of the existence of essential genes.

In the sperm–egg fusion process, only two key factors, a member of the immuno-globulin superfamily, IZUMO1, in the spermatozoon and a member of the tetraspanin family, CD9, in the oocyte, have been identified to be essential (Inoue et al. 2005; Miyado et al. 2000; Le Naour et al. 2000; Kaji et al. 2000). Loss of these factors results in sterility, but currently there is no evidence that they act as direct fusogenic proteins. However, the fact that the requirement for these factors can be restored by bypassing the fusion step via intracytoplasmic sperm injection (ICSI) suggests that they make specific contributions to egg–sperm fusion (Inoue et al. 2005).

IZUMO1 was initially identified by a sperm–egg fusion inhibitory monoclonal antibody. IZUMO1-deficient spermatozoa appear morphologically normal, and bind and penetrate the zona pellucida surrounding the egg, but are not capable of fusing with eggs (Inoue et al. 2005). IZUMO1 is initially hidden in the acrosomal organelle under the plasma membrane. After the acrosome reaction, it relocates to the surface of the sperm head, suggesting that redistribution of IZUMO1 is essential for fusion (Inoue et al. 2005; Satouh et al. 2012).

CD9 was shown to be required for fusion on the egg plasma membrane (Miyado et al. 2000; Le Naour et al. 2000; Kaji et al. 2000). It was proposed that exosome-like CD9-containing vesicles are secreted from unfertilized eggs, thereby confer-ring fusion competence to the spermatozoon (Miyado et al. 2008). In addition to the sterility phenotype, the length, thickness, and density of the microvilli of the eggs of CD9-deficient mice are altered, suggesting that CD9 participates in microvilli for-mation and that microvilli are important for sperm–egg fusion (Runge et al. 2007). CD9 has also been suggested to be involved in adhesion of the membrane to trigger sperm–egg fusion (Jégou et al. 2011).

There is no evidence indicating that IZUMO1 and CD9 directly interact during sperm–egg fusion. Even though IZUMO1 and CD9 are essential for sperm–egg fusion, it is still unclear how they participate in the process and if other proteins are involved.

32.2   IZUMO1 Is an Essential Protein for Sperm–Egg Fusion

We produced the anti-mouse spermatozoon monoclonal antibody, OBF13, that inhibits the fusion process (Okabe et al. 1987; Okabe et al. 1988). The antigen rec-ognized by OBF13 was not identified for many years. However, it was recently identified by two-dimensional gel electrophoresis and subsequent immunoblotting and liquid chromatography-tandem mass spectroscopy analysis. We named the anti-gen "Izumo" after a Japanese shrine dedicated to marriage. The gene encodes a novel immunoglobulin superfamily type I membrane protein with an extracellular Ig domain. Recently, according to Ellerman et al. (Ellerman et al. 2009), IZUMO

Fig. 32.1 Gamete fusion-related factor IZUMO1. Accumulation of many spermatozoa in the perivitelline space of the eggs recovered from the female mice mated with *Izumo1-deficient* male mice. Spermatozoa in the perivitelline space were labeled with the acrosome-reacted sperm-specific monoclonal antibody MN9

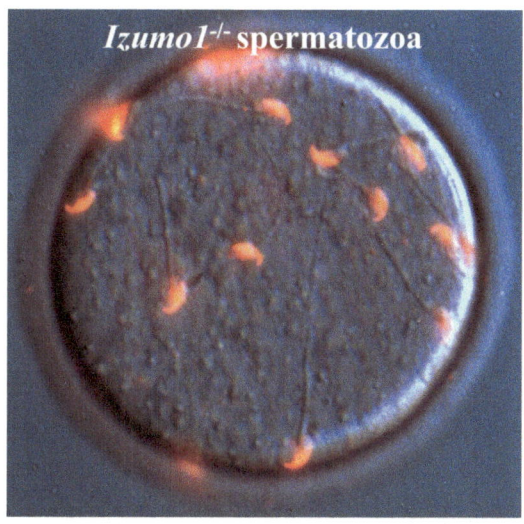

proteins consist of four family proteins (IZUMO1 to IZUMO4). The N-terminal domain between signal peptides and the Ig domain showed a significant homology to each other and was termed the Izumo domain.

After producing *Izumo1*$^{-/-}$ mutant mice, we found that they were healthy and showed no overt developmental abnormalities. As expected, *Izumo1*$^{-/-}$ male mice became sterile although they exhibited normal mating behavior and ejaculated to form normal vaginal plugs. Moreover, the sperm penetrated the zona pellucida without any problem but failed to fuse with the eggs, causing an accumulation of spermatozoa in the perivitelline space of the eggs (Inoue et al. 2005) (Fig. 32.1).

We also examined the acrosomal status of *Izumo1*$^{-/-}$ spermatozoa. To verify the acrosome reaction, we stained the spermatozoa with an MN9 monoclonal antibody, which stains only to the equatorial segment of acrosome-reacted spermatozoa (Toshimori et al. 1998) (Fig. 32.1). As shown in Fig. 32.1, *Izumo1*$^{-/-}$ spermatozoa were clearly stained for MN9. This finding indicated that *Izumo1*$^{-/-}$ spermatozoa had undergone the acrosome reaction but failed to fuse with eggs.

We further examined whether the defect of *Izumo1*$^{-/-}$ spermatozoa is limited to their fusing ability with eggs or whether it extends to later developmental stages. To address this question, we injected *Izumo1*$^{-/-}$ spermatozoa directly into the cytoplasm of wild-type eggs and observed the ability of later development. Eggs injected with *Izumo1*$^{-/-}$ spermatozoa were successfully activated and implanted normally. The embryos developed to term in a normal ratio (Inoue et al. 2005).

32.3   The Role of *N*-Glycosylation in IZUMO1

IZUMO1 possesses an *N*-glycosylation site, which is well conserved among species, in the middle of an Ig loop. This site must be glycosylated because if we incubated mouse IZUMO1 from spermatozoa with *N*-glycosidase, the molecular weight of IZUMO1 decreased from its original size of 56 to 50 kDa. Because glycan composition is known to be involved in many molecular interaction mechanisms (Ohtsubo and Marth 2006), we attempted to examine the role of N-glycan on IZUMO1. To answer this question, we produced mouse lines expressing mutated IZUMO1. In particular, residue 204, asparagine, in the putative N-glycosylation site, was substituted with glutamine by site-directed mutagenesis under the testis-specific *Calmegin* promoter with the rabbit beta-globin polyadenylation signal. After we established N204Q-IZUMO1 male mice, we crossed this transgenic mouse line with *Izumo1*$^{-/-}$ mice and produced a mouse line that has spermatozoa with no N-glycosylation site in IZUMO1. Although the litter sizes were smaller compared to wild-type IZUMO1 mice, the infertile phenotype was rescued in N204Q-IZUMO1 mice. The efficiency was low, but spermatozoa from N204Q-IZUMO1 mice could fuse with eggs. We extracted proteins from testis and spermatozoa from N204Q-IZUMO1 mice and analyzed IZUMO1 by Western blot analysis. N204Q-IZUMO1 from testis appeared in the 50-kDa area because of the lack of N-linked glycan. However, severe fragmentation was observed for N204Q-IZUMO1 from spermatozoa, which was not observed in wild-type IZUMO1. The majority of fragmented bands was observed in the ~30- and ~35-kDa areas. Although N204Q-IZUMO1 could rescue the infertile phenotype, the amount of intact N204Q-IZUMO1 present in spermatozoa was significantly smaller compared to that of wild-type IZUMO1, in spite of an abundance of N204Q-IZUMO1 in testis (Inoue et al. 2008). This finding indicates that glycosylation is not essential for the function of IZUMO1 but that it has a role in protecting IZUMO1 from fragmentation in the cauda epididymis.

32.4 IZUMO1-Interacting Protein

Because IZUMO1 has no "fusogenic" peptide or "SNARE"-like structure in it, we considered the possibility that IZUMO1 might be one of the components forming a fusogenic machinery on spermatozoa. To search for IZUMO1-interacting proteins, we made a transgenic mouse line producing IZUMO1-His on spermatozoa and introduced it to an *Izumo1*$^{-/-}$ background, which allowed us to immunoprecipitate IZUMO1 using an anti-His antibody. The IZUMO1-interacting protein was purified from acrosome-intact sperm lysates by using anti-His microbeads. We could detect a specific 80-kDa band in the purified fraction by silver staining. After liquid chromatography-tandem mass spectroscopy analysis, the protein was identified as ACE3 (angiotensin I-converting enzyme 3) (Rella et al. 2007).

We generated *Ace3*-deficient mice by homologous recombination. Differing from our expectation, *Ace3*$^{-/-}$ mice showed signs of infertility in both males and females. We analyzed the fertilizing ability of *Ace3*$^{-/-}$ spermatozoa in an in vitro fertilization system. Again, *Ace3*$^{-/-}$ spermatozoa showed normal fertilizing ability in our in vitro fertilization system using both cumulus-intact and cumulus-free eggs. These results suggest that ACE3 binds to IZUMO1, but this characteristic nature is not required for spermatozoa to fertilize eggs (Inoue et al. 2010).

32.5   Dynamic Translocation of IZUMO1

IZUMO1 is not exposed on the freshly prepared sperm surface, and it appears on the plasma membrane after acrosome reaction. It is not known how IZUMO1 appears on the plasma membrane or how it behaves at the moment of sperm-egg fusion. To visualize the behavior of IZUMO1 in living spermatozoa, we generated a transgenic mouse line expressing mCherry-tagged IZUMO1 (Red-IZUMO1) using a testis-specific promoter and optimized a confocal microscope system with ultralow invasiveness. By using Red-IZUMO1 transgenic mice in combination with $Izumo1^{-/-}$ mice and Green-acrosome transgenic mice that overexpress GFP specifically in the acrosome (Nakanishi et al. 1999), we examined the precise localization of IZUMO1 before and after acrosome reaction as well as subsequent fertilization processes. The physiological function of Red-IZUMO1 was proven in a gene rescue experiment. Red-IZUMO1 was localized on both inner and outer acrosomal membranes in freshly prepared spermatozoa. At the time of acrosome reaction, Red-IZUMO1 demonstrated rapid diffusion onto the whole sperm head and then tended to localize in the medial region of the sperm head, called the equatorial segment (Fig. 32.2a,b). The moment of sperm–egg fusion was also imaged live through Red-IZUMO1, demonstrating that the diffusion of Red-IZUMO1 onto the egg membrane starts at the equatorial segment (Fig. 32.2c). Further experimentation using CD9-GFP eggs indicated that IZUMO1 on the inner acrosomal membrane is engulfed into the egg cytoplasm. In this study, we visualized the dynamic movement of the fusion-related protein IZUMO1 during fertilization processes. We found that IZUMO1 migrated out through fusion pores formed between the plasma membrane and outer acrosomal membrane and that it diffused away from the equatorial segment to the egg plasma membrane at the beginning of sperm–egg fusion (Satouh et al. 2012).

32.6   Perspective

A partial molecular mechanism of fertilization has been clarified by gene-manipulated animal experiments. However, to elucidate detailed mechanism leading to a new schema, it is necessary to analyze the properties of each factor as well. Concerning the membrane fusion of spermatozoon and oocyte that is the central event of fertilization, targeted deletion studies have revealed only two proteins, CD9 on oocyte and IZUMO1 on spermatozoon, to be necessary in sperm–egg fusion. If all the essential components and molecular behaviors of the fusion machinery are sufficiently identified, elucidation of the sperm–egg fusion mechanism will be greatly promoted. Also, this result might prompt us to thoroughly understand the fundamental principle of wider areas of the cell–cell fusion process such as the formation of myotubes, placenta, multinucleated osteoclasts, and macrophages. Clarification of the molecular mechanism of fertilization will benefit clinical treatment of sterility and will support the potential development of novel contraceptive strategies in the future.

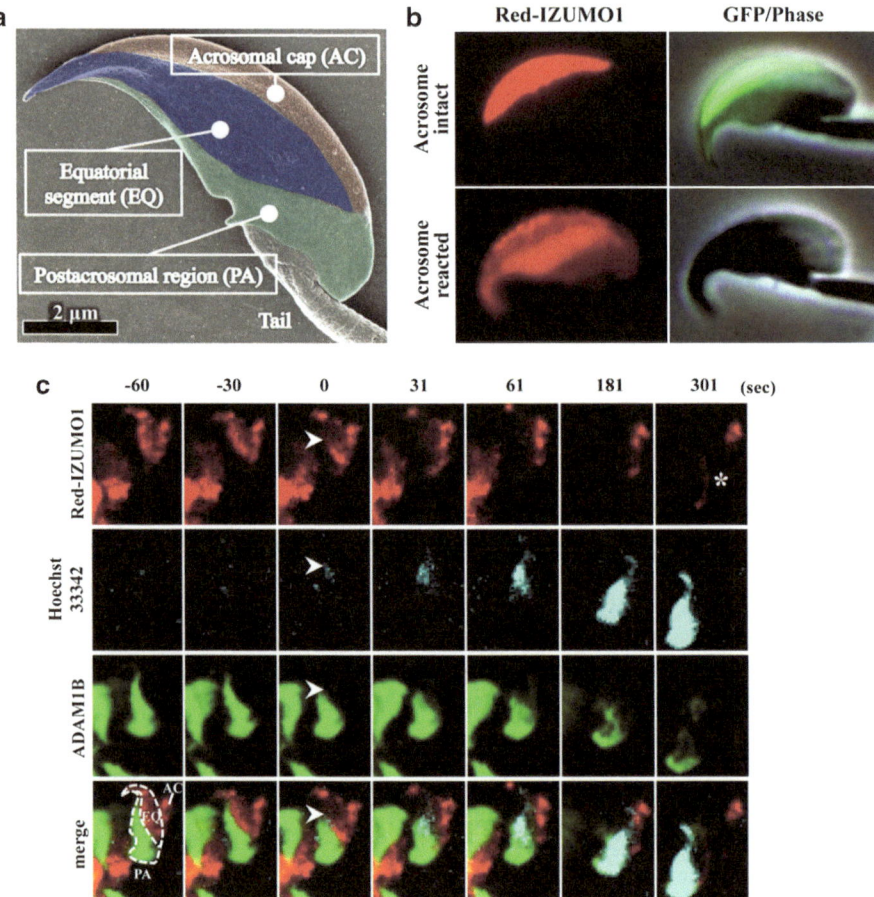

Fig. 32.2 Visualization of sperm–egg fusion. (**a**) The sperm head consists mainly of three subcellular regions. Scanning electron micrographic view of the sperm head indicates each region: acrosomal cap (*AC*), equatorial segment (*EQ*), and postacrosomal region (*PA*). (**b**) Fluorescent images of spermatozoa from Red-IZUMO1 and Green-acrosome double-transgenic mouse. IZUMO1 was initially localized in both the outer membrane and inner acrosomal membrane in the acrosomal cap but not in the equatorial segment. Thus, acrosome-intact spermatozoa have Red-IZUMO1 in the acrosomal cap area with GFP, whereas Red-IZUMO1 spread to the entire head, including the equatorial segment, after acrosome reaction (GFP negative). (**c**) A representative time-lapse view of sperm–egg fusion. Initiation of fusion (time 0) was detected by diffusion of Red-IZUMO1 from the equatorial segment and the concomitant transfer of Hoechst 33342 dye to spermatozoa in the same area (*arrowheads*). Diffusion of the membrane in the postacrosomal area identified by ADAM1B (*green*) started at 60 s after the initiation of sperm–egg fusion, and this was accompanied by expansion of the Hoechst 33342 staining area toward the posterior head. After the fusion process, IZUMO1 is only present in the inner acrosomal membrane (*asterisk*)

Acknowledgments This work was supported by grants from the Ministry of Education, Culture, Sports, Science, and Technology of Japan (21112006 and 21687018).

References

Ellerman DA, Pei J, Gupta S, Snell WJ, Myles D, Primakoff P (2009) Izumo is part of a multiprotein family whose members form large complexes on mammalian sperm. Mol Reprod Dev 76:1188–1199

Inoue N, Ikawa M, Isotani A, Okabe M (2005) The immunoglobulin superfamily protein Izumo is required for sperm to fuse with eggs. Nature (Lond) 434:234–238

Inoue N, Ikawa M, Okabe M (2008) Putative sperm fusion protein IZUMO and the role of N-glycosylation. Biochem Biophys Res Commun 377:910–914

Inoue N, Kasahara T, Ikawa M, Okabe M (2010) Identification and disruption of sperm-specific angiotensin converting enzyme-3 (ACE3) in mouse. PLoS One 5:e10301

Jégou A, Ziyyat A, Barraud-Lange V, Perez E, Wolf JP, Pincet F, Gourier C (2011) CD9 tetraspanin generates fusion competent sites on the egg membrane for mammalian fertilization. Proc Natl Acad Sci USA 108:10946–10951

Kaji K, Oda S, Shikano T, Ohnuki T, Uematsu Y, Sakagami J, Tada N, Miyazaki S, Kudo A (2000) The gamete fusion process is defective in eggs of Cd9-deficient mice. Nat Genet 24:279–282

Le Naour F, Rubinstein E, Jasmin C, Prenant M, Boucheix C (2000) Severely reduced female fertility in CD9-deficient mice. Science 287:319–321

Miyado K, Yamada G, Yamada S, Hasuwa H, Nakamura Y, Ryu F, Suzuki K, Kosai K, Inoue K, Ogura A, Okabe M, Mekada E (2000) Requirement of CD9 on the egg plasma membrane for fertilization. Science 287:321–324

Miyado K, Yoshida K, Yamagata K, Sakakibara K, Okabe M, Wang X, Miyamoto K, Akutsu H, Kondo T, Takahashi Y, Ban T, Ito C, Toshimori K, Nakamura A, Ito M, Miyado M, Mekada E, Umezawa A (2008) The fusing ability of sperm is bestowed by CD9-containg vesicles released from eggs in mice. Proc Natl Acad Sci USA 105:1292–1296

Nakanishi T, Ikawa M, Yamada S, Parvinen M, Baba T, Nshimune Y, Okabe M (1999) Real-time observation of acrosomal dispersal from mouse sperm using GFP as a marker protein. FEBS Lett 449:277–283

Ohtsubo K, Marth JD (2006) Glycosylation in cellular mechanisms of health and disease. Cell 126:855–867

Okabe M, Adachi T, Takada K, Oda H, Yagasaki M, Kohama Y, Mimura T (1987) Capacitation-related changes in antigen distribution on mouse sperm heads and its relation to fertilization rate in vitro. J Reprod Immunol 11:91–100

Okabe M, Yagasaki M, Oda H, Matzno S, Kohama Y, Mimura T (1988) Effect of a monoclonal anti-mouse sperm antibody (OBF13) on the interaction of mouse sperm with zona-free mouse and hamster eggs. J Reprod Immunol 13:211–219

Rella M, Elliot JL, Revett TJ, Lanfear J, Phelan A, Jackson RM, Turner AJ, Hooper NM (2007) Identification and characterisation of the angiotensin converting enzyme-3 (ACE3) gene: a novel mammalian homologue of ACE. BMC Genomics 8:194

Runge KE, Evans JE, He ZY, Gupta S, McDonald KL, Stahlberg H, Primakoff P, Myles DG (2007) Oocyte CD9 is enriched on the microvillar membrane and required for normal microvillar shape and distribution. Dev Biol 304:317–325

Satouh Y, Inoue N, Ikawa M, Okabe M (2012) Visualization of the moment of mouse sperm-egg fusion and dynamic localization of IZUMO1. J Cell Sci 125:4985–4990

Toshimori K, Saxena DK, Tanii I, Yoshinaga K (1998) An MN9 antigenic molecule, equatorin, is required for successful sperm-oocyte fusion in mice. Biol Reprod 59:22–29

Chapter 33
A ZP2 Cleavage Model of Gamete Recognition and the Postfertilization Block to Polyspermy

Jurrien Dean

Abstract The molecular basis of gamete recognition and the corresponding block to polyspermy in mammals have intrigued investigators for decades. Taking advantage of the fastidious nature of human sperm, which will not bind to the mouse zona pellucida, gain-of-function assays have been established in transgenic mice by replacing endogenous mouse proteins with the corresponding human homologue. In the presence of human ZP2, by itself or with the three other human zona proteins, human sperm bind and penetrate the 'humanized' zona pellucida but do not fuse with the mouse egg. Using recombinant ZP2 peptides in a bead binding assay, the gamete recognition site was located to a ~115 amino-acid N-terminal domain. Following fertilization, egg cortical granules exocytose ovastacin, an oocyte-specific metalloendoprotease, that cleaves the N-terminus of ZP2 and prevents sperm binding to the zona surrounding the preimplantation embryo. Genetic ablation of the enzyme or mutation of the ZP2 cleavage site prevents cleavage and sperm bind de novo to the surface of the zona pellucida even after fertilization and cortical granule exocytosis. These observations form the basis of the ZP2 cleavage model of gamete recognition in which mammalian sperm bind to an N-terminal domain of ZP2. Following penetration through the zona matrix and gamete fusion, the egg cortical granules exocytose ovastacin, which cleaves ZP2 and provides a definitive block to sperm binding at the surface of the zona pellucida.

Keywords Block to polyspermy • Fertilization • Ovastacin • Zona pellucida • ZP2

J. Dean (✉)
Laboratory of Cellular and Developmental Biology, NIDDK,
National Institutes of Health, Bethesda, MD 20892, USA
e-mail: jurriend@helix.nih.gov

H. Sawada et al. (eds.), *Sexual Reproduction in Animals and Plants*,
DOI 10.1007/978-4-431-54589-7_33, © The Author(s) 2014

33.1   Introduction

The fertilizing sperm is capacitated by passage through the female reproductive tract. As sperm approach ovulated eggs in the ampulla of the oviduct, the acrosome, a subcellular organelle underlying the sperm surface, remains intact. The ovulated egg(s) is surrounded by two investments: a gelatinous cumulus mass composed of hyaluronan interspersed with cumulus cells and an insoluble, extracellular zona pellucida. The successful sperm navigates the cumulus mass, and binds and penetrates the zona pellucida before entry into the perivitelline space between the zona matrix and egg. Only acrosome-reacted sperm are observed in the perivitelline space, but the site and molecular basis of induction of acrosome exocytosis remain controversial and the function(s) of acrosomal contents remains an area of active investigation (Avella and Dean 2011). Acrosome-reacted sperm fuse with the egg plasma membrane and fertilization activates the egg.

The zona pellucida is an extracellular matrix surrounding eggs and early embryos that plays critical roles in gamete recognition required for fertilization and in ensuring monospermy necessary for the successful onset of development (Fig. 33.1a). The mouse and human zona pellucida are composed of three and four (ZP1, ZP2, ZP3, ZP4) glycoproteins, respectively (Bleil and Wassarman 1980; Shabanowitz and O'Rand 1988; Lefievre et al. 2004). Although each zona protein is well conserved (62–71 % amino-acid identity), human sperm are fastidious and will not bind to the mouse zona pellucida (Bedford 1977). Following fertilization, the zona pellucida is modified and mouse sperm do not bind to two-cell embryos. Although a number of changes in the zona matrix have been inferred, only biochemical cleavage of ZP2 has been experimentally documented (Bleil et al. 1981; Bauskin et al. 1999). The zona pellucida also plays a critical role in ensuring passage of the early embryo through the oviduct before implantation. The biochemical removal of the zona matrix leads to resorption of the embryo into the epithelia lining of the oviduct and is lethal (Bronson and McLaren 1970; Modlinski 1970).

Using mouse genetics, we have investigated the molecular basis of gamete recognition, the postfertilization block to polyspermy, and the role of the zona in protecting the early embryo. Based on experimental results, we propose a ZP2 cleavage model of gamete recognition in which sperm bind to the N-terminus of ZP2, which is cleaved by ovastacin to prevent postfertilization sperm binding (Gahlay et al. 2010; Baibakov et al. 2012; Burkart et al. 2012). The absence of ZP2 or ZP3 precludes formation of a zona pellucida surrounding ovulated eggs and leads to resorption and embryonic lethality (Liu et al. 1996; Rankin et al. 1996; Rankin et al. 2001).

33.2   The Structure of the Zona Pellucida

By scanning electron microscopy, the mouse and human zonae pellucidae appear similar, with fenestration thought to provide passage for sperm penetration (Familiari et al. 2006; Familiari et al. 2008). The mouse egg and surrounding zona pellucida

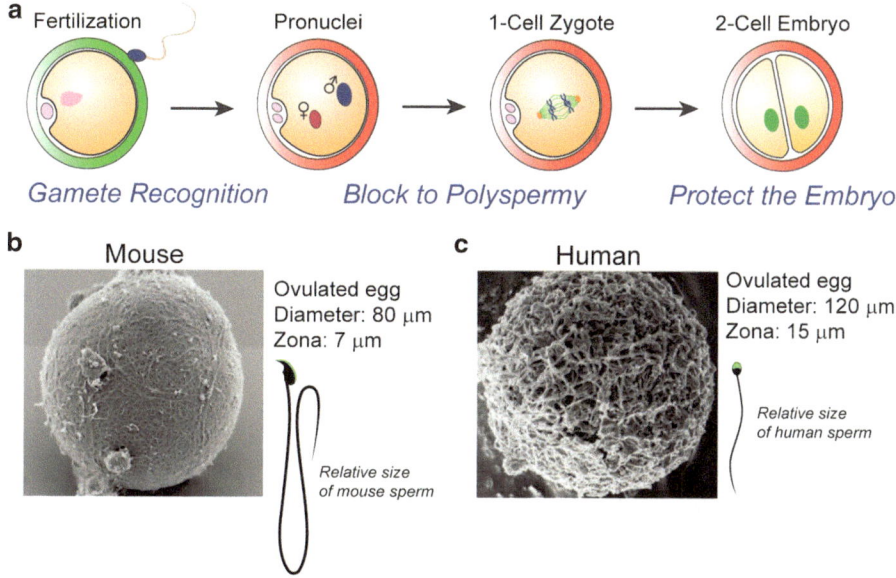

Fig. 33.1 Role of the zona pellucida in fertilization and early development. (**a**) Ovulated eggs in the cumulus mass are fertilized by a single sperm in the ampulla of the oviduct. Each haploid gamete forms a pronucleus and after syngamy develops into a one-cell zygote that divides within 24 h to form the two-cell embryo. The extracellular zona pellucida that surrounds the egg is permissive for sperm binding and penetration (*green*). However, following fertilization the zona matrix is modified (*red*) and sperm do not bind. The zona pellucida is critically important for gamete recognition, a postfertilization block to sperm binding, and for protecting the embryo as it passes through the oviduct. (Modified from Li et al. 2013). (**b**) An insoluble, extracellular zona pellucida (~7 μm wide) surrounds mouse eggs (~80 μm diameter) (Familiari et al. 2008). Mouse sperm are ~125 μm long with a thin acrosome overlying a distinctive falciform (hook-like) head; the zona matrix has multiple pores that may facilitate sperm penetration. (**c**) The human zona pellucida (~15 μm) surrounds a larger egg (~120 μm) and has a structure in scanning electron microscopy (EM) similar to that of mouse (Familiari et al. 2006). Human sperm are half as long (~ 60 μm) as mouse sperm with a smaller, flattened, spatulate head

are smaller than those of the human, although the mouse sperm is considerably longer than the diminutive human sperm (Fig. 33.1b,c). As noted, the human zona pellucida has four glycoproteins, but the mouse zona pellucida has only three. Each zona pellucida protein is encoded by a single-copy gene that is found on syntenic chromosomes in mouse and human (Hoodbhoy et al. 2005). However, mouse *Zp4* is an expressed pseudogene that is not translated into protein because of multiple missense and stop codons (Lefièvre et al. 2004).

Each of the other three mouse zona genes has been ablated in embryonic stem cells to establish mouse lines lacking the cognate protein. *Zp1* null mice form a zona pellucida with a more loosely woven matrix than normal. Female mice are fertile, albeit with decreased fecundity (Rankin et al. 1999). However, mice lacking either ZP2 (Rankin et al. 2001) or ZP3 (Rankin et al. 1996; Liu et al. 1996) do not form a zona pellucida surrounding ovulated eggs, and female mice are sterile. Thus, ZP1 is not essential for fertility, but in the absence of a zona matrix, these loss-of-function observations do not provide insight into the role of either ZP2 or ZP3 in gamete recognition.

33.3   ZP2 Cleavage Model of Gamete Recognition

In more recent genetic studies, we have taken advantage of two physiological dichotomies. The first is that human sperm are fastidious and bind to the human, but not the mouse, zona pellucida. The second is that homologous mouse sperm bind to eggs but not two-cell embryos. To exploit the first dichotomy, we have used mouse genetics to replace each endogenous mouse protein with the corresponding human protein; we have established transgenic mice expressing human ZP4; and we have crossed the four lines together to establish the quadruple rescue line that expresses the four human zona proteins and none of the three mouse zona proteins (Rankin et al. 1998; Rankin et al. 2003; Yauger et al. 2011; Baibakov et al. 2012).

Each of these lines formed a zona pellucida and was fertile when mated with normal male mice. Eggs from each line were tested for their ability to support capacitated human sperm binding in vitro using eggs in cumulus (Fig. 33.2a). Only a zona matrix with human ZP2, either by itself or in conjunction with the three other human zona proteins was able to support human sperm binding. In the presence of all four human proteins, sperm penetrated the zona matrix and accumulated in the perivitelline space, unable to fuse with mouse eggs. Following fertilization with mouse sperm, an effective postfertilization block to polyspermy was established and human sperm were unable to bind to the zona pellucida. The binding site on ZP2 was further defined by assaying sperm binding to beads coated with recombinant human and mouse N-terminal ZP2 peptides. Human sperm bound to beads coated with human ZP2$^{39\text{-}154}$, but not to the corresponding mouse ZP2 peptide, and binding was dependent on structural constraints imposed by disulfide bonds (Baibakov et al. 2012).

Following fertilization, ZP2 is cleaved and the cleavage site was defined by microscale Edman degradation as immediately upstream of a diacidic motif, a known recognition site for the astacin family of metalloendoproteases. When ZP2 was mutated in transgenic mice to prevent postfertilization cleavage (^{166}LA$\downarrow$DE^{169} → ^{166}LG$\downarrow$AA^{169}), mouse sperm bound de novo to two-cell embryos despite fertilization and cortical granule exocytosis (Fig. 33.2b) (Gahlay et al. 2010). Ovastacin is an oocyte-specific metalloendoprotease that is conserved in mouse and human (Quesada et al. 2004). Using a peptide-specific antibody, mouse ovastacin was detected in cortical granules of normal eggs and was absent in two-cell embryos, consistent with its postfertilization exocytosis. The single-copy gene *Astl* encodes ovastacin. When *Astl* was ablated in transgenic mice, homozygous null female were fertile with a modest decrease in fecundity. However, ZP2 remained uncleaved following fertilization and mouse sperm bound to the zona pellucida surrounding two-cell embryos (Burkart et al. 2012).

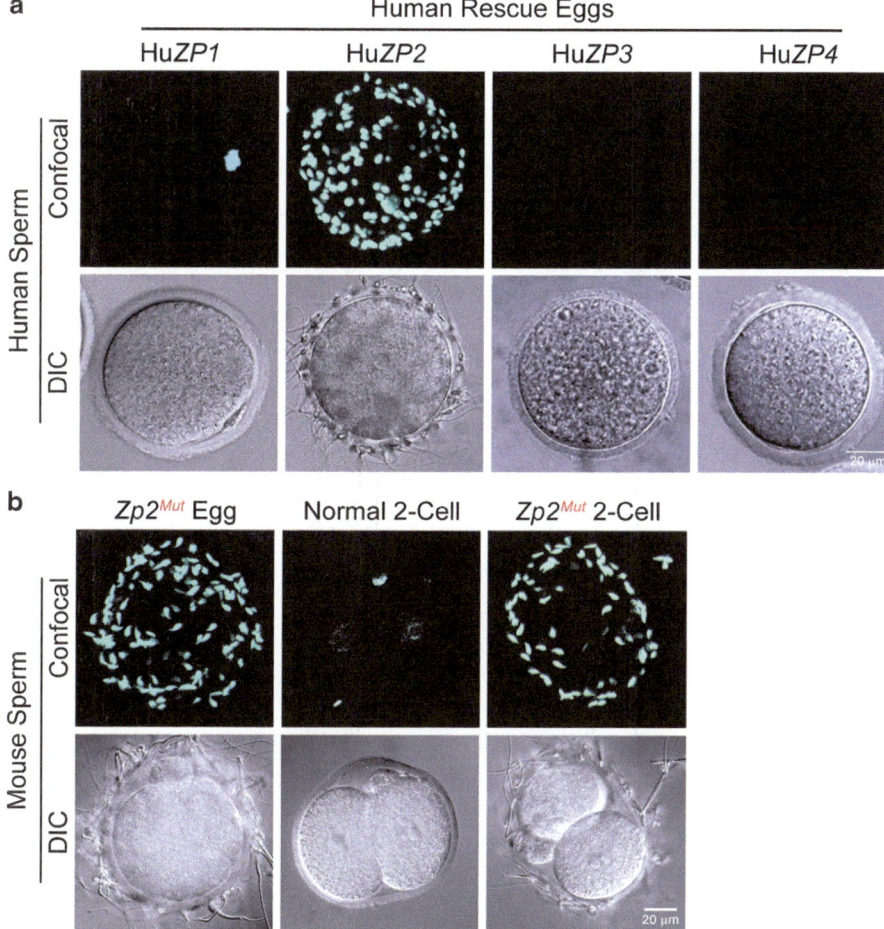

Fig. 33.2 Human and mouse sperm binding to genetically altered zonae pellucidae. (**a**) Confocal and differential interference contrast (DIC) images of capacitated human sperm binding to the zona pellucida of mice expressing human ZP1, human ZP2, human ZP3, or human ZP4, in the absence of the corresponding mouse protein. Human sperm bind only to human ZP2 in transgenic mice. (Modified from Baibakov et al. 2012). (**b**) Capacitated mouse sperm binding to two-cell embryos in which ZP2 remains intact because ZP2 was mutated to prevent postfertilization cleavage ($Zp2^{Mut}$). $Zp2^{Mut}$ eggs serve as positive controls and normal two-cell embryos as negative controls. (Modified from Gahlay et al. 2010)

Cortical granule exocytosis is associated with a postfertilization block to polyspermy that occurs via multiple mechanisms and is critically important to ensure monospermy essential for normal development. The most rapid block prevents supernumerary sperm already present in the perivitelline space from fusing with the egg plasma membrane. Although attributed to membrane depolarization in other species, the molecular basis of this block remains to be determined in mammals (Jaffe et al. 1983; Horvath et al. 1993). There is an additional block to penetration of the zona pellucida that appears to occur within the first hour after fertilization (Sato 1979; Stewart-Savage and Bavister 1988). During the ensuing several hours, ovastacin cleaves ZP2 after which sperm do not bind to the zona pellucida (Baibakov et al. 2007). Although slow compared to the first two blocks and probably of less immediate importance, cleavage of ZP2 provides a definitive and irreversible block to polyspermy.

These observations are consistent with earlier observations in *Xenopus laevis* in which gp69/64, the homologue of mouse ZP2, inhibited primary sperm binding to eggs in vitro (Tian et al. 1997). Following fertilization, gp69/64 was cleavage at the conserved ^{155}FD$^{\downarrow}$DD^{158} site and the C-terminal native gp69/64 glycopeptide did not inhibit sperm binding. Although a short recombinant peptide gp69/64$^{130-156}$ also did not affect sperm binding, the longer gp69/64$^{34-156}$ N-terminal domain (homologous to mouse ZP2$^{35-149}$) was not tested. Following fertilization, pg69/64 is cleaved by a zinc metalloprotease (Lindsay and Hedrick 2004), and postfertilization proteolysis of the N-terminal domain of gp69/64 could account for the lack of sperm binding in *Xenopus* embryos. Thus, the ZP2 cleavage model of gamete recognition may apply broadly among vertebrates.

33.4   The Model and Future Validation

The ZP2 cleavage model provides a simple, unifying explanation to account for these experimental observations in transgenic mice (Fig. 33.3). To wit, capacitated sperm bind to the N-terminus of ZP2 in the extracellular zona pellucida surrounding ovulated eggs. After penetration of the zona matrix, the gamete membranes are fused at fertilization. Postfertilization cortical granule exocytosis releases ovastacin, which diffuses through the zona matrix. This metalloendoprotease cleaves mouse ZP2 at ^{166}LA$^{\downarrow}$DE^{169} and provides a definitive block to polyspermy. The zona pellucida that surrounds the early embryo is essential to ensure passage through the oviduct for implantation and the successful onset of development.

This model makes predictions that can be tested experimentally with transgenic mice expressing chimeric human–mouse or truncated forms of ZP2. If the model remains robust, a more precise definition of the sperm-binding site may provide reagents to identify the long elusive cognate sperm receptor. Particularly satisfying would be the replacement of the mouse sperm receptor with the human homologue, after which 'humanized' sperm should not bind to the mouse zona pellucida unless it contains human ZP2.

Acknowledgments The critical reading of the manuscript by Dr. Matteo Avella is appreciated. This research was support by the Intramural Research Program of the National Institutes of Health, NIDDK.

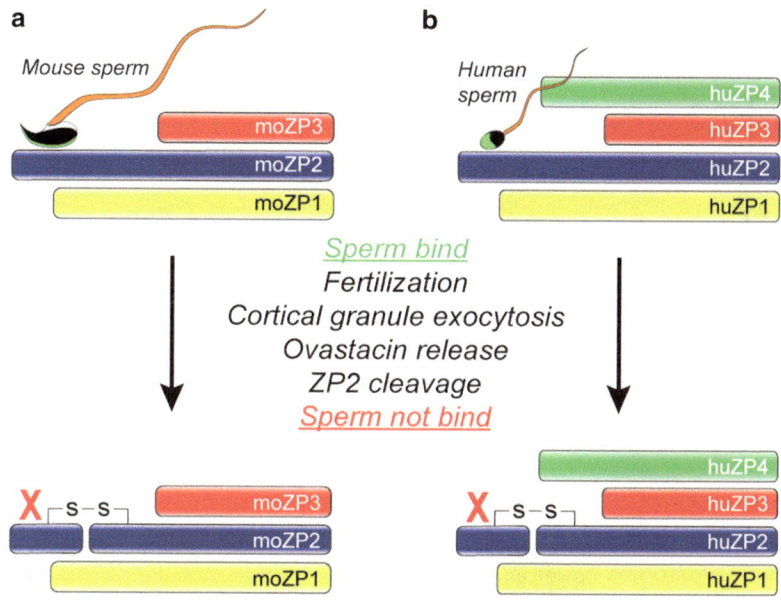

Fig. 33.3 The ZP2 cleavage model of gamete recognition. (**a**) The mouse zona pellucida is composed of three glycoproteins, ZP1, ZP2, and ZP3. Capacitated sperm bind to an N-terminal domain of ZP2, penetrate through the zona matrix, and fertilize the egg by fusing with its membrane: this activates the egg and leads to cortical granule exocytosis, which releases ovastacin, a metalloendoprotease. Ovastacin cleaves ZP2 within the zona pellucida and provides a definitive block to sperm binding in the early embryo. (**b**) Same as **a**, but with the human zona pellucida composed of four glycoproteins

References

Avella MA, Dean J (2011) Fertilization with acrosome-reacted mouse sperm: implications for the site of exocytosis. Proc Natl Acad Sci USA 108:19843–19844

Baibakov B, Gauthier L, Talbot P et al (2007) Sperm binding to the zona pellucida is not sufficient to induce acrosome exocytosis. Development (Camb) 134:933–943

Baibakov B, Boggs NA, Yauger B et al (2012) Human sperm bind to the N-terminal domain of ZP2 in humanized zonae pellucidae in transgenic mice. J Cell Biol 197:897–905

Bauskin AR, Franken DR, Eberspaecher U et al (1999) Characterization of human zona pellucida glycoproteins. Mol Hum Reprod 5:534–540

Bedford JM (1977) Sperm/egg interaction: the specificity of human spermatozoa. Anat Rec 188:477–488

Bleil JD, Wassarman PM (1980) Structure and function of the zona pellucida: identification and characterization of the proteins of the mouse oocyte's zona pellucida. Dev Biol 76:185–202

Bleil JD, Beall CF, Wassarman PM (1981) Mammalian sperm–egg interaction: fertilization of mouse eggs triggers modification of the major zona pellucida glycoprotein, ZP2. Dev Biol 86:189–197

Bronson RA, McLaren A (1970) Transfer to the mouse oviduct of eggs with and without the zona pellucida. J Reprod Fertil 22:129–137

Burkart AD, Xiong B, Baibakov B et al (2012) Ovastacin, a cortical granule protease, cleaves ZP2 in the zona pellucida to prevent polyspermy. J Cell Biol 197:37–44

Familiari G, Heyn R, Relucenti M et al (2006) Ultrastructural dynamics of human reproduction, from ovulation to fertilization and early embryo development. Int Rev Cytol 249:53–141

Familiari G, Heyn R, Relucenti M et al (2008) Structural changes of the zona pellucida during fertilization and embryo development. Front Biosci 13:6730–6751

Gahlay G, Gauthier L, Baibakov B et al (2010) Gamete recognition in mice depends on the cleavage status of an egg's zona pellucida protein. Science 329:216–219

Hoodbhoy T, Joshi S, Boja ES et al (2005) Human sperm do not bind to rat zonae pellucidae despite the presence of four homologous glycoproteins. J Biol Chem 280:12721–12731

Horvath PM, Kellom T, Caulfield J et al (1993) Mechanistic studies of the plasma membrane block to polyspermy in mouse eggs. Mol Reprod Dev 34:65–72

Jaffe LA, Sharp AP, Wolf DP (1983) Absence of an electrical polyspermy block in the mouse. Dev Biol 96:317–323

Lefievre L, Conner S, Salpekar A et al (2004) Four zona pellucida glycoproteins are expressed in the human. Hum Reprod 19:1580–1586

Lefièvre L, Conner SJ, Salpekar A et al (2004) Four zona pellucida glycoproteins are expressed in the human. Hum Reprod 19:1580–1586

Li L, Lu X, Dean J (2013) The maternal to zygotic transition in mammals. Mol Aspects Med. doi:10.1016/j.mam.2013.01.003

Lindsay LL, Hedrick JL (2004) Proteolysis of *Xenopus laevis* egg envelope ZPA triggers envelope hardening. Biochem Biophys Res Commun 324:648–654

Liu C, Litscher ES, Mortillo S et al (1996) Targeted disruption of the mZP3 gene results in production of eggs lacking a zona pellucida and infertility in female mice. Proc Natl Acad Sci USA 93:5431–5436

Modlinski JA (1970) The role of the zona pellucida in the development of mouse eggs in vivo. J Embryol Exp Morphol 23:539–547

Quesada V, Sanchez LM, Alvarez J et al (2004) Identification and characterization of human and mouse ovastacin: a novel metalloproteinase similar to hatching enzymes from arthropods, birds, amphibians, and fish. J Biol Chem 279:26627–26634

Rankin T, Familari M, Lee E et al (1996) Mice homozygous for an insertional mutation in the Zp3 gene lack a zona pellucida and are infertile. Development (Camb) 122:2903–2910

Rankin TL, Tong Z-B, Castle PE et al (1998) Human ZP3 restores fertility in Zp3 null mice without affecting order-specific sperm binding. Development (Camb) 125:2415–2424

Rankin T, Talbot P, Lee E et al (1999) Abnormal zonae pellucidae in mice lacking ZP1 result in early embryonic loss. Development (Camb) 126:3847–3855

Rankin TL, O'Brien M, Lee E et al (2001) Defective zonae pellucidae in *Zp2* null mice disrupt folliculogenesis, fertility and development. Development (Camb) 128:1119–1126

Rankin TL, Coleman JS, Epifano O et al (2003) Fertility and taxon-specific sperm binding persist after replacement of mouse 'sperm receptors' with human homologues. Dev Cell 5:33–43

Sato K (1979) Polyspermy-preventing mechanisms in mouse eggs fertilized in vitro. J Exp Zool 210:353–359

Shabanowitz RB, O'Rand MG (1988) Characterization of the human zona pellucida from fertilized and unfertilized eggs. J Reprod Fertil 82:151–161

Stewart-Savage J, Bavister BD (1988) A cell surface block to polyspermy occurs in golden hamster eggs. Dev Biol 128:150–157

Tian J, Gong H, Thomsen GH et al (1997) Gamete interactions in *Xenopus laevis*: identification of sperm binding glycoproteins in the egg vitelline envelope. J Cell Biol 136:1099–1108

Yauger B, Boggs N, Dean J (2011) Human ZP4 is not sufficient for taxon-specific sperm binding to the zona pellucida in transgenic mice. Reproduction 141:313–319

Chapter 34
Involvement of Carbohydrate Residues of the Zona Pellucida in In Vitro Sperm Recognition in Pigs and Cattle

Naoto Yonezawa

Abstract The zona pellucida (ZP), which surrounds the mammalian oocyte, plays roles in species-selective sperm–oocyte interactions. In pigs and cattle, the ZP consists of ZP2, ZP3, and ZP4. Nonreducing terminal β-galactosyl (Gal) residues of neutral N-linked carbohydrate chains of the ZP are necessary for porcine sperm–ZP binding, and nonreducing terminal α-mannosyl (Man) residues of high-mannose-type chains are necessary for bovine sperm–ZP binding. Acrosome-intact porcine sperm prefer β-Gal, whereas acrosome-intact bovine sperm prefer α-Man, as shown using glycolipid analogues. The major N-linked chains of recombinant porcine and bovine ZP glycoproteins expressed using the baculovirus-Sf9 cell expression system are pauci- and high-mannose-type chains that are different in structure from the major neutral N-linked chains of the native porcine ZP but similar to those of the native bovine ZP. Porcine and bovine ZP3/ZP4 complexes coexpressed in Sf9 cells bind to bovine sperm but not to porcine sperm. Hybrid complexes consisting of native porcine ZP4 and recombinant porcine ZP3 bind to porcine sperm, whereas complexes consisting of native porcine ZP3 and recombinant porcine ZP4 do not bind to porcine sperm. These data indicate that the sugar preference of sperm is consistent with the nonreducing terminal residues of N-linked chains of sperm binding-active ZP glycoproteins and suggest that, in the in vitro assay system, the nonreducing terminal sugar residues are essential for species-selective recognition of sperm in pigs and cattle.

Keywords Baculovirus-Sf9 cell • N-linked chains • Sperm ligand • Zona pellucida

N. Yonezawa (✉)
Graduate School of Science, Chiba University, 1-33 Yayoi-cho,
Inage-ku, Chiba 263-8522, Japan
e-mail: nyoneza@faculty.chiba-u.jp

H. Sawada et al. (eds.), *Sexual Reproduction in Animals and Plants*,
DOI 10.1007/978-4-431-54589-7_34, © The Author(s) 2014

34.1   Introduction

Mammalian oocytes are surrounded by a transparent envelope called the zona pellucida (ZP), which is involved in several critical aspects of fertilization, including species-selective recognition of sperm. In many mammalian species, there are four ZP glycoproteins (ZPGs) (ZP1, ZP2, ZP3, ZP4) (Stetson et al. 2012), although there are three ZPGs (ZP1, ZP2, ZP3) in mice. The porcine and bovine ZPs are composed of three ZPGs (ZP2, ZP3, ZP4). All ZPGs contain the ZP domain, which consists of about 260 amino acids and contains eight conserved Cys residues (Jovine et al. 2005).

Many studies have proposed that in mice the carbohydrate chains of ZP3 play an essential role in sperm recognition (the glycan model). However, a series of studies using transgenic mice do not support the glycan model (Hoodbhoy and Dean 2004). Instead, the supramolecular complex model, in which the supramolecular structure of the ZP polypeptides is necessary for sperm recognition in mice, was proposed based on studies using transgenic mice rescued by the human *ZP2* gene (Rankin et al. 2003). Furthermore, the domain-specific model, in which both protein and carbohydrate moieties in the domain(s) of ZPGs are involved in sperm recognition, was recently proposed (Clark 2011). More recently still, it was shown using transgenic mice that sperm–ZP binding is not necessary for fertilization of oocytes surrounded by the *cumulus oophorus* (Tokuhiro et al. 2012). Whether this finding is applicable to mammals other than mice remains to be determined.

One of our research questions is whether carbohydrate chains are involved in sperm–ZP binding in pigs and cattle. Here, we discuss this subject mainly by reviewing our biochemical studies.

34.2   Polypeptides of Porcine and Bovine ZPGs

The porcine ZP is approximately 16 μm in width and contains 30–33 ng glycoproteins. The bovine ZP is of similar size to the porcine ZP. The estimated ZP2/ZP3/ZP4 protein molar ratio is 1:6:6 in the porcine ZP and 1:2:1 in the bovine ZP. In mice, the estimated ZP1/ZP2/ZP3 molar ratio is 1:4:4, and ZP2 and ZP3 form a filamentous equimolar complex, whereas ZP1 crosslinks the ZP2/ZP3 complex (Greve and Wassarman 1985). Based on the similarity in the protein molar ratios, it appears that ZP architecture is similar in pigs and mice but bovine ZP architecture appears to differ from that of porcine and murine ZP.

Cys positions of the ZP domain are classified into two patterns: the ZP3 pattern and the ZP1/ZP2/ZP4 pattern (Fig. 34.1). Porcine ZP3 and ZP4 exhibit disulfide patterns different from the ZP3 and ZP1/ZP2/ZP4 patterns reported for mice, rats, humans, and fish (Kanai et al. 2008). The chick homologue of the mammalian ZP3 precursor protein has a pig-type ZP3 pattern (Han et al. 2010), whereas betaglycan has a pig-type ZP1/ZP2/ZP4 pattern (Lin et al. 2011), as revealed by X-ray crystallography. The two ZP3 disulfide bond patterns cause only subtle structural differences (Han et al. 2010). The subtle structural differences may affect the specificity of

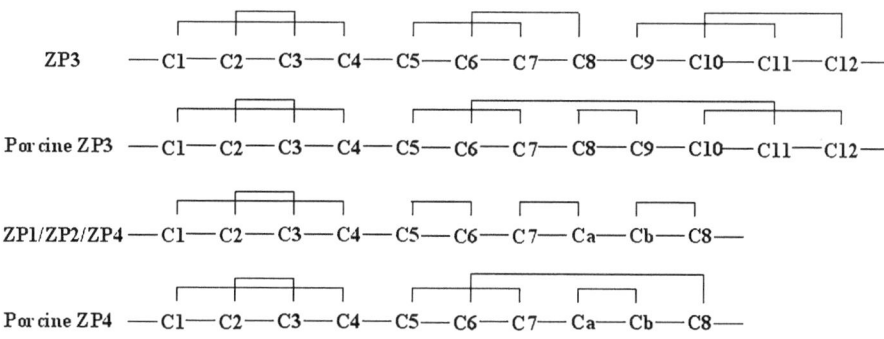

Fig. 34.1 Schematic representations of disulfide bond patterns in zona pellucida (ZP)3 and ZP1/ZP2/ZP4. The disulfide bond pattern of ZP3 from mice, human, rat and fish, the pattern of porcine ZP3, the pattern in the ZP domain of ZP1/ZP2/ZP4-type proteins from mice, rat and fish, the pattern in the ZP domain of porcine ZP4 are shown schematically. The patterns of C1–C4 are completely conserved in the ZP domain proteins. *C* Cys

the interaction between ZP1/ZP2/ZP4 and ZP3 (Darie et al. 2004). When the disulfide bond pattern of ZP3 is of the pig type, the ZP1/ZP2/ZP4 pattern of the same species is also the pig type, and vice versa. The disulfide bond patterns might change during zona matrix formation, following processing at the consensus furin cleavage site.

34.3   *N*-Linked Carbohydrate Chains from Porcine and Bovine ZPGs

The *N*-linked chains of porcine ZP comprise neutral and acidic chains at a molar ratio of about 1:3. Both neutral chains and acidic chains comprise di-, tri-, and tetra-antennary complex-type chains with an α-fucosyl (Fuc) residue at the innermost *N*-acetylglucosamine (GlcNAc) (Nakano and Yonezawa 2001). A large proportion of neutral chains have nonreducing terminal β-galactosyl (Gal) residues.

In the bovine ZP, the major neutral *N*-linked chain consists of only one structure: the high-mannose-type chain containing five mannosyl (Man) residues ($Man_5GlcNAc_2$) (Katsumata et al. 1996). Sialylation is predominant in acidic chains of bovine ZP; in porcine ZP, sulfation of the *N*-acetyl lactosamine unit is more dominant.

34.4   Involvement of Carbohydrate Chains from Porcine and Bovine ZPGs in Sperm–ZP Binding

Neutral *N*-linked carbohydrate chains released from the porcine ZP3/ZP4 mixture retain their sperm-binding activities, with the tri- and tetra-antennary complex-type chains binding more strongly than the di-antennary complex-type chains

(Nakano and Yonezawa 2001). In one study, the nonreducing terminal β-Gal residues of the complex-type *N*-linked chains were shown to be involved in sperm binding (Yonezawa et al. 2005a). Conversely, another study reported that the *O*-linked carbohydrate chains, and not the *N*-linked chains, released from the ZP3/ZP4 mixture inhibit sperm–ZP binding (Yurewicz et al. 1991). Therefore, both the *N*- and *O*-linked carbohydrate chains are thought to be sperm ligands. A polypeptide scaffold is necessary for the sperm-binding activity of carbohydrate chains, because the sperm-binding activities of carbohydrate chains released from polypeptides are much weaker than the activity of the ZP3/ZP4 mixture.

Porcine ZP3 and ZP4 have three *N*-linked chains, at Asn124, Asn146, and Asn271 in ZP3 and at Asn203, Asn220, and Asn333 in ZP4. Tri- and tetra-antennary chains are localized at Asn271 in ZP3 and at Asn220 in ZP4 (Nakano and Yonezawa 2001). ZP2 has five *N*-glycosylation sites: Asn84, Asn268, Asn316, Asn323, and Asn530 (von Witzendorff et al. 2005). A remarkable difference between the *N*-linked chain structures of ZP2 and ZP3/ZP4 is that ZP2 has a high-mannose-type chain containing five Man residues, probably located at Asn268.

Nonreducing terminal α-Man residues of the high-mannose-type chain $Man_5GlcNAc_2$ play an essential role in bovine sperm–ZP binding (Amari et al. 2001). Sialic acid residues at the nonreducing ends of acidic *N*-linked or *O*-linked chains of bovine ZPGs are also involved in sperm binding (Velásquez et al. 2007). The *N*-glycosylation sites of bovine ZP3 and ZP4 have not yet been determined. The *N*-glycosylation sites of bovine ZP2 are Asn83, Asn191, and Asn527 (Ikeda et al. 2002).

Glycolipid analogue-possessing monosaccharides adsorbed on plastic wells show sperm-binding activity, but monosaccharide solution does not inhibit sperm–ZP binding (Takahashi et al. 2013). Two-dimensional coating of plastic wells with glycolipid analogue may cause the formation of a cluster of sugars, which is necessary for the high avidity for sperm binding. Acrosome-intact porcine sperm exhibit their strongest affinity for plastic wells coated with β-Gal and the second strongest affinity for β-*N*-acetylgalactosamine (GalNAc), whereas acrosome-intact bovine sperm exhibit their strongest binding affinities for α-Man and their second strongest affinity for β-glucose (Glc) and β-GlcNAc. These specificities are consistent with the nonreducing terminal sperm ligand sugar residues of porcine and bovine ZPs.

34.5   Sperm Binding-Active Regions of Porcine and Bovine ZPGs

We identified the sperm binding-active fragment of porcine ZP4 as an N-terminal region (Asp137 to Lys247) containing two *N*-linked chains (Nakano and Yonezawa 2001). Porcine ZP4 purified by reverse-phase HPLC is actually contaminated with ZP3, and formation of a heterocomplex of ZP3 and ZP4 is essential for the sperm-binding activity of the glycoproteins (Yurewicz et al. 1998). We established an expression system for recombinant porcine ZPGs (rpZPGs) using baculovirus-Sf9

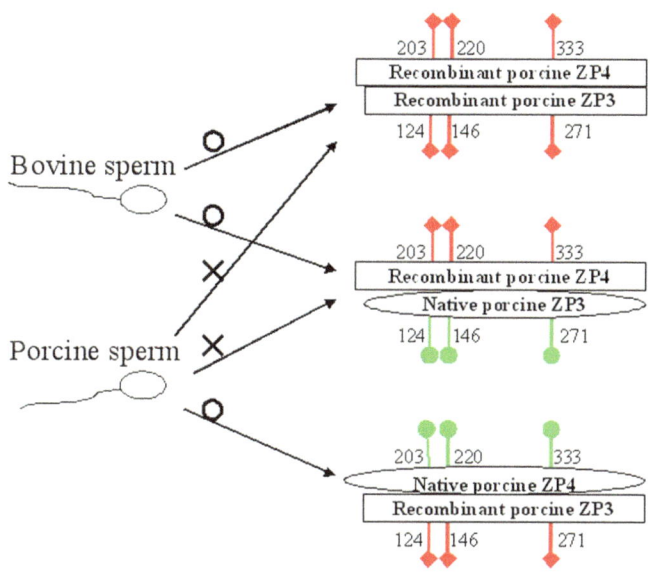

Fig. 34.2 Schematic representation of the specificity in recognition between sperm and zona pellucida (ZP) glycoproteins. Porcine sperm bind to the heterocomplex of porcine ZP3 and ZP4 in which ZP4 has complex-type *N*-linked chains (*closed circles in green*) but not to the heterocomplexes in which ZP4 has pauci- and high-mannose-type chains (*closed diamonds in red*). On the other hand, bovine sperm bind to the heterocomplexes in which ZP4 has pauci- and high-mannose-type chains

cells to obtain ZP4 without contamination of ZP3 (Yonezawa et al. 2005b). The major structures of the *N*-linked chains of rpZPGs are believed to be pauci- and high-mannose-type chains with or without a Fuc residue at the innermost GlcNAc. They have nonreducing terminal α-Man residues and are thus similar in structure to the sperm ligand *N*-linked chain of bovine ZPGs. The rpZP3/rpZP4 mixture coexpressed by Sf9 cells shows binding activity toward bovine sperm but not to porcine sperm (Fig. 34.2). The sperm-binding activity of rpZP4 is much weaker than that of rpZP3/rpZP4 (Yonezawa et al. 2005b). This study supports the previous report by Yurewicz et al. (1998).

To reexamine the sperm binding-active region of porcine ZP4, rpZP4 mutants with an Asp substitution at Asn203, Asn220, or Asn333 were coexpressed with rpZP3 in Sf9 cells (Yonezawa et al. 2005b). The mutation at Asn333 does not have an effect on the sperm-binding activity of the rpZP3/rpZP4 mixture, whereas the mutations at Asn203 and Asn220 reduce the sperm-binding activity secondarily and primarily among the three sites, respectively. These results are in agreement with our previous finding that the N-terminal fragment containing two *N*-linked chains at Asn203 and Asn220 exhibits sperm-binding activity, and that the tri- and tetra-antennary chains are localized at Asn220 in porcine ZP4.

Porcine sperm bind to native porcine ZPGs, but not to rpZPGs, probably because of differences in carbohydrate structures between native and recombinant ZPGs. A mixture of rpZP3 and native porcine ZP4 binds to the acrosomal region of porcine sperm and inhibits porcine sperm–ZP binding (Fig. 34.2) (Yonezawa et al. 2012). A mixture of native porcine ZP3 and rpZP4 does not inhibit the binding, although the mixture inhibits bovine sperm–ZP binding. This study indicated that native ZP4, but not rZP4, is necessary for the binding activity of the porcine ZP3/ZP4 complex toward porcine sperm and further suggested that the carbohydrate structures of ZP4 in the porcine ZP3/ZP4 complex are responsible for the porcine sperm-binding activity of the complex. Thus, in the present model for porcine sperm–ZP binding, the N-linked chains with nonreducing terminal β-Gal residues linked to the N-terminal region of ZP4 form a sperm-binding domain in the ZP3/ZP4 heterocomplex.

ZP4 has the highest sperm-binding activity among bovine ZPGs purified from ovaries, while ZP2 and ZP3 have weak but significant sperm-binding activities (Yonezawa et al. 2001). In the case of bovine ZPGs, ZP4 is not completely purified by reverse-phase HPLC and is contaminated with ZP3. To obtain ZP3 and ZP4 without contamination of other ZPGs, we established an expression system for bovine ZPGs using the baculovirus-Sf9 cell system (Kanai et al. 2007). Recombinant bovine ZP3 (rbZP3) and rbZP4 do not show sperm-binding activity, whereas the rbZP3/rbZP4 complex does show sperm-binding activity. The region from the N-terminus to the trefoil domain of rbZP4 is dispensable for formation of the complex with rbZP3 and for the sperm-binding activity of the complex. More detailed analysis of the sperm-binding region of bovine ZP3/ZP4 has not been accomplished.

34.6   Conclusion

Our studies indicate that porcine and bovine sperm proteins recognize β-Gal and α-Man residues, respectively, at the nonreducing ends of carbohydrate chains linked to the ZP3/ZP4 complex. In pigs and cattle, ZP3 and ZP4 form a framework on which carbohydrate chains active in sperm binding can exhibit high avidity for sperm. To prove the significance of the carbohydrate chains in sperm binding, the sperm proteins that bind specifically to β-Gal in pigs and α-Man in cattle need to be identified and characterized.

References

Amari S, Yonezawa N, Mitsui S et al (2001) Essential role of the nonreducing terminal α-mannosyl residues of the N-linked carbohydrate chain of bovine zona pellucida glycoproteins in sperm-egg binding. Mol Reprod Dev 59:221–226

Clark GF (2011) Molecular models for mouse sperm-oocyte binding. Glycobiology 21:3–5

Darie CC, Biniossek ML, Jovine L et al (2004) Structural characterization of fish egg vitelline envelope proteins by mass spectrometry. Biochemistry 43:7459–7478

Greve JM, Wassarman PM (1985) Mouse egg extracellular coat is a matrix of interconnected filaments possessing a structural repeat. J Mol Biol 181:253–264

Han L, Monné M, Okumura H et al (2010) Insights into egg coat assembly and egg-sperm interaction from the X-ray structure of full-length ZP3. Cell 143:404–415

Hoodbhoy T, Dean J (2004) Insights into the molecular basis of sperm-egg recognition in mammals. Reproduction 127:417–422

Ikeda K, Yonezawa N, Naoi K et al (2002) Localization of N-linked carbohydrate chains in glycoprotein ZPA of the bovine egg zona pellucida. Eur J Biochem 269:4257–4266

Jovine L, Darie CC, Litscher ES et al (2005) Zona pellucida domain proteins. Annu Rev Biochem 74:83–114

Kanai S, Yonezawa N, Ishii Y et al (2007) Recombinant bovine zona pellucida glycoproteins ZP3 and ZP4 coexpressed in Sf9 cells form a sperm-binding active hetero-complex. FEBS J 274:5390–5405

Kanai S, Kitayama T, Yonezawa N et al (2008) Disulfide linkage patterns of pig zona pellucida glycoproteins ZP3 and ZP4. Mol Reprod Dev 75:847–856

Katsumata T, Noguchi S, Yonezawa N et al (1996) Structural characterization of the N-linked carbohydrate chains of the zona pellucida glycoproteins from bovine ovarian and fertilized eggs. Eur J Biochem 240:448–453

Lin SJ, Hu Y, Zhu J et al (2011) Structure of betaglycan zona pellucida (ZP)-C domain provides insights into ZP-mediated protein polymerization and TGF-beta binding. Proc Natl Acad Sci USA 108:5232–5236

Nakano M, Yonezawa N (2001) Localization of sperm ligand carbohydrate chains in pig zona pellucida glycoproteins. Cells Tissues Organs 168:65–75

Rankin TL, Coleman JS, Epifano O et al (2003) Fertility and taxon-specific sperm binding persist after replacement of mouse sperm receptors with human homologs. Dev Cell 5:33–43

Stetson I, Izquierdo-Rico MJ, Moros C et al (2012) Rabbit zona pellucida composition: a molecular, proteomic and phylogenetic approach. J Proteomics 75:5920–5935

Takahashi K, Kikuchi K, Uchida Y et al (2013) Binding of sperm to the zona pellucida mediated by sperm carbohydrate-binding proteins is not species-specific in vitro between pigs and cattle. Biomolecules 3:85–107

Tokuhiro K, Ikawa M, Benham AM et al (2012) Protein disulfide isomerase homolog PDILT is required for quality control of sperm membrane protein ADAM3 and male fertility. Proc Natl Acad Sci USA 109:3850–3855

Velásquez JG, Canovas S, Barajas P et al (2007) Role of sialic acid in bovine sperm-zona pellucida binding. Mol Reprod Dev 74:617–628

von Witzendorff D, Ekhlasi-Hundrieser M, Dostalova Z et al (2005) Analysis of N-linked glycans of porcine zona pellucida glycoprotein ZPA by MALDI-TOF MS: a contribution to understanding zona pellucida structure. Glycobiology 15:475–488

Yonezawa N, Fukui N, Kuno M et al (2001) Molecular cloning of bovine zona pellucida glycoproteins ZPA and ZPB and analysis for sperm-binding component of the zona. Eur J Biochem 268:3587–3594

Yonezawa N, Amari S, Takahashi K et al (2005a) Participation of the nonreducing terminal beta-galactosyl residues of the neutral N-linked carbohydrate chains of porcine zona pellucida glycoproteins in sperm-egg binding. Mol Reprod Dev 70:222–227

Yonezawa N, Kudo K, Terauchi H et al (2005b) Recombinant porcine zona pellucida glycoproteins expressed in Sf9 cells bind to bovine sperm but not to porcine sperm. J Biol Chem 280:20189–20196

Yonezawa N, Kanai-Kitayama S, Kitayama T et al (2012) Porcine zona pellucida glycoprotein ZP4 is responsible for the sperm-binding activity of the ZP3/ZP4 complex. Zygote 20:389–397

Yurewicz EC, Pack BA, Sacco AG (1991) Isolation, composition, and biological activity of sugar chains of porcine oocyte zona pellucida 55K glycoproteins. Mol Reprod Dev 30:126–134

Yurewicz EC, Sacco AG, Gupta SK et al (1998) Hetero-oligomerization-dependent binding of pig oocyte zona pellucida glycoproteins ZPB and ZPC to boar sperm membrane vesicles. J Biol Chem 273:7488–7494

Part V
Organella, Proteolysis, and New Techniques

Chapter 35
The Role of Peroxisomes in Plant Reproductive Processes

Shino Goto-Yamada, Shoji Mano, and Mikio Nishimura

Abstract Peroxisomes are ubiquitous organelles found in eukaryotic cells. These organelles perform important metabolic functions required for normal development. Peroxisomes play crucial roles in flowering plants, functioning in such processes as lipid metabolism, photorespiration, and the production of phytohormones, rendering these organelles indispensable for normal plant growth. In addition to functioning in vegetative tissues, recent studies have shown that peroxisomes play essential roles in reproductive processes. This chapter describes the analysis of peroxisomes in gametophytes, using the approach of organelle visualization in reproductive tissues, which reveals organelle distribution in gametophytes and the dynamics of peroxisomes in pollen cells. Studies using mutants defective in peroxisomal functions demonstrate the importance of peroxisomes to pollen fertility and pollen tube elongation in pistils. We also summarize the relationship between peroxisomal function and reproductive processes in plants. Peroxisomes may contribute to pollen fertility by generating energy for pollen tube growth and producing signaling molecules required for pollen tube elongation and orientation. Moreover, peroxisomes may be involved in male–female gametophyte recognition.

Keywords Fluorescent protein • Peroxisome • Plant • Reproduction

S. Goto-Yamada
Department of Cell Biology, National Institute for Basic Biology, Okazaki 444-8585, Japan

S. Mano (✉) • M. Nishimura
Department of Cell Biology, National Institute for Basic Biology, Okazaki 444-8585, Japan

Department of Basic Biology, School of Life Science, The Graduate University for Advanced Studies, Okazaki 444-8585, Japan
e-mail: mano@nibb.ac.jp

H. Sawada et al. (eds.), *Sexual Reproduction in Animals and Plants*,
DOI 10.1007/978-4-431-54589-7_35, © The Author(s) 2014

35.1   Introduction

Peroxisomes are single membrane-bound organelles that are ubiquitous in eukaryotic cells. As peroxisomes contribute to lipid metabolism via β-oxidation activities in various organisms, including animals and plants, these organelles are essential for survival. Peroxisomes have a variety of functions, such as polyamine metabolism (Nishikawa et al. 2000; Wu et al. 2003; Kamada-Nobusada et al. 2008), the synthesis of various secondary metabolites (Imazaki et al. 2010; Meijer et al. 2010), and vitamin production (Babujee et al. 2010; Tanabe et al. 2011), in addition to lipid metabolism, which demonstrates that peroxisomes are involved in biological processes at various developmental stages. Recent analyses using various mutants showed that defects in peroxisomal function result in the failure of reproductive processes (Footitt et al. 2007; Boisson-Dernier et al. 2008; Wu et al. 2010). This chapter focuses on the relationship between peroxisomal functions and reproductive processes in plants.

35.2   Peroxisomes in Higher Plants

35.2.1   Peroxisomal Functions During Plant Development

In plants, peroxisomes appear in various tissues as globular organelles of approximately 1–1.5 μm. Several types of peroxisomes are found in plants, including glyoxysomes and leaf peroxisomes (Kamada et al. 2003). Plant peroxisomes adapt to environmental and developmental changes (Nishimura et al. 1993). Peroxisomes in etiolated cotyledons are referred to as glyoxysomes. These organelles are responsible for the conversion of fatty acids to sucrose, which provides energy for seedling growth after germination (Fig. 35.1a). In photosynthetic leaves, a functional transition from glyoxysomes to leaf peroxisomes takes place (Fig. 35.1b) (Nishimura et al. 1993). Leaf peroxisomes contain enzymes that function in the glycolate pathway and are active during photorespiration. This function is indispensable for scavenging by-products produced by photosynthesis and is required for normal plant growth (Tolbert 1981). Moreover, plant peroxisomes contribute to the generation and detoxification of reactive oxygen species (ROS), the biosynthesis of plant hormones, such as jasmonate and auxin, and the metabolism of polyamines (Kamada-Nobusada et al. 2008). Studies of various Arabidopsis mutants with defective peroxisomes have demonstrated that peroxisomal functions and biogenesis are essential for plant growth (Hayashi et al. 1998, 2002; Mano et al. 2004, 2006; Goto et al. 2011).

Because all peroxisomal proteins are encoded by the nuclear genome, peroxisomal proteins are translated on free ribosomes in the cytosol. These proteins are then transported to peroxisomes when their peroxisomal targeting signals 1 or 2 (PTS1 or PTS2) are recognized by the cytosolic receptors PEROXIN 5 (PEX5) and PEX7, respectively, which direct the proteins to the peroxisomal membrane. Constitutive expression of the gene encoding the PTS1 fused green fluorescent protein (GFP) construct under the control of the CaMV 35S promoter results in the

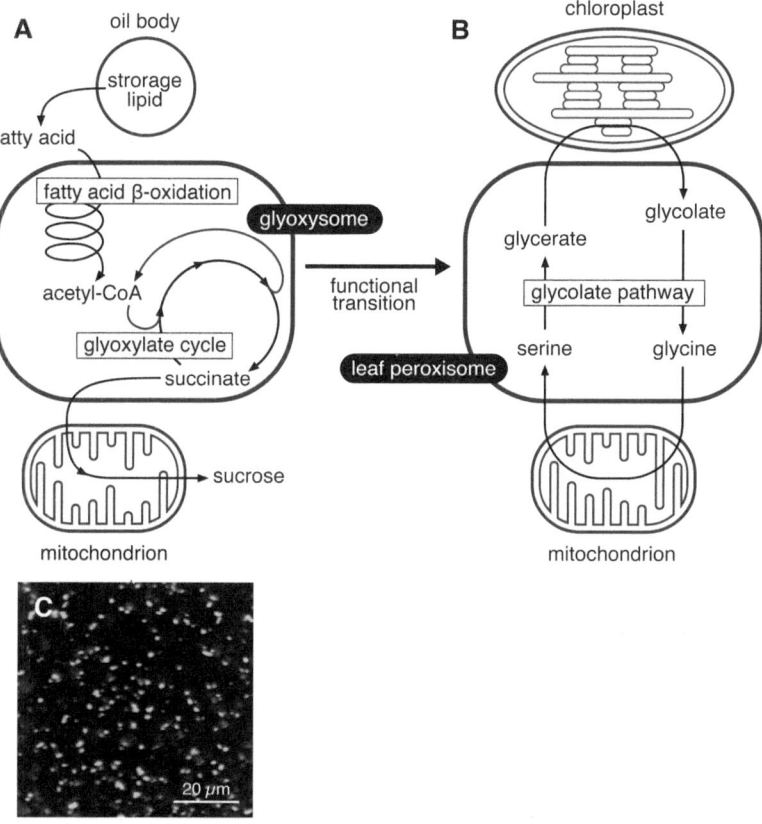

Fig. 35.1 Plant peroxisomes. (**a, b**) Metabolic pathways in peroxisomes. Lipid metabolism in glyoxysomes (**a**) and photorespiration in leaf peroxisomes (**b**). After the plant receives light, a functional transition from glyoxysomes to leaf peroxisomes occurs. (**c**) GFP-labeled peroxisomes in leaves of transgenic Arabidopsis expressing *GFP-PTS1* under the control of the CaMV 35S promoter

transport of this protein to the peroxisomes; GFP in the peroxisomes is detectable in almost all cells when visualized under a fluorescence microscope (Fig. 35.1c) (Mano et al. 2002). This visualization system is useful for a variety of experiments, including the observation of peroxisome dynamics, such as morphology and movement, and the isolation of mutants (Mano et al. 2004, 2006, 2011; Goto et al. 2011).

35.2.2   Peroxisomes in Plant Reproductive Tissues

Electron microscopic observation has revealed that peroxisomes exist in both vegetative and reproductive tissues. Many vesicles thought to be oil bodies are present in pollen grains, and peroxisomes are found among them (Fig. 35.2a, e). Pollen peroxisomes appear as spherical structures, and they are smaller than peroxisomes in vegetative tissues, with an average size of approximately 500 nm. To understand the

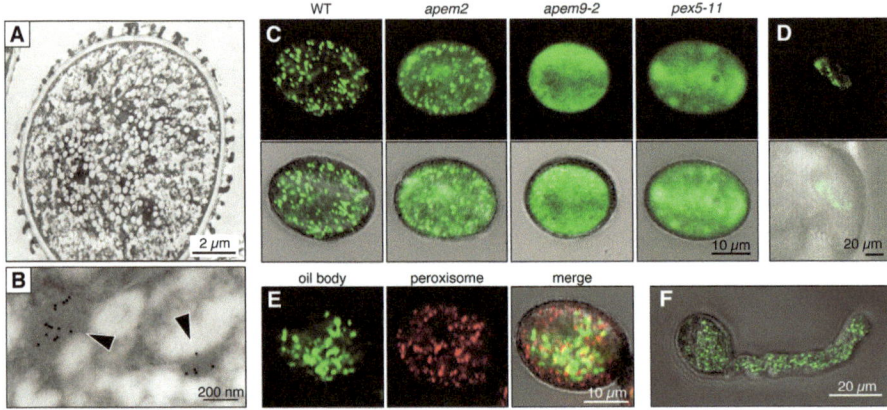

Fig. 35.2 Peroxisomes in reproductive tissues. (**a**) Electron micrograph of Arabidopsis pollen. (**b**) Immunoelectron micrograph of pollen from transgenic Arabidopsis harboring *Lat52p:GFP-PTS1*. Arrowheads indicate peroxisomes, which were immunogold labeled with anti-GFP antibodies. (**c**) GFP fluorescence pattern in wild-type and peroxisome-defective mutants *apem2*, *pex5-11*, and *apem9-2*, which express *GFP-PTS1*. The *pex5-11* and *apem9-2* represent the T-DNA insertion lines SALK_018579 and SALK-132193, respectively. Both mutants are unable to produce homozygotes, and heterozygotes produce two types of pollen that exhibit different phenotypes, including wild-type-like spherical structures and mutant structures exhibiting cytosolic fluorescence; only pollen grains of the mutant are shown in this figure. (**d**) Peroxisomes in egg cell of female gametophyte. Peroxisomes were visualized using *DD45p:GFP-PTS1*, with the expression of *GFP-PTS1* under the control of the *DD45* promoter. (**e**) Localization of oil bodies and peroxisomes in pollen. Oil bodies were visualized with GFP-labeled oleosin, a membrane protein present in oil bodies. Peroxisomes were visualized with TagRFP-PTS1. (**f**) Subcellular localization of peroxisomes in elongating pollen tube

temporal and spatial behavior of peroxisomes during reproductive processes, we generated transgenic Arabidopsis-expressing peroxisome-targeted fluorescent protein in pollen and egg cells using the *Lat52* and *DD45* (Twell et al. 1991; Steffen et al. 2007) promoters, respectively. Because the CaMV 35S promoter does not function properly in these cells, *Lat52* and *DD45* promoter-controlled *GFP-PTS1* (Lat52p:GFP-PTS1 or DD45p:GFP-PTS1) was used to visualize peroxisomes in pollen and egg cell, respectively. Using these constructs, we could easily observe peroxisomal movement and distribution during the reproductive processes (Fig 35.2c,d). In mature pollen grains, the peroxisomes are quite immobile, but after pollen germination, the peroxisomes move in the direction of pollen tube growth (Fig. 35.2f). The peroxisomes then move through the pollen tube that reaches the female gametophyte.

35.3 Key Genes of Peroxisome Biogenesis Also Function in Pollen Cells

In flowering plants and in animals, peroxisomal functions are essential for normal development and survival. In humans, peroxisome biogenesis disorders, including Zellwerger syndrome, are fatal genetic disorders (Fujiki 2000). In plants, defects of peroxisomal biogenesis and function result in various defects, including the failure to germinate without sucrose (Hayashi et al. 1998; Hayashi et al. 2002), dwarfism (Mano et al. 2004, 2006; Goto et al. 2011), and embryonic lethality (Schumann et al. 2003; Sparkes et al. 2003; Fan et al. 2005). Moreover, recent studies have revealed that defects in peroxisomes result in a decrease of pollen fertility and failures in gametophyte recognition (Boisson-Dernier et al. 2008), indicating that explication of peroxisomal functions is important for understanding reproductive processes in plants.

As reported previously, we isolated a number of Arabidopsis *apem* (*aberrant peroxisome morphology*) mutants expressing *35Sp:GFP-PTS1* that had different patterns of GFP fluorescence from the parental GFP-PTS1 plants (Mano et al. 2004, 2006, 2011; Goto et al. 2011). To understand the role of peroxisomes in pollen fertility, we expressed the *Lat52p:GFP-PTS1* transgene in various mutants defective in peroxisomal biogenesis and function, including the *apem* mutants. Of these mutants, both *apem2* and *apem9* have mutations in the *PEX* genes, which are key genes of peroxisome biogenesis. *APEM2* and *APEM9* encode PEX13 and plant-specific factor, respectively, and both proteins are involved in peroxisomal protein transport. Therefore, in both mutants, GFP-PTS1 protein is accumulated in the cytosol as well as in peroxisomes (Mano et al. 2006; Goto et al. 2011). As shown in Fig. 35.2c, the pattern of GFP fluorescence in the pollen cells of both mutants is similar to that in vegetative tissues; GFP fluorescence is observed in the cytosol, whereas GFP is observed only in peroxisomes in the wild-type background. A similar pattern is observed in the pollen of a mutant defective in *PEX5*, which encodes a receptor of peroxisomal matrix protein and is essential for peroxisomal protein transport. As shown in Fig. 35.2c, GFP-PTS1 protein is accumulated in the cytosol in pollen cells of the *pex5-11* mutant. These results show that peroxisomal protein transport functions in pollen peroxisomes and that peroxisome biogenesis factors, such as PEXs, act in pollen cells.

35.4   Effects of Peroxisome Defects on Male Gametes

To examine the contribution of peroxisomal function to reproductive process, the peroxisome-defective *apem9* mutant was employed. First, to investigate the fertility of gametes in peroxisome-defective mutants, reciprocal crosses of *apem9-2/+* with wild-type Columbia (Col) were performed. When *apem9-2/+* pollen was used to pollinate Col pistils, only 14.6 % [transmission efficiency (TE)=17.1, $n=96$] of the seeds had the *apem9-2/+* genotype, whereas when *apem9-2/+* pistils were pollinated with Col pollen, 45.5 % (TE=83.3, $n=44$) of the seeds had the *apem9-2/+* genotype (Table 35.1). These results show that the transmission of the mutation through the female gametophyte is not impaired, and *apem9* ovules are fertile. However, pollen development and function are severely affected in the *apem9* mutant.

Table 35.1 Reciprocal crosses between peroxisome mutants and wild-type plants

Mutant name	Gene	TE *female* (%)[a]	TE *male* (%)[a]	Reference
pex5_11[b]	PEROXIN5	110.0	30.2	This study
apem9-2[c]	APEM9	83.3	17.1	This study
amc/apem2	PEROXIN13	82.8	45.3	Boisson-Dernier et al. (2008)
cts-1[d]	COMATOSE	93.9	21.6	Footitt et al. (2007)
cts-2[d]	COMATOSE	111.3	10.3	Footitt et al. (2007)

[a]Transmission efficiencies (TE) were calculated according to Howden et al. (1998)
[b]*pex5_11* represents SALK_018579
[c]*apem9-2* represents SALK-132193
[d]*cts-1* and *cts-2* are allelic to *ped3/pxa1*, which is defective in peroxisomal ABC transporter

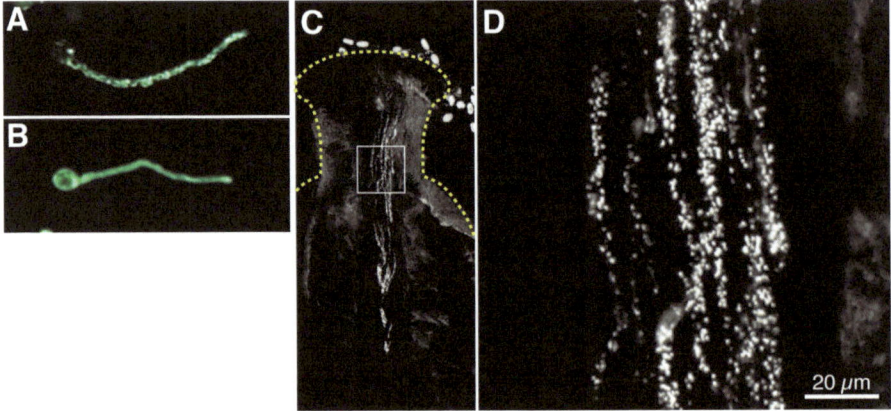

Fig. 35.3 Analysis of pollen tube elongation in *apem9-2* mutant and wild-type plant. (**a, b**) Fluorescence images of elongating wild-type (**a**) and *apem9-2* (**b**) pollen tubes grown in vitro. (**c**) In vivo pollen germination assay. Mutant *apem9-2/+* pollens were used to pollinate a wild-type pistil, and GFP fluorescence was observed with a laser-scanning confocal microscope 1 h after pollination. (**d**) Magnified image of the area enclosed by the *white box* in **c**

Pollen fertilization occurs via the following steps: (1) pollen maturation, (2) pollen germination and elongation of the pollen tube, (3) guidance of the pollen tube to the ovule, and (4) fertilization with the ovule. To determine whether the reduced male transmission of the *apem9-2* mutation was caused by a pollen germination defect, and to identify which step(s) in fertilization are disturbed in *apem9-2* pollen, an in vitro pollen germination assay was performed, and the numbers of germinated wild-type (+) and *apem9-2* pollen were counted. There was no obvious difference in the number of germinated pollen grains between wild-type and *apem9-2* pollen, indicating that *apem9-2* pollen matured successfully, as did the wild-type pollen, and had the ability to germinate under artificial germination conditions (Fig. 35.3a,b). However, in the pollen germination assay in vivo, in which pollen was germinated on wild-type stigmas, the *apem9-2* pollen tubes, unlike the wild-type pollen tubes, failed to elongate in the top parts of the pistils within 1 h of pollination (Fig. 35.3c,d). Based on these results, we determined that the defect in the *APEM9* gene may decrease the efficiency of pollen fertility caused by a delay in pollen tube elongation in vivo or a failure of pollen tube penetration into the papilla cells, indicating that peroxisome biogenesis and function are required for pollen fertility.

35.5 Relationship Between Peroxisomal Function and Reproduction

In addition to the *apem9* mutant, other peroxisome mutants also show defects in pollen fertility. Table 35.1 shows the results of reciprocal crosses between peroxisome mutants. Although the female TEs are almost normal in these mutants, the male TEs are obviously decreased. These results indicate that peroxisomal functions are required for pollen fertility. This section reviews the relationship between peroxisomal functions and reproductive processes, including pollen fertility.

35.5.1 Peroxisomal β-Oxidation and Pollen Fertility

Fatty acid β-oxidation is one of the major metabolic processes that occur in peroxisomes. Fatty acid β-oxidation is required for the metabolism of lipids, as well as the conversion of fatty acid derivates into phytohormones, such as auxin and jasmonic acid (JA) (Wain and Wightman 1954; Sanders et al. 2000; Stintzi and Browse 2000). JA plays a crucial role as a signaling molecule during plant growth and development and in plant responses to biotic and abiotic stress. Moreover, JA contributes to reproductive processes. Peroxisome-targeted OPR3 (12-oxophytodienoic acid reductase 3) is one of the enzymes required for JA biosynthesis (Sanders et al. 2000; Stintzi and Browse 2000). The Arabidopsis *opr3* mutant shows the male-sterility phenotype, resulting from defects of sufficient elongation of the anther filament, normal anther dehiscence, and the production of viable pollen (Sanders et al. 2000; Stintzi and Browse 2000). These results indicate the importance of peroxisomal β-oxidation in plant reproductive processes.

Another important aspect of fatty acid β-oxidation is lipid metabolism in seeds. Oilseed plants, such as Arabidopsis and pumpkin, convert seed-storage lipids to sucrose during postgerminative growth as energy sources before the plant begins photosynthesis (Fig. 35.1a). Triacylglycerols are accumulated as storage lipids in organelles known as oil bodies, which are surrounded by oleosin in seeds. Similarly, numerous oil bodies and peroxisomes are observed in pollen grains when examined by electron (Fig. 35.2a) and fluorescence microscopy (Fig. 35.2e). The similarity between seeds and pollen suggests that fatty acid β-oxidation provides energy during pollen germination and/or pollen tube elongation. Indeed, the lack of CTS (COMATOSE), an ABC transporter that contributes to the uptake of fatty acids into peroxisomes, results in the suppression of pollen tube elongation (Footitt et al. 2007), whereas this mutant pollen tube can elongate in artificial pollen germination medium containing sucrose. The phenotype for suppression of pollen tube elongation is also seen in other mutants defective in fatty acid β-oxidation, such as KAT2 (3-keto acyl-CoA thiolase), LACS6, and LACS7 (long-chain acyl-CoA synthetase 6 and 7), which demonstrates the importance of peroxisomal fatty acid β-oxidation in pollen fertility and pollen tube elongation (Footitt et al. 2007).

35.5.2 Peroxisome Biogenesis and Male–Female Gametophyte Recognition

A recent study has revealed that peroxisomes are involved in male–female gametophyte recognition (Boisson-Dernier et al. 2008). The *amc* (*abstinence by mutual consent*) mutant was originally identified as a mutant with defective in male–female gamete recognition. In addition to decreased pollen fertility, the *amc* mutant exhibits impaired fertilization only when both male and female gameto-phytes have mutations in *AMC*. In wild-type plants, pollen germinates after it attaches to a pistil. The pollen tube then elongates, reaches one of the two synergid cells of the female gametophyte, and ruptures to release two sperm cells. However, the *amc* mutant shows impaired pollen tube reception when an *amc* pollen tube reaches an *amc* female gametophyte, resulting in the continued growth of the pollen tube in the synergid cell. As the pollen tube fails to rupture, the sperm cells are not released, and homozygous mutants never develop. Interestingly, the *AMC* gene is allelic to *apem2* (Mano et al. 2006). Therefore, the *amc* mutant is defective in one of the peroxisome biogenesis factors, PEX13, which is required for peroxisomal protein transport. The *AMC/PEX13* gene is transiently expressed in both male and female gametophytes during fertilization (Boisson-Dernier et al. 2008). As PEX13 is one of the peroxisome biogenesis factors, these results indicate that peroxisome biogenesis plays an unexpected key role in gametophyte recognition.

35.5.3 Peroxisome-Derived ROS and NO and Reproduction

ROS and nitric oxide (NO) are signaling molecules; peroxisomes are involved in the production of these compounds. A recent study has revealed that peroxisome-targeted PAO (polyamine oxidase) contributes to pollen tube elongation in associa-tion with ROS production (Wu et al. 2010). This study illustrates that PAO regulates pollen tube elongation during the oxidation of spermidine, one of the polyamines, and generates hydrogen peroxide (H_2O_2), which triggers the opening of the hyper-polarization-activated Ca^{2+}-permeable channels in pollen. The Arabidopsis genome has five isoforms of PAO, and three of these (PAO2, PAO3, and PAO4) are localized to peroxisomes (Kamada-Nobusada et al. 2008). The *pao3* mutant exhibits suppressed pollen tube elongation and decreased seed numbers, indicating the importance of peroxisome function, combined with ROS production, in plant reproduction. The details of how PAO-mediated H_2O_2 production contributes to pollen fertility will become clearer in future studies. In addition, Prado et al. reported that NO is generated in peroxisomes in the pollen tube, and NO regulates pollen tube orientation in vitro (Prado et al. 2004). Although the activity of NOS (nitrogen oxide synthase) is detected in peroxisomes isolated from pea (*Pisum sativum*) (Barroso et al. 1999), peroxisome-targeted NOS has not been identified yet. The metabolic process leading to NO production is still largely unclear and should be investigated in the future.

35.6   Perspectives

This chapter describes how plant peroxisomes contribute to pollen development, pollen tube elongation, and male–female gametophyte recognition. Although some details about how peroxisomes function during reproduction remain to be elucidated, the importance of peroxisomes in reproductive processes has become clear through the analysis of peroxisome mutants. Defects in peroxisomal biogenesis and metabolism impair sperm development and fertility in *Drosophila* and other animals as well as in plants (Huyghe et al. 2006; Nenicu et al. 2007; Chen et al. 2010; Nakayama et al. 2011), suggesting that peroxisomes may function in the reproductive processes of various organisms in a similar manner.

In mammalian cells, the β-oxidation is localized in both peroxisomes and mitochondria. Long- and medium-chain fatty acids were metabolized in peroxisomes, whereas short-chain fatty acids were metabolized in mitochondria but not in peroxisomes. In plants, however, the activity of β-oxidation is present only in peroxisomes. Therefore, we have to consider the difference along with the similarity of peroxisomal functions between plants and animals. Transdisciplinary investigation that concerns both plant and mammalian peroxisomes will clarify the similarities or differences of peroxisomal functions in the reproductive process among various organisms.

Acknowledgments   We thank Dr. Tomoko Igawa in Chiba University for providing the *DD45* clone, and Dr. Sheila McCormick in University of California, Berkeley, for providing the *Lat52* clone. This work was supported by MEXT KAKENHI of Grant-in-aid for Scientific Research on Innovative Areas, "Elucidating Common Mechanisms of Allogenic Authentication" (No. 22112523), to S.M., and the Japan Society for the Promotion of Science for a Research Fellowship for Young Scientists (No. 22.9) to S.G.-Y. We thank Chinami Yamaguchi, Azusa Matsuda, Masami Araki, Chihiro Nakamori, Maki Kondo, and Dr. Makoto Hayashi from the Division of Cell Mechanisms at the National Institute for Basic Biology for supporting this work, and staff from the Model Plant Research Facility at the National Institute for Basic Biology for plant care.

References

Babujee L, Wurtz V, Ma C, Lueder F, Soni P, van Dorsselaer A, Reumann S (2010) The proteome map of spinach leaf peroxisomes indicates partial compartmentalization of phylloquinone (vitamin K_1) biosynthesis in plant peroxisomes. J Exp Bot 61(5):1441–1453

Barroso JB, Corpas FJ, Carreras A, Sandalio LM, Valderrama R, Palma JM, Lupianez JA, del Rio LA (1999) Localization of nitric-oxide synthase in plant peroxisomes. J Biol Chem 274(51): 36729–36733

Boisson-Dernier A, Frietsch S, Kim T-H, Dizon MB, Schroeder JI (2008) The peroxin loss-of-function mutation *abstinence by mutual consent* disrupts male–female gametophyte recognition. Curr Biol 18(1):63–68

Chen H, Liu Z, Huang X (2010) *Drosophila* models of peroxisomal biogenesis disorder: peroxins are required for spermatogenesis and very-long-chain fatty acid metabolism. Hum Mol Genet 19(3):494–505

Fan J, Quan S, Orth T, Awai C, Chory J, Hu J (2005) The Arabidopsis PEX12 gene is required for peroxisome biogenesis and is essential for development. Plant Physiol 139(1):231–239

Footitt S, Dietrich D, Fait A, Fernie AR, Holdsworth MJ, Baker A, Theodoulou FL (2007) The COMATOSE ATP-binding cassette transporter is required for full fertility in Arabidopsis. Plant Physiol 144(3):1467–1480

Fujiki Y (2000) Peroxisome biogenesis and peroxisome biogenesis disorders. FEBS Lett 476(1-2):42–46

Goto S, Mano S, Nakamori C, Nishimura M (2011) Arabidopsis ABERRANT PEROXISOME MORPHOLOGY9 is a peroxin that recruits the PEX1-PEX6 complex to peroxisomes. Plant Cell 23(4):1573–1587

Hayashi M, Toriyama K, Kondo M, Nishimura M (1998) 2,4-Dichlorophenoxybutyric acid-resistant mutants of Arabidopsis have defects in glyoxysomal fatty acid β-oxidation. Plant Cell 10(2):183–195

Hayashi M, Nito K, Takei-Hoshi R, Yagi M, Kondo M, Suenaga A, Yamaya T, Nishimura M (2002) Ped3p is a peroxisomal ATP-binding cassette transporter that might supply substrates for fatty acid β-oxidation. Plant Cell Physiol 43(1):1–11

Howden R, Park SK, Moore JM, Orme J, Grossniklaus U, Twell D (1998) Selection of T-DNA-tagged male and female gametophytic mutants by segregation distortion in Arabidopsis. Genetics 149(2):621–631

Huyghe S, Schmalbruch H, De Gendt K, Verhoeven G, Guillou F, Van Veldhoven PP, Baes M (2006) Peroxisomal multifunctional protein 2 is essential for lipid homeostasis in sertoli cells and male fertility in mice. Endocrinology 147(5):2228–2236

Imazaki A, Tanaka A, Harimoto Y, Yamamoto M, Akimitsu K, Park P, Tsuge T (2010) Contribution of peroxisomes to secondary metabolism and pathogenicity in the fungal plant pathogen *Alternaria alternata*. Eukaryot Cell 9(5):682–694

Kamada T, Nito K, Hayashi H, Mano S, Hayashi M, Nishimura M (2003) Functional differentiation of peroxisomes revealed by expression profiles of peroxisomal genes in *Arabidopsis thaliana*. Plant Cell Physiol 44(12):1275–1289

Kamada-Nobusada T, Hayashi M, Fukazawa M, Sakakibara H, Nishimura M (2008) A putative peroxisomal polyamine oxidase, AtPAO4, is involved in polyamine catabolism in *Arabidopsis thaliana*. Plant Cell Physiol 49(9):1272–1282

Mano S, Nakamori C, Hayashi M, Kato A, Kondo M, Nishimura M (2002) Distribution and characterization of peroxisomes in Arabidopsis by visualization with GFP: dynamic morphology and actin-dependent movement. Plant Cell Physiol 43(3):331–341

Mano S, Nakamori C, Kondo M, Hayashi M, Nishimura M (2004) An *Arabidopsis* dynamin-related protein, DRP3A, controls both peroxisomal and mitochondrial division. Plant J 38(3): 487–498

Mano S, Nakamori C, Nito K, Kondo M, Nishimura M (2006) The Arabidopsis pex12 and *pex13* mutants are defective in both PTS1- and PTS2-dependent protein transport to peroxisomes. Plant J 47(4):604–618

Mano S, Nakamori C, Fukao Y, Araki M, Matsuda A, Kondo M, Nishimura M (2011) A defect of peroxisomal membrane protein 38 causes enlargement of peroxisomes. Plant Cell Physiol 52(12):2157–2172

Meijer WH, Gidijala L, Fekken S, Kiel JA, van den Berg MA, Lascaris R, Bovenberg RA, van der Klei IJ (2010) Peroxisomes are required for efficient penicillin biosynthesis in *Penicillium chrysogenum*. Appl Environ Microbiol 76(17):5702–5709

Nakayama M, Sato H, Okuda T, Fujisawa N, Kono N, Arai H, Suzuki E, Umeda M, Ishikawa HO, Matsuno K (2011) *Drosophila* carrying *pex3* or *pex16* mutations are models of Zellweger syndrome that reflect its symptoms associated with the absence of peroxisomes. PLoS One 6(8):e22984

Nenicu A, Luers GH, Kovacs W, David M, Zimmer A, Bergmann M, Baumgart-Vogt E (2007) Peroxisomes in human and mouse testis: differential expression of peroxisomal proteins in germ cells and distinct somatic cell types of the testis. Biol Reprod 77(6):1060–1072

Nishikawa M, Hagishita T, Yurimoto H, Kato N, Sakai Y, Hatanaka T (2000) Primary structure and expression of peroxisomal acetylspermidine oxidase in the methylotrophic yeast *Candida boidinii*. FEBS Lett 476(3):150–154

Nishimura M, Takeuchi Y, De Bellis L, Hara-Nishimura I (1993) Leaf peroxisomes are directly transformed to glyoxysomes during senescence of pumpkin cotyledons. Protoplasma 175(3–4):131–137

Prado AM, Porterfield DM, Feijo JA (2004) Nitric oxide is involved in growth regulation and reorientation of pollen tubes. Development (Camb) 131(11):2707–2714

Sanders PM, Lee PY, Biesgen C, Boone JD, Beals TP, Weiler EW, Goldberg RB (2000) The Arabidopsis *DELAYED DEHISCENCE1* gene encodes an enzyme in the jasmonic acid synthesis pathway. Plant Cell 12(7):1041–1061

Schumann U, Wanner G, Veenhuis M, Schmid M, Gietl C (2003) AthPEX10, ariuclear gene essential for peroxisome and storage organelle formation during *Arabidopsis* embryogenesis. Proc Natl Acad Sci USA 100(16):9626–9631

Sparkes IA, Brandizzi F, Slocombe SP, El-Shami M, Hawes C, Baker A (2003) An Arabidopsis pex10 null mutant is embryo lethal, implicating peroxisomes in an essential role during plant embryogenesis. Plant Physiol 133(4):1809–1819

Steffen JG, Kang IH, Macfarlane J, Drews GN (2007) Identification of genes expressed in the Arabidopsis female gametophyte. Plant J 51(2):281–292

Stintzi A, Browse J (2000) The *Arabidopsis* male-sterile mutant, *opr3*, lacks the 12-oxophytodienoic acid reductase required for jasmonate synthesis. Proc Natl Acad Sci USA 97(19):10625–10630

Tanabe Y, Maruyama J, Yamaoka S, Yahagi D, Matsuo I, Tsutsumi N, Kitamoto K (2011) Peroxisomes are involved in biotin biosynthesis in *Aspergillus* and *Arabidopsis*. J Biol Chem 286(35):30455–30461

Tolbert NE (1981) Metabolic pathways in peroxisomes and glyoxysomes. Annu Rev Biochem 50:133–157

Twell D, Yamaguchi J, Wing RA, Ushiba J, McCormick S (1991) Promoter analysis of genes that are coordinately expressed during pollen development reveals pollen-specific enhancer sequences and shared regulatory elements. Gene Dev 5(3):496–507

Wain RL, Wightman F (1954) The growth regulating activity of certain ω-substituted alkyl carboxylic acids in relation to their β-oxidation within the plant. Proc R Soc Lond B Biol Sci 142(909):525–536

Wu T, Yankovskaya V, McIntire WS (2003) Cloning, sequencing, and heterologous expression of the murine peroxisomal flavoprotein, N^1-acetylated polyamine oxidase. J Biol Chem 278(23): 20514–20525

Wu J, Shang Z, Jiang X, Moschou PN, Sun W, Roubelakis-Angelakis KA, Zhang S (2010) Spermidine oxidase-derived H_2O_2 regulates pollen plasma membrane hyperpolarization-activated Ca^{2+}-permeable channels and pollen tube growth. Plant J 63(6):1042–1053

Chapter 36
Regulation of Vacuole-Mediated Programmed Cell Death During Innate Immunity and Reproductive Development in Plants

Tomoko Koyano, Takamitsu Kurusu, Shigeru Hanamata, and Kazuyuki Kuchitsu

Abstract Programmed cell death (PCD), organized destruction of cells, is essential in development, maintenance of cellular homeostasis, and innate immunity in multicellular organisms. In most angiosperms, development of male and female organs involves spatial and temporal regulation of PCD. The tapetum, the innermost layer of the anther, provides both nutrient and lipid components to developing microspores, and has been proposed to be degraded by PCD during the later stages of pollen maturation. Plants lack homologues of most apoptosis-related genes in animals and have evolved specific mechanisms for PCD. PCD is also a crucial event in plant immune responses against microbial infection that prevents the spread of pathogens. Recent live cell imaging techniques have revealed the dynamic features and significant roles of the vacuole during defense responses and PCD. Disintegration or collapse of the vacuolar membrane has been suggested to trigger the final step of

T. Koyano and T. Kurusu contributed equally to this work.

T. Koyano • S. Hanamata
Department of Applied Biological Science, Tokyo University of Science, 2641 Yamazaki, Noda, Chiba 278-8510, Japan

T. Kurusu
Department of Applied Biological Science, Tokyo University of Science, 2641 Yamazaki, Noda, Chiba 278-8510, Japan

Research Institute for Science and Technology, Tokyo University of Science, 2641 Yamazaki, Noda, Chiba 278-8510, Japan

School of Bioscience and Biotechnology, Tokyo University of Technology, 1404-1 Katakura, Hachioji, Tokyo 192-0982, Japan

K. Kuchitsu (✉)
Department of Applied Biological Science, Tokyo University of Science, 2641 Yamazaki, Noda, Chiba 278-8510, Japan

Research Institute for Science and Technology, Tokyo University of Science, 2641 Yamazaki, Noda, Chiba 278-8510, Japan
e-mail: kuchitsu@rs.noda.tus.ac.jp

H. Sawada et al. (eds.), *Sexual Reproduction in Animals and Plants*,
DOI 10.1007/978-4-431-54589-7_36, © The Author(s) 2014

PCD in several cell types. We here overview spatiotemporal dynamic changes of the vacuole triggered by signals from pathogens and comparatively discuss PCD during innate immunity and reproductive development in plants.

Keywords Innate immunity • Programmed cell death • Reproductive development • Tapetum • Vacuole

36.1 Programmed Cell Death in Plants

Programmed cell death (PCD) is a genetically regulated process of cellular suicide that is well known to play a fundamental role in a wide variety of developmental and physiological functions in multicellular organisms (Bozhkov and Lam 2011; Fuchs and Steller 2011; Teng et al. 2011). In plants, PCD plays a critical role in the control of developmental processes such as xylogenesis, embryogenesis, pollen maturation, seed development, seed germination, and leaf senescence, as well as various stress responses including innate immunity against pathogen attack (Pennell and Lamb 1997). Reproductive development in angiosperms involves PCD in a variety of cells in reproductive organs, such as reproductive primordium abortion, style transmitting tissue, nonfunctional megaspores, synergids, antipodals, endosperm, anther tapetum, and abortive pollen in male sterility (Greenberg 1996; Pennell and Lamb 1997; Wei et al. 2002). Plants lack homologues of most apoptosis-related genes in animals and have evolved specific mechanisms for PCD. We here comparatively discuss mechanisms for PCD in innate immunity and reproductive development with special reference to the roles of the vacuole.

36.2 PCD in Plant Immunity Against Pathogen Attack

Plants lack immune systems based on antibodies or phagocytosis. Instead, they have evolved multiple layers of active defense responses including the deliberate production of reactive oxygen species (ROS), pathogenesis-related (PR) proteins, and antimicrobial secondary metabolites called phytoalexins. The dynamic reorganization of plant cells is triggered at the site of infection and often accompanies localized PCD, known as the hypersensitive response (HR), which is effective in preventing the spread of pathogens (Heath 2000; Mur et al. 2008).

The plant cells show different characteristics from animal cells, most notably the presence of the cell wall, the plastids/chloroplasts, and the vacuole, all of which play crucial roles in the regulation of plant immunity and PCD. This difference suggests that execution of PCD takes place with different morphological features from typical animal PCD such as apoptosis. Cellular morphological changes in animal cells undergoing apoptosis, including cell shrinkage and nuclear fragmentation, are followed by the fragmentation of cells and formation of apoptotic bodies, which are then phagocytosed. Although the similarities and differences between PCD in plants and animals have been extensively discussed (Cacas 2010; Reape and McCabe 2010), the mechanisms for execution and regulation of plant PCD including HR still largely unclear.

Cell biological aspects of immune responses accompanying PCD have been studied in various experimental systems using a combination of plants and microbes. A major experimental approach is immunostaining of plant tissues infected with microbes (Kobayashi et al. 1994; Skalamera and Heath 1996; Kobayashi et al. 1997). However, the deformation of endomembrane systems by chemical fixation with this technique should be noted. Recently, green fluorescent protein (GFP)-based *in vivo* imaging has allowed time-sequential observations of the endomembrane systems in plant–microbe interactions (Takemoto et al. 2003; Koh et al. 2005).

The execution of cell death in a regulated fashion should accompany dynamic reorganization of the cellular architecture. In plants, cellular morphological changes are often governed by cytoskeletons such as actin microfilaments and microtubules, as well as the vacuole, an organelle occupying most of the cell volume. In recent years, the rapid development of live cell imaging techniques has provided novel aspects on the dynamics of intracellular structures. We here focus on the dynamics of the vacuole and discuss its functional significance in innate immunity and PCD.

36.2.1 Tobacco BY-2 Cells as an Excellent Model System to Study Immune Responses Accompanying Localized Plant PCD

Plant cell cultures are useful simple model systems for the monitoring of cellular events. Defense responses including intracellular reorganization and gene expression upon fungal infection are basically similar between cultured cells and *in planta* (Gross et al. 1993). Treatment of cultured cells with purified signal molecules from pathogens can mimic most defense responses, including the expression of defense-related genes, as effectively as microbial infections (Lecourieux et al. 2002; Kadota et al. 2004a; Kadota and Kuchitsu 2006).

Treatment of tobacco (*Nicotiana tabacum*) BY-2 cells with cryptogein, a protein from the oomycete *Phytophthora cryptogea*, induces various immune responses such as membrane potential changes, ion fluxes, biphasic ROS production, and MAP kinase activation in a cell cycle-dependent manner (Kadota et al. 2005; Kadota et al. 2006), followed by cell-cycle arrest and PCD (Kadota et al. 2004b; Ohno et al. 2011). The slow prolonged phase, not the rapid transient phase, of ROS production shows strong correlation with downstream events including expression of defense-related genes and PCD (Kadota et al. 2005; Kadota et al. 2006).

NADPH oxidase-mediated deliberate ROS production has been suggested to play a crucial role in triggering and regulating PCD (Torres et al. 2005; Suzuki et al. 2011). Respiratory burst oxidase homologue (Rboh) proteins show ROS-producing activity synergistically activated by binding of Ca^{2+} to their EF-hand motifs and

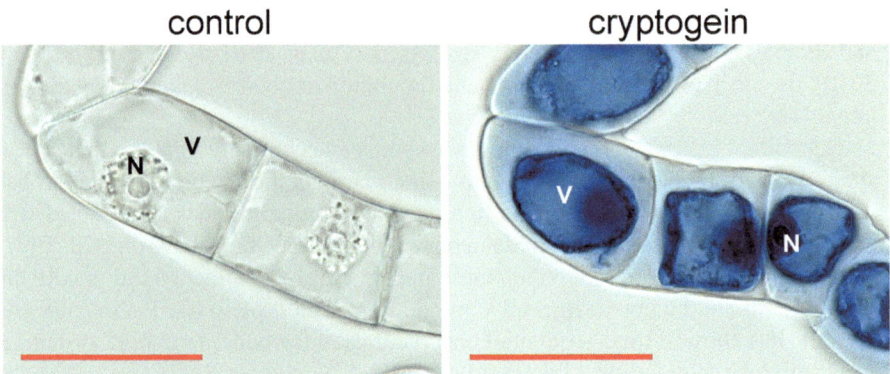

Fig. 36.1 Cryptogein-induced programmed cell death in suspension-cultured tobacco BY-2 cells.
Three-day-old cultured BY-2 cells were treated with 1 μM cryptogein or distilled water (control)
for 24 h. Dead cells were stained with Evans blue 24 h after cryptogein treatment (Higaki et al.
2007). *V* vacuole, *N* nucleus. *Bar* 50 μm.

protein phosphorylation (Ogasawara et al. 2008; Takeda et al. 2008; Kimura et al.
2012). Potato StRBOHB has been shown to be activated by phosphorylation by
calcium-dependent protein kinases StCDPK4 and StCDPK5 (Kobayashi et al. 2007).
Arabidopsis AtRbohF has recently been shown to bind CIPK26, a protein kinase
activated by binding of calcineurin B-like Ca^{2+} sensor proteins CBLs, *in planta*
(Kimura et al. 2013), and be activated in the presence of CBL1/CBL9 and CIPK26
(Drerup et al. 2013).

Transgenic tobacco BY-2 cell lines expressing the GFP-markers for vacuolar
membranes (VM; Kutsuna and Hasezawa 2002) have been effective in monitoring
dynamic changes in vacuoles during cryptogein-induced defense responses in vivo
(Higaki et al. 2007).

36.2.2 Regulation of Vacuole-Mediated PCD in Innate Immunity

The plant vacuole is an organelle that occupies most of the cell volume and contains
many hydrolytic enzymes for digestive processes, similar to lysosomes in animal
cells. The vacuole performs various functions essential for plant growth, development, and adaptation to both abiotic and biotic stresses (Marty 1999).

Disintegration or collapse of the VM has been suggested to trigger the final step
of PCD in several cell types (Jones 2001; Hara-Nishimura and Hatsugai 2011;
Higaki et al. 2011). At the final stage of cryptogein-induced PCD in tobacco BY-2
cells, disintegration of the VM is followed by the irreversible loss of plasma
membrane integrity and cell shrinkage (Higaki et al. 2007; Fig. 36.1). Tobacco
mosaic virus-induced hypersensitive cell death in tobacco leaves involves vacuolar
rupture, in which a vacuolar-localized protease called vacuolar processing enzyme

(VPE) exhibiting caspase-1-like activity is involved (Hatsugai et al. 2004, 2006). Disintegration of the VM or vacuolar collapse needs to be strictly regulated to accomplish PCD with appropriate timing. Another mechanism involving fusion of the VM with the plasma membrane, resulting in the discharge of vacuolar antibacterial proteins to the outside of the cells where bacteria proliferate, has also been proposed for biotrophic bacteria-induced hypersensitive cell death (Hatsugai et al. 2009).

Cryptogein-induced PCD in BY-2 cells accompanies dynamic reorganization of the vacuole before the execution of cell death (Higaki et al. 2008). Cryptogein induces decrease in the transvacuolar strands (TVS), tubular regions of the cytoplasm connecting the nucleus to the cell periphery, and formation of a spherical intravacuolar structure called the 'bulb' (Higaki et al. 2007) that has been observed in a wide range of plant tissues (Saito et al. 2002). The bulb-like structure could be derived from the excess VM comprising the TVS. At the later stage of the PCD, the bulb-like structure disappears and the structure of the large central vacuole becomes simpler. Architecture of the vacuole is governed by actin filaments. Indeed, an actin polymerization inhibitor facilitates both the disappearance of the bulb-like VM structures and induction of PCD (Higaki et al. 2007). These findings suggest that the elicitor-induced reorganization of actin microfilaments followed by modification of the vacuolar structure to induce VM disintegration plays a key role in induction of the PCD.

36.3   Vacuole-Mediated PCD in Plant Vegetative Development

Vacuolar collapse plays a critical role at the final stage of execution of PCD during differentiation of the tracheary element in *Zinnia elegans* (Kuriyama 1999; Obara et al. 2001). PCD in the inner integument cell layers of developing *Arabidopsis* seeds also involves vacuolar rupture, in which a vacuolar-localized protease called VPE exhibiting caspase-1-like activity is involved (Nakaune et al. 2005).

The structural simplification of vacuoles has been commonly observed in various PCD processes including tracheary element differentiation (Obara et al. 2001), gibberellin-mediated PCD in central aleurone cells (Guo and Ho 2008), embryogenesis in a gymnosperm (Smertenko et al. 2003), and leaf formation in a lace plant (Gunawardena 2008). Decreases in the number of TVS and VM reorganization were confirmed in the process of developmental PCD in the lace plant *Aponogeton madagascariensis* (Wright et al. 2009), which suggests a general role of VM reorganization in vacuolar rupture-mediated PCD in plants.

36.4 PCD in Plant Reproductive Development

During plant reproduction, proper induction of PCD is prerequisite for reproductive success. Failure in PCD often results in sterility. In most angiosperms, development of male and female organs involves spatial and temporal regulation of PCD. For example, tapetal cells, the innermost layer of the anther, supply fundamental nutrients necessary for normal pollen development. Shortly after the microspore release from the tetrad and before mitosis, tapetal cells begin to degenerate by PCD. In female gamete development within the ovules, a typically diploid megaspore mother cell undergoes meiosis, giving four megaspores, and three of these megaspores undergo PCD.

Besides development of reproductive organs, PCD also has an important role in fertilization. For example, in self-incompatibility response in *Papaver*, an incompatible pollen tube is stopped by interactions with the pistil *S*-determinant by PCD (Bosch and Franklin-Tong 2008). In this way, PCD plays important roles in multiple steps of normal reproduction in plants from development of gametes to fertilization.

36.4.1 PCD as a Key System for Tapetum Degradation During Pollen Development

During pollen development, the tapetum is degraded to supply metabolites, nutrients, and sporopollenin precursors to developing microspores. Defects in the degradation of the tapetum cause the abnormal formation of the pollen coat and pollen grains and result in severe male sterility (Ku et al. 2003; Li et al. 2006; Zhang et al. 2008; Ariizumi and Toriyama 2011). The degradation of the tapetum is highly regulated and exhibits the hallmark features of PCD such as cell shrinkage, condensation of chromatin, swelling of the endoplasmic reticulum (ER), and persistence of the mitochondria (Rogers et al. 2005) as well as nuclear fragmentation (Wang et al. 1999; Vardar and Unal 2012).

The correct timing of tapetal PCD is important for normal pollen development. The initiation signal for tapetal PCD has been proposed to commence as early as the tetrad stage (Kawanabe et al. 2006). The *ms1* mutant showed delayed tapetal breakdown and a switch from PCD degradation to necrotic-based breakdown (Vizcay-Barrena and Wilson 2006). A rice mutant *tapetal degeneration retardation* (*Ostdr*) shows significantly delayed tapetal breakdown and PCD, resulting in the failure of pollen wall deposition and subsequent microspore degeneration (Li et al. 2006; Zhang et al. 2008). Degradation of the tapetum was also regulated by a plant hormone gibberellin (Cheng et al. 2004; Aya et al. 2009).

36.4.2   Regulation of Vacuole-Mediated PCD in Tapetum Degradation

Abnormal vacuolization in the tapetum during the tetrad stage causes abnormal tapetal PCD and results in male sterility (Wan et al. 2010). The vacuoles in the tapetum may supply tetrad wall-degrading enzymes before their secretion into the anther locules (Wu and Yang 2005). These results suggest important roles of tapetal vacuoles during anther development and PCD.

In the tapetum of *Lathyrus undulatus* L. at the vacuolated microspore stage, rupture of the VM and vacuolar collapse is induced, followed by possible release of hydrolytic enzymes and the degradation of cell components (Vardar and Unal 2012). Interestingly, the increment of VPE in the *Arabidopsis* anther has also been reported (Hatsugai et al. 2006), suggesting a possible role of VPE-mediated proteolysis during tapetal PCD. However, the molecular mechanisms for the dynamic reorganization of the vacuoles during tapetal PCD have not yet been elucidated.

36.5   Conclusions and Future Perspectives

Vacuolar collapse seems to be important in the triggering of plant PCD during innate immunity and development in both reproductive and vegetative tissues. Recent findings have shed light on the machinery for various plant PCD. Further live cell imaging may reveal novel dynamic aspects of the vacuole. In light of these circumstances, automated microscopy and image analysis techniques for the quantitative evaluation of cellular dynamics are of increasing importance.

Besides apoptosis, autophagy is also a participant in cell degeneration and PCD in animals (Shimizu et al. 2004; Tsujimoto and Shimizu 2005). Autophagic cell death is characterized by the occurrence of double-membrane autophagosomes within the dying cells that remove the cell remnants (Gump and Thorburn 2011). In *Drosophila melanogaster*, the destruction of the salivary glands and digestive tract was shown to be mediated by a marked upregulation of autophagy before and during cell death during metamorphosis (Melendez and Neufeld 2008).

Autophagy has recently been shown to be involved in various processes such as recycling of nutrients and senescence in plants (Yoshimoto 2012). Although autophagy-deficient mutants of *Arabidopsis* have been reported to show a normal life cycle, autophagy may also play roles in PCDs in development or stress responses in plants (van Doorn and Woltering 2010). A simple easy method to quantitatively analyze autophagic fluxes has recently been developed in plants (Hanamata et al. 2013). Such technical advances in combination with genetic analyses may reveal novel aspects of autophagy in PCD in plants.

References

Ariizumi T, Toriyama K (2011) Genetic regulation of sporopollenin synthesis and pollen exine development. Annu Rev Plant Biol 62:437–460

Aya K, Ueguchi-Tanaka M, Kondo M et al (2009) Gibberellin modulates anther development in rice via the transcriptional regulation of GAMYB. Plant Cell 21:1453–1472

Bosch M, Franklin-Tong VE (2008) Self-incompatibility in *Papaver*: signalling to trigger PCD in incompatible pollen. J Exp Bot 59:481–490

Bozhkov PV, Lam E (2011) Green death: revealing programmed cell death in plants. Cell Death Differ 18:1239–1240

Cacas JL (2010) Devil inside: does plant programmed cell death involve the endomembrane system? Plant Cell Environ 33:1453–1473

Cheng H, Qin L, Lee S et al (2004) Gibberellin regulates *Arabidopsis* floral development via suppression of DELLA protein function. Development (Camb) 131:1055–1064

Drerup M, Schlücking K, Hashimoto K et al (2013) The calcineurin B-like calcium sensors CBL1 and CBL9 together with their interacting protein kinase CIPK26 regulate the *Arabidopsis* NADPH oxidase RBOHF. Mol Plant 6:559–569

Fuchs Y, Steller H (2011) Programmed cell death in animal development and disease. Cell 147:742–758

Greenberg JT (1996) Programmed cell death: a way of life for plants. Proc Natl Acad Sci USA 93:12094–12097

Gross P, Julius C, Schmelzer E et al (1993) Translocation of cytoplasm and nucleus to fungal penetration sites is associated with depolymerization of microtubules and defense gene activation in infected, cultured parsley cells. EMBO J 12:1735–1744

Gump JM, Thorburn A (2011) Autophagy and apoptosis: what is the connection? Trends Cell Biol 21:387–392

Gunawardena AH (2008) Programmed cell death and tissue remodelling in plants. J Exp Bot 59:445–451

Guo WJ, Ho TH (2008) An abscisic acid-induced protein, HVA22, inhibits gibberellin-mediated programmed cell death in cereal aleurone cells. Plant Physiol 147:1710–1722

Hanamata S, Kurusu T, Okada M et al (2013) In vivo imaging and quantitative monitoring of autophagic flux in tobacco BY-2 cells. Plant Signal Behav 8:e22510

Hara-Nishimura I, Hatsugai N (2011) The role of vacuole in plant cell death. Cell Death Differ 18:1298–1304

Hatsugai N, Kuroyanagi M, Yamada K et al (2004) A plant vacuolar protease, VPE, mediates virus-induced hypersensitive cell death. Science 305:855–858

Hatsugai N, Kuroyanagi M, Nishimura M et al (2006) A cellular suicide strategy of plants: vacuole-mediated cell death. Apoptosis 11:905–911

Hatsugai N, Iwasaki S, Tamura K et al (2009) A novel membrane fusion-mediated plant immunity against bacterial pathogens. Genes Dev 23:2496–2506

Heath MC (2000) Hypersensitive response-related death. Plant Mol Biol 44:321–334

Higaki T, Goh T, Hayashi T et al (2007) Elicitor-induced cytoskeletal rearrangement relates to vacuolar dynamics and execution of cell death: in vivo imaging of hypersensitive cell death in tobacco BY-2 cells. Plant Cell Physiol 48:1414–1425

Higaki T, Kadota Y, Goh T et al (2008) Vacuolar and cytoskeletal dynamics during elicitor-induced programmed cell death in tobacco BY-2 cells. Plant Signal Behav 3:700–703

Higaki T, Kurusu T, Hasezawa S et al (2011) Dynamic intracellular reorganization of cytoskeletons and the vacuole in defense responses and hypersensitive cell death. J Plant Res 124: 315–324

Jones AM (2001) Programmed cell death in development and defense. Plant Physiol 125:94–97

Kadota Y, Kuchitsu K (2006) Regulation of elicitor-induced defense responses by Ca^{2+} channels and cell cycle in tobacco BY-2 cells. In: Nagata T, Matsuoka K, Inze D (eds) Biotechnology in agriculture and forestry 58 tobacco BY-2 cells: from cellular dynamics to omics. Springer, Berlin, pp 207–221

Kadota Y, Goh T, Tomatsu H et al (2004a) Cryptogein-induced initial events in tobacco BY-2 cells: pharmacological characterization of molecular relationship among cytosolic Ca^{2+} transients, anion efflux and production of reactive oxygen species. Plant Cell Physiol 45:160–170

Kadota Y, Watanabe T, Fujii S et al (2004b) Crosstalk between elicitor-induced cell death and cell cycle regulation in tobacco BY-2 cells. Plant J 40:131–142

Kadota Y, Watanabe T, Fujii S et al (2005) Cell cycle dependence of elicitor-induced signal transduction in tobacco BY-2 cells. Plant Cell Physiol 46:156–165

Kadota Y, Fujii S, Ogasawara Y et al (2006) Continuous recognition of the elicitor signal for several hours is prerequisite for induction of cell death and prolonged activation of signaling events in tobacco BY-2 cells. Plant Cell Physiol 47:1337–1342

Kawanabe T, Ariizumi T, Kawai-Yamada M et al (2006) Abolition of the tapetum suicide program ruins microsporogenesis. Plant Cell Physiol 47:784–787

Kimura S, Kaya H, Kawarazaki T et al (2012) Protein phosphorylation is a prerequisite for the Ca^{2+}-dependent activation of *Arabidopsis* NADPH oxidases and may function as a trigger for the positive feedback regulation of Ca^{2+} and reactive oxygen species. Biochim Biophys Acta 1823:398–405

Kimura S, Kawarazaki T, Nibori H et al (2013) The CBL-interacting protein kinase CIPK26 is a novel interactor of *Arabidopsis* NADPH oxidase AtRbohF that negatively modulates its ROS-producing activity in a heterologous expression system. J Biochem (Tokyo) 153:191–195

Kobayashi I, Kobayashi Y, Hardham AR (1994) Dynamic reorganization of microtubules and microfilaments in flax cells during the resistance response to flax rust infection. Planta (Berl) 195:237–247

Kobayashi Y, Kobayashi I, Funaki Y et al (1997) Dynamic reorganization of microfilaments and microtubules is necessary for the expression of non-host resistance in barley coleoptile cells. Plant J 11:525–537

Kobayashi M, Ohura I, Kawakita K et al (2007) Calcium-dependent protein kinases regulate the production of reactive oxygen species by potato NADPH oxidase. Plant Cell 19:1065–1080

Koh S, Andre A, Edwards H et al (2005) *Arabidopsis thaliana* subcellular responses to compatible *Erysiphe cichoracearum* infections. Plant J 44:516–529

Ku S, Yoon H, Suh HS et al (2003) Male-sterility of thermosensitive genic male-sterile rice is associated with premature programmed cell death of the tapetum. Planta (Berl) 217:559–565

Kuriyama H (1999) Loss of tonoplast integrity programmed in tracheary element differentiation. Plant Physiol 121:763–774

Kutsuna N, Hasezawa S (2002) Dynamic organization of vacuolar and microtubule structures during cell cycle progression in synchronized tobacco BY-2 cells. Plant Cell Physiol 43:965–973

Lecourieux D, Mazars C, Pauly N et al (2002) Analysis and effects of cytosolic free calcium increases in response to elicitors in *Nicotiana plumbaginifolia* cells. Plant Cell 14:2627–2641

Li N, Zhang DS, Liu HS et al (2006) The rice tapetum degeneration retardation gene is required for tapetum degradation and anther development. Plant Cell 18:2999–3014

Marty F (1999) Plant vacuoles. Plant Cell 11:587–600

Melendez A, Neufeld TP (2008) The cell biology of autophagy in metazoans: a developing story. Development (Camb) 135:2347–2360

Mur LA, Kenton P, Lloyd AJ et al (2008) The hypersensitive response; the centenary is upon us but how much do we know? J Exp Bot 59:501–520

Nakaune S, Yamada K, Kondo M et al (2005) A vacuolar processing enzyme, δVPE, is involved in seed coat formation at the early stage of seed development. Plant Cell 17:876–887

Obara K, Kuriyama H, Fukuda H (2001) Direct evidence of active and rapid nuclear degradation triggered by vacuole rupture during programmed cell death in *Zinnia*. Plant Physiol 125:615–626

Ogasawara Y, Kaya H, Hiraoka G et al (2008) Synergistic activation of *Arabidopsis* NADPH oxidase AtrbohD by Ca^{2+} and phosphorylation. J Biol Chem 283:8885–8892

Ohno R, Kadota Y, Fujii S et al (2011) Cryptogein-induced cell cycle arrest at G_2 phase is associated with inhibition of cyclin-dependent kinases, suppression of expression of cell cycle-related genes and protein degradation in synchronized tobacco BY-2 cells. Plant Cell Physiol 52:922–932

Pennell RI, Lamb C (1997) Programmed cell death in plants. Plant Cell 9:1157–1168

Reape TJ, McCabe PF (2010) Apoptotic-like regulation of programmed cell death in plants. Apoptosis 15:249–256

Rogers LA, Dubos C, Surman C et al (2005) Comparison of lignin deposition in three ectopic lignification mutants. New Phytol 168:123–140

Saito C, Ueda T, Abe H et al (2002) A complex and mobile structure forms a distinct subregion within the continuous vacuolar membrane in young cotyledons of *Arabidopsis*. Plant J 29:245–255

Shimizu S, Kanaseki T, Mizushima N et al (2004) Role of Bcl-2 family proteins in a non-apoptotic programmed cell death dependent on autophagy genes. Nat Cell Biol 6:1221–1228

Skalamera D, Heath MC (1996) Cellular mechanisms of callose deposition in response to fungal infection or chemical damage. Can J Bot 74:1236–1242

Smertenko AP, Bozhkov PV, Filonova LH et al (2003) Re-organisation of the cytoskeleton during developmental programmed cell death in *Picea abies* embryos. Plant J 33:813–882

Suzuki N, Miller G, Morales J et al (2011) Respiratory burst oxidases: the engines of ROS signaling. Curr Opin Plant Biol 14:691–699

Takeda S, Gapper C, Kaya H et al (2008) Local positive feedback regulation determines cell shape in root hair cells. Science 319:1241–1244

Takemoto D, Jones DA, Hardham AR (2003) GFP-tagging of cell components reveals the dynamics of subcellular re-organization in response to infection of *Arabidopsis* by oomycete pathogens. Plant J 33:775–792

Teng X, Cheng WC, Qi B et al (2011) Gene-dependent cell death in yeast. Cell Death Dis 2:e188

Torres MA, Jones JD, Dangl JL (2005) Pathogen-induced, NADPH oxidase-derived reactive oxygen intermediates suppress spread of cell death in *Arabidopsis thaliana*. Nat Genet 37: 1130–1134

Tsujimoto Y, Shimizu S (2005) Another way to die: autophagic programmed cell death. Cell Death Differ 12(2):1528–1534

van Doorn WG, Woltering EJ (2010) What about the role of autophagy in PCD? Trends Plant Sci 15:361–362

Vardar F, Unal M (2012) Ultrastructural aspects and programmed cell death in the tapetal cells of *Lathyrus undulatus* Boiss. Acta Biol Hung 63:52–66

Vizcay-Barrena G, Wilson ZA (2006) Altered tapetal PCD and pollen wall development in the *Arabidopsis ms1* mutant. J Exp Bot 57:2709–2717

Wan L, Xia X, Hong D et al (2010) Abnormal vacuolization of the tapetum during the tetrad stage is associated with male sterility in the recessive genic male sterile *Brassica napus* L. line 9012A. J Plant Biol 53:121–133

Wang M, Hoekstra S, Van-Bergen S et al (1999) Apoptosis in developing anthers and the role of ABA in this process during androgenesis in *Hordeum vulgare* L. Plant Mol Biol 39:489–501

Wei CX, Lan SY, Xu ZX (2002) Ultrastructural features of nucleus degradation during programmed cell death of starchy endosperm cells in rice. Acta Bot Sin 44:1396–1402

Wright H, van Doorn WG, Gunawardena AH (2009) In vivo study of developmental programmed cell death using the lace plant (*Aponogeton madagascariensis*; Aponogetonaceae) leaf model system. Am J Bot 96:865–876

Wu H, Yang M (2005) Reduction in vacuolar volume in the tapetal cells coincides with conclusion of the tetrad stage in *Arabidopsis thaliana*. Sex Plant Reprod 18:173–178

Yoshimoto K (2012) Beginning to understand autophagy, an intracellular self-degradation system in plants. Plant Cell Physiol 53:1355–1365

Zhang DS, Liang WQ, Yuan Z et al (2008) Tapetum degeneration retardation is critical for aliphatic metabolism and gene regulation during rice pollen development. Mol Plant 1:599–610

Chapter 37
Sperm Proteasome as a Putative Egg Coat Lysin in Mammals

Edward Miles and Peter Sutovsky

Abstract During animal and human fertilization, the fertilizing spermatozoon creates and passes through a fertilization slit in the vitelline coat (VC) of the oocyte. It has been hypothesized that the penetration of the mammalian VC, the zona pellucida (ZP), is aided by a proteolytic enzyme capable of locally degrading ZP proteins. This putative "zona lysin" is predicted to reside within the sperm head acrosome and be released or exposed by ZP-induced acrosomal exocytosis. Evidence has been accumulating in favor of the 26S proteasome, the ubiquitin-dependent multi-subunit protease acting as the putative vitelline coat/zona lysin in humans and animals. To confirm this hypothesis, three criteria must be met: (1) the sperm receptor on the ZP must be ubiquitinated, (2) proteasomes must be present, exposed, and enzymatically active in the sperm acrosome, and (3) sperm proteasomes must be able to degrade the sperm receptor on the egg coat/ZP during fertilization. This review discusses recent data from a number of mammalian and nonmammalian models addressing these predictions. These data shed light on the mechanisms controlling sperm interactions with VC and on the evolutionary conservation of the proteasome-assisted fertilization mechanisms.

Keywords Egg coat lysin • Fertilization • Oocyte • Proteasome • Proteolysis • Sperm • Ubiquitin • Vitelline coat • Zona pellucida

E. Miles • P. Sutovsky (✉)
Division of Animal Science and Departments of Obstetrics Gynecology and Women's Health,
University of Missouri, Columbia, MO 65211-5300, USA
e-mail: SutovskyP@missouri.edu

H. Sawada et al. (eds.), *Sexual Reproduction in Animals and Plants*,
DOI 10.1007/978-4-431-54589-7_37, © The Author(s) 2014

37.1   Introduction: Zona Lysin and the Elusive Mechanism of Egg Coat Penetration by Sperm

The process of mammalian fertilization has long been studied, but many fundamental questions still remain unanswered. One that continues to be debated, despite decades of research, concerns the mechanism utilized by the fertilizing spermatozoa to penetrate the oocyte vitelline coat (VC), the zona pellucida (ZP). When a capacitated, fertilization-competent spermatozoon binds to the sperm receptor on the ZP, it undergoes acrosomal exocytosis (AE), which causes vesiculation of the acrosomal membrane and exocytosis of the acrosomal cap (Yurewicz et al. 1998; Bleil and Wassarman 1980). This step exposes the acrosome-borne proteolytic enzymes and results in the formation of the acrosomal shroud, which allows the spermatozoa to create a local microenvironment that supports the opening of the fertilization slit and penetration of the sperm through the ZP (Yurewicz et al. 1998). There are two schools of thought about the mechanism of sperm–ZP penetration: (1) the mechanical force of the sperm tail motility is sufficient to fully penetrate the ZP (Green and Purves 1984; Bedford 1998); and (2) an enzyme originating from the acrosome of the fertilizing spermatozoa acts as a putative zona lysin that digests the fertilization slit in the ZP and allows sperm to penetrate through the ZP (Austin and Bishop 1958; Austin 1975). The latter theory was first proposed as early as 1958 by Austin and Bishop (1958).

Several candidates have been proposed for this hypothetical zona lysin. Originally, the acrosomal protease acrosin was favored as the most likely zona lysin, but was ruled out when acrosin knockout mice remained fertile with only delayed ZP penetration (Baba et al. 1994). Acrosin has since been considered to be involved in proteolysis or processing of proteins in the acrosome and on acrosomal membranes (Honda et al. 2002). The serine protease Tesp5/Prss21 on the mouse sperm surface was identified as a plausible candidate. Double-knockout studies have shown reduced fertility in vitro, which was rescued by exposure of the spermatozoa to the uterine microenvironment or by treatment with uterine fluids (Yamashita et al. 2008; Kawano et al. 2010). These candidates have since been proposed to be involved in initial sperm–ZP binding (Yamashita et al. 2008; Kawano et al. 2010). Evidence has been accumulating in favor of the 26S proteasome as the candidate zona lysin in mammals, ascidians, and invertebrates (Sakai et al. 2004; Yi et al. 2007a; Sutovsky et al. 2004). This review highlights the significant studies and current research examining the possible role of the 26S proteasome as the putative mammalian zona lysin.

37.2   Ubiquitin-Proteasome System

37.2.1   Ubiquitin

Through a multistep enzymatic process, the ubiquitin-proteasome system (UPS) tags outlived or damaged intracellular proteins with a small chaperone protein ubiquitin. This process, referred to as ubiquitination, typically targets the ubiquitinated substrates for proteolytic degradation by a multi-subunit protease, the 26S proteasome, (Glickman and Ciechanover 2002). Ubiquitin, an 8.5-kDa, 76-amino-acid protein, was first isolated and purified by the Goldstein laboratory in 1975. They found that this polypeptide induced the differentiation of bovine thymus-derived and bone marrow-derived immunocytes in vitro (Goldstein et al. 1975). Consequently, they named this newly discovered polypeptide ubiquitous immunopoietic polypeptide (UBIP) because of its high degree of evolutionary conservation, exhibiting close structural, functional, and immunological similarity when isolated from species as diverse as protozoans to mammals and plants. In 1976, Etlinger and Goldberg first described a novel ATP-dependent proteolytic system responsible for the rapid degradation of abnormal proteins separate from the lysosomal degradation pathway (Etlinger and Goldberg 1977). The discovery that metabolic energy is required for intracellular protein degradation opposed the commonly accepted idea that cellular proteolysis was an entirely exergonic process occurring in the lysosome. In a joint effort, Aaron Ciechanover, Avram Hershko, and Ervin Rose described the ubiquitin-mediated protein degradation, a discovery for which they shared the Nobel Prize in Chemistry in 2004. Ubiquitin is not just a housekeeping protein that helps recycle outlived or damaged protein molecules. It is also involved in a number of cellular mechanisms and pathologies, including but not limited to antigen presentation in the immune system, apoptosis, Alzheimer's disease, cell-cycle control, endocytosis of membrane receptors, human immunodeficiency virus (HIV) particle internalization, protein quality control in the endoplasmic reticulum, reticulocyte differentiation, signaling, and transcriptional control (Pines 1994; Hershko and Ciechanover 1998; Glickman and Ciechanover 2002).

37.2.2 Ubiquitination

Ubiquitin is typically a cytoplasmic and nuclear protein, but it can be found in the extracellular space in some mammalian, lower vertebrate, and invertebrate systems and is highly substrate specific. Polymerization of substrate-bound ubiquitin molecules into multi-ubiquitin chains serves as a degradation signal for numerous target proteins. The degradation of substrate protein is initiated by the covalent attachment of an isopeptide chain of four or more ubiquitin molecules. These ubiquitin molecules are linked to each other through one of seven lysine residues (K6, K11, K27, K29, K33, K48, K63) to form the poly-ubiquitin chains. All seven of these Lys residues are potential ubiquitin-chain initiation sites, but K48 is the most common linkage site for poly-ubiquitin chains recognized by the 26S proteasome (Walczak et al. 2012; Iwai 2012). The K63 site is the most common site of di-ubiquitination, targeting membrane receptors for lysosomal degradation, chromatin remodeling, and DNA repair (Walczak et al. 2012; Kim et al. 2006).

The formation and ligation of the poly-ubiquitin chain to its target protein occurs through a concerted series of ATP-dependent enzymatic reactions (Fig. 37.1). Ubiquitination starts with the activation of a single ubiquitin molecule with a

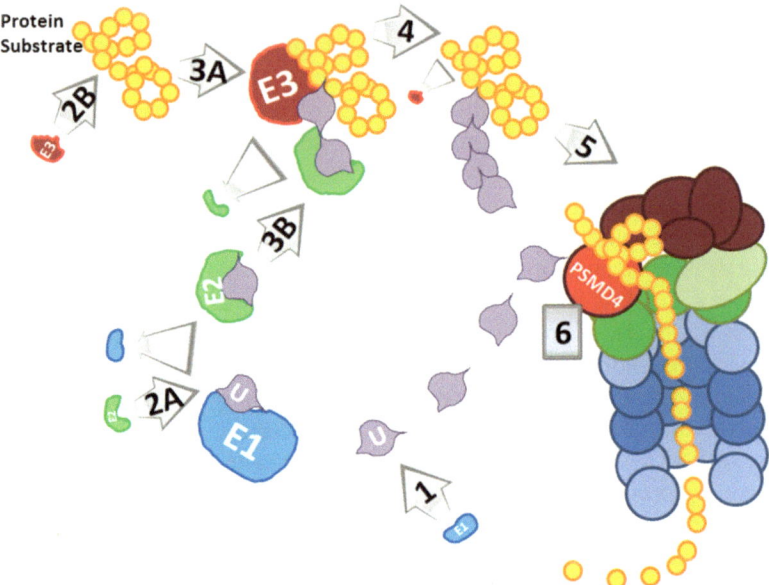

Fig. 37.1 Diagram of protein ubiquitination and degradation by the 26S proteasome. Step *1*. Monoubiquitin is activated by a phosphorylated ubiquitin-activating enzyme E1 (*UBA1*). Step *2A*. UBE1 is supplanted by ubiquitin carrier enzyme/ubiquitin-conjugating enzyme E2 (*UBE2*). Step *2B*. Concurrently, the substrate protein is engaged by an E3-type ubiquitin ligase. Step *3A*. Ubiquitin ligase E3 covalently links an activated monoubiquitin to an internal Lys residue of the substrate protein. Step *3B*. A second activated ubiquitin molecule is linked to the substrate-bound ubiquitin. Step *4*. The ensuing tandem ligation of additional activated ubiquitin molecules results in the formation of a multi-ubiquitin chain. Step *5*. Multi-ubiquitin chain of four or more ubiquitin molecules is recognized and engaged by subunit *PSMD4* of the 19S proteasomal regulatory complex. Step *6*. The substrate protein is deubiquitinated (liberated ubiquitin molecules re-enter the cycle), unfolded, threaded through the 20S core, and cleaved into small peptides, released from the 20S core lumen. [From Sutovsky 2011, © Society for Reproduction and Fertility (2013). Reproduced by permission]

phosphorylation-dependent ubiquitin-activating enzyme E1 (UBE1 or UBA1 in HUGO nomenclature). The UBA1 is then supplanted by ubiquitin carrier/ubiquitin-conjugating enzyme E2. Simultaneously, an E3-type ubiquitin ligase seeks out and engages the target protein to be ubiquitinated and degraded (Hershko 2005). The E3-type ligases are highly diverse and responsible for the substrate specificity of protein ubiquitination as there are about two E1 proteins, approximately 30 E2 proteins, and more than 500 different E3 proteins in humans (Tanaka 2009). The E3-ligases catalyze the covalent ligation of the C-terminal glycine/Gly residue (G76) of ubiquitin to an internal lysine/Lys residue of the target protein. Next, the ubiquitin chain elongates from the mono-ubiquitinated substrate protein resulting in a poly-ubiquitinated protein. A poly-ubiquitin tail of four or more ubiquitin molecules is needed to signal proteasomal degradation.

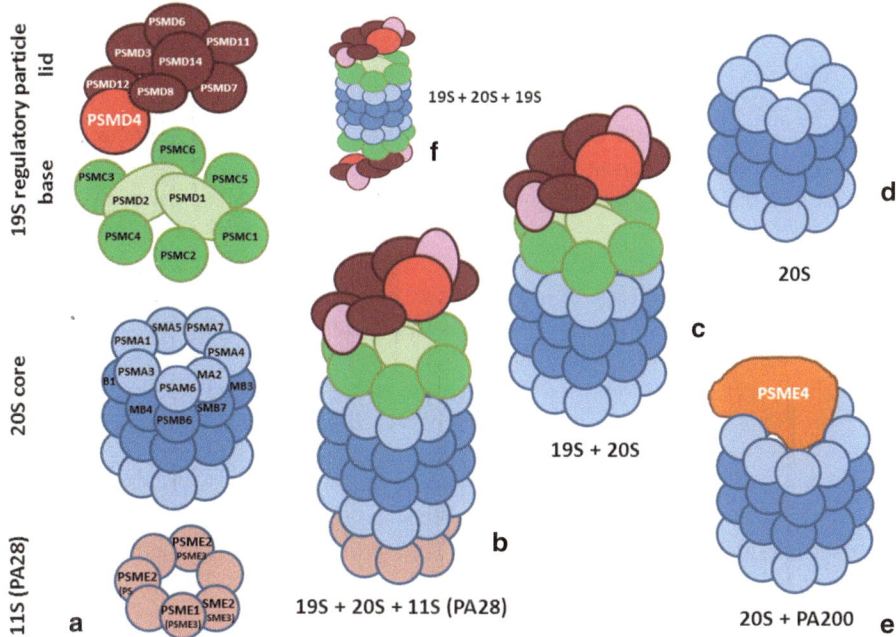

Fig. 37.2 Variations on the subunit composition of the 26S proteasome. (**a**) Subunit makeup of (*top* to *bottom*) 19S regulatory particle (lid + base), 20S core, and 11S particle (PA28). (**b**) 20S core capped with 19S particle on one side and an 11S complex/PA28 on the other. (**c**) 20S core capped with one 19S particle. (**d**) Uncapped 20S proteasome. (**e**) 20S core capped with proteasome activator PA200. (**f**) Canonical 26S proteasome, with one 19S particle/cap on each side of 20S core. [From Sutovsky 2011, © Society for Reproduction and Fertility (2013). Reproduced by permission]

37.2.3   26S Proteasome

The 26S proteasome is a 2.5-MDa multi-catalytic canonical protease localized in the cell cytosol and nucleus (Fig. 37.2). The 26S proteasome is responsible for degrading ubiquitinated proteins in the cell, although nonubiquitinated proteins could be substrates as well. The typical enzymatically active proteasome consists of two subcomplexes. One is a hollow catalytic 20S core particle (20S CP) capped by one or two terminal 19S regulatory particle(s) (19S RP). The 19S RP is responsible for poly-ubiquitin chain recognition and binding, deubiquitination, and substrate protein priming/linearization/unfolding. The 19S RP consists of at least 17 different subunits between the two subcomplexes: the lid and the base. The lid complex contains 14 regulatory particle non-ATPase subunits (PSMD1-14) (Hanna and Finley 2007). Subunit PSMD4 (Rpn10) is the main 19S subunit responsible for substrate recognition. It binds the poly-ubiquitin chains on the target protein during poly-ubiquitin chain recognition by the 26S proteasome (Yi et al. 2010a, b). Subunit PSMD14 (Rpn11), then deubiquitinates the captured substrate by cleaving the poly-ubiquitin chain at a proximal site, which is then further cleaved into single ubiquitin molecules by deubiquitinases (DUBs) that can associate with the 19S RP (Verma et al. 2002). The cleaved ubiquitin molecules can be reused for ubiquitination of other protein molecules. New functions are still being discovered for subunits composing the 19S lid. Table 37.1 lists the known 26S proteasomal subunits and their proposed functions by species.

The 19S RP base subcomplex is responsible for (1) capturing target proteins via poly-ubiquitin chain recognition, (2) unfolding the substrate protein, and (3) opening the channel into the 20S CP. The base subcomplex contains six homologous AAA-ATPase subunits (PSMC1–6) and three non-ATPase subunits (PSMD1, -2, -4). These subunits create a narrow gated channel, which only allows unfolded proteins to enter the 20S core. PSMD1 and -2 have been proposed to work together as a functional unit and are required for substrate translocation and opening of the 20S channel. This compartmentalized design of the 26S proteasome separates proteolysis from the cellular milieu and restricts degradation to unfolded and imported proteins. Subunit PSMD2 attaches to the 20S CP, whereas subunit PSMD1 is located on top of PSMD2 and serves as a docking site for substrate recruitment factors (Rosenzweig et al. 2008). Subunit PSMD4 functions as an integral ubiquitin receptor to trap poly-ubiquitinated proteins; PSMD4 does this by its C-terminal ubiquitin-interacting motif (Deveraux et al. 1995). Six ATPase subunits (PSMC1-6) surrounding the base are organized into a hexameric ring that controls the opening of the channel and allows the protein to reach the catalytic sites of the 20S CP. The base ATPases are required not only to open the channel into the 20S CP but to also unfold the substrate protein: this allows the protein to be threaded through the narrow channel of the 20S CP where the catalytic protease subunits are located. The details of the ATP-dependent mechanisms behind these processes are still unknown, but it is certain that these subunits work through a process that requires ATP hydrolysis (Liu et al. 2006).

The 20S CP is a 750-kDa cylindrical protein complex responsible for the proteolysis of the substrate protein. The 20S CP contains 28 subunit molecules of 14 types formed from the stacking of two outer α-rings and two inner β-rings. Each ring is composed of seven structurally similar α- and β-subunits, respectively. The rings are heptamerically stacked in an $\alpha_{1-7}\beta_{1-7}\beta_{1-7}\alpha_{1-7}$ pattern. The subunits of the α-ring (PSMA1-7) connect the 20S CP to the 19S base and act as a gate that opens in the presence of an ubiquitinated protein. The two inner rings are each composed of seven β-type subunits (PSMB1-7) and confer the threonine protease activities of the 26S proteasome. There are three catalytic β-type subunits: PSMB6 ($\beta1$), PSMB7 ($\beta2$), and PSMB5 ($\beta5$). The $\beta1$-subunit is associated with caspase-like/PGPH (peptidylglutamyl-peptide hydrolyzing) activity, -$\beta2$ with trypsin-like activity, and $\beta5$ with chymotrypsin-like activities, which confer the ability to cleave peptide bonds at the C-terminal side of acidic, basic, and hydrophobic amino-acid residues, respectively (Tanaka 2009). The 20S proteasome then degrades the target protein into 3 to 15 amino-acid oligopeptides that are released into the cytosol and are further hydrolyzed into single amino acids by cytosolic oligopeptidases or amino-carboxyl peptidases (Tanaka 2009).

Table 37.1 26S proteasome subunit function and nomenclature by species

HUGO/GDB symbol	Homo sapiens	Bos taurus	Sus scrofa	Rattus norvegicus	Mus musculus	Caenorhabditis elegans	Saccharomyces cerevisiae	Function
20S α-type								
α1	Pros27/Iota/p27k	PSMA6	PSMA6	Psma6	Psma6/Pros-27	PAS-1	Scl1/YC7	
α2	HC3	PSMA2	PSMA2	Psma2	Psma2	PAS-2	Pre8/Y7	
α3	HC9	PSMA4	PSMA4	Psma4	Psma4	PAS-3	Pre9/Y13	
α4	XAPC7-S	PSMA7	PSMA7	Psma7	Psma7	PAS-4	Pre6	
α5	Zeta	PSMA5	PSMA5	Psma5	Psma5	PAS-5	Pup2/Doa5	
α6	HC2/Pros30	PSMA1	PSMA1	Psma1	Psma1/Pros-30	PAS-6	Pre5	
α7	HC8	PSMA3	PSMA3	Psma3	Psma3	PAS-7	Pre10/YC1	
20S β-type								
β1	Y/Delta	PSMB6	PSMB6	Psmb6	Psmb6/Lmp19/Mpnd	PBS-1	Pre3	Caspase-like peptidase activity
β2	Z	PSMB7	PSMB7	Psmb7	Psmb7/MC14	PBS-2	Pup1	Trypsin-like protease activity
β3	HC10	PSMB3	PSMB3	Psmb3	Psmb3	PBS-3	Pup3	
β4	HC7	PSMB2	PSMB2	Psmb2	Psmb2	PBS-4	Pre1	
β5	X/MB1	PSMB5	PSMB5	Psmb5	Psmb5	PBS-5	Pre2/Doa3	Chymotrypsin-like protease activity
β6	HC5	PSMB1	PSMB1	Psmb1	Psmb1	PBS-6	Pre7	
β7	HN3	PSMB4	PSMB4	Psmb4	Psmb4	PBS-7	Pre4	
β1i	LMP2/RING12	PSMB9	PSMB9	Psmb9/Lmp2	Psmb9/Lmp2	–	–	Caspase-like activity/immunoproteasome

(continued)

Table 37.1 (continued)

	HUGO/GDB symbol	Homo sapiens	Bos taurus	Sus scrofa	Rattus norvegicus	Mus musculus	Caenorhabditis elegans	Saccharomyces cerevisiae	Function
β2i	PSMB10	MECL1	PSMB10	PSMB10	Psmb10	Psmb10/Mecl-1/Lmp7	–	–	Trypsin-like activity/immunoproteasome
β5i	PSMB8	LMP7/RING10	PSMB8/LMP7	PSMB8	Psmb8/Ring10	Psmb8	–	–	Chymotrypsin-like activity/immunoproteasome
19S ATPase									
Rpt1/S7	PSMC2	MSS1/S7	PSMC2/MSS1	PSMC2	Psmc2/Mss1	Psmc2/Mss1	RPT-1	Yta3/Cim5	ATPase
Rpt2/S4	PSMC1	p56/S4	PSMC1/p56	PSMC1	Psmc1/S4	Psmc1/P26S4	RPT-2	Yta5/mts2	ATPase, Gate-opening
Rpt3/S6b	PSMC4	TBP7/S6/p48	PSMC4/p48	PSMC4	Psmc4/Tbp7	Psmc4	RPT-3	Yta2	ATPase, Gate-opening
Rpt4/S10b	PSMC6	p42/CADP44/SUG2	PSMC6/p42	PSMC6	Psmc6	Psmc6	RPT-4	Sug2/Pcs1/Crl13	ATPase
Rpt5/S6a	PSMC3	TBP1/S6a	PSMC3/p50	PSMC3	Psmc3/Tbp1	Psmc3	RPT-5	Yta1	ATPase, Gate-opening
Rpt6/S8	PSMC5	p45/S8/TRIP1/SUG1	PSMC5/p45	PSMC5	Psmc5/Sug1	Psmc5/FZA-B/Sug1/M56	RPT-6	Sug1/Cim3/let1/Crl3	ATPase
19S non-ATPase									
Rpn1/S2	PSMD2	p97/TRAP-2/55.11/S2	PSMD2/p97	PSMD2	Psmd2	Psmd2/Tex190	RPN-1	Nas1/mts4/Hrd2	PIPs scaffolding
Rpn2/S1	PSMD1	p112/S1	PSMD1/p112	PSMD1	Psmd1	Psmd1/P112	RPN-2	Sen3	PIPs scaffolding

	HUGO	GDB							
Rpn3/S3	PSMD3	p58/S3/TSTA2	PSMD3/p58	PSMD3	Psmd3	Psmd3/P91A	RPN-3	Sun2	
Rpn4								Son1/Ufd5	Proteasome gene, transcription
Rpn5/p55	PSMD12	p55	PSMD13/p55	PSMD12	Psmd12	Psmd12/P55	RPN-5	Nas5	
Rpn6/S9	PSMD11	p44.5/S9	PSMD11/p44.5	PSMD11	Psmd11	Psmd11/P44.5	RPN-6	Nas4	
Rpn7/S10a	PSMD6	p44/S10a	PSMD6/p44	PSMD6	Psmd6	Psmd6	RPN-7		
Rpn8/S12	PSMD7	p40/S12/MOV34	PSMD7/p40	PSMD7	Psmd7	Psmd7/Mov-34	RPN-8	Nas3	
Rpn9/S11	PSMD13	p40.5/S11	PSMD13/p40.5	PSMD13	Psmd13	Psmd13	RPN-9	Nas7/mts1	
Rpn10/S5a	PSMD4	S5a/AF	PSMD4	PSMD4	Psmd4	Psmd4/S5a/Mcb1	RPN-10	Mcb1/pus1/Sun1	Ub receptor
S5b	PSMD5	p50.5/S5b	PSMD5/p50.5	PSMD5	Psmd5	Psmd5	–	–	
Rpn11/S13	PSMD14	S13/POH1/PAD1	PSMD14	PSMD14	Psmd14	Psmd14/Pad1/Poh1	RPN-11	Mpr1/pad1/mts5	DUB
Rpn12/S14	PSMD8	p31/S14/HIP6/Nin1p	PSMD8/p31	PSMD8	Psmd8	Psmd8	RPN-12	Nin1/mts3	
Rpn13	ADRM1	ADRM1						DAQ1	Ub receptor, UCH37 recruitment
p28	PSMD10	p28/S15	PSMD10/p28	PSMD10	Psmd10/p28Gank	Psmd10/P28/gankyrin	–	Nas6	
p27/S15	PSMD9	p27/S15	PSMD10/p27	PSMD9	Psmd9/Bridge	Psmd9/P27	–	Nas2	PSM modulator

HUGO human genome organization, *GDB* human genome database, *DUB* deubiquitinating enzyme, *Ub* ubiquitin, *PSM* proteasome, *PIPs* proteasome-interacting proteins/proteasome substrates (Voges et al. 1999; Chen et al. 2008; Tanaka 2009)

37.3   Sperm Proteasome During Plant, Ascidian, and Echinoderm Fertilization

37.3.1   Plant Fertilization

The sexual reproduction cycle in higher plants occurs in the reproductive organs of the flower. Similar to animal spermatozoon, pollen, the male gametophyte, is a highly specialized structure, consisting of two or three cells. Pollen is released from the anthers and adheres to the surface of the female stigma, equipped with a cytoplasmic extension known as the pollen tube (Boavida et al. 2005). The pollen tube rapidly grows through the style to the ovules (Boavida et al. 2005). Pollen tube adhesion and growth guides the pollen toward the ovary where it fertilizes the embryo sac, the female gametophyte (Boavida et al. 2005). For a more in-depth account of fertilization in higher plants, please refer to the review by Boavida et al. (2005).

Evolutionary conservation of the components and roles of the UPS appear to reach beyond the animal kingdom to plants, where pollen adhesion and guidance seem to depend on ubiquitin. For example, during maize pollen development, the levels of ubiquitin and ubiquitin–protein conjugates in young pollen precursors, the microspores without vacuoles were 10 to 50 times lower than compared to mature pollen grains (Callis and Bedinger 1994). Treatment with 26S proteasomal inhibitors, MG132 and epoxomicin, significantly reduced pollen tube growth in kiwifruit (*Actinidia deliciosa*) (Speranza et al. 2001). Kim et al. (2006) showed that pollen adhesion to styles of lily (*Lilium longiflorum* Thunb.) can be enhanced in vitro with the addition of exogenous ubiquitin. Furthermore, in *Arabidopsis thaliana* the proper conformation of SCF E3 ubiquitin-ligases that direct ubiquitination is required for male fertility (Devoto et al. 2002). Mutations in genes encoding 26S proteasomal subunits, ubiquitin-specific proteases, deubiquitinating enzymes, and 19S regulatory particle non-ATPases result in infertility from defects in pollen maturation or transmission (Doelling et al. 2007; Book et al. 2009). Several E3s have been shown to play key roles in self/non-self pollen recognition in snapdragons (*Antirrhinum*) (Qiao et al. 2004). In snapdragons, the SCF[AhSLF2] complex acts as an E3 that targets non-self S-RNases for ubiquitination and destruction by the 26S proteasome; however, it leaves self-S-RNases active during the self-incompatibility response, thus functioning as a cytotoxin that degrades RNA and terminates pollen tube growth (Qiao et al. 2004).The current data strongly suggest a role for the UPS in plant germination and gametophyte development, but more research is needed to elucidate how the UPS is utilized in these processes.

37.3.2   Ascidian Fertilization

Ascidians are hermaphrodites that reproduce by releasing sperm and eggs simultaneously into the surrounding seawater during the spawning season; thus, large quantities of readily fertilizable gametes can be easily obtained for research. The exposure to alkaline seawater activates the spermatozoa, very similar to capacitation in mammals, suggesting that the proteasome is then secreted to the sperm surface. The spermatozoa then undergo species-specific binding to the proteinaceous vitelline coat, which they must penetrate via sperm surface proteases that act as a VC-lysin to complete the fertilization process. It has been shown that the proteasomes are exposed on the ascidian sperm surface; proteasomal proteolytic activity was specifically detected in the sperm head region and was clearly increased upon sperm activation (Sawada et al. 2002a, b). Anti-proteasome antibodies, and proteasomal inhibitors MG115 and MG132, inhibit ascidian fertilization (Sawada et al. 2002b).

Studies in two ascidian species, *Ciona intestinalis* and *Halocynthia roretzi*, have shown a strict self-sterility in fertilization (Sawada et al. 2002a). After self/non-self recognition via interactions between the sperm and VC, the sperm-borne egg coat lysin system is activated (Sawada et al. 2002a). Data from Sawada's laboratory have provided evidence that VC penetration by ascidian sperm involves the sequential ubiquitination and proteasomal degradation of the sperm receptor protein HrVC70, a major protein component of the ascidian VC. Ubiquitination of HrVC70 is accomplished by a 700-kDa ubiquitin-conjugating enzyme complex released during sperm activation (Sawada et al. 2002a; Sakai et al. 2003). A 930-kDa proteasome (26S-like proteasome) seems to function as the egg coat lysin that directly degrades the VC. These data suggest that an extracellular, sperm-borne, ubiquitin-conjugating enzyme system is essential for the formation of these poly-ubiquitin chains on HrVC70 and for ascidian fertilization.

The HrVC70 is a 70-kDa component of the VC shown to be a novel sperm receptor that bears significant similarity to components of the mammalian zona pellucida, particularly the proposed mammalian/murine sperm receptor ZPC (Sawada et al. 2002a). Three N-terminal cystine/Cys residues in the ZP domain of the 120-kDa HrVC70 precursor (HrVC120) share conserved positions within mammalian and frog ZPC (Sawada et al. 2002a). HrVC70 contains 12 EGF (epidermal growth factor)-like repeats that confer its role in sperm binding to VC (Sawada et al. 2002b). This protein has been shown to be degraded by purified ascidian sperm 26S-like proteasomes only in the presence of ubiquitin, ATP, and the ubiquitin-conjugating enzymes purified from a rabbit reticulocyte lysate (Sawada et al. 2002b; Sakai et al. 2004). These results reveal that an extracellular UPS is essential for ascidian fertilization, particularly in the degradation of the proteinaceous vitelline coat.

Recently, Yokota and Sawada investigated an extracellular transport signal and have discovered a novel posttranslational modification of the ascidian sperm proteasome. They found that the 20S core PSMA1/α6-subunit of purified sperm proteasomes is distinct from purified egg and muscle proteasomes (Yokota et al. 2011). Tissue specific α-subunits are not commonplace, but several are expressed in the testis of *Drosophila* (α3T, α4T2, and α6T) (Belote and Zhong 2009). Among these testis-specific α-subunits, α6T is reported to be crucial for spermatogenesis and fertility (Belote and Zhong 2009). The α6-subunit of *H. roretzi* contains a cluster of acidic amino acid residues and the removal of this cluster may mimic the state of dephosphorylation of the α6-subunit which may affect function and localization of the sperm proteasome (Yokota et al. 2011). Alternatively, it is also possible that the conserved sequence of the α6-subunit in *H. roretzi* and *C. intestinalis* may function as a transport signal to the acrosome during ascidian spermiogenesis. It is not known at present if similar modifications exist in the mammalian sperm proteasome.

37.3.3   Echinoderm Fertilization

Most echinoderms can reproduce sexually by spawning eggs and spermatozoa into the seawater, much as do ascidians, and asexually by regenerating body parts. For the purpose of this review, only sexual reproduction is examined. It has been confirmed that the proteasomes are present in sea urchin spermatozoa and that the sperm proteasome may be involved in the acrosome reaction; but differing from ascidians, the proteasome is not essential for sperm binding to the echinoderm VC (Yokota and Sawada 2007; Matsumura and Aketa 1991). The role of sperm proteasomes in echinoderm fertilization has been investigated in the sea urchin *Pseudocentrotus depressus* by Yokota and Sawada (2007), who examined the effects of proteasomal inhibitors and synthetic peptide substrates for the proteasome on different steps of the fertilization process. The inhibition of fertilization by proteasomal inhibitors suggests that the echinoderm sperm proteasomes are important in the acrosome reaction and in sperm penetration through the VC, most likely as an egg coat lysin (Yokota and Sawada 2007). Among the examined proteasomal substrates and inhibitors, a caspase substrate (Z-Leu-Leu-Glu-MCA) competitively inhibited the caspase-like catalytic center (β1-subunit) of the proteasome, which decreased fertilization in a concentration-dependent manner (Yokota and Sawada 2007); this suggests that the caspase-like activity of the proteasome β1-subunit must play a key role in sea urchin fertilization.

Echinoderm spermatozoa undergo acrosome reaction/acrosomal exocytosis when they enter the jelly coat surrounding the VC. Protein kinase A (PKA) has been implicated in the acrosome reaction in sea urchins (Su et al. 2005), and shown to stimulate the function of mammalian 26S proteasome (Zhang et al. 2007). Therefore, the sperm proteasome may act to degrade certain unknown PKA modulators, which would result in the irreversible activation of PKA during the acrosome reaction. During the acrosome reaction, the acrosomal contents containing an egg coat lysin are exposed and released, allowing spermatozoa to penetrate the VC (Yokota and Sawada 2007). For the sperm proteasome to act as egg coat/VC lysin, the VC must be ubiquitinated before fertilization and extracellular ATP must be present (Yokota and Sawada 2007). Yokota and Sawada used Western blotting with a monoclonal anti-ubiquitin antibody FK2, which recognizes both mono- and multi-ubiquitinated proteins, to reveal a band pattern indicative of VC ubiquitination. The identity of the ubiquitinated VC protein(s) remains to be determined, but current data support the hypothesis that the echinoderm sperm 26S proteasome is released during the acrosome reaction and degrades the ubiquitinated VC proteins in an ATP-dependent fashion.

37.4 Proteasome Localization and Activity in Mammalian Spermatozoa

Mammalian fertilization is a unique cellular-recognition event that comprises sequential interactions between the sperm and the oocyte vitelline coat (Yanagimachi 1994). First, a capacitated spermatozoon penetrates the cumulus oophorus, facilitated by a sperm membrane-bound hyaluronidase, and then binds to the zona pellucida, a protective glycoprotein matrix that surrounds the ovulated oocyte. The ZP plays a pivotal role in the species specificity of sperm–oocyte recognition, and binding, as well as in the induction of acrosomal exocytosis (AE), anti-polyspermy defense, and protection of the embryo until implantation. In most of the mammals studied, only spermatozoa with an intact acrosome bind to the ZP, in a species-specific manner. The fertilizing spermatozoan interaction with the zona glycoprotein that serves as sperm receptor (ZP3/ZPC in mouse/human and ZPB/ZPC complex in pig) triggers the AE via signal transduction events in the sperm acrosome, such as the opening of T-type, low voltage-activated Ca^{2+} channels, causing an influx of Ca^{2+} and the activation of heterotrimeric G proteins (Pasten et al. 2005; Ikawa et al. 2010). An increase in intracellular pH, resulting in sustained Ca^{2+} influx, which drives AE, causes the release or exposure of acrosomal enzymes and, when combined with vigorous sperm motility, allows the spermatozoon to penetrate the ZP and fuse with the oocyte plasma membrane, the oolemma (Pasten et al. 2005; Ikawa et al. 2010).

The role of the UPS has been well documented in the reproductive processes of several mammalian species. For example, multi-enzymatic sperm protease complexes with similar sedimentation coefficients as purified proteasomes from other tissues have been isolated from mouse (Pasten et al. 2005), pig (Zimmerman and Sutovsky 2009), and human (Morales et al. 2003; Morales et al. 2004; Baker et al. 2007) spermatozoa. Accordingly, the substrate-specific enzymatic activities (trypsin-like, chymotrypsin-like, and peptidylglutamyl peptide-hydrolyzing activities) of these sperm proteasomes have been recorded in mice (Pasten et al. 2005; Bedard et al. 2011; Rivkin et al. 2009), pigs (Sutovsky et al. 2004; Yi et al. 2007a; Yi et al. 2009), and humans (Morales et al. 2003; Chakravarty et al. 2008; Kong et al. 2009).

The sperm-associated proteasomes are tethered to the acrosomal surface in mice (Pasten et al. 2005), pigs (Sutovsky et al. 2004; Yi et al. 2009), and humans (Morales et al. 2004). The proteasomes probably become associated with the spermatozoa inner and outer acrosomal membranes during acrosomal biogenesis at the spermatid stage (Rivkin et al. 2009). It appears that the extracellular proteasomes exposed on the acrosomal surface of mammalian spermatozoa are involved in the ZP-induced AE and are able to directly interact with the ZP during fertilization (Sutovsky et al. 2004; Yi et al. 2009). The proteasomes located on the cell surface could participate in sperm–ZP binding in some mammalian species, such as humans (Chakravarty et al. 2008; Naz and Dhandapani 2010) and mice (Pasten et al. 2005). Then, the intracellular and extracellular proteasomes would blend into the acrosomal shroud and participate during the AE. Sperm proteasomes have been shown to be essential for acrosomal function and sperm–zona penetration during fertilization in several mammalian species such as mouse, pig, and human (Pasten et al. 2005; Kong et al. 2009; Zimmerman et al. 2011). It is possible that the proteasomes modulating these cellular events could participate in other steps of fertilization as well.

In addition to proteasomes, other UPS components have been implicated in sper-matogenesis, epididymal sperm maturation, and fertilization. For example, ubiqui-tinated substrates have been detected in the epididymal fluid, the seminal plasma, on the surface of defective epididymal spermatozoa, and on the outer surface of the ZP (Sutovsky et al. 2001, 2004; Zimmerman et al. 2011). Anti-proteasome antibodies have been reported to be present in the seminal plasma of infertile men (Bohring et al. 2001; Bohring and Krause 2003). Addition of the deubiquitinating enzyme inhibitor, ubiquitin-aldehyde, to boar in vitro fertilization assays promotes poly-spermy, which might be related to an increase in ubiquitination and degradation of the ZP because this C-terminally modified ubiquitin species accelerates protea-somal proteolysis (Yi et al. 2007b). More recently, sperm deubiquitinating enzymes have been explored by the Wing laboratory in Canada. They reported that the sper-matozoa of deubiquitinating enzyme *Usp2* knockout mice are severely subfertile, which appeared to be caused by a defect in ZP binding or penetration, even though the USP2 protein in wild-type mice does not seem to be specifically localized in the acrosome. In elongating spermatids, USP2 was localized perinuclearly in a thin layer between the outer acrosomal membrane and the plasma membrane, but absent from the nucleus (Lin et al. 2000). The *Usp2* null spermatozoa are motile and undergo the AE but fail to penetrate the ZP (Bedard et al. 2011). These data suggest that the deubiquitination of sperm or spermatid proteins by USP2 is an important regulatory mechanism required for the acquisition of fertilizing ability by mamma-lian spermatozoa. It is not clear if the observed severe subfertility is a result of impaired deubiquitinating activity of the acrosomal USP2 enzyme during fertiliza-tion, or if it is rather a result of abnormal assembly of the sperm acrosome during spermiogenesis. Either way, the failure of motile *Usp2* null spermatozoa to pene-trate the zona supports the view that sperm motility alone is not sufficient to propel the sperm head through the zona matrix.

The role of UPS in sperm penetration through the ZP has been well documented. Proteasomal inhibitors and anti-proteasome antibodies effectively block mouse, pig, and human sperm ability to penetrate the ZP of their respective species (Morales et al. 2003; Sutovsky et al. 2004; Pasten et al. 2005). The coincubation of pig sper-matozoa with proteasomal inhibitors prevents sperm–ZP penetration without affect-ing sperm motility and binding (Sutovsky et al. 2003, 2004; Yi et al. 2007a; Zimmerman and Sutovsky 2009). However, if the zona is removed before fertiliza-tion, fertilization proceeds even with the addition of proteasomal inhibitors (Sutovsky et al. 2004), which suggests a role for the proteasome in sperm–ZP pen-etration. Chakravarty et al. reported that the human sperm-associated proteasomal activity is not stimulated by binding of recombinant ZP proteins, but that it remains steady after AE, most likely caused by the association of the proteasomes with the inner acrosomal membrane, which is not removed by AE (Chakravarty et al. 2008).

The ZP3/ZPC in mice and humans has been identified as the primary sperm-binding receptor on the ZP and perhaps the inducer of AE (Naz and Dhandapani 2010). The human sperm ubiquitin-associated protein UBAP2L was identified as a ZP-interacting protein by the Naz Lab (Naz and Dhandapani 2010) using a yeast two-hybrid screen (Y2H) and was further tested when UBAP2L antibodies were shown to inhibit sperm–ZP binding when tested in vitro. The Y2H procedure is used to identify proteins that interact with a target protein expressed in yeast as a hybrid with a DNA-binding domain and has been widely used to examine protein–protein interactions (Naz and Dhandapani 2010). Human sperm–oocyte recognition, binding, and AE have been proposed to be mediated by sugar residues (*O*- and *N*-linked) and peptide moieties of ZP3 (Chakravarty et al. 2005; Gupta et al. 2009). Results from the Naz lab suggest that ubiquitination, in addition to glycosylation of the ZP proteins, may regulate the sperm–ZP interactions (Naz and Dhandapani 2010).

In agreement with the foregoing study, Aitken and colleagues found that the proteasome is a component of a multimeric ZP-binding complex found in human spermatozoa (Redgrove et al. 2011). They propose that sperm–ZP binding requires the concerted action of several sperm proteins that form multimeric recognition complexes on the sperm surface, which is an alternate and novel view different from the traditional simple lock-and-key mechanism of one receptor, one ligand. The formation of these complexes on the sperm surface is purported to depend upon posttesticular maturation driven by the environmental changes to which the spermatozoa are exposed in the epididymis and within the female reproductive system. They report that human spermatozoa express a number of high molecular weight protein complexes on their surface and that subsets of these complexes display affinity for homologous ZP (Redgrove et al. 2011). Two of these complexes were revealed to be a chaperonin-containing TCP-1 (CCT), which harbors a putative ZP-binding protein, ZPBP2, and several components of the 20S proteasome that were found previously in an analysis of the human sperm proteome (Redgrove et al. 2011; Johnston et al. 2005). The role of these protein complexes in sperm–ZP interactions was further confirmed when antibodies against the individual components of the complex, including proteasomal subunits, inhibited sperm binding to zona-intact oocytes (Redgrove et al. 2011). Because many sperm proteins exhibit ZP-binding affinity, it is possible that complexes containing these ZP-binding proteins may participate in sequential or hierarchical molecular interactions or act synergistically to ensure successful sperm–ZP binding. It is important to note that the proteasome complex was shown as three large bands when examined by blue native polyacrylamide gel electrophoresis. Aitken et al. theorize that this is caused by posttranslational modifications of certain subunits of the complex (α1–7, β1, and β4) that displayed charge shift signatures characteristic of tyrosine phosphorylation (Redgrove et al. 2011). This finding is consistent with the proteasomal subunit phosphorylation reported in the acrosome that, according to Diaz et al., modulates the fertilizing capacity of human spermatozoa by inducing AE (Diaz et al. 2007), suggesting that proteasome complexes may be differentially activated during the individual steps of fertilization.

37.5   Is the Mammalian Egg Coat Ubiquitinated
Before Fertilization?

To confer with the hypothesis that the 26S proteasome acts as the putative egg coat lysin in mammalian fertilization, there must be an ubiquitinated sperm receptor on porcine ZP that is degraded by the sperm acrosome-borne proteasomes during porcine fertilization. These three prerequisites are required: (1) the sperm receptor on the mammalian egg coat is ubiquitinated; (2) proteasomes are present, exposed, and enzymatically active in the sperm acrosomal cap; and (3) sperm proteasomes degrade the sperm receptor on the egg coat during fertilization.

There is evidence that the sperm receptor complex ZPB-ZPC on the porcine ZP is ubiquitinated. In porcine oocytes, the ZPB-ZPC complex has been shown to be responsible for sperm–ZP binding (Yurewicz et al. 1998). Furthermore, the porcine ZPC homologues in mouse (ZP3) and human (ZP3) have been implicated in sperm binding and induction of acrosomal exocytosis (Shur et al. 2006; Gupta et al. 2009). Sutovsky et al. (2004) have shown the presence of ubiquitinated proteins in the unfertilized porcine egg coat. They reported that ubiquitin-immunoreactive proteins can be detected on the outer surface of porcine ZP, visualized ZP digestion with immunofluorescence microscopy with anti-ubiquitin antibodies, and recorded the presence of ubiquitinated proteins in ZP preparations from high-quality metaphase II ova. More recently, they have observed Gly-Gly modifications, a fingerprint of ubiquitinated internal Lys-residues, on all three components of porcine ZP (ZPA, ZPB, and ZPC) using Nanospray LC-MS/MS spectroscopy (Peng et al. 2003; Zimmerman et al. 2011). These findings corroborate the pattern of sequential ubiquitination and proteasomal degradation of the ascidian sperm-receptor protein HrVC70, an analogue of mammalian ZP. Similarly, Yokota and Sawada (2007) reported that the vitelline envelopes of sea urchin eggs are already ubiquitinated before fertilization (Yokota and Sawada 2007).

Enzymatically active proteasomes have been found in the boar sperm acrosome. There is evidence that sperm proteasomes are exposed onto the sperm surface and remain associated with acrosomal membranes during sperm–ZP penetration. Boar sperm proteasomes are associated with acrosomal membranes and matrix before AE and remain associated with the inner acrosomal membrane after AE (Morales et al. 2004; Zimmerman and Sutovsky 2009; Yi et al. 2009; Yi et al. 2010a, b). Proteasomal activity is present in motile boar spermatozoa (Yi et al. 2009). Adenosine triphosphate (ATP) is essential for the integrity of the 26S proteasome, which is maintained by the six 19S ATPase subunits. Depletion of sperm surface ATP by *Solanum tuberosum* apyrase inhibits sea urchin fertilization (Yokota and Sawada 2007) and porcine in vitro fertilization (IVF) (Yi et al. 2009). Enzymatic activity of boar sperm proteasomes was confirmed in whole, motile spermatozoa, sperm acrosomal fractions, and affinity purified proteasomes (Yi et al. 2009; Miles et al. 2013). These purified proteasomes were also tested for functionality via casein degradation (Manandhar and Sutovsky, unpublished data).

37.6 Sperm Proteasomes Can Degrade Zona Pellucida Proteins in Solution and In Situ

Studies in porcine and avian models suggest that sperm proteasomes can degrade the sperm receptor on porcine ZP and on the vitelline coat of Japanese quail egg, respectively (Zimmerman et al. 2011; Sasanami et al. 2012). It is difficult to capture the action of sperm acrosomal enzymes during fertilization, but the Sutovsky lab developed an in vitro assay using live, freshly collected, and capacitated boar spermatozoa coincubated with the ZP proteins solubilized from 100 meiotically mature, fertilization-competent porcine oocytes (Zimmerman et al. 2011). The soluble ZP proteins are thus enabled to bind to sperm acrosomal surface as the zona matrix would during fertilization and induce acrosomal exocytosis, resulting in the formation of acrosomal shrouds similar to those seen on the surface of ZP during IVF. Because there is no solid zona matrix in this assay, the acrosomal shrouds are easily separated from spermatozoa and interrogated for the degradation of the bound ZP-proteins. A distinct degradation pattern of the porcine sperm-receptor component ZPC was observed, similar to the degradation pattern of ZPC with purified sperm proteasomes (Zimmerman et al. 2011). The observed, rapid degradation of ZPC by spermatozoa and purified sperm proteasomes was inhibited by ATP depletion with *S. tuberosum* apyrase and with proteasomal inhibitors (MG132, CLBL, and Epoxomicin); it was accelerated by ubiquitin-aldehyde, a C-terminally modified ubiquitin protein that stimulates proteasomal proteolysis (Zimmerman et al. 2011). Furthermore, they were able to record that purified boar sperm proteasomes can digest intact pig ZP of in vitro maturing ova and reduce the rate of polyspermic fertilization after IVF (Zimmerman et al. 2011). These results were corroborated by a study in Japanese quail model that used a similar in vitro system to demonstrate that the sperm proteasome is important for avian fertilization and helps sperm penetration through the perivitelline membrane, which is homologous to the ZP in mammals. Japanese quail spermatozoa contain proteasomes localized in the acrosomal region and can degrade the ZP1 protein in a fashion similar to the degradation of ZP3/ZPC in the porcine model; this degradation is also inhibited by proteasomal inhibitor MG132 and by extracellular ATP depletion by apyrase (Sasanami et al. 2012).

37.7 Recent Advances in the Study of Sperm Ubiquitin Proteasome System

There is a growing acceptance of the involvement of the ubiquitin proteasome system in the reproductive process in plants, ascidians, echinoderms, and mammals, but many outstanding questions remain. How is sperm proteasomal activity regulated during sperm storage, capacitation, AE, and ZP penetration? How are proteasomes inserted in the acrosome during spermiogenesis? What is the role of deubiquitinating enzymes in fertilization? Can we target sperm proteasomes for a contraceptive effect? Novel animal models and assays will pave the way to elucidate these mechanisms.

A novel transgenic boar model with green fluorescent protein (GFP) fused to the C-terminus of 20S proteasomal core subunit alpha-type 6 (PSMA1) has been developed through a joint effort between the Sutovsky, Prather, and Wells laboratories at the University of Missouri–Columbia (Miles et al. 2013). Functional GFP-tagged proteasomes have been shown to be incorporated in not only fertilization-competent spermatozoa but in other tissues and cell types (Miles et al. 2013). Using cross-immunoprecipitation experiments, the authors identified various proteins interacting with the GFP-PSMA1 subunit such as lactadherin/MFGE8, spermadhesins, and disintegrins/ADAM metalloproteinases; these proteins may regulate sperm proteasomal activity or may be the substrates of proteasomal proteolysis during fertilization (Miles et al. 2013). They have also proposed a method of isolating enzymatically active GFP proteasomes through GFP affinity purification (Miles et al. 2013). This novel model will be useful for studies of fertilization and wherever UPS plays a role in cellular function or pathology.

A protection assay utilizing proteasomal inhibitors was used as an alternative method of identifying sperm–proteasome interacting proteins. The treatment of porcine spermatozoa with proteasomal inhibitors during coincubation with solubilized ZP proteins led to the accumulation of sperm acrosomal surface-associated proteins that would otherwise be degraded during AE (Zimmerman et al. 2011). The identified proteins that were protected from degradation by proteasomal inhibitors included sperm adhesion molecule 1 (SPAM1), lactadherin/MFGE8, zona pellucida-binding protein 2 (ZPBP2), and fragments of acrosin-binding protein ACRBP (SP32) (Zimmerman et al. 2011). Proteasomal degradation of these acrosomal zona-binding proteins could facilitate the ZP-induced acrosomal exocytosis or serve to terminate primary sperm–ZP binding as the fertilizing spermatozoon starts to move forward and penetrate deeper into the ZP. Furthermore, proteasomal inhibitors suppressed the induction of AE when human spermatozoa were incubated with recombinant human ZP3 and ZP4 (Chakravarty et al. 2008). Current research also implicates the ubiquitin proteasome system in sperm capacitation and AE, and that these sperm transformations may be regulated and reversed by deubiquitinating enzymes that also regulate oocyte anti-polyspermy defense and oocyte maturation. Inhibition of the ubiquitin-activating enzyme UBA1 (E1) during boar sperm capacitation alters proteasomal subunit properties and sperm-fertilizing ability in vitro in a dose-dependent manner (Yi et al. 2012). Ubiquitin-conjugating enzyme, the ubiquitin-ligase UBR7, has been detected in the boar sperm acrosome, and de novo ubiquitination of UBB+1 has been achieved using sperm acrosomal extract (S.W. Zimmerman and P. Sutovsky, unpublished data). Deubiquitinating enzymes from the ubiquitin C-terminal hydrolase (UCHL) family have been shown to regulate anti-polyspermy defense and oocyte cortex and meiotic spindle formation. Block of sperm UCHL3 increases porcine polyspermy in vitro while supplementation of recombinant UCHs to IVF media reduced polyspermy and increased the rate of monospermic fertilization (Yi et al. 2007b). Interference with bovine oocyte

UCHL1 alters cortical granule maturation and causes polyspermy (Susor et al. 2010). Block of murine oocyte UCHL1 and UCHL3 prevents sperm incorporation in the ooplasm and causes meiotic spindle anomalies and polar body extrusion defects (Mtango et al. 2012a, b). Subfertility, in vitro polyspermy, and failed morula compaction have been reported in the gad mutant mouse expressing an inactive form of UCHL1 (Kwon et al. 2003; Mtango et al. 2012a). These data suggest that ubiquitination plays a key role in sperm function and that the deubiquitinating enzymes (UCHL1 and UCHL3) are important during fertilization and pre-implantation embryo development.

37.8 Conclusions

Accumulating evidence suggests that the putative mammalian egg coat lysin is the 26S proteasome. There are data supporting all the prerequisites in the porcine fertilization model: (1) all three pig zona components (ZPA, ZPB, ZPC) are ubiquitinated, (2) enzymatically active proteasomes are present in the boar sperm acrosome and exposed on the acrosomal surface, and (3) motile boar spermatozoa and isolated boar sperm proteasomes are capable of degrading the sperm receptor components of the zona pellucida in vitro.

Recent advances have elucidated more ways in which the UPS affects mammalian reproduction that could possibly be manipulated to develop novel forms of nonhormonal contraceptives. Sperm proteasomes have been shown to copurify with various acrosomal sperm membrane-binding proteins. Sperm capacitation and acrosome reaction have been reported to be altered by interference with UPS enzymes responsible for ubiquitin–substrate conjugation, deubiquitination, and proteasomal proteolysis. Acrosomal extracts can ubiquitinate exogenous substrates. Therefore, the properties of sperm-borne proteasomes are consistent with their proposed role of mammalian zona lysin. Evidence from several laboratories, animal models, and human gamete studies supports the participation of sperm proteasomes in multiple aspects of the fertilization process, including sperm capacitation, sperm zona binding, acrosomal exocytosis, and sperm–zona/vitelline coat penetration.

Acknowledgments We thank Ms. Kathryn Craighead for editorial and clerical assistance. Our gratitude belongs to our colleagues/associates Dr. Young-Joo Yi, Ms. Miriam Sutovsky, Ms. Wonhee Song, Dr. Peter Vargovic, Dr. Gaurishankar Manandhar, and Dr. Shawn Zimmerman, as well as to our collaborators Drs. Randal Prather, Kevin Wells, Richard Oko, Satish Gupta, and David Miller. Some of the work reviewed in this manuscript was supported by Agriculture and Food Research Initiative Competitive Grants no. 2011-67015-20025 and 2013-67015-20961 from the USDA National Institute of Food and Agriculture, and by the seed funding from the Food for the Twenty-first Century Program of the University of Missouri.

References

Austin CR, Bishop MW (1958) Role of the rodent acrosome and perforatorium in fertilization. Proc R Soc Lond B Biol Sci 149(935):241–248

Baba T, Azuma S et al (1994) Sperm from mice carrying a targeted mutation of the acrosin gene can penetrate the oocyte zona pellucida and effect fertilization. J Biol Chem 269(50): 31845–31849

Baker MA, Reeves G et al (2007) Identification of gene products present in Triton X-100 soluble and insoluble fractions of human spermatozoa lysates using LC-MS/MS analysis. Proteomics Clin Appl 1(5):524–532

Bedard N, Yang Y et al (2011) Mice lacking the USP2 deubiquitinating enzyme have severe male subfertility associated with defects in fertilization and sperm motility. Biol Reprod 85(3):594–604

Bedford JM (1998) Mammalian fertilization misread? sperm penetration of the eutherian zona pellucida is unlikely to be a lytic event. Biol Reprod 59(6):1275–1287

Belote JM, Zhong L (2009) Duplicated proteasome subunit genes in *Drosophila* and their roles in spermatogenesis. Heredity (Edinb) 103(1):23–31

Bleil JD, Wassarman PM (1980) Mammalian sperm–egg interaction: identification of a glycoprotein in mouse egg zonae pellucidae possessing receptor activity for sperm. Cell 20(3): 873–882

Boavida LC, Vieira AM et al (2005) Gametophyte interaction and sexual reproduction: how plants make a zygote. Int J Dev Biol 49(5-6):615–632

Bohring C, Krause W (2003) Characterization of spermatozoa surface antigens by antisperm antibodies and its influence on acrosomal exocytosis. Am J Reprod Immunol 50(5):411–419

Bohring C, Krause E et al (2001) Isolation and identification of sperm membrane antigens recognized by antisperm antibodies, and their possible role in immunological infertility disease. Mol Hum Reprod 7(2):113–118

Book AJ, Smalle J et al (2009) The RPN5 subunit of the 26s proteasome is essential for gametogenesis, sporophyte development, and complex assembly in *Arabidopsis*. Plant Cell 21(2): 460–478

Byrne K, Leahy T et al (2012) Comprehensive mapping of the bull sperm surface proteome. Proteomics 12(23-24):3559–3579

Callis J, Bedinger P (1994) Developmentally regulated loss of ubiquitin and ubiquitinated proteins during pollen maturation in maize. Proc Natl Acad Sci USA 91(13):6074–6077

Chakravarty S, Suraj K et al (2005) Baculovirus-expressed recombinant human zona pellucida glycoprotein-B induces acrosomal exocytosis in capacitated spermatozoa in addition to zona pellucida glycoprotein-C. Mol Hum Reprod 11(5):365–372

Chakravarty S, Bansal P et al (2008) Role of proteasomal activity in the induction of acrosomal exocytosis in human spermatozoa. Reprod Biomed Online 16(3):391–400

Chen C, Huang C et al (2008) Subunit-subunit interactions in the human 26S proteasome. Proteomics 8(3):508–520

Deveraux Q, van Nocker S et al (1995) Inhibition of ubiquitin-mediated proteolysis by the *Arabidopsis* 26 S protease subunit S5a. J Biol Chem 270(50):29660–29663

Devoto A, Nieto-Rostro M et al (2002) COI1 links jasmonate signalling and fertility to the SCF ubiquitin-ligase complex in *Arabidopsis*. Plant J 32(4):457–466

Diaz ES, Kong M et al (2007) Effect of fibronectin on proteasome activity, acrosome reaction, tyrosine phosphorylation and intracellular calcium concentrations of human sperm. Hum Reprod 22(5):1420–1430

Doelling JH, Phillips AR et al (2007) The ubiquitin-specific protease subfamily UBP3/UBP4 is essential for pollen development and transmission in *Arabidopsis*. Plant Physiol 145(3): 801–813

Etlinger JD, Goldberg AL (1977) A soluble ATP-dependent proteolytic system responsible for the degradation of abnormal proteins in reticulocytes. Proc Natl Acad Sci USA 74(1):54–58

Glickman MH, Ciechanover A (2002) The ubiquitin-proteasome proteolytic pathway: destruction for the sake of construction. Physiol Rev 82(2):373–428

Goldstein G, Scheid M et al (1975) Isolation of a polypeptide that has lymphocyte-differentiating properties and is probably represented universally in living cells. Proc Natl Acad Sci USA 72(1):11–15

Green DP, Purves RD (1984) Mechanical hypothesis of sperm penetration. Biophys J 45(4): 659–662

Gupta SK, Bansal P et al (2009) Human zona pellucida glycoproteins: functional relevance during fertilization. J Reprod Immunol 83(1-2):50–55

Hanna J, Finley D (2007) A proteasome for all occasions. FEBS Lett 581(15):2854–2861

Hershko A (2005) The ubiquitin system for protein degradation and some of its roles in the control of the cell division cycle. Cell Death Differ 12(9):1191–1197

Hershko A, Ciechanover A (1998) The ubiquitin system. Annu Rev Biochem 67:425–479

Honda A, Siruntawineti J et al (2002) Role of acrosomal matrix proteases in sperm-zona pellucida interactions. Hum Reprod Update 8(5):405–412

Ikawa M, Inoue N et al (2010) Fertilization: a sperm's journey to and interaction with the oocyte. J Clin Invest 120(4):984–994

Iwai K (2012) Synthesis and analysis of linear ubiquitin chains. Methods Mol Biol 832:229–238

Johnston DS, Wooters J et al (2005) Analysis of the human sperm proteome. Ann N Y Acad Sci 1061:190–202

Kawano N, Kang W et al (2010) Mice lacking two sperm serine proteases, ACR and PRSS21, are subfertile, but the mutant sperm are infertile in vitro. Biol Reprod 83(3):359–369

Kierszenbaum AL, Rivkin E et al (2011) Cytoskeletal track selection during cargo transport in spermatids is relevant to male fertility. Spermatogenesis 1(3):221–230

Kim ST, Zhang K et al (2006) Exogenous free ubiquitin enhances lily pollen tube adhesion to an in vitro stylar matrix and may facilitate endocytosis of SCA. Plant Physiol 142(4):1397–1411

Kong M, Diaz ES et al (2009) Participation of the human sperm proteasome in the capacitation process and its regulation by protein kinase A and tyrosine kinase. Biol Reprod 80(5): 1026–1035

Kwon J, Kikuchi T et al (2003) Characterization of the testis in congenitally ubiquitin carboxy-terminal hydrolase-1 (Uch-L1) defective (gad) mice. Exp Anim 52(1):1–9

Lin H, Keriel A et al (2000) Divergent N-terminal sequences target an inducible testis deubiquitinating enzyme to distinct subcellular structures. Mol Cell Biol 20(17):6568–6578

Liu CW, Li X et al (2006) ATP binding and ATP hydrolysis play distinct roles in the function of 26S proteasome. Mol Cell 24(1):39–50

Matsumura K, Aketa K (1991) Proteasome (multicatalytic proteinase) of sea urchin sperm and its possible participation in the acrosome reaction. Mol Reprod Dev 29(2):189–199

Miles EL, O'Gorman C, Zhao J, Samuel M, Walters E, Yi YJ, Sutovsky M, Prather RS, Wells K, Sutovsky P (2013) Transgenic pig carrying green fluorescent proteasomes. Proc Natl Acad Sci USA 110(16):6334–6339

Morales P, Kong M et al (2003) Participation of the sperm proteasome in human fertilization. Hum Reprod 18(5):1010–1017

Morales P, Pizarro E et al (2004) Extracellular localization of proteasomes in human sperm. Mol Reprod Dev 68(1):115–124

Mtango NR, Sutovsky M et al (2012a) Essential role of maternal UCHL1 and UCHL3 in fertilization and preimplantation embryo development. J Cell Physiol 227(4):1592–1603

Mtango NR, Sutovsky M et al (2012b) Essential role of ubiquitin C-terminal hydrolases UCHL1 and UCHL3 in mammalian oocyte maturation. J Cell Physiol 227(5):2022–2029

Naz RK, Dhandapani L (2010) Identification of human sperm proteins that interact with human zona pellucida3 (ZP3) using yeast two-hybrid system. J Reprod Immunol 84(1):24–31

Pasten C, Morales P et al (2005) Role of the sperm proteasome during fertilization and gamete interaction in the mouse. Mol Reprod Dev 71(2):209–219

Peng J, Schwartz D et al (2003) A proteomics approach to understanding protein ubiquitination. Nat Biotechnol 21(8):921–926

Pines J (1994) Cell cycle. Ubiquitin with everything. Nature 371(6500):742–743

Qiao H, Wang H et al (2004) The F-box protein AhSLF-S2 physically interacts with S-RNases that may be inhibited by the ubiquitin/26S proteasome pathway of protein degradation during compatible pollination in *Antirrhinum*. Plant Cell 16(3):582–595

Redgrove KA, Anderson AL et al (2011) Involvement of multimeric protein complexes in mediating the capacitation-dependent binding of human spermatozoa to homologous zonae pellucidae. Dev Biol 356(2):460–474

Rivkin E, Kierszenbaum AL et al (2009) Rnf19a, a ubiquitin protein ligase, and Psmc3, a component of the 26S proteasome, tether to the acrosome membranes and the head–tail coupling apparatus during rat spermatid development. Dev Dyn 238(7):1851–1861

Rosenzweig R, Osmulski PA et al (2008) The central unit within the 19S regulatory particle of the proteasome. Nat Struct Mol Biol 15(6):573–580

Sakai N, Sawada H et al (2003) Extracellular ubiquitin system implicated in fertilization of the ascidian, *Halocynthia roretzi*: isolation and characterization. Dev Biol 264(1):299–307

Sakai N, Sawada MT et al (2004) Non-traditional roles of ubiquitin-proteasome system in fertilization and gametogenesis. Int J Biochem Cell Biol 36(5):776–784

Sasanami T, Sugiura K et al (2012) Sperm proteasome degrades egg envelope glycoprotein ZP1 during fertilization of Japanese quail (*Coturnix japonica*). Reproduction 144(4):423–431

Sawada H, Sakai N et al (2002a) Extracellular ubiquitination and proteasome-mediated degradation of the ascidian sperm receptor. Proc Natl Acad Sci USA 99(3):1223–1228

Sawada H, Takahashi Y et al (2002b) Localization and roles in fertilization of sperm proteasomes in the ascidian *Halocynthia roretzi*. Mol Reprod Dev 62(2):271–276

Shur BD, Rodeheffer C et al (2006) Identification of novel gamete receptors that mediate sperm adhesion to the egg coat. Mol Cell Endocrinol 250(1-2):137–148

Speranza A, Scoccianti V et al (2001) Inhibition of proteasome activity strongly affects kiwifruit pollen germination. Involvement of the ubiquitin/proteasome pathway as a major regulator. Plant Physiol 126(3):1150–1161

Su YH, Chen SH et al (2005) Tandem mass spectrometry identifies proteins phosphorylated by cyclic AMP-dependent protein kinase when sea urchin sperm undergo the acrosome reaction. Dev Biol 285(1):116–125

Susor A, Liskova L et al (2010) Role of ubiquitin C-terminal hydrolase-L1 in antipolyspermy defense of mammalian oocytes. Biol Reprod 82(6):1151–1161

Sutovsky P (2011) Sperm proteasome and fertilization. Soc Reprod Fertil 142:1–14

Sutovsky P, Moreno R et al (2001) A putative, ubiquitin-dependent mechanism for the recognition and elimination of defective spermatozoa in the mammalian epididymis. J Cell Sci 114(pt 9): 1665–1675

Sutovsky P, McCauley TC et al (2003) Early degradation of paternal mitochondria in domestic pig (*Sus scrofa*) is prevented by selective proteasomal inhibitors lactacystin and MG132. Biol Reprod 68(5):1793–1800

Sutovsky P, Manandhar G et al (2004) Proteasomal interference prevents zona pellucida penetration and fertilization in mammals. Biol Reprod 71(5):1625–1637

Tanaka K (2009) The proteasome: overview of structure and functions. Proc Jpn Acad Ser B Phys Biol Sci 85(1):12–36

Verma R, Aravind L et al (2002) Role of Rpn11 metalloprotease in deubiquitination and degradation by the 26S proteasome. Science 298(5593):611–615

Voges D, Zwickl P et al (1999) The 26S proteasome: a molecular machine designed for controlled proteolysis. Annu Rev Biochem 68:1015–1068

Walczak H, Iwai K et al (2012) Generation and physiological roles of linear ubiquitin chains. BMC Biol 10:23

Yamashita M, Honda A et al (2008) Reduced fertility of mouse epididymal sperm lacking Prss21/ Tesp5 is rescued by sperm exposure to uterine microenvironment. Genes Cells 13(10): 1001–1013

Yanagimachi R (1994) Fertility of mammalian spermatozoa: its development and relativity. Zygote 2(4):371–372

Yi YJ, Manandhar G et al (2007a) Mechanism of sperm-zona pellucida penetration during mammalian fertilization: 26S proteasome as a candidate egg coat lysin. Soc Reprod Fertil Suppl 63:385–408

Yi YJ, Manandhar G et al (2007b) Ubiquitin C-terminal hydrolase-activity is involved in sperm acrosomal function and anti-polyspermy defense during porcine fertilization. Biol Reprod 77(5):780–793

Yi YJ, Park CS et al (2009) Sperm-surface ATP in boar spermatozoa is required for fertilization: relevance to sperm proteasomal function. Syst Biol Reprod Med 55(2):85–96

Yi YJ, Manandhar G et al (2010a) Inhibition of 19S proteasomal regulatory complex subunit PSMD8 increases polyspermy during porcine fertilization in vitro. J Reprod Immunol 84(2): 154–163

Yi YJ, Manandhar G et al (2010b) Interference with the 19S proteasomal regulatory complex subunit PSMD4 on the sperm surface inhibits sperm-zona pellucida penetration during porcine fertilization. Cell Tissue Res 341(2):325–340

Yi YJ, Zimmerman SW et al (2012) Ubiquitin-activating enzyme (UBA1) is required for sperm capacitation, acrosomal exocytosis and sperm-egg coat penetration during porcine fertilization. Int J Androl 35(2):196–210

Yokota N, Sawada H (2007) Sperm proteasomes are responsible for the acrosome reaction and sperm penetration of the vitelline envelope during fertilization of the sea urchin *Pseudocentrotus depressus*. Dev Biol 308(1):222–231

Yokota N, Kataoka Y et al (2011) Sperm-specific C-terminal processing of the proteasome PSMA1/alpha6 subunit. Biochem Biophys Res Commun 410(4):809–815

Yurewicz EC, Sacco AG et al (1998) Hetero-oligomerization-dependent binding of pig oocyte zona pellucida glycoproteins ZPB and ZPC to boar sperm membrane vesicles. J Biol Chem 273(13):7488–7494

Zhang F, Hu Y et al (2007) Proteasome function is regulated by cyclic AMP-dependent protein kinase through phosphorylation of Rpt6. J Biol Chem 282(31):22460–22471

Zimmerman S, Sutovsky P (2009) The sperm proteasome during sperm capacitation and fertilization. J Reprod Immunol 83(1-2):19–25

Zimmerman SW, Manandhar G et al (2011) Sperm proteasomes degrade sperm receptor on the egg zona pellucida during mammalian fertilization. PLoS One 6(2):e17256

Chapter 38
Germline Transformation in the Ascidian *Ciona intestinalis*

Yasunori Sasakura

Abstract Genetic technologies are necessary for understanding the molecular mechanisms of wide-ranging biological phenomena. However, genetic approaches are limited in the so-called model organisms. For example, genetics are not available in most marine invertebrates. *Ciona intestinalis* is a marine invertebrate chordate that provides excellent systems for studying allorecognition. We have established germline transformation with transposable elements, enhancer detection, and mutagenesis with transposons and customized nucleases. These genetic technologies are invaluable tools for uncovering the genetic functions underlying fertilization and allorecognition in this ascidian. In this chapter, achievements in the genetics of *C. intestinalis* are discussed.

Keywords Engineered nuclease • Fluorescent protein • Mutagenesis • Transgenic line • Transposon

38.1 Introduction

Ascidians, or sea squirts, are sessile invertebrates that live in the ocean (Fig. 38.1) (Satoh 1994). The larvae of ascidians are typical free-swimming tadpoles (Fig. 38.1a). As the larval tadpole body suggests, ascidians are closely related to vertebrates (Satoh 2003). Ascidians are members of the subphylum Urochordata, and belong to the phylum Chordata with cephalochordates and vertebrates. There are some basic body features that characterize chordates in addition to the tadpole body: a central nervous system consisting of a dorsally located neural tube, possession of a notochord at

Y. Sasakura (✉)
Shimoda Marine Research Center, University of Tsukuba,
Shimoda, Shizuoka 415-0025, Japan
e-mail: sasakura@kurofune.shimoda.tsukuba.ac.jp

H. Sawada et al. (eds.), *Sexual Reproduction in Animals and Plants*, 465
DOI 10.1007/978-4-431-54589-7_38, © The Author(s) 2014

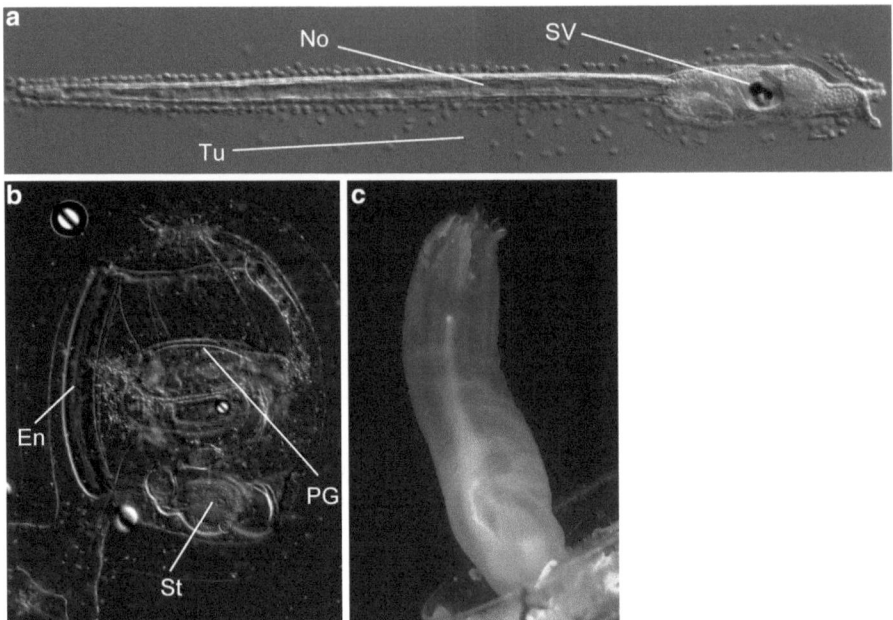

Fig. 38.1 The ascidian *Ciona intestinalis*. (**a**) A larva. *No*, notochord; *SV*, sensory vesicle (brain); *Tu*, tunic. (**b**) A juvenile. Larvae convert their tadpole body into sessile juveniles through metamorphosis. *En*, endostyle; *PG*, pharyngeal gill; *St*, stomach. (**c**) An adult. Body size is approximately 10 cm

certain developmental stages, gill slits at the pharynx, and an endostyle/thyroid gland. Among chordates, urochordates are thought to be the most closely related to vertebrates (Delsuc et al. 2006).

Reflecting their phylogenetic position, ascidians have a gene set that specifies the chordate body with less redundancy than vertebrates (Dehal et al. 2002). Vertebrates are thought to have experienced genome duplication twice during evolution (Holland et al. 1994) and, as a consequence, they usually have several genes that are similar with respect to sequence, expression, and functions. These similar genes sometimes compensate functions for each other. Because of this redundancy of genes, disrupting a gene does not always cause strong phenotypes, and multiple paralogous genes must be analyzed to determine their functions. Genome duplication is not thought to have occurred in the ascidian lineage, and ascidians usually have one orthologous gene to their vertebrate counterparts (Sasakura et al. 2003a, b). Therefore, it is necessary to study one orthologous gene to determine such functions in ascidians. A knowledge of ascidian genes is useful in deducing functions of vertebrate orthologues that may be hidden by their redundancy. Ascidians provide an excellent experimental system because of their simple genome.

The cosmopolitan ascidian *Ciona intestinalis* holds a representative position among ascidians. The genome sequence of *C. intestinalis* was determined in 2002 (Dehal et al. 2002), and the genome has been well assembled and annotated

(Satou et al. 2008). Accompanying the genome sequence, abundant expressed sequence tag (EST) and cDNA information and expression profiles of many developmentally relevant genes are available (Satou et al. 2002; Imai et al. 2006). Omic analyses of *C. intestinalis* have been conducted with the aid of the genome sequence, which provides gene and protein expression profiles of the germ cells (Hozumi et al. 2004; Yamada et al. 2009). Basic experimental systems analyzing genetic functions have been established in this ascidian. Forced expression of exogenous genes can be carried out by electroporation of plasmid DNAs into hundreds of embryos at once (Corbo et al. 1997). Additionally, gene functions have been disrupted by microinjection of antisense morpholino oligonucleotides (MOs) (Satou et al. 2001). The life cycle of *C. intestinalis* is about 2–3 months, and inland culturing systems have been developed (Joly et al. 2007). The short generation time facilitates genetic studies that are necessary for studying gene functions at later developmental stages, as discussed here (Nakatani et al. 1999; Sordino et al. 2000; Harada et al. 2008). With these characteristics, *C. intestinalis* is extensively studied as a splendid experimental model of ascidians.

38.2 Transposon-Mediated Germline Transgenesis

In *C. intestinalis*, methods of germline transformation with Tc1/*mariner* superfamily transposable elements have been well established (Sasakura et al. 2003c; Hozumi et al. 2013). Two transposons, *Minos* and *Sleeping Beauty*, have been used for germline transformation of *C. intestinalis* (Franz and Savakis 1991; Ivics et al. 1997). Transposon vectors can be modified to contain the DNA elements that a researcher wishes to introduce into the *C. intestinalis* genome without losing transposon activity. The transposon vectors are introduced with their transposase mRNA into *C. intestinalis* embryos through microinjection or electroporation (Matsuoka et al. 2005). In experimental *Ciona*, transposon vectors are inserted into the genome with the aid of transposase, and the transposon insertions are inherited stably by subsequent generations (Fig. 38.2). Approximately 30 % of transposon-introduced *Ciona* become founders (Sasakura et al. 2007). The transgenic lines are useful because they express exogenous genes in a non-mosaic fashion. The expression of exogenous genes can be controlled by selecting appropriate *cis* elements for the purposes of experiments. For example, the *cis* element of a gene expressing in a tissue-specific manner can be used to express the exogenous gene specifically in the tissue (Fig. 38.2).

Many kinds of exogenous genes can be utilized for molecular and cellular studies in *C. intestinalis*. Genes encoding fluorescent proteins are good markers for labeling cells that are alive. These fluorescent proteins can be fused to signal peptides to specifically label the organelle. Various researchers have developed multiple indicators that monitor cellular features such as calcium, voltage, apoptosis, and cell cycle (Miyawaki et al. 1997; Nakai et al. 2001; Takemoto et al. 2003; Tsutsui et al. 2008; Sakaue-Sawano et al. 2008). Transgenic lines are valuable resources for

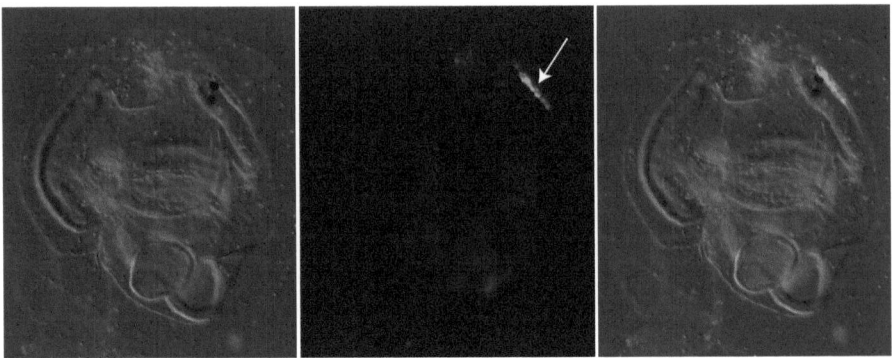

Fig. 38.2 Transgenic line of *Ciona intestinalis*. A juvenile of the transgenic line expressing green fluorescent protein (GFP) in the central nervous system (*arrow* in *middle image*). *Left*, a differential contrast image; *middle*, a fluorescence image; *right*, a merged image

analyzing the cellular features of *C. intestinalis*. *C. intestinalis* transgenic lines are databased, and the information is available on the *C. intestinalis* Transgenic line RESources (CITRES) website (http://marinebio.nbrp.jp/ciona/). Transgenic lines are provided to researchers upon request (Sasakura et al. 2009).

38.3   Enhancer Detection

Enhancers are DNA elements that control the spatial and temporal expression patterns of genes. Enhancers are usually flexible in terms of their location and orientation with respect to the genes they regulate. When a transposon vector with a reporter gene is inserted near enhancer elements, the expression pattern of the reporter gene is altered by the enhancers (O'Kane and Gehring 1987). This phenomenon is called "enhancer detection" or "enhancer trap."

In *C. intestinalis*, both *Minos* and *Sleeping Beauty* cause enhancer detection (Awazu et al. 2004; Hozumi et al. 2013). Large-scale enhancer detection has been conducted with *Minos* (Sasakura et al. 2008; Hozumi et al. 2010). Through enhancer detection, many marker transgenic lines that express fluorescent proteins in a tissue- or organ-specific manner have been created. Because the *C. intestinalis* genome is compact and on average one gene per 10-kilobase pair is located, enhancers are also densely present in the genome; this results in high frequency enhancer detection in this organism. With the aid of fluorescent markers of the enhancer detection lines, a detailed characterization of tissues and organs has been achieved in *Ciona* (Ohta et al. 2010). For example, the collection of enhancer detection lines that emit green fluorescent protein (GFP) in the digestive organs revealed a novel subdivision of the digestive tract (Yoshida and Sasakura 2012).

38.4   Insertional Mutagenesis

Transposons have the chance to insert themselves into the genomic regions and disrupt the functions of genes. For this reason, transposons can be utilized as a tool for mutagenesis. In *C. intestinalis*, *Minos*-mediated insertional mutants have been reported. In a mutant named *swimming juvenile*, the gene encoding cellulose synthase is mutated (Sasakura et al. 2005). Ascidians have in common the characteristic that they produce cellulose. Ascidian cellulose is found in the mantle layer, named the tunic (Fig. 38.1a), which surrounds the body to protect it from predators. In larvaceans, cellulose is found in the 'house,' which acts as a filter for collecting small particles in the seawater to feed (Sagane et al. 2010). The *swimming juvenile* mutants of *C. intestinalis* have a malformed and soft tunic compared to that of wild types, suggesting that cellulose gives tunics the physical strength to withstand attack from predators. In addition to this defect, *swimming juvenile* mutants exhibit abnormality in their process of metamorphosis. The wild-type larvae start metamorphosis after they adhere to substrates. The *swimming juvenile* larvae start some metamorphic events at the trunk region without adhesion to substrates, suggesting that cellulose or cellulose synthase is responsible for the pathway triggering certain metamorphic events (Nakayama-Ishimura et al. 2009).

Reflecting the compact genome of *C. intestinalis*, enhancer detection lines of this organism frequently have transposon insertions inside or close to genes (Hozumi et al. 2013), which suggests that enhancer detection lines are a good resource of insertional mutants. Indeed, an insertional mutant line of *Ci-Hox1*, which encodes a transcription factor widely conserved among metazoans, has been isolated (Sasakura et al. 2012). Detailed analyses of the *Ci-Hox1* mutant have revealed that this gene is essential for forming the atrial siphon primordia, structures in the epidermis thought to be homologous to the otic placodes in vertebrates (Manni et al. 2004; Kourakis et al. 2010).

To the best of my knowledge, no mutant has yet been reported in *C. intestinalis* that shows defects during fertilization. Future studies will screen mutants focusing on fertilization to isolate mutants in this step, and characterization of the mutants will reveal novel molecular and genetic mechanisms responsible for allorecognition in *C. intestinalis*.

38.5   Engineered Nuclease-Mediated Genome Editing

In *C. intestinalis*, one major approach for disrupting the functions of genes-of-interest is to introduce antisense MOs into embryos (Satou et al. 2001). The MOs that are designed to have a sequence complementary to that of the target genes can be bound to the RNAs of the target genes to disrupt their splicing or translation. A technical limitation of this method is that MOs are lost during development and therefore their knockdown effects do not persist to the later stages. For this reason, disruption of genes in the genome through genetic modifications, namely knockout, is desirable to observe functions of genes in the germ cells because these cells are produced at later (adult) stages.

Attempts at such a reverse-genetic approach have been made in *C. intestinalis* with engineered nucleases (Bibikova et al. 2003; Kawai et al. 2012). The engineered nucleases are fusions of the nuclease domain of the restriction enzyme *FokI* and zinc-finger DNA-binding motifs (ZFNs). We can change the target sequences of ZFNs by customizing the zinc-finger motifs through substitutions of amino-acid sequences. When the ZFNs are bound to the target DNA sequences, the *FokI* domain introduces double-strand breaks to the target site. The double-strand breaks are repaired through two mechanisms: homologous recombination and nonhomologous end-joining. During the latter repair system, insertions and/or deletions are introduced to the target sequence, which results in the mutation of the target genes.

In *C. intestinalis*, the mutation frequency of ZFNs has been estimated with the ZFNs targeting enhanced GFP genes (EGFP) of the transgenic lines (Kawai et al. 2012). When 10 ng/µl EGFP-ZFN mRNA solution was introduced into embryos of an EGFP-transgenic line created by *Minos*, almost all the embryos became negative in their EGFP fluorescence. Genomic analyses of the EGFP-ZFN-introduced embryos indicated that the mutation frequency of EGFP is nearly 100 %. Mutations were observed in germ cells as well as in somatic cells, and the mutations can be inherited by subsequent generations. Therefore, mutagenesis with ZFNs is feasible in *C. intestinalis*.

Future studies will achieve knockout of endogenous genes with engineered nucleases to characterize their functions. One difficulty with ZFNs is their complicated way of construction. Recently, another DNA-binding motif derived from the transcription activator-like (TAL) effector of a plant pathogen has been used for engineered nucleases named TALENs. The DNA-binding motifs of TAL effectors can be customized more easily than those of zinc-fingers (Christian et al. 2010; Cermak et al. 2011). Therefore, knockout of genes with TALENs will be a better choice for the easy disruption of genes-of-interest.

38.6   Conclusions

As stated here, *C. intestinalis* provides a splendid experimental system for analyzing genetic functions in the chordate body. Genetic and transgenic technologies are helpful for studying fertilization and allorecognition mechanisms in this organism, and provide a superb opportunity to obtain considerable novel knowledge in future studies.

Acknowledgments The author thanks the members of the Shimoda Marine Research Center at the University of Tsukuba for their kind cooperation during our study. I am also grateful to the National Bioresource Project, MEXT, Dr. Nobuo Yamaguchi, Dr. Kunifumi Tagawa, and Dr. Shigeki Fujiwara and his colleagues for providing me with *Ciona* adults.

References

Awazu S, Sasaki A, Matsuoka T, Satoh N, Sasakura Y (2004) An enhancer trap in the ascidian *Ciona intestinalis* identifies enhancers of its *Musashi* orthologous gene. Dev Biol 275: 459–472

Bibikova M, Beumer K, Trautman JK, Carroll D (2003) Enhancing gene targeting with designed zinc finger nucleases. Science 300(5620):764

Cermak T, Doyle EL, Christian M, Wang L, Zhang Y, Schmidt C, Baller JA, Somia NV, Bogdanove AJ, Voytas DF (2011) Efficient design and assembly of custom TALEN and other TAL effector-based constructs for DNA targeting. Nucleic Acids Res 39(12):e82

Christian M, Cermak T, Doyle EL, Schmidt C, Zhang F, Hummel A, Bogdanove A, Voytas DF (2010) Targeting DNA double-strand breaks with TAL effector nucleases. Genetics 186: 757–761

Corbo JC, Levine M, Zeller RW (1997) Characterization of a notochord-specific enhancer from the *Brachyury* promoter region of the ascidian, *Ciona intestinalis*. Development (Camb) 124: 589–602

Dehal P, Satou Y, Campbell RK, Chapman J, Degnan B, De Tomaso A, Davidson B, Di Gregorio A, Gelpke M, Goodstein DM, Harafuji N, Hastings KE, Ho I, Hotta K, Huang W, Kawashima T, Lemaire P, Martinez D, Meinertzhagen IA, Necula S, Nonaka M, Putnam N, Rash S, Saiga H, Satake M, Terry A, Yamada L, Wang HG, Awazu S, Azumi K, Boore J, Branno M, Chin-Bow S, DeSantis R, Doyle S, Francino P, Keys DN, Haga S, Hayashi H, Hino K, Imai KS, Inaba K, Kano S, Kobayashi K, Kobayashi M, Lee BI, Makabe KW, Manohar C, Matassi G, Medina M, Mochizuki Y, Mount S, Morishita T, Miura S, Nakayama A, Nishizaka S, Nomoto H, Ohta F, Oishi K, Rigoutsos I, Sano M, Sasaki A, Sasakura Y, Shoguchi E, Shin-I T, Spagnuolo A, Stainier D, Suzuki MM, Tassy O, Takatori N, Tokuoka M, Yagi K, Yoshizaki F, Wada S, Zhang C, Hyatt PD, Larimer F, Detter C, Doggett N, Glavina T, Hawkins T, Richardson P, Lucas S, Kohara Y, Levine M, Satoh N, Rokhsar DS (2002) The draft genome of *Ciona intestinalis*: insight into chordate and vertebrate origins. Science 298(5601):2157–2167

Delsuc F, Brinkmann H, Chourrout D, Philippe H (2006) Tunicates and not cephalochordates are the closest living relatives of vertebrates. Nature (Lond) 439(7079):965–968

Franz G, Savakis CC (1991) *Minos*, a new transposable element from *Drosophila hydei*, is a member of the *Tc1*-like family of transposons. Nucleic Acids Res 19:6646

Harada Y, Takagaki Y, Sunagawa M, Saito T, Yamada L, Taniguchi H, Shoguchi E, Sawada H (2008) Mechanism of self-sterility in a hermaphroditic chordate. Science 320(5875):548–550

Holland PW, Garcia-Fernàndez J, WIlliams NA, Sidow A (1994) Gene duplications and the origins of vertebrate development. Dev Suppl:125–133

Hozumi A, Satouh Y, Ishibe D, Kaizu M, Konno A, Ushimaru Y, Toda T, Inaba K (2004) Local database and the search program for proteomic analysis of sperm proteins in the ascidian *Ciona intestinalis*. Biochem Biophys Res Commun 319:1241–1246

Hozumi A, Kawai N, Yoshida R, Ogura Y, Ohta N, Satake H, Satoh N, Sasakura Y (2010) Efficient transposition of a single *Minos* transposon copy in the genome of the ascidian *Ciona intestinalis* with a transgenic line expressing transposase in eggs. Dev Dyn 239:1076–1088

Hozumi A, Mita K, Miskey C, Mates L, Izsvak Z, Ivics Z, Satake H, Sasakura Y (2013) Germline transgenesis of the chordate *Ciona intestinalis* with hyperactive variants of sleeping beauty transposable element. Dev Dyn 242:30–43

Imai KS, Levine M, Satoh N, Satou Y (2006) Regulatory blueprint for a chordate embryo. Science 312:1183–1187

Ivics Z, Hackett PB, Plasterk RH, Izsvak Z (1997) Molecular reconstruction of Sleeping Beauty, a Tc1-like transposon from fish, and its transposition in human cells. Cell 91:501–510

Joly JS, Kano S, Matsuoka T, Auger H, Hirayama K, Satoh N, Awazu S, Legendre L, Sasakura Y (2007) Culture of *Ciona intestinalis* in closed systems. Dev Dyn 236:1832–1840

Kawai N, Ochiai H, Sakuma T, Yamada L, Sawada H, Yamamoto T, Sasakura Y (2012) Efficient targeted mutagenesis of the chordate *Ciona intestinalis* genome with zinc-finger nucleases. Dev Growth Differ 54:535–545

Kourakis MJ, Newman-Smith E, Smith WC (2010) Key steps in the morphogenesis of a cranial placode in an invertebrate chordate, the tunicate *Ciona savignyi*. Dev Biol 340:134–144

Manni L, Lane N, Joly J, Gasparini F, Tiozzo S, Caicci F, Zaniolo G, Burighel P (2004) Neurogenic and non-neurogenic placodes in ascidians. J Exp Zool B Mol Dev Evol 302:483–504

Matsuoka T, Awazu S, Shoguchi E, Satoh N, Sasakura Y (2005) Germline transgenesis of the ascidian *Ciona intestinalis* by electroporation. Genesis 41:61–72

Miyawaki A, Llopis J, Heim R, McCaffery JM, Adams JA, Ikura M, Tsien RY (1997) Fluorescent indicators for Ca^{2+} based on green fluorescent proteins and calmodulin. Nature (Lond) 388: 882–887

Nakai J, Ohkura M, Imoto K (2001) A high signal-to-noise Ca(2+) probe composed of a single green fluorescent protein. Nat Biotechnol 19:137–141

Nakatani Y, Moody R, Smith WC (1999) Mutations affecting tail and notochord development in the ascidian *Ciona savignyi*. Development 126:3293–3301

Nakayama-Ishimura A, Chambon JP, Horie T, Satoh N, Sasakura Y (2009) Delineating metamorphic pathways in the ascidian *Ciona intestinalis*. Dev Biol 326:357–367

Ohta N, Horie T, Satoh N, Sasakura Y (2010) Transposon-mediated enhancer detection reveals the location, morphology and development of the cupular organs, which are putative hydrodynamic sensors, in the ascidian *Ciona intestinalis*. Zool Sci 27:842–850

O'Kane C, Gehring WJ (1987) Detection in situ of genomic regulatory elements in *Drosophila*. Proc Natl Acad Sci USA 84:9123–9127

Sagane Y, Zech K, Bouquet JM, Schmid M, Bal U, Thompson EM (2010) Functional specialization of cellulose synthase genes of prokaryotic origin in chordate larvaceans. Development (Camb) 137:1483–1492

Sakaue-Sawano A, Kurokawa H, Morimura T, Hanyu A, Hama H, Osawa H, Kashiwagi S, Fukami K, Miyata T, Miyoshi H, Imamura T, Ogawa M, Masai H, Miyawaki A (2008) Visualizing spatiotemporal dynamics of multicellular cell-cycle progression. Cell 132:487–498

Sasakura Y, Yamada L, Takatori N, Satou Y, Satoh N (2003a) A genomewide survey of developmentally relevant genes in *Ciona intestinalis*. VII. Molecules involved in the regulation of cell polarity and actin dynamics. Dev Genes Evol 213:273–283

Sasakura Y, Shoguchi E, Takatori N, Wada S, Meinertzhagen IA, Satou Y, Satoh N (2003b) A genomewide survey of developmentally relevant genes in *Ciona intestinalis*. X. Genes for cell junctions and extracellular matrix. Dev Genes Evol 213:303–313

Sasakura Y, Awazu S, Chiba S, Satoh N (2003c) Germ-line transgenesis of the Tc1/*mariner* superfamily transposon *Minos* in *Ciona intestinalis*. Proc Natl Acad Sci USA 100(13):7726–7730

Sasakura Y, Nakashima K, Awazu S, Matsuoka T, Nakayama A, Azuma J, Satoh N (2005) Transposon-mediated insertional mutagenesis revealed the functions of animal cellulose synthase in the ascidian *Ciona intestinalis*. Proc Natl Acad Sci USA 102:15134–15139

Sasakura Y, Oogai Y, Matsuoka T, Satoh N, Awazu S (2007) Transposon-mediated transgenesis in a marine invertebrate chordate, *Ciona intestinalis*. Genome Biol 8(suppl 1):S3

Sasakura Y, Konno A, Mizuno K, Satoh N, Inaba K (2008) Enhancer detection in the ascidian *Ciona intestinalis* with transposase-expressing lines of *Minos*. Dev Dyn 237:39–50

Sasakura Y, Inaba K, Satoh N, Kondo M, Akasaka K (2009) *Ciona intestinalis* and *Oxycomanthus japonicus*, representatives of marine invertebrates. Exp Anim 58:459–469

Sasakura Y, Kanda M, Ikeda T, Horie T, Kawai N, Ogura Y, Yoshida R, Hozumi A, Satoh N, Fujiwara S (2012) Retinoic acid-driven Hox1 is required in the epidermis for forming the otic/atrial placodes during ascidian metamorphosis. Development (Camb) 139:2156–2160

Satoh N (1994) Developmental biology of ascidians. Cambridge University Press, New York

Satoh N (2003) The ascidian tadpole larva: comparative molecular development and genomics. Nat Rev Genet 4:285–295

Satou Y, Imai KS, Satoh N (2001) Action of morpholinos in *Ciona* embryos. Genesis 30: 103–106

Satou Y, Yamada L, Mochizuki Y, Takatori N, Kawashima T, Sasaki A, Hamaguchi M, Awazu S, Yagi K, Sasakura Y, Nakayama A, Ishikawa H, Inaba K, Satoh N (2002) A cDNA resource from the basal chordate *Ciona intestinalis*. Genesis 33:153–154

Satou Y, Mineta K, Ogasawara M, Sasakura Y, Shoguchi E, Ueno K, Yamada L, Matsumoto J, Wasserscheid J, Dewar K, Wiley GB, Macmil SL, Roe BA, Zeller RW, Hastings KE, Lemaire P, Lindquist E, Endo T, Hotta K, Inaba K (2008) Improved genome assembly and evidence-based global gene model set for the chordate *Ciona intestinalis*: new insight into intron and operon populations. Genome Biol 9:R152

Sordino P, Heisenberg CP, Cirino P, Toscano A, Giuliano P, Marino R, Pinto MR, De Santis R (2000) A mutational approach to the study of development of the protochordate *Ciona intestinalis* (Tunicata, Chordata). Sarsia 85:173–176

Takemoto K, Nagai T, Miyawaki A, Miura M (2003) Spatio-temporal activation of caspase revealed by indicator that is insensitive to environmental effects. J Cell Biol 160:235–243

Tsutsui H, Karasawa S, Okamura Y, Miyawaki A (2008) Improving membrane voltage measurements using FRET with new fluorescent proteins. Nat Methods 5:683–685

Yamada L, Saito T, Taniguchi H, Sawada H, Harada Y (2009) Comprehensive egg-coat proteome of an ascidian *Ciona intestinalis* reveals gamete recognition molecules involved in self-sterility. J Biol Chem 284:9402–9410

Yoshida R, Sasakura Y (2012) Establishment of enhancer detection lines expressing GFP in the gut of the ascidian *Ciona intestinalis*. Zool Sci 29:11–20

Index

H. Sawada et al. (eds.), *Sexual Reproduction in Animals and Plants*,
DOI 10.1007/978-4-431-54589-7, © The Author(s) 2014